The Geography of Phytochemical Races

Bruce A. Bohm

The Geography of Phytochemical Races

Springer

THE NEW YORK BOTANICAL GARDEN

Dr. Bruce A. Bohm
3685 West 15th Avenue
Vancouver, British Columbia
V6R 2Z6
Canada
e-mail: brucebohm@shaw.ca

ISBN: 978-1-4020-9051-6 e-ISBN: 978-1-4020-9052-3

Library of Congress Control Number: 2008934053

Cover design: Boekhorst Design BV

Printed on acid-free paper

9 8 7 6 5 4 3 2 1

springer.com

This book is dedicated to Wilfred (Wilf) B. Schofield, Emeritus Professor of Botany, University of British Columbia, bryologist extraordinaire, indomitable field botanist for whom the force of gravity is a mere inconvenience, scholar, teacher, colleague, and, above all, good friend.

Acknowledgments

I thank many workers, too numerous to list, who have provided reprints, preprints, and general information on the subject of chemical races. Special thanks go to Prof. Dr. Tod Stuessy, Institut für Botanik, Universität Wien, for his help with the literature on this and other projects, and for encouragement and support over the years. I am also very grateful to Prof. Dr. Eckhard Wollenweber, Institut für Botanik der TU Darmstadt, for his friendship and collegiality over the years, for access to unpublished data, and for rendering the Abstract in German, thus saving everyone from having to suffer my abuse of his beautiful language. I also thank Mapthematics Ltd. of Renton, Washington who assisted me in acquiring the mapping program Geocart Version 2.6 with which the maps in this review were prepared.

Abstract

The discovery of chemical races was one of the important outcomes of the "chemotaxonomic age," providing systematists with another tool in the search for relationships. The subject was discussed in a few review articles at the time, but, by and large, receded into the background as more powerful techniques, generally referred to as molecular biology, came on stream. This review is an attempt to update the subject of chemical races by bringing together an assortment of examples from the literature where geographically distinct profiles in the occurrence of secondary metabolites have been described.

The examples are sorted into several general categories, within which examples are further arranged according to geographical areas. The first major category involves disjunctions that occur more or less within continents, with some obvious exceptions such as some African taxa that not only occur primarily on the continent but are also represented on Madagascar or on the Canary Islands. The second category is similar to the first, but specifically addresses distributions likely caused by retreat of glaciers. The third category treats intercontinental disjunctions; category four is similar, but treats examples characterized by wide disjunctions that do not fit easily into the intercontinental group. Category four also includes several examples from the lichen chemistry literature. Category five features disjunctions involving oceanic islands, while polar disjunctions are covered in the final section.

Inclusion of an example in a particular category is admittedly arbitrary. Examples in the postglacial group, for instance, could just as easily have been included within the continental group, but were given special consideration because most of the workers involved with those taxa discussed their results in terms of refugia and postglacial migration. Some taxa appear under two or more categories. For example, some taxa within *Chrysosplenium* are disjunct between Asia and North America, others between North America and Europe, and still others between Asia and South America. At closer focus, one finds sectional disjunctions between eastern and western North America, and in one case, populational differences within a species. Unfortunately, few systems have been examined this thoroughly, as will be noted in several places below.

In a few cases, information obtained from macromolecular methods is included where such data might have a bearing on interpretation of how a given secondary metabolic profile could have come about. Specifically, it is possible to speculate on which biosynthetic step in a given pathway might have been eliminated or altered in the establishment of a new chemical profile.

Examples involve a wide assortment of chemical types, and range from the simplest approach—a comparison of chromatograms—to application of cladistic methodology to complex sets of chemical structural data. Equally wide is the sampling of the plant kingdom, with examples including lichens, bryophytes, ferns, flowering plants, and conifers. Although the study of conifer chemistry is often limited to simple compounds of widespread occurrence, we see that statistical analysis of quantitative data (GLC) provides a powerful means of assessing relationships, as well as the history of postglacial migrations. Application of statistical analysis to flavonoid data has also revealed relationships in a few cases.

Zusammenfassung

Die Entdeckung chemischer Rassen war eines der wichtigsten Ergebnisse des "Zeitalters der Chemotaxonomie", hat sie doch den Systematikern ein weiteres Werkzeug für die Suche nach Verwandtschaftsbeziehungen in die Hand gegeben. Das Thema wurde seinerzeit in mehreren Review-Artikeln diskutiert, aber nach und nach ist es gegenüber leistungsfähigeren Techniken, die allgemein unter dem Stichwort Molekularbiologie zusammengefaßt werden können, in den Hintergrund gerückt. Dieser Übersichtsartikel ist ein Versuch, das Thema Chemische Rassen zu aktualisieren, indem eine Zusammenstellung von Beispielen aus der Literatur gegeben wird, wo für pflanzliche Sekundärstoffen geographisch distinkte Muster beschrieben wurden.

Die Beispiele sind nach mehreren Kategorien geordnet und innerhalb dieser Kategorien nach geographischen Gebieten. Die erste Kategorie umfaßt Disjunktionen, die mehr oder weniger nur innerhalb eines Kontinents vorkommen. Sie umfaßt auch einige Ausnahmen, wie etwa afrikanische Sippen, die zwar vorwiegend auf dem Kontinent, aber auch auf Madagaskar oder auf den Kanaren vertreten sind. Die zweite Gruppe ähnelt der ersten, betrifft aber speziell Verbreitungsmuster, die vermutlich durch den Rückzug von Gletschern bedingt sind. Die dritte Gruppe behandelt interkontinentale Disjunktionen. Die vierte Gruppe enthält Beispiele für weiträumige Disjunktionen, die nicht ohne weiteres in die interkontinentale Gruppe passen. Diese Gruppe enthält auch einige Beispiele aus der Literatur zur Flechtenchemie. Gruppe fünf behandelt Disjunktionen, die vor allem die ozaenischen Inseln betreffen, während polare Disjunktionen im letzten Kapitel behandelt werden.

Die Aufnahme eines Beispiels in eine bestimmte Gruppe ist zugegebenermaßen etwas willkürlich. So könnten etwa Beispiele aus der nach-eiszeitlichen Kategorie ebensogut in der kontinentalen Kategorie stehen. Hier wurde aber besonders berücksichtigt, daß die meisten Autoren, die sich mit solchen Sippen beschäftigen, Ihre Ergebnisse im Zusammenhang mit Rückzugsgebieten und nach-eiszeitlichen Wanderungen diskutieren. Einige Sippen erscheinen unter zwei oder mehr Kategorien. Manche Sippen innerhalb von *Chrysosplenium* beipielsweise sind disjunkt zwischen Asien und Nordamerika verbreitet, andere zwischen Nordamerika und Europa, wieder andere zwischen Asien und Südamerika. Bei näherer Betrachtung findet man Diskunktionen auf Sektionsebene zwischen dem östlichen und dem westlichen Teil Nordamerikas, und in einem Fall unterscheiden sich Populationen innerhalb einer Art.

Leider wurden nur wenige Verwandtschaftskreise derart gründlich untersucht, wie an mehreren Stellen gezeigt werden wird.

In einigen Fällen wurden Informationen aus makromolekularen Untersuchungen mit einbezogen, wenn diese Daten bei der Interpretation der Sekundärstoffmuster hilfreich sind. Insbesondere kann man Vermutungen darüber anstellen, welche Schritte eines bestimmten Biosynthesewegs beim Aufbau eines neuen Sekundärstoffmusters ausgefallen oder verändert worden sein könnten.

Die Beispiele umfassen einen weiten Bereich Chemotypen, von einfachsten Fällen (Vergleich von Chromatogrammen) bis hin zur cladistischen Methodik für komplexe Sätze chemischer Merkmale. Ähnlich weit ist die Auswahl aus dem Pflanzenreich; sie umfaßt Flechten, Moose, Farne, Angiospermen und Gymnospermen. Obwohl sich die chemischen Daten bei Koniferen oft auf einfache Verbindungen mit weitverbreitetem Vorkommen beschränken, werden wir doch sehen, daß statistische Analysen quantitativer Daten (GLC) ein wertvolles Hilfsmittel zur Aufdeckung von Verwandschaftsbeziehungen ebenso wie der nach-eiszeitlichen Wanderungen darstellen. Die Anwendung statistischer Analysen auf Flavonoid-Daten hat in einigen Fällen ebenfalls zur Klärung von Verwandschaftbeziehungen geführt.

Contents

1 Introduction 1

2 Examples Within Continents 3
2.1 Africa (Including Madagascar) 3
2.1.1 A Comparison of Rain Forests 3
2.1.2 The *Senecio radicans* Complex (Asteraceae) 5
2.1.3 *Pearsonia* (Fabaceae) 7
2.1.4 *Phytolacca dodecandra* (Phytolaccaceae) 7
2.1.5 *Aloë* (Asphodelaceae) 8
2.1.6 *Aspalathus linearis*—Rooibos tea (Fabaceae) 10
2.1.7 Mangrove Waxes 11
2.1.8 *Papaver aculeatum* (Papaveraceae) 11
2.1.9 *Ocimum gratissimum* (Lamiaceae) 12
2.2 Australia and New Zealand 12
2.2.1 *Chamelaucium uncinatum*—The Geraldton Waxflower (Myrtaceae) 12
2.2.2 *Duboisia myoporoides* (Solanaceae) 14
2.2.3 *Zieria* (Rutaceae) 16
2.2.4 *Melaleuca* (Myrtaceae) 17
2.2.5 *Isotachis lyallii* (Balantiopsidaceae) 19
2.2.6 *Leptospermum scoparium* (Myrtaceae) 20
2.2.7 *Chionochloa* (Poaceae) 23
2.3 Europe 24
2.3.1 *Ramalina siliquosa* (Ramalinaceae) 24
2.3.2 Cyanogenesis in Clover (*Trifolium,* Fabaceae) 25
2.3.3 *Lotus corniculatus* (Fabaceae) 28
2.3.4 *Silene latifolia (= alba, = pratensis)* (Caryophyllaceae) 31
2.3.5 *Achillea* (Asteraceae) 36
2.3.6 *Pulicaria dysenterica* (Asteraceae) 38
2.3.7 *Thalictrum minus* (Ranunculaceae) 38
2.3.8 *Valeriana officinalis* (Valerianaceae) 40
2.3.9 *Cistus* (Cistaceae) 41
2.3.10 Three Examples from Finland (Fig. 2.25) 44

2.3.11 *Thymus* (Lamiaceae) 47
2.3.12 *Pinus uncinata* (Pinaceae) 49
2.4 The Mediterranean Basin 51
2.4.1 *Withania somnifera* (Solanaceae) 51
2.4.2 *Foeniculum vulgare* (Apiaceae) 52
2.4.3 *Genista* (Fabaceae) 53
2.4.4 *Pistacia lentiscus* (Anacardiaceae) 55
2.4.5 *Thymus* (Lamiaceae) 56
2.4.6 *Lavandula stoechas* (Lamiaceae) 58
2.4.7 *Mentha suaveolens* (Lamiaceae) 58
2.4.8 *Teucrium polium* (Lamiaceae) 59
2.4.9 *Juniperus phoenicea* (Cupressaceae) 60
2.4.10 *Phlomis lychnitys* (Lamiaceae) 61
2.5 Asia 61
2.5.1 *Datura* and *Berberis* — Alkaloids and Elevation 61
2.5.2 *Cymbopogon distans* (Poaceae) 63
2.5.3 *Cupressus* or *Chamaecyparis?* (Cupressaceae) 64
2.5.4 *Jungermannia vulcanicola* (Jungermanniaceae) 65
2.5.5 *Biebersteinia* (Geraniaceae) 66
2.5.6 *Rhodiola crenulata* (Crassulaceae) 67
2.5.7 *Hippophae rhamnoides* (Elaeagnaceae) 67
2.5.8 *Thujopsis dolabrata* (Cupressaceae) 68
2.6 South America 68
2.6.1 *Vanillosmopsis erythropappa* (Asteraceae) 68
2.6.2 *Cunila galioides* (Lamiaceae) 70
2.6.3 *Cinchona* (Rubiaceae) 70
2.6.4 *Lycopersicon pennellii* (Solanaceae) 72
2.6.5 C_3 and C_4 Grasses in Argentina 74
2.6.6 *Tagetes minuta* (Asteraceae) 76
2.6.7 *Hymenaea* (Fabaceae) 76
2.6.8 *Minthostachys verticillata* (Lamiaceae) 77
2.7 North and Central America 78
2.7.1 *Notholaena standleyi* (Adiantaceae) 78
2.7.2 *Chenopodium fremontii* (Chenopodiaceae) 81
2.7.3 *Asclepias tuberosa* (Asclepiadaceae) 85
2.7.4 *Phlox carolina* (Polemoniaceae) 86
2.7.5 *Lasthenia* (Asteraceae) 88
2.7.6 *Brickellia cylindracea* (Asteraceae) 93
2.7.7 *Helianthus maximiliani* (Asteraceae) 94
2.7.8 *Ambrosia* (Asteraceae) 94
2.7.9 *Gaillardia pulchella* (Asteraceae) 99
2.7.10 *Dudleya* (Crassulaceae) 100
2.7.11 *Delphinium variegatum* (Ranunculaceae) 101
2.7.12 *Isomeris arborea* (Capparaceae) 102

2.7.13 Soils and Asters ... 103
2.7.14 *Lupinus sericeus* (Fabaceae) ... 103
2.7.15 *Satureja douglasii* (Lamiaceae) ... 105
2.7.16 *Pentagramma* (*Pityrogramma*) *triangularis* (Adiantaceae) ... 108
2.7.17 *Encelia farinosa* (Asteraceae) ... 111
2.7.18 *Larrea tridentata* (Zygophyllaceae) ... 113
2.7.19 *Chrysosplenium glechomaefolium* (Saxifragaceae) ... 113
2.7.20 *Quercus rubra* (Fagaceae) ... 116
2.7.21 *Cnidoscolus spinosus* (Euphorbiaceae) ... 117
2.7.22 *Cupressus* (Cupressaceae) ... 118
2.7.23 *Mimulus aurantiacus* (Scrophulariaceae) ... 120
2.7.24 *Cladonia cervicornis* (Cladoniaceae) ... 120
2.7.25 *Echinacea angustifolia* (Asteraceae) ... 121
2.7.26 *Lippia dulcis* (Verbenaceae) ... 122

3 After the Ice ... 125
3.1 North America ... 126
3.1.1 *Luetkea pectinata* (Rosaceae) ... 126
3.1.2 *Tolmiea menziesii* (Saxifragaceae) ... 126
3.1.3 *Arnica* (Asteraceae), a Note on *Sedum* (Crassulaceae), and a Comment on *Antennaria* (Asteraceae) ... 128
3.1.4 *Menziesia* (Ericaceae) ... 134
3.1.5 *Thuja plicata*-Western Red Cedar (Cupressaceae) ... 137
3.1.6 *Liriodendron tulipifera* (Magnoliaceae) ... 138
3.1.7 *Ambrosia psilostachya* (Asteraceae) ... 140
3.2 North American Conifers ... 141
3.2.1 *Abies* (Pinaceae) ... 141
3.2.2 *Cupressus bakeri* (Cupressaceae) ... 144
3.2.3 *Picea* (Pinaceae) ... 146
3.2.4 *Pinus* (Pinaceae) ... 147
3.2.5 *Pseudotsuga* ... 157
3.2.6 *Sequoia sempervirens* (Cupressaceae) ... 159
3.2.7 *Juniperus* (Cupressaceae), and a Comment on *Marshallia graminiifolia* (Asteraceae) ... 161
3.3 Europe ... 162
3.3.1 *Pinus halepensis* (Pinaceae) ... 162
3.3.2 *Pinus brutia* var. *brutia* (Pinaceae) ... 164
3.3.3 *Pinus sylvestris* (Pinaceae) ... 164
3.3.4 *Abies alba* (Pinaceae) ... 166
3.4 South America ... 167
3.4.1 *Fitzroya cupressoides* (Cupressaceae) ... 167
3.4.2 *Austrocedrus chilensis* (Cupressaceae) ... 170
3.4.3 *Araucaria araucana* (Araucariaceae) ... 171

4 Intercontinental Disjunctions 173
4.1 Across the Atlantic Ocean 173
4.1.1 Datiscaceae 173
4.1.2 *Armeria maritima* (Plumbaginaceae) 174
4.1.3 *Cakile* (Brassicaceae) 176
4.1.4 *Cypripedium calceolus* (Orchidaceae) 179
4.1.5 *Ephedra* (Ephedraceae) 180
4.1.6 Cactaceae and Aizoaceae 181
4.2 Across the Indian Ocean (Primarily) 182
4.2.1 *Hymenaea verrucosa* (Fabaceae) 182
4.2.2 Restionaceae 183
4.2.3 *Villarsia* (Menyanthaceae) 187
4.2.4 Icacinaceae 188
4.3 North Pacific 190
4.3.1 *Glehnia* (Apiaceae) 191
4.3.2 *Agastache* (Lamiaceae) 192
4.3.3 Cladothamneae (Ericaceae) 193
4.3.4 *Tiarella* (Saxifragaceae) 194
4.3.5 *Chrysosplenium* (Saxifragaceae) 195
4.3.6 *Iris setosa* (Iridaceae) 197
4.3.7 *Humulus lupulus* (Cannabaceae) 198
4.4 South Pacific 199
4.4.1 *Eucryphia* (Eucryphiaceae) 199
4.4.2 Blennospermatinae (Asteraceae) 202
4.4.3 *Alexa* and *Castanospermum* (Fabaceae) 203
4.4.4 *Scoparia* (Scrophulariaceae) 205
4.4.5 *Fuchsia* (Onagraceae) 205
4.4.6 *Nothofagus* (Nothofagaceae) 208
4.5 Western Pacific: Eastern Asia, Japan, and the Philippines 210
4.5.1 *Euchresta* (Fabaceae) 210
4.5.2 *Heterotropa* (=*Asarum*)(Aristolochiaceae) 212
4.5.3 *Rhododendron* (Ericaceae) 214
4.5.4 Gesneriaceae 215

5 Wide Disjunctions 217
5.1 Vascular Plants 217
5.1.1 Coriariaceae 217
5.1.2 Gentianaceae 217
5.1.3 *Hordeum* (Poaceae) 219
5.1.4 *Pteridium aquilinum* sens. lat. (Pteridaceae) 219
5.2 Nonvascular Plants 221
5.2.1 *Pseudevernia furfuracea* (Parmeliaceae) 221
5.2.2 *Rhizoplaca melanophthalma* (Lecanoraceae) 221
5.2.3 *Thamnolia vermicularis* (Icmadophilaceae) 223
5.2.4 *Lobaria pulmonaria* (Stictaceae) 223

5.2.5 *Bazzania trilobata* (Lepidoziaceae) 223
5.2.6 *Conocephalum conicum* (Conocephalaceae) 225
5.2.7 *Adelanthus decipiens* (Adelanthaceae) 228
5.2.8 *Plagiochila* (Plagiochilaceae) 229
5.2.9 *Marchantia* (Marchantiaceae) 231
5.2.10 *Mastigophora diclados* (Mastigopharaceae) 232
5.2.11 *Makinoa crispata* (Makinoaceae) 233
5.2.12 Brown Algae 233
5.2.13 *Cystoseira amentacea* (Cystoseiraceae) 234
5.2.14 *Plocamium* (Plocamiaceae) 235
5.2.15 *Portieria hornemanii* (Rhodophyllidaceae) 239
5.2.16 *Laurencia* (Rhodomelaceae) 240
5.2.17 *Codium* (Codiaceae) 244

6 Oceanic Islands 245
6.1 The Galapagos Islands 245
6.1.1 Cyanogenic Plants 246
6.1.2 *Lycopersicon cheesmanii* (Solanaceae) 248
6.1.3 *Scalesia* (Asteraceae) 249
6.2 The Hawaiian Islands 249
6.2.1 *Bidens* (Asteraceae) 251
6.2.2 *Cyperus rotundus* (Cyperaceae) 253
6.2.3 *Hesperomannia* (Asteraceae) 254
6.2.4 *Metrosideros* (Myrtaceae) 255
6.2.5 Silversword alliance (Asteraceae) 255
6.2.6 *Vaccinium* (Ericaceae) 257
6.2.7 *Zanthoxylum* 257
6.3 Southern Pacific 259
6.3.1 *Piper methysticum* (Piperaceae) 259
6.3.2 *Degeneria* (Degeneriaceae) 262
6.4 Juan Fernandez Islands 263
6.4.1 *Dendroseris* (Asteraceae) 264
6.4.2 *Robinsonia* (Asteraceae) 265
6.4.3 *Erigeron* (Asteraceae) 266
6.4.4 *Gunnera* (Gunneraceae) 268
6.4.5 *Peperomia* (Piperaceae) 271
6.4.6 *Sophora* (Fabaceae) 272
6.5 Lord Howe Island 273
6.5.1 *Carmichaelia* (Fabaceae) 273
6.6 Ullung Island, Korea 275
6.6.1 *Acer* (Sapindaceae) 275
6.6.2 *Cotoneaster wilsonii* (Rosaceae) 275
6.7 Tristan da Cunha and the Falklands 276
6.7.1 *Empetrum* (Empetraceae) 276

6.8 Kerguelen Islands ... 277
6.8.1 *Ranunculus* (Ranunculaceae) ... 277
6.9 Macaronesia ... 279
6.9.1 *Anomalographis madeirensis* (Graphidaceae) ... 280
6.9.2 *Dactylis glomerata* (Poaceae) ... 280
6.9.3 *Laurus azorica* (Lauraceae) ... 282
6.9.4 *Todaroa aurea* (Apiaceae) ... 282

7 Polar Disjunctions ... 285
7.1 *Ambrosia chamissonis* (Asteraceae) ... 285
7.2 *Blennosperma* (Asteraceae) ... 288
7.3 *Chrysosplenium* (Saxifragaceae) ... 289
7.4 *Encelia canescens* (Asteraceae) ... 290
7.5 *Eschscholzia californica* (Papaveraceae) ... 291
7.6 *Krameria* (Krameriaceae) ... 291
7.7 *Larrea* (Zygophyllaceae) ... 292
7.8 *Lasthenia* (Asteraceae) ... 295
7.9 *Perityle emoryi* (Asteraceae) ... 295
7.10 *Tradescantia* (Commelinaceae) ... 296

8 Conclusions ... 299

Bibliography ... 301

Plant Family Index ... 339

Plant Genus Index ... 341

Subject Index ... 345

Chapter 1
Introduction

Patterns of occurrence of morphological features of a species across its natural range can often provide insights into evolutionary relationships within the taxon. In principal, at least, this should be no less true when the differences involve secondary metabolites. That secondary metabolites indeed have played and continue to play a role in systematics is borne out by the very significant impact that chemotaxonomy has had on the subject. An added feature of the use of secondary metabolites is the extensive literature documenting the biosynthetic steps by which they are formed, and in several cases, the genetic basis for many of the steps involved. Thus, if the details of a given biosynthetic pathway are sufficiently well understood, it is often possible to infer differences in the genetic composition of species (or populations) that differ by the presence or absence of a particular compound or set of compounds. This application is most useful in the case of compound types that have lent themselves to straightforward biosynthetic analysis, flavonoids, and simple terpene derivatives in particular. In the case of simple terpenes, where quantitative information is more reliably available, differences in chemical structures are less important.

Although it is possible to infer genetic differences between species, or among populations, as mentioned above, the direction of evolutionary change in the chemistries involved is not at all straightforward. For example, a very simple flavonoid profile may be interpreted as either representing the primitive (or pleisiomorphic) condition or an advanced (apomorphic) condition where loss of one or more members of a pathway represents a derived condition. Only when detailed phylogenetic analyses have been performed is it possible to say which of these positions the profile represents. In a few cases described below, phylogenetic analysis of the taxa involved has been carried out, making it possible to comment on the evolutionary significance, if any, of the secondary compounds involved.

The natural product literature is vast, with only a passing comment on chemical differences between plants from different sources occasionally included. It is virtually impossible to track down all of these examples; thus, the examples below represent only a sampling of chemical variation in the plant kingdom. Although not exhaustive, the sampling does include most classes of natural products and a fairly wide sampling of major taxa, including lichens, algae, bryophytes, ferns, flowering plants, and conifers.

B.A. Bohm, *The Geography of Phytochemical Races*,

It is useful to mention at this point a few citations from the earlier literature that might be of interest to readers. Historical aspects of secondary metabolite distributions were commented upon by Ralph Alston and Billie Turner in their groundbreaking book *Biochemical Systematics* published in 1963. The general topic of geographic patterning was revisited by Ralph Alston (1967) in a comprehensive review. Soon thereafter, Tom Mabry (1973) reviewed the chemistry of geographic races. Prof. Otto Gottlieb (1986) discussed microchemical evolution, including geographic and ecological aspects of the subject, and reviews on other aspects of chemical geographic patterns were treated by the present author (Bohm 1998a, b), the first being a more general discussion and the second a more specialized one focusing on evolutionary relationships of island plants. All of these provide detailed citation lists.

Information presented in this review appears under five major headings: examples within continents; examples involving glacial refugia and postglacial migrations; examples involving oceanic disjunctions; examples involving island species; and examples that exhibit north–south (polar) disjunctions. In many cases, larger taxa have representatives that could fit into two or more categories; hence, the assignments to some sections are quite arbitrary. For example, species of *Chrysosplenium* occur in eastern Asia; eastern, western, and boreal North America; northern Europe; and in extreme southern South America. It is also worth bearing in mind the idea that all plant distributions have been influenced to a greater or lesser extent by glaciation, and that assigning an example to a particular category may be seen by some readers as arbitrary.

Chapter 2
Examples Within Continents

The survey of geographical distribution patterns of secondary plant constituents starts with examples that come, more or less, from within continents. In many examples, the patterns of variation involve comparatively small areas, a few counties perhaps, and could be referred to as local; whereas in others, considerably larger areas are involved, often several states or provinces, or even large pieces of a country. In most cases, however, the regions under consideration lie within a single continental landmass. A few examples come from island systems (e.g., New Zealand) where the existence of major landmasses separated by a small stretch of water does not appear to have had an effect upon the distribution reported, or at least none has been reported.

2.1 Africa (Including Madagascar)

2.1.1 A Comparison of Rain Forests

The first example from the African continent represents one of the more wide-ranging projects that we will meet. Ninety species were sampled in a study comparing aspects of the phytochemistry of two African rain forests (Gartlan et al., 1980). The object of the study was to compare different plant communities from the perspective of their production of secondary metabolites (presumably) employed in allelochemical defense. The areas studied are located in the Kibale Forest in Uganda, East Africa, and the Douala-Edea Forest Reserve in Cameroon, West Africa (Fig. 2.1). The two areas differ in several respects: rainfall, soil chemistry, terrain, and vegetation. The Douala-Edea site receives over twice the amount of annual rainfall (3500–4000 mm over ca. 220 days) than does the Kibale site (ca. 1500 mm over ca. 166 days). Temperatures in Douala-Edea show comparatively smaller seasonal variation (23–32°C), whereas average temperatures in Kibale are somewhat lower and show a greater range (12.7–25.5°C). This is expected in view of the coastal, low elevation of Douala-Edea compared to the mid-elevation (1300–1500 m) location of the Kibale site. The soil at the Douala-Edea site is sandy, low in nutrients, and has an average pH = 3.89. In contrast, the soil in Kibale is dark gray to red sandy loam of moderate fertility with an average pH = 5.64. The Douala-Edea site vegetation is

B.A. Bohm, *The Geography of Phytochemical Races*,

Fig. 2.1 Map of Africa showing general locations of the Douala-Edea site in Cameroon and the Kibale Forest (Uganda). Abbreviations of countries referred to in other African examples: Bots = Botswana, Cam = Cameroon, C.A.R. = Central African Republic, Gh = Ghana, Ken = Kenya, Mad = Madagascar, Moz = Mozambique, Tanz = Tanzania, Zam = Zambia, and Zim = Zimbabwe

typical lowland, evergreen rain forest characterized by 11.5% tree species; 7% of its species are deciduous. Kibale Forest is also an evergreen rain forest but is characterized by 27% tree species and 18% deciduous species. Of the 93 species examined in this study, only one, *Symphonia globulifera* L. f. (Guttiferae), was present in both study areas. Limited space precludes listing of the species; suffice it to say that the vegetation of the two areas is quite different, even at the family level.

The allelochemistry of the two forests also shows striking differences. Analyses showed that the Douala-Edea site has a higher proportion of tannin-producing taxa, whereas species at the Kibale site exhibited a much higher frequency of alkaloid producers. Only eight species appeared to lack both tannins and alkaloid-positive compounds. One of these exhibited a high concentration of a biflavonoid, a polyphenolic compound based on the coupling of two flavonoid monomeric units; another was shown to be positive for iridoids (15-carbon terpenoid derivatives). As Gartlan et al. (1980) discussed, these observations are in agreement with the idea that taxa from habitats where plants undergo rapid growth (i.e., stem elongation and leaf replacement) invest a comparatively smaller amount of resources in production of defensive chemicals (allelochemics) than species in habitats where growth is slower. Growth would be expected to be slower in habitats where soil nutrients are in short supply, as observed in the Douala-Edea site in this study. It has been debated that plants in nutrient-challenged habitats are more likely to produce tannins or other polyphenolic compounds, rather than alkaloids or other "qualitative" allelochemics, as defensive compounds. This wide-ranging study would seem to offer support for the idea.

A few years earlier, Levin (1976) discussed, in some detail, the distribution of alkaloid-bearing plants in relation to geography. The basic premise was that plants from tropical regions are under more concentrated attack by predators of one sort or another, and thus invest more biosynthetic effort in chemical defense systems. His survey of the literature provided information on alkaloid occurrence in 110 families, representing 38 orders of dicots, and showed that: (1) nearly twice as many annuals as perennials were alkaloid bearing; (2) tropical floras had almost twice the level of alkaloid-bearing taxa as temperate floras; (3) Magnoliales and Ranales had higher levels of alkaloid-bearing components than other dicot families; and (4) there is a latitudinal gradient with higher percentages of alkaloid-bearing taxa nearer the equator. Three values representing the extremes of the latter point are 40% alkaloid-bearing taxa in Kenya (0°), 24.9% in taxa sampled from Mexico (ca. 25°N), and 10.8% in taxa from New Zealand (ca. 43°S).

In a subsequent paper, Levin and York (1978) assessed the alkaloid situation in greater detail — again based on an extensive search of the literature — by calculating "alkaloid toxicity indices." Analysis at the generic level revealed that herbs, shrubs, and trees consistently showed higher levels of toxicity in tropical floras (based on 159 genera), lesser levels in subtropical floras (based on 109 genera), and the smallest levels in temperate floras (based on 210 genera). The same trend was evident when the data were assessed at the family level (analysis was based upon 24 tropical families, 14 cosmopolitan families, and 14 temperate families).

2.1.2 The **Senecio radicans** *Complex (Asteraceae)*

The *Senecio radicans* complex, a somewhat amorphous taxonomic assemblage, is a group of succulent groundsels, most of which are native to Africa. Glennie et al. (1971) examined 25 "species" of this group for their flavonoids. Although the overall flavonoid profile of the group was comparatively simple, some interesting patterns of distribution did emerge. The pigment profile of the complex consisted of kaempferol [1] (See Fig. 2.2 for structures 1–5) and quercetin [2] 3-O-mono- and diglycosides,

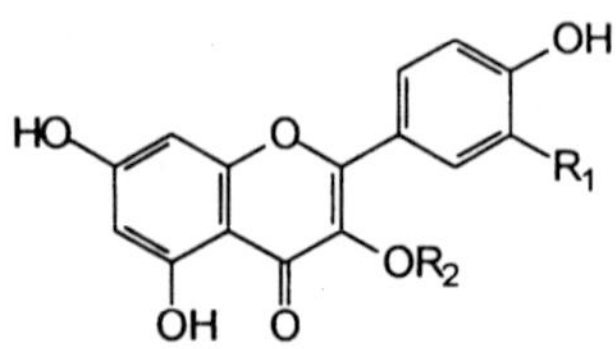

[1] $R_1 = H, R_2 = Sugar$
[2] $R_1 = OH, R_2 = Sugar$
[3] $R_1 = H, R_2 = CH_3$

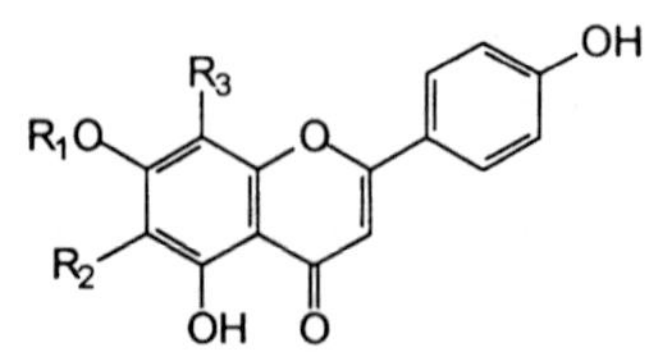

[4] $R_1 = Glc, R_2 = R_3 = H$
[5] $R_1 = H, R_2 = R_3 = Glc$

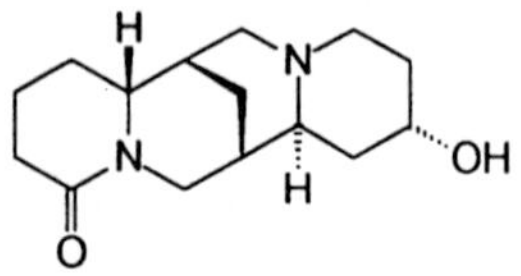

[6]

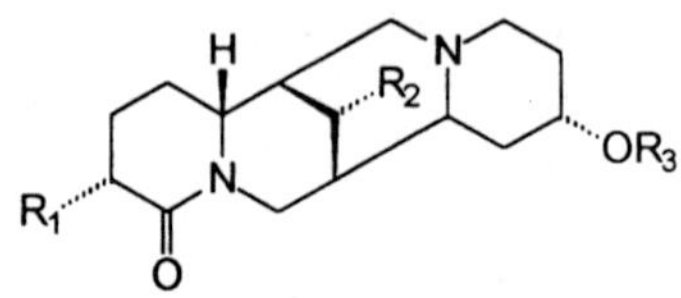

[7] $R_1 = R_3 = H, R_2 = OH$
[8] $R_1 = OH, R_2 = H, R_3 =$ Angelate
[9] $R_1 = R_2 = OH, R_3 =$ Angelate

Fig. 2.2 Compounds 1–5 from *Senecio radicans* complex; Compounds 6–9 from *Pearsonia*

quercetin 3-methyl ether [3], apigenin 7-O-glucoside [4], and 6,8-di-C-rhamnosyl-apigenin [5]. The first point of interest is that flavonoid profiles of individual taxa did not correlate with ploidy level. Individuals with 2*n* values of 20, 30, 40, 60, 80, 100, and 102 were included in the survey as well as *S. angulatus* L. f. that has $2n = 180$. The most striking profile difference reported involved *S. longiflorus* Sch. Bip. var. *madagascariensis* Rowl. from Madagascar, *S. longiflorus* var. *violacea* Berg. from Kenya, and *S. kleinia* Sch. Bip. from the Canary Islands. The flavonoid profiles of these taxa consisted solely of quercetin 3-O-glucoside, a compound not otherwise detected in the study. Quercetin 3-O-rutinoside, on the other hand, was observed in all but a scant few other specimens. Kaempferol 3-O-glucoside and 3-O-rutinoside were more scattered in their occurrence as was quercetin 3-methyl ether and the flavone derivatives. *Senecio angulatus*, with $2n = 180$, was unique within the taxa in having only apigenin 7-O-glucoside and 6,8-di-C-rhamnosylapigenin. Other geographic correlations include the observation that apigenin 7-O-glucoside occurs uniformly in taxa from South West Africa, but only sporadically in South African ones. 6,8-Di-C-rhamnosylapigenin, conversely, was not observed in taxa from South West Africa but was present in most specimens from South Africa.

Perhaps the most unusual observation in this study, other than the unique pigment profile in *S. angulatus*, is the simple and identical flavonoid profile in the Kenyan, Madagascaran, and Canary Islands specimens. A close relationship between the two varieties from East Africa is not difficult to appreciate. The occurrence of this profile in specimens from the Canary Islands, however, points to a closer relationship than the distance between these areas might suggest. There is no way to know, at least from the data presented, whether this represents a case of convergence of flavonoid biosynthetic capacities involving unrelated species, whether it points to a relationship based upon a long-distance dispersal event followed by differentiation of a daughter species, or whether the two extremes represent relics of a once broader distribution. Other features of these taxa would have to be studied in order to resolve that question.

2.1.3 **Pearsonia** ***(Fabaceae)***

Pearsonia (Fabaceae, Crotalarieae) is a genus of 12 species in tropical and Southern Africa with one in Madagascar (Mabberley, 1997, p. 535). Leaves of 59 specimens representing nine species were examined for alkaloids by Van Wyk and Verdoorn (1991). In all, eight compounds were detected. High levels of both qualitative and quantitative variation were observed in this study and attributed to four phenomena: (1) interspecific and intersubspecific differences; (2) developmental differences; (3) organ differences; and (4) interpopulational (geographic) differences. As an indication of the degree of geographic differences in alkaloid concentration in *P. cajanifolia* (Harv.), Polhill subsp. *cajanifolia,* and subsp. *cryptantha* (Bak.) Polhill collected in northeastern South Africa in the vicinity of Pretoria. Their data can be found in Table 2.1. The four characteristic compounds identified in this study are 13α-hydroxylupanine [6]; 8α, 13α-dihydroxylupanine [7], cajanifoline [8], and pearsonine [9] (See Fig. 2.2 for structures 6–9).

2.1.4 **Phytolacca dodecandra** ***(Phytolaccaceae)***

Interest in chemical variation in *Phytolacca dodecandra* L'Hert. stems from its potentially important biologically active constituents, perhaps the most significant of which is the molluscicidal compound "lemmatoxin." Attempts to measure variation within this taxon, which occurs widely on the African continent, have involved a study of morphological features (Adams et al., 1989) and the comparative chemical work to be described here (Adams et al., 1990). This more recent study involved analysis of nonpolar constituents from three populations from Ethiopia, and one

Table 2.1 Percentage of alkaloids in leaves of *Pearsonia cajanifolia* from sites in South Africa (after Van Wyk and Verdoom, 1991)

Taxa and Sites	Cmpd-[6][a]	Cmpd-[7]	Cmpd-[8]	Cmpd-[9]
Subsp. *cajanifolia*				
Kensington (no data)	25 22 54[b]	–	8 12 8	–
Northcliff (26°9′S, 27°58′E)	38 16 13	–	16 20 19	–
Magaliesberg (26°S, 27°33′E)	35 53 22	65 47 27	t – t	–
Subsp. *cryptantha*				
Blydepoort (no data)	t t t	– t t	–	99 85 99
Pilgrims Rest (24°55′S, 30°44′E)	8 40 23	–	5 t –	t – –
Lydenburg (25°10′S, 30°29′E)	14 t 6	– 13 –	– 7 1	–

[a] Compound numbers refer to structures in Fig. 2.1.
[b] Values are percent (%) for three plants from each population; t = trace.

each from Ghana, Kenya, Madagascar, Nigeria, Zambia, and Zimbabwe. The list of identified compounds comprises myristic, palmitic, and oleic acids, phytol palmitate, phytol oleate, and phytol linoleate, α-tocopherol and the long-chain alkanes eicosane (C_{20}), henicosane (C_{21}), tricosane (C_{23}), nonacosane (C_{29}), hentriacontane (C_{31}), tritriacontane (C_{33}), and pentatriacontane (C_{35}). In addition to the named molecules, 22 unidentified compounds were used in a statistical analysis. Interpretation of the contour diagrams representing the data set within this species is left to readers familiar with this analytical approach. Suffice it say, as the authors concluded, that there is a significant level of chemical differentiation evident on the continent. Dealing strictly with average values, the authors noted that northeastern African populations yielded 15.8–22.8% oleic acid compared to 1.8% in Zambian material. Hentriacontane (C_{31}) concentration varied from 52.3% in Zambian populations to 11.3% for plants collected in Madagascar. The level of α-tocopherol in plants from Zimbabwe was 5.8%, but only 1.4% in plants collected in Kenya (where the species appears to be rare). A further distinction was noted in the concentration of the phytol esters, which varied from 6.3 to 9.7% in Ghanaian plants, but reached no higher than 4.5% (phytol oleate in Kenyan material) in any other specimen. For the purpose of germplasm conservation, four areas were defined: (1) northeast Africa comprising Ethiopia and Kenya; (2) Madagascar; (3) Southern Africa comprising Zambia and Zimbabwe; and (4) West Africa comprising Ghana and Nigeria. It did not go without notice that dispersal of seed of different chemotypes of this species through human movement has been (and may still be) a contributing factor.

2.1.5 Aloë *(Asphodelaceae)*

Aloë is a genus of respectable size, with some 365 species distributed widely in Africa, Madagascar, the Arabian Peninsula, and the Canary Islands (Mabberley, 1997, p. 26). The taxa of special interest here are the shrubby species native to East Africa. Cutler et al. (1980) undertook a multidisciplinary study of the group with an interest in, among other things, the nature of the cuticular patterns on the epidermis of mature leaves. This feature offers a useful means of identifying most species, with closely related taxa exhibiting similar patterns. Twelve species were examined representing sites in Tanzania, Kenya, and Uganda. Six of these were diploids ($2n = 14$), exhibiting a range of morphological features, including the cuticular pattern. The other six, all tetraploids ($2n = 27, 28, 29$), were morphologically very similar, which were taken to suggest that they may have originated from a single source, possibly a diploid of the form of *A. morijensis* Carter and Brandham. The choice of shrubby diploid progenitors was narrowed from six to two on the basis of morphological features, *A. morijensis* and *A. fibrosa* Lavranos and Newton. It was also seen as relevant that these two taxa, along with *A. babatiensis* Christian and Verdoorn, yielded identical chromatographic profiles. *Aloë fibrosa* was eliminated as a candidate owing to its large size. On comparison of *A. morijensis* and the tetraploid *A. kedongensis,* Reynolds revealed such a level of similarity that formation of the latter from *A. morijensis* by straightforward autotetraploidization seemed a reasonable suggestion (Cutler et al., 1980). This event was thought to have occurred in the Morijo region of southwestern Kenya, followed by colonization northward up the Rift Valley, with branches to the east in the vicinity of Mt. Kenya and to the west into Uganda.

Chromatographic analyses in this genus have taken advantage of the presence of a class of leaf exudate compounds known as aloins. Barbaloin [10] (See Fig. 2.3 for structures 10–17), nataloin [11], and homonataloin [12] are typical members of this group of anthrone derivatives. Accompanying the anthrones in these plants are compounds such as aloesin [13] and aloenin [14]. Comparative chromatographic studies of the diploid and tetraploid shrubby aloes described by Reynolds (1986, 1990) support the origin of the tetraploids as suggested by Cutler et al. (1980).

The second study of *Aloë* involved the distribution of the naphthalene derivative "plicataloside" [15], first identified by Wessels et al. (1996) as a component

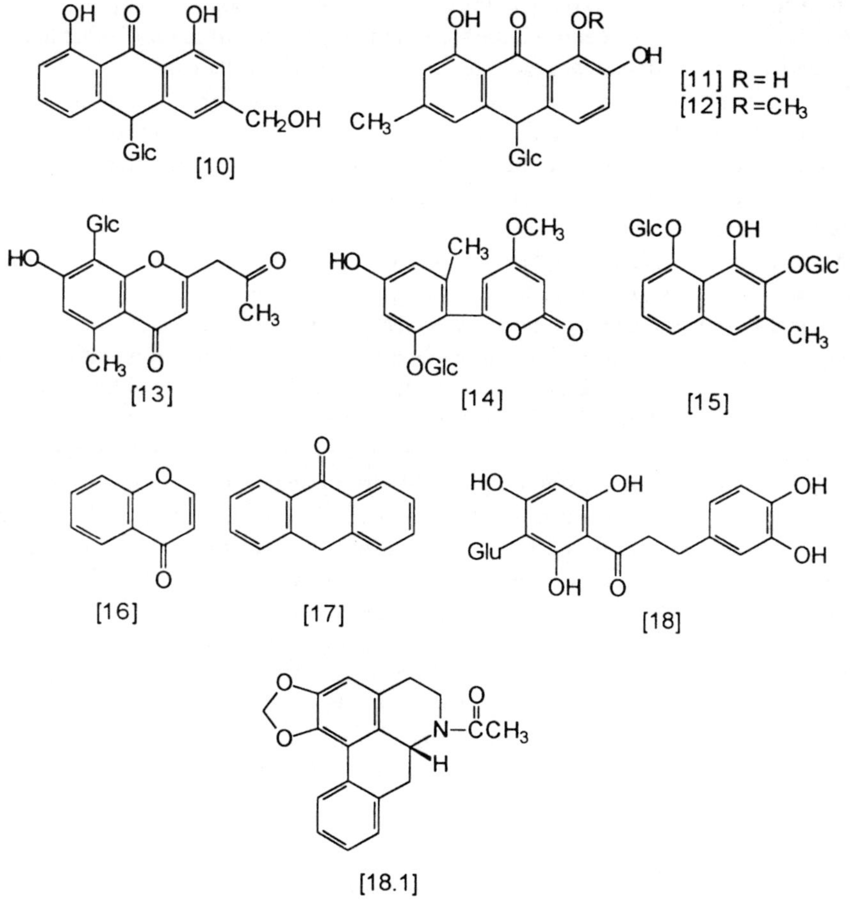

Fig. 2.3 Compounds 10–17 from *Aloë*; Compound 18 from *Aspalathus linearis*; Compound 18.1 from *Papaver aculeatum*

of *Aloë plicatilis* (L.) Miller leaf exudate. Those workers speculated that it might be restricted to this anomalous species, the sole member of section *Kumura*. A more recent study involving a much larger sampling of species revealed a very interesting geographic distribution pattern. Viljoen et al. (1999) investigated the leaf exudate chemistry of 380 taxa of *Aloë* using high-performance column chromatography coupled with a diode array detector. Twenty species were found to have plicataloside, often as the only detectable compound. It was accompanied in a few species by trace amounts of compounds normally encountered in *Aloë*, namely, chromones [16] and anthrones [17], or by unidentified substances. Most of the plicataloside-positive species occur in Kenya, Tanzania, and Uganda with outliers in extreme southwestern Angola (*A. palmiformis* Baker) and extreme southwestern South Africa (*A. plicatilis*, the original source). Other positive taxa originated in Mozambique, Zimbabwe, and northeastern South Africa. *Aloe schweinfurthii* Baker was also found to contain plicataloside. This taxon occurs in a narrow band running from southern Sudan through Central African Republic, southern Chad, northern Cameroon, and through central Nigeria, Benin, Togo, and east-central Ghana. Species endemic to Madagascar available for study lacked plicataloside.

The authors were very conservative about any phylogenetic speculations based upon the single chemical apomorphy. The genus is known to have relatively few morphological apomorphies at the infrageneric level, and no morphological feature was held in common by the plicataloside-positive taxa. The workers' (Viljoen et al., 1999) concluding remarks are best appreciated directly: "We do not wish to suggest that chemical characters should enjoy preference over the morphological characters as all problems (e.g., convergence) encountered with morphological characters are prevalent for chemical characters. However, we do believe that the presence of this unique compound should not be completely dismissed as chemotaxonomic coincidence, and we should at least explore further the possibility of taxonomic affinity between [sic] these plicataloside-containing taxa."

It is obvious in situations like this that a clear phylogenetic picture is necessary in order to gain some understanding of the distributional history of the taxa in question, in this case the plicataloside-positive ones. This would appear to be a situation where gene sequence information might well provide the needed framework. It would also be of interest to learn where the biosynthesis of plicataloside diverges from the pathway that leads to the anthrones normally seen in *Aloë*.

Viljoen et al. (2002) have described the occurrence of a new anthrone derivative, which they named "homonataloside B," from 14 species of *Aloë* (of the 380 tested). Species containing this compound, a hitherto unknown diglycoside, were obtained from scattered sites in East Africa, the Horn of Africa, and southern parts of the Arabian Peninsula. The compound was not observed in any of the species native to Madagascar.

2.1.6 Aspalathus linearis—*Rooibos tea (Fabaceae)*

Aspalathus is an endemic South African legume genus, particularly well represented in the Cape District. According to Mabberley (1997, p. 61), 278 species are recognized.

Leaves of *A. linearis* (Burm. F.) Dahlgren, provide the caffeine-free herbal "rooibos tea." Although cultivated commercially, this species occurs naturally. As pointed out in their paper on chemical variants (van Heerden et al., 2003), the species is highly variable with respect to habit, fire survival, vegetative structures, reproductive morphology, electrophoretic characteristics, and flavonoids (see their paper for leading references). Despite the availability of a variety of features, no infraspecific classification has been agreed upon. The chemical study by those workers revealed several profiles based upon, among others, the C-glycosides orientin, isoörientin, and vitexin, quercetin glycosides, and aspalathin [18] (See Fig. 2.3). Aspalathin belongs to one of the smaller groups of unusual flavonoids—the C-glycosyldihydrochalcones. Perhaps the most noteworthy finding was the observation that the combination of aspalathin and one of the unidentified flavonoids appears to characterize certain populations from the more northerly part of the taxon's range. However, some plants from southern populations did exhibit trace amounts of the dihydrochalcone along with major amounts of the other compounds.

2.1.7 Mangrove Waxes

Preliminary studies of leaf waxes from species of *Rhizophora* (Rhizophoraceae) in West Africa (Dodd et al., 1995) revealed a significant level of variation in both aliphatic hydrocarbons and triterpene derivatives. In the following year the work was expanded to a comparison of wax components of mangrove species from Gabon in West Africa, and Guyana on the northeastern coast of South America (Rafii et al., 1996). Most striking was the comparison between *Avicennia germinans* (L.) L. (Avicenniaceae) from the two coasts. Triterpenes in the Guyana material amounted to less than 2%, whereas these compounds in the plants from Gabon comprised 10–23% of total extract. The hydrocarbon fraction also showed large differences with the C_{28} fraction ranging between 38.6% and 66.7% for Guyanan plants and between 10.9% and 23.1% for those from Gabon. The ranges were reversed for C_{32} hydrocarbons, with the values for Guyanan specimens ranging from 1.1% to 2.0% and the Gabonese specimens weighing in the range of 16.8–28.4%. It is interesting to note that some taxonomists have recognized the plants from Gabon as *Avicennia africana* Beauv. in spite of the absence of clear morphological differences between them and specimens from South America.

2.1.8 **Papaver aculeatum** *(Papaveraceae)*

The genus *Papaver* consists of 80 species (Mabberley, 1997, p. 525), nearly all of which are Northern Hemisphere in occurrence. One species occurs in the Cape Verde Islands and another, the subject of this entry, is found in South Africa and is naturalized in Australia. *Papaver aculeatum* Thunb. (*P. gariepinun* Burch. ex DC; *P. horridum* DC) occurs widely in South Africa in areas characterized by summer rainfall and can be found from sea level to 2800 m in appropriate habitats. There have been suggestions that it is most closely allied to taxa in the Mediterranean region. A recent paper by Langlois et al. (2004) examined the alkaloid components of *P. aculeatum* in an attempt to see if chemotaxonomic information might assist in pointing to relationships within the genus, particularly with regard to sectional affinities. The major compound present in all parts of the plant was identified as *N*-acetylanonaine [18.1, Fig. 2.3], a noraporphine alkaloid not otherwise known in the genus, although members of sect. *Pilosa* have yielded members of this group of compounds.

2.1.9 Ocimum gratissimum *(Lamiaceae)*

Ocimum gratissimum L., the tree basil, is a highly variable taxon with a center of origin in Africa. Two subspecies have been recognized, subsp. *gratissimum* and subsp. *macrophyllum* Briq., based upon stem, leaf, and inflorescence characteristics, but levels of variation in these morphological traits are significant. In an effort to develop additional sources of data to address the problem, Vieira et al. (2001) undertook a study of volatile components, flavonoids, and random amplified polymorphic DNA (RAPD) markers. The number of samples used in the study was small—only a dozen accessions—but the amount of variation in both sets of secondary metabolites suggests that a detailed examination of a much larger set representing the geographical range of the taxon, as well as cultivated material, would repay the effort. Both qualitative and quantitative differences were observed, which would allow detailed statistical analysis, an approach that the reader will see below to have revealed significant groupings in several taxa.

2.2 Australia and New Zealand

2.2.1 Chamelaucium uncinatum—*The Geraldton Waxflower (Myrtaceae)*

Chamelaucium is a genus of a dozen species endemic to southwestern Australia. The species under investigation here, *C. uncinatum* Schauer, occurs along the coast from somewhat north of Perth (ca. 29°S) to north of Geraldton (ca. 27°30′S), over a distance of approximately 350 km (Fig. 2.4). This species has provided several cultivars for the cut-flower trade, where it is known as "waxflower" or "Geraldton waxflower." A recent study of the species sampled from its entire range demonstrated the existence of clear-cut monoterpene races (Egerton-Warburton et al., 1998).

Leaf material representing 38 sites was collected from plants cultivated under common garden conditions (University of Western Australia). Gas chromatographic analysis of the leaf oil revealed 33 components, nine of which accounted for 80.5%

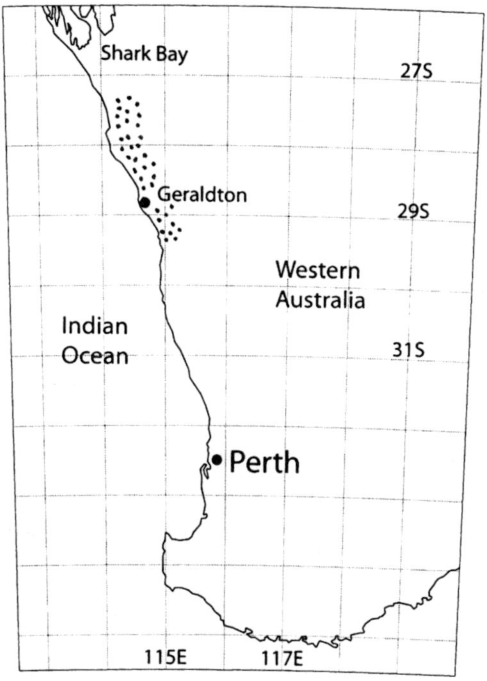

Fig. 2.4 Map of Western Australia showing location of Geraldton after which the Geraldton waxflower (*Chamelaucium uncinatum*) is named. Stippled area represents an approximation of the taxon's range

of the total terpene variation as shown by principal-components analysis. Monoterpenes identified included the acyclic compounds geraniol [19] (See Fig. 2.5 for compounds 19–25), linaloöl [20], and citronellal [21], and cyclic compounds such as α-pinene [22], ß-pinene [23], limonene [24], and α-terpenyl acetate [25]. Cluster analysis based on the mean percentage of leaf oils revealed four terpene races: (1) citronellal dominant, (2) limonene dominant, (3) α-pinene dominant, and

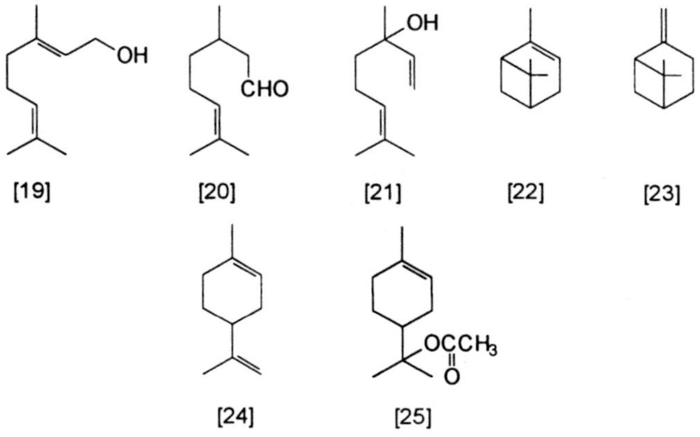

Fig. 2.5 Compounds 19–25 from *Chamelaucium uncinatum*, the Geraldton waxflower

(4) a group consisting of limonene and α-pinene as codominant constituents. The northern-most region was dominated by a race, represented by six sites, in which citronellal was the major component, with concentrations ranging from 36.98% to 62.24% (Ave. = 48.32%). No other site yielded oil with higher than 25.63% citronellal. The southernmost race (six sites) had limonene as the dominant terpene, with concentrations falling in the range 43.18–56.92% (Ave. = 49.91%). The highest concentration of limonene observed otherwise was 42.98% in one of the codominant races. Populations in the middle of the range of this species tended to exhibit mixed terpene compositions. The workers concluded that the citronellal race should best be considered as an ecotype that has only recently evolved, a view supported in floral bud and leaf morphology.

2.2.2 Duboisia myoporoides *(Solanaceae)*

The genus *Duboisia* consists of three species limited in occurrence to eastern Australia (Fig. 2.6) and parts of New Guinea. *Duboisia myoporoides* R. Br., the taxon of interest to us in the present context, is solely Australian in its occurrence. Evidence for the existence of chemical variation within the species was first described by Loftus Hills et al. (1953, 1954) who observed that scopolamine [26] (see Fig. 2.7 for structures 26–30) was the dominant alkaloid in plants collected north of Gosford, New South Wales, whereas hyoscyamine [27] was the dominant alkaloid in collections south of Gosford. (Gosford, NSW, lies roughly between Newcastle and Sydney at 33°25′S, 151°18′E.) Later, a third variant was discovered in a region known as the Acacia Plateau, near Killarney, Queensland (southwest of Brisbane), that was characterized by the possession of the pyridine-based alkaloids, nicotine [28] and anabasine [29] (see also Mortimer and Wilkinson, 1957). Other minor components reported, according to Gritsanapan and Griffin (1991), included valeroidine, valtropine, tigloidine, tropine, butropine, norscopoline, poroidine, isoporoidine, acetyltropine, aposcopolamine, noratropine, atropine, and tetramethylputrescine. None of these appears to figure significantly in the geographic patterning.

The most recent work, by Gritsanapan and Griffin (1991), extended the range of study to more northerly sites in Queensland and included populations at different elevations. Four sites were studied: (1) northern Queensland at elevation 1000 m; (2) near Tinaroo Dam on the Atherton Plateau (near Cairns, Qld.) at 2200 m; (3) Mt. Glorious (near Brisbane) at 700 m; and (4) a revisited site at Acacia Plateau at an elevation of about 1000 m. Scopolamine was the major alkaloid present in plant material from Tinaroo Dam, from Mt. Glorious, and from coastal northern Queensland. Accompanying scopolamine in the Mt. Glorious plants were nicotine and anabasine. Plants from Acacia Plateau yielded a small amount of scopolamine, which was the only tropane alkaloid detected, but major quantities of nicotine, anabasine, and hygrine [30]. Hygrine was also observed as a lesser component of plant material from the Mt. Glorious site. It is instructive to recognize that hygrine, or a biologically active form thereof, lies on the biosynthetic pathway leading to the

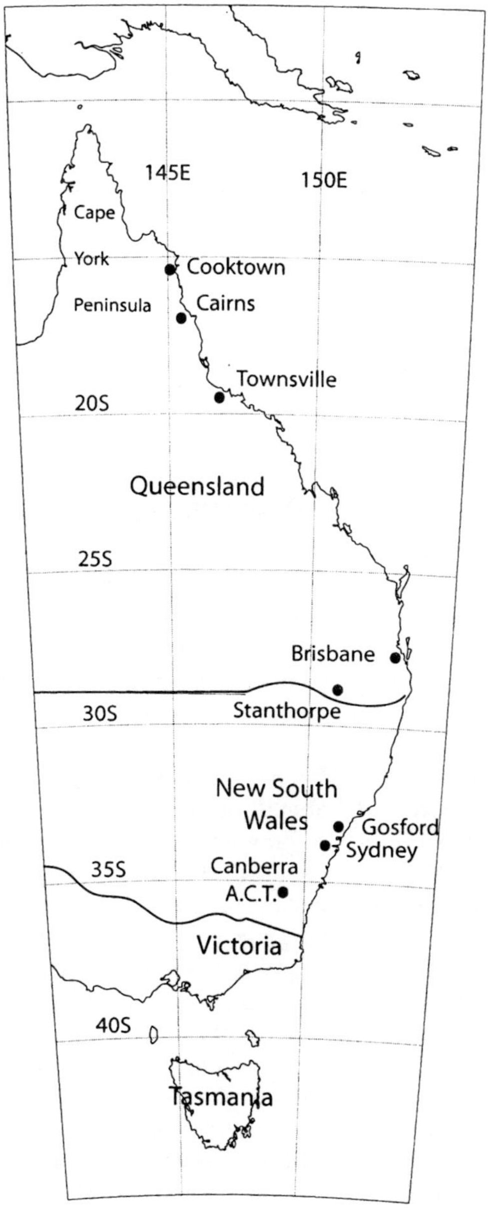

Fig. 2.6 Map of Eastern Australia and Tasmania showing places referred to in the discussion of *Duboisia myoporoides*

tropane alkaloids. Thus, the presence of hygrine as a major alkaloid, replacing scopolamine in this instance, represents not so much the replacement of one alkaloid by another, as the accumulation of an intermediate owing, presumably, to the absence or reduced functionality of at least one enzyme in the pathway. It would be of interest to determine which enzyme is involved.

Fig. 2.7 Compounds 26–30 from *Duboisia myoporoides*

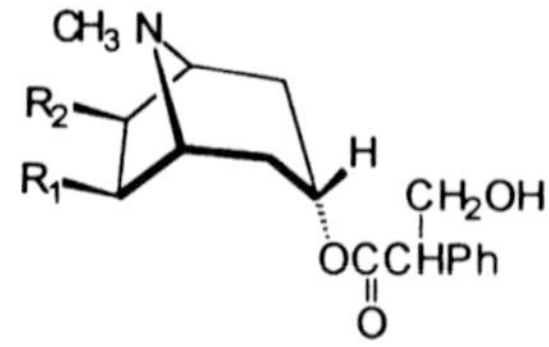

[26] $R_1 + R_2$ = epoxide
[27] $R_1 = R_2$ = H

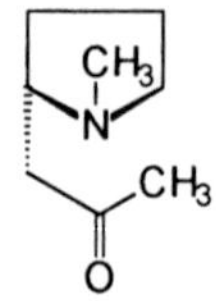

[30]

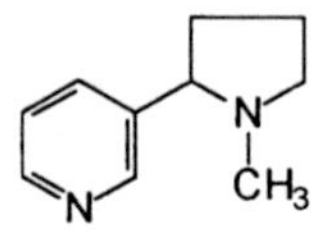

[28]

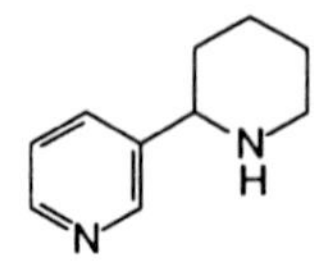

[29]

2.2.3 Zieria *(Rutaceae)*

Zieria, a genus of about two dozen species in Rutaceae, is endemic to eastern Australia, including Tasmania, except for one species, *Z. chevalieri* Virot, which is a native of New Caledonia. A detailed study of the genus has shown it to be a rich source of secondary compounds, including a cyanogenic glucoside (based on *p*-hydroxybenzaldehyde) (Flynn and Southwell, 1987a) and an array of simple phenolic and terpene-derived volatile compounds (Flynn and Southwell, 1987b). We are interested here is an examination of chemical variation within the genus studies carried out by Southwell and Armstrong (1987). Collections of the *Z. arborescens* Sims aff. *arborescens-smithii* group from Tasmania and along the southern coast of Victoria, north to the area around Cooktown, Queensland (Fig. 2.6) provided an interesting picture of biosynthetic specialization over a range of 25° of latitude. There is a significant decline in the zierone [31] (see Fig. 2.8 for structures 31–40) content and a corresponding increase in the phenolic ethers, eugenol [32], safrole [33], elemicin [34], 2,3,4,6-tetramethoxystyrene [35], and 2,6-dimethoxy-3,4-methylenedioxystyrene [36], as one goes northward from Tasmania to northern coastal Queensland. Although there is a moderate degree of variation among populations in any given area, the overall trend in carbon utilization in the biosynthesis of these different classes of compounds is impressive.

Other observations on chemical variation within *Zieria* can be mentioned at this point. *Zieria cytisoides* "form b" occurs along the coast of New South Wales and in similar habitats in Tasmania. Five populations concern us here, two from the former area and three from the latter. Significant differences in the concentration of the monoterpene derivative (–)-chrysanthenone [37] were observed by Southwell and Armstrong (1987). The populations on the mainland exhibited concentrations of the ketone of 15% and 18%, whereas those in Tasmania had higher amounts, that is, 42%, 49%, and 55%. The New Caledonian *Z. chevalieri* was shown to exhibit an unusual profile of compounds, relative to the Australian species, with 2,4,6-trimethoxystyrene [38],

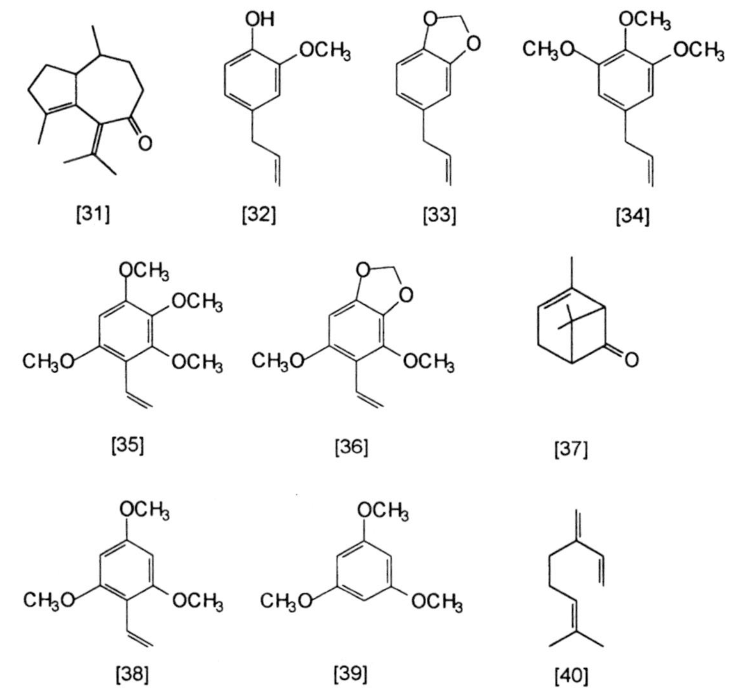

Fig. 2.8 Compounds 31–40 from *Zieria*

1,3,5-trimethoxybenzene [39], and the open-ring monoterpene myrcene [40] as the prominent constituents.

2.2.4 Melaleuca *(Myrtaceae)*

Melaleuca alternifolia (Maiden and Betche) Cheel, locally known as the Australian "tea tree" or as "paperbark," is a commercially important oil plant native to northeastern New South Wales and southern Queensland. Since the concentration of 1,8-cineole [41] (see Fig. 2.9 for structures 41–43) figures significantly in the usefulness, and hence value of the oil, considerable interest has been shown in locating 1,8-cineole-low populations. Over the years, studies have revealed the existence of several chemotypes throughout the range of the species (Penfold et al., 1948a, b; Southwell et al., 1992; Butcher et al., 1994). The most extensive survey of the tea tree for oil composition is the recent study by Homer et al. (2000), which involved 615 individual trees, 15 each from 41 sites covering the entire range of the species. This range with the coordinates 28°34′48″–31°22′23″S and 151°34′53″–153°19′36″E is centered roughly near Stanthorpe, Queensland (on the border with New South Wales) (Fig. 2.6).

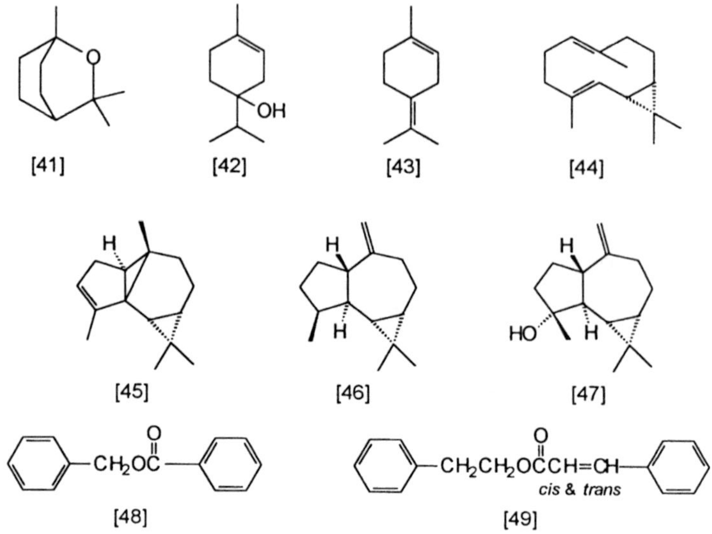

Fig. 2.9 Compounds 41–43 from *Melaleuca alternifolia*; Compounds 44–49 from *Isotachis lyallii*

The existence of six chemotypes was documented by the work of Homer et al. (2000) who also established a set of formulae for comparing results obtained by steam distillation versus static headspace gas chromatography (GC). Although several terpenes are regularly observed in tea tree oils, three compounds were clearly the most significant and were used by the present workers to define the chemotypes, terpinen-4-ol [42], terpinolene [43], and 1,8-cineole. Chemotype **1** is defined by the presence of a high concentration of terpinen-4-ol, accompanied by lesser amounts of α-pinene, α-terpinene, p-cymene, and γ-terpinene (see Fig. 2.35 for cymene and γ-terpinene structures). Chemotype **2** is characterized by high levels of terpinolene. Chemotype **5** is marked by high concentrations of 1,8-cineole and has the associated compounds limonene and α-terpineol in lesser amounts. Chemotypes **3**, **4**, and **6** are dominated by 1,8-cineole, but differ with regard to the relative amounts of terpinen-4-ol or terpinolene. (Chemotype numbering is an historical artifact.) Geographical differentiation of the populations was observed nearer the coast, with a predominance of chemotype **1** (high terpeinen-4-ol) in the north (south of Casino, NSW) and a predominance of chemotypes **4** and **5** (higher 1,8-cineole) in the south (south of Grafton, NSW), but no clear-cut line of demarcation appears to exist. Chemotype **2**, however, which occurs in the vicinity of Stanthorpe, Queensland (southwest of Brisbane), is separated by a distance of 40–50 km from the coastal groups. It has also been noted that the Queensland specimens afford smaller yields than do trees from New South Wales, and that this difference is likely due to the drier conditions in which the former occur (Butcher et al., 1994).

Homer et al. (2000) speculated that differences in oil composition may reflect localized inbreeding leading "some stands or areas becoming more pure-breeding for the different oil chemotypes." They went on to conclude that low frequencies of different chemotypes in a given area may "indicate low incidence of recessive gene expression, occasional inflow of genetic material from wider afield (e.g., during flooding), or some other genetic control mechanism over oil chemotypes." Genetic differentiation between the NSW and Queensland populations, using other techniques, has also been revealed (Butcher et al., 1992; Aitken et al., 1998; Rossetto et al., 1999). As in numerous other situations encountered in the literature, a population phylogeny of *M. alternifolia* would likely provide valuable insights into the evolutionary relationships within this species.

A detailed statistical analysis of chemical composition and morphological differences of trees from 40 populations was recently described by L. S. Lee et al. (2002). Three drainage catchments were represented, the Clarence and Richmond Rivers in NSW and the Severn River in Queensland. The two NSW sites represent warm and moist environments; the Severn site is higher and drier. The results of this set of analyses pointed to the existence of significant differences in chemotype representation in the different catchments. Also, this is the first report that there is a relationship between chemotype and tree size, with the trees from the Queensland site being the smaller. The higher level of chemotype differentiation in the Queensland population lies in opposition to the lower level of genetic variation recorded for trees from that area as pointed out above (Rossetto et al., 1999). Whether the differences recorded in the recent paper are the result of genetic divergence or to environmental differences has not been firmly established.

Melaleuca quinquenervis also ranges along the eastern coast of Australia from near Sydney in New South Wales, north to the Cape York Peninsula, Queensland and along the southern coast of Papua New Guinea and in New Caledonia. Ireland et al. (2002) made extensive collections of leaves of this species for study of essential oil content. Two clear-cut chemotypes emerged, one characterized by *E*-nerolidol (74–95%) and linaloöl (14–30%), both acylic C-10 terpene alcohols, representing the "southern" race, and one characterized by combinations of 1,8-cineole (10–75%), viridiflorol (13–66%), α-terpineol (0.5–14%), and β-caryophyllene (0.5–28%) [98] comprising the "northern" race. In addition to the compositional differences, oil yield differed significantly between the two areas with yields less than 0.5% from northern race plants and yields of 1.5–3.0% in southern populations. The southern race extended from Sydney northward to Selection Flat, NSW with a disjunct population farther north near Maryborough in Queensland. It was possible to discern sub-races within the southern chemotype based upon linaloöl concentration. The northern form occurred throughout the range of the species, but was only rarely seen in the south. Variation of the major components appeared to vary continuously. The authors offered no explanation for the existence of several chemotypes within this species, but did note that the time of year during which collection took place did not influence the results.

2.2.5 Isotachis lyallii *(Balantiopsidaceae)*

Isotachis lyallii Mitt. is a liverwort that occurs on both the north and south islands of New Zealand. A study of the lipid fraction of this species collected from both islands revealed different secondary metabolite profiles involving sesquiterpenes and aromatic esters (Asakawa et al., 1997). The sesquiterpene array consisted of

Table 2.2 Occurrence of sesquiterpenes and aromatic esters in the liverwort, *Isotachis lyallii*, in New Zealand (from Asakawa et al., 1997)

Compounds[a]	South Island	North Island
Bicyclogermacrene [44]	++[b]	++
Anastreptene [45]	+++	+++
Aromadendrene [46]	+	–
Spathulenol [47]	+	++
Benzyl benzoate [48]	–	++
Phenylethyl *cis*-cinnamate	++	–
Phenylethyl *trans*-cinnamate	++++	–

[a] Structures shown in Fig. 2.9.
[b] Relative concentrations.

bicyclogermacrene [44] (see Fig. 2.9 for structures 44–49), anastreptene [45], aromadendrene [46], and spathulenol [47], while the ester array consisted of benzyl benzoate [48] and both isomers of β-phenylethyl cinnamate [49]. The occurrence of these compounds from the North Island and South Island are given in Table 2.2. The authors suggested that spathulenol might be an artifact formed from bicyclogermacrene by autooxidation, as shown by Toyota et al. (1996). Nonetheless, spathulenol might also be derived from aromadendrene by enzymatic oxidation. If that were the case, the absence of aromadendrene in the profile of North Island plants might also be explained by its efficient conversion to the hydroxylated derivative, spathulenol. The authors reported that the concentration of spathulenol was higher in the North Island plants, which is in accord with the more efficient utilization of its precursor. It is not possible to conclude much on the ester differences other than that it is possible that either or both of the precursor fragments could be lacking. The authors noted as well that plants from the South Island also had the phytosterols campestrol and stigmastrol; no comments on that apparent difference were offered.

2.2.6 Leptospermum scoparium *(Myrtaceae)*

Leptospermum scoparium J.R. & F. Forst, known by the Maori name "manuka," occurs widely throughout New Zealand. Owing to the potential of manuka as an oil-producing species, Perry et al. (1997a) investigated the steam-volatile components of specimens representing much of its range in New Zealand as well as specimens grown in a common garden. Gas chromatographic analysis of the distillates from 15 sites showed some 50 or so peaks that represented three groups of compounds, monoterpenes, sesquiterpenes, and "triketones." Three chemotypes emerged from these analyses that agreed reasonably well with morphological differences that had been noted earlier (unpublished observations of W. Harris cited in the current paper). A northern chemotype was distinguished on the basis of its high concentration of α-pinene, an East Cape chemotype was characterized by the presence of high concentrations of the triketones leptospermone [50], isoleptospermone [51], and flavesone [52] (see Figs. 2.10 and 2.11 for structures 50–52), and a southern

Fig. 2.10 Triketones, compounds 50–52, and flavonoids, compounds 53–56 from *Leptospermum scoparium*

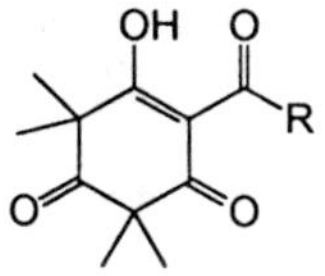

[50] R = isobutyl
[51] R = sec-butyl
[52] R = isopropyl

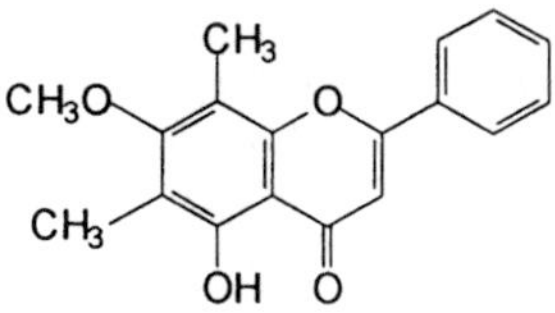

[53]

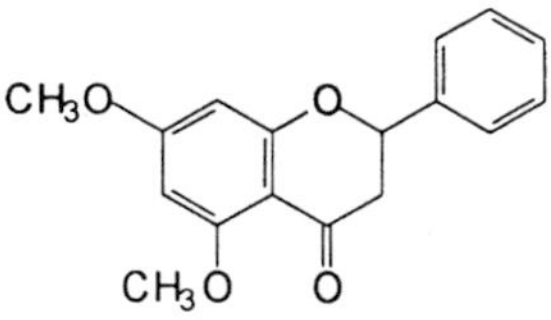

[54]

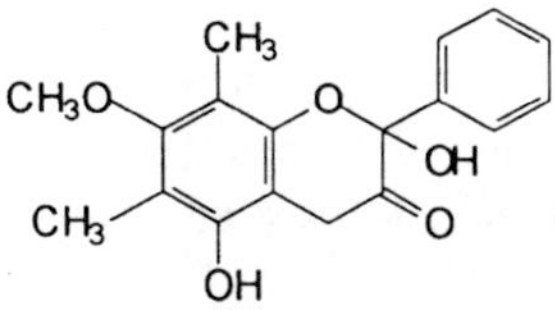

[55]

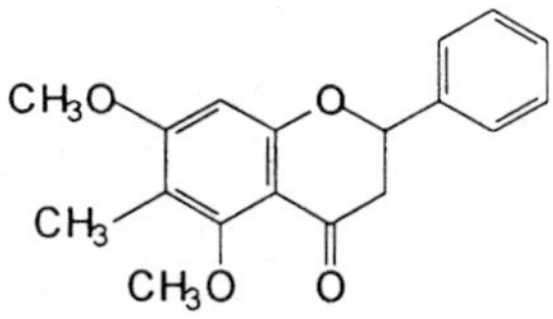

[56]

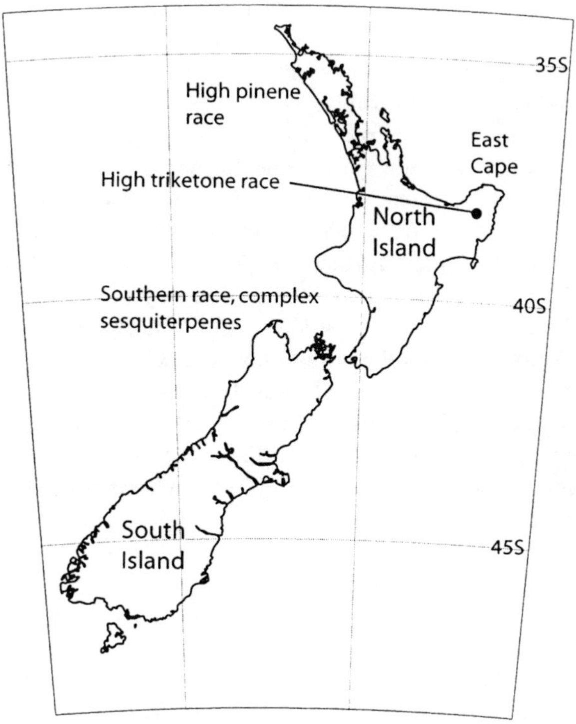

Fig. 2.11 Map of New Zealand showing general locations of the three chemotypes of *Leptospermum scoparium*

Table 2.3 Levels of selected compounds and total monoterpenes, sesquiterpenes, and triketones from *Leptospermum scoparium* (from Perry et al., 1997a)

Compound	Australian	N.Z. North.	N.Z. East Cape	N.Z. South
α-Pinene	17.3(7.3)[a]	22.5(3.7)	0.7(0.9)	2.9(3.8)
p-Cymene	4.2(3.6)	0.8(0.8)	0.2(0.0)	0.6(0.7)
1, 8-Cineole	19.9(16.6)	1.6(0.3)	0.3(0.1)	0.7(1.0)
Leptospermone	0.4(0.6)	0.8(0.5)	18.9(0.7)	1.7(1.1)
Isoleptospermone	0.4(0.6)	0.0(0.1)	5.3(0.9)	0.5(0.6)
Flavesone	0.1(0.2)	0.1(0.1)	8.3(0.3)	0.1(0.2)
Total monoterpenes	50.6(24.4)	39.9(4.6)	3.0(1.1)	11.8(9.1)
Total sesquiterpenes	24.7(16.0)	42.1(6.0)	53.7(3.0)	64.7(10.3)
Total triketones	1.2(0.9)	1.0(0.5)	32.5(1.9)	2.4(1.6)

[a] Percentage of total gas chromatography (GC) peak area (standard deviation).

chemotype characterized by the highest level of sesquiterpenes observed in the study, along with an array of monoterpenes (Fig. 2.11). Values for the levels of the defining compounds of these chemotypes, along with those of the three classes of compounds, are summarized in Table 2.3.

Included in the table are data for *L. scoparium* obtained from five sites in Australia. The Australian material exhibited a level of α-pinene approaching that of the northern race from New Zealand, but was clearly characterized by the highest level of 1,8-cineole observed in the study, along with the highest value for total monoterpenes and lowest value for total sesquiterpenes. The authors noted the need for further taxonomic study, particularly of the Australian material.

Häberlein and Tschiersch (1998) have documented geographical variation in the external leaf flavonoids of *L. scoparium.* Plants representing a range of climatic and geological situations were collected from seven sites from the north and three from the northern part of the south islands of New Zealand (Fig. 2.11). Detailed isolation and structural studies afforded an array of O- and C-methylated B-ring deoxyflavones and flavanones, including a very unusual member of the latter class (see Fig. 2.11 for structures 53–56). Characteristic of these are 5-hydroxy-6,8-di-*C*-methyl-7-methoxyflavone [53], 5,7-dimethoxyflavanone [54], and the unusual flavanone 2,5-dihydroxy-6,8-di-*C*-methyl-7-methoxyflavan-3-one [55]. Plants from the four most northerly sites, Auckland, Coromandel, Rawhiti, and Whangaruru North, exhibited the highest concentrations of flavonoids, as well as the full range of compounds identified, four flavones and five flavanones, including the unusual compound [54]. Plants from the other six locations afforded lower amounts of flavonoid material and simpler profiles. The least complex profiles, comprising only two compounds, 5,7-dimethoxy-6-*C*-methylflavanone [56] and 5,7-dimethoxyflavanone, were two from the North Island (Te Urewara and Tongariro) and one from the South Island (Marlborough Administrative Region). The unusual flavanone was observed in five of the northern plants and one of the southern ones, which excluded it as a marker. The possibility that these pigment profiles represent adaptations to local conditions was not pursued, although the authors clearly indicated that they were aware that environmental factors might play an important part in this system.

Returning to the terpenes for a moment, it is interesting to note that Perry et al. (1997b) also examined the essential oils of *Kunzea ericoides* (A. Rich.) J. Thompson

(Myrtaceae), known locally as "kanuka." Fifty-one specimens from New Zealand along with six from Australia (and other species of *Kunzea*) were included in the study. Oils from Australian and New Zealand specimens were quite similar with α-pinene as the most prominent component in both, along with lower but similar levels of *p*-cymene and 1,8-cineole. Although two chemotypes can be defined within this taxon, they occur within populations and do not show the geographic patterning seen in manuka.

2.2.7 Chionochloa *(Poaceae)*

Chionochloa, known locally as "snow tussock," consists of 22 species, 21 of which are native to New Zealand. The remaining species occurs in southeastern Australia and does not figure in the discussion. Studies of triterpene methyl ethers (TMEs) have revealed the existence of different "chemodemes" (Connor and Purdie, 1976, 1981). In some instances, it was shown that TMEs were present in populations of a given species in one part of its range but absent from others: arundoin [57] (see Fig. 2.12 for structures 57–59) was identified from plants collected at Harpers

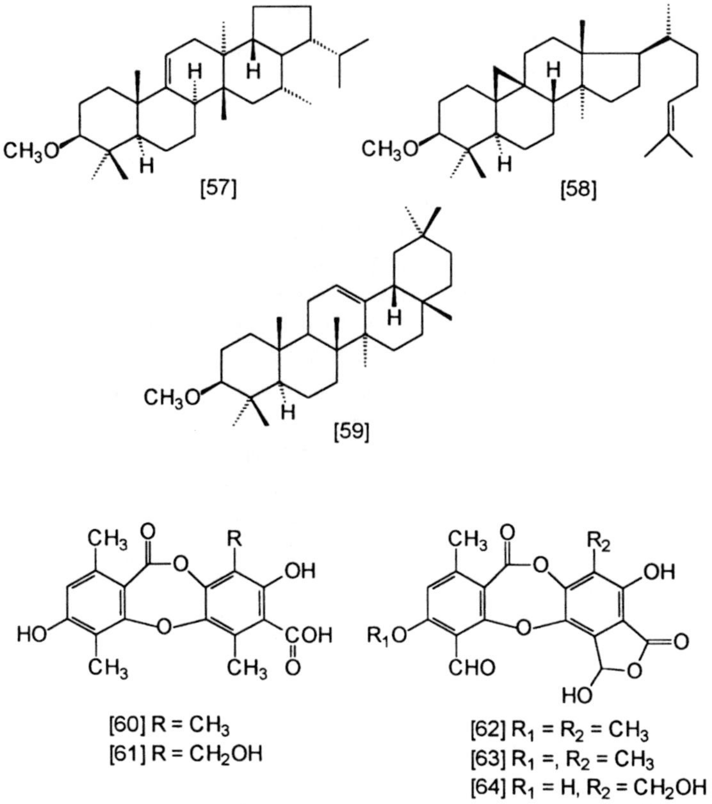

Fig. 2.12 Compounds 57–59, triterpene methyl ethers (TMEs), from *Chionochloa*. Compounds 60–64, lichen-acid derivatives from European *Ramalina siliquosa*

Knob (southern Canterbury Administrative Region; see Fig. 2.11) and cycloartenol methyl ether [58] was found in plants from Clarke River, whereas four other collections failed to yield any TMEs. Similarly, populations of *C. rigida* (Raoul) Zotov from eastern South Island do not synthesize TMEs, whereas those from western South Island were shown to accumulate β-amyrin methyl ether [59]. An additional difference between populations from these two sites lies in the capacity of the former to produce short-chain wax components, as opposed to long-chain compounds from the latter (Cowlishaw et al., 1983). Different chemodemes were also described for *C. cheesemanii* (Hackel ex Cheesem.) Zotov; plants from North Island exhibited lupeol methyl ether as the dominant compound along with lesser amounts of arundoin and two unidentified compounds, whereas populations from the South Island had arundoin as the major compound with lesser amounts of lupeol methyl ether.

The study by Cowlishaw et al. (1983) involved an examination of the epicuticular wax chemistry of four species of the genus, *C. flavescens* Zotov, *C. pallens* Zotov, *C. rigida*, and *C. rubra* Zotov, with regard to the chain lengths of acids, alcohols, aldehydes, alkanes, and wax esters. These compounds consist of long-chain, unbranched hydrocarbons with the appropriate functional group at one end. Wax esters involve a long-chain alcohol and a long-chain acid in a normal ester linkage. The major acids, aldehydes, and alcohols identified ranged between C_{24} and C_{32}, the alkanes ranged between C_{29} and C_{33}, and the esters ranged between C_{36} and C_{52} (recall that these larger numbers represent sums of alcohol and acid carbons). Differences in the profiles of wax components led to the recognition of distinct chemical regions for the studied taxa. For example, *C. rigida* assorted into eastern and western South Island types (mentioned above with respect to TMEs). Three wax chemical types were evident in *C. flavescens*, one encompassing populations from Canterbury and near Nelson, a second consisting of populations from the western part of South Island, and a third consisting of populations from the southern part of North Island. Chemodemes were also evident within both *C. pallens* and *C. rubra*, both of which sorted into four types. The overall outcome of these studies was the suggestion that the northwest Nelson district could be the center of biosynthetic diversity with diversification having occurred along three routes, one to the western part of the South Island, one to eastern South Island, and one to the southern part of North Island. As differentiation proceeded, enzymes involved in the elongation steps of fatty acid biosynthesis and those that catalyzed decarboxylation underwent changes in their specificity in response to selection pressures in the respective environments. It would be of considerable interest to see what gene sequence studies within this system would yield.

2.3 Europe

2.3.1 Ramalina siliquosa *(Ramalinaceae)*

Although this example represents one of the most thoroughly studied examples of local chemical differentiation, a little background involving a broader area is useful to put the specific study in perspective. The *Ramalina siliquosa* (Huds.) A. L. Sm. species complex consists of moderately large, fruticose, sexually reproductive organisms widely distributed on acid substrates along the coasts of western and northern Europe. The chemistry of this species complex has been described in considerable detail (Culberson and Culberson, 1967; Culberson, 1967, 1969, 1970). In all, six chemical races have been described, five differing in the nature of their β-orcinol derivative, with the sixth characterized by the absence of medullary substances. Structures of the compounds identified in this study are protocetraric acid [60], hypoprotocetraric acid [61], stictic acid [62], norstictic acid [63], and salazanic acid [64] (see Fig. 2.12 for structures 60–64).

The salazanic acid race is the most northerly, occupying suitable habitats in arctic Norway, Iceland, and throughout the Baltic region. The norstictic acid race and medullary negative races occur in southwestern Norway, western Great Britain, and in the Brittany Peninsula. The hypoprotocetraric acid race is common in western Great Britain and France, and extends south to Portugal. Finally, the protocetraric and stictic acid races are not only most common in Portugal, but also occur northwards to southern Norway. All six races occur "sympatrically" in western coastal Wales. In the strict sense, however, use of the term sympatric is incorrect. Careful analysis of individuals along several transects on a rocky headland (Culberson and Culberson, 1967; see for photograph) revealed a significant level of specialization of chemical race in (apparent) response to the degree of harshness of the particular aspect of the site. Thus, the southern (Portuguese) race, characterized by hypoprotocetraric acid, occurs on the northern-most, sheltered side of the promontory. The stictic acid race occurs on the most exposed faces in this locale, the seaward faces to the west and south sides. Plants characterized by norstictic acid occupy intermediate sites. These studies revealed that an otherwise apparently homogeneous area can harbor chemically different races separated by no more than a few centimeters. How the different chemicals elaborated by these races aid in the survival of individuals in their respective habitats, if in fact they do, remains an unanswered question, at least so far as this writer is aware.

2.3.2 Cyanogenesis in Clover (Trifolium, Fabaceae)

The study of cyanogenesis in European clover involves both latitudinal and elevational transects and represents one of the most frequently cited examples of pattern variation in the production of a secondary metabolite by a flowering plant. We can introduce this example with an interesting tale, related by Briggs and Walters (1997), involving the discovery of cyanogenic glycosides in species of *Lotus* that were toxic to transport animals used by British forces in the Sudan campaign at the end of the nineteenth century. It was subsequently demonstrated that some plants of *Lotus corniculatus* L., one of the species responsible for the poisonings, gave a positive reaction for hydrocyanic acid (HCN), whereas others did not (Armstrong et al., 1912). Subsequently, Dawson (1941) reported an HCN polymorphism within

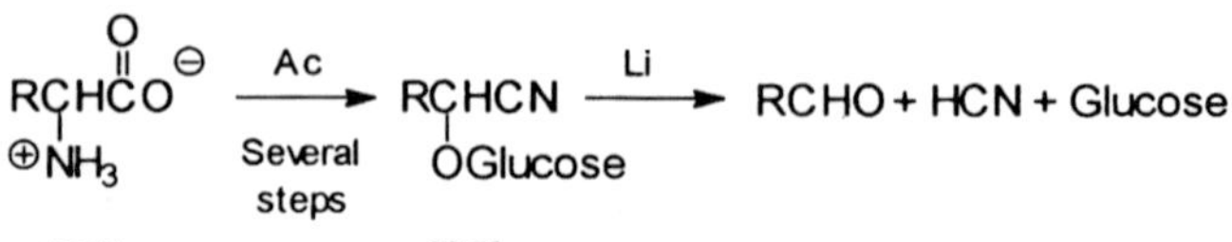

Fig. 2.13 Generalized reaction sequence from starting amino acid [65] to product cyanogenic glycoside [66]

L. corniculatus in southern England, and demonstrated that presence of cyanogenic glycosides is dominant over their absence. Further work on inheritance of cyanogenesis in this species was complicated by the fact that this clover is a tetraploid. Work with other species led to more definitive results, however.

The genetics of cyanogenic glycoside formation were determined using *Trifolium repens* L. (Corkill, 1942; Atwood and Sullivan, 1943). Two loci were identified, one that controls formation of the cyanogenic glycoside (*Ac*) and one, (*Li*), that controls synthesis of the glycohydrolase (also referred to as linamarinase). The reactions are represented in the conversion of generalized amino acid [65] to the cyanogenic glucoside [66] (see Fig. 2.13), which requires several chemical steps, with subsequent breakdown of the latter to HCN, glucose, and an aldehyde. In order for a plant to be cyanogenic, it must have dominant alleles at both loci, that is, *Ac__Li__*. Three different combinations yield acyanogenic plants: *Ac_lili*, which has cyanogenic glycosides but lacks the glycohydrolase, *acacLi_*, which has the enzyme but lacks the glycoside; and *acaclili*, which lacks both. In the case of *Lotus corniculatus*, which exhibits the same phenomenon, plants that were cyanogenic were referred to as variety (or forma) *amara*, whereas acyanogenic plants were referred to as variety *dulcis*, which language scholars will recognize as the terms, respectively, for bitter and sweet.

Attention was refocused on cyanogenic polymorphism in the early 1950s, owing to extensive studies of the phenomenon in clover (*Trifolium repens* L.) in Europe by Daday (1954a, b). In the first of a series of papers, he reported that populations of clover in extreme southern and southwestern Europe had much higher frequencies of HCN-positive individuals than populations located in northeastern Europe, some of which lacked cyanogenic glycosides altogether (Fig. 2.14). Populations in central Europe exhibited intermediate frequencies. The second report described the frequency of occurrences as a function of elevation, wherein one sees the frequency of cyanogenic individuals in populations decrease as one progresses to higher and higher sites in the Alps. Whereas a population situated at 580 m had predominantly HCN-positive individuals, plants in a population at 1950 m were acyanogenic. de Araujo (1976) reported a similar relationship between elevation and cyanogenesis in clover growing in Wales.

Further confirmation that the HCN polymorphism is a general phenomenon came from the work of a colleague at the University of British Columbia, Fred Ganders (1990), who studied white clover from six sites in southwestern British Columbia and adjacent Washington. In all but one of the sites (the exception being a particularly

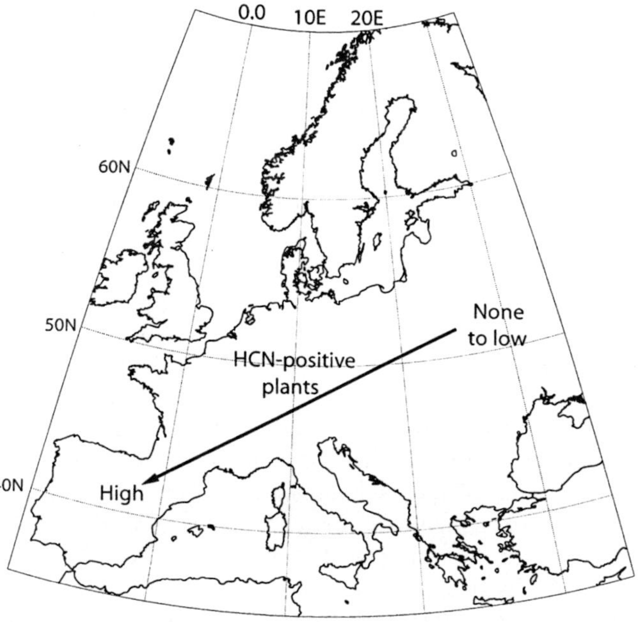

Fig. 2.14 Generalized pattern of distribution of cyanogenic glycosides in Europe

badly disturbed ski development) the frequency of cyanogenic plants was highest in the low-level populations (53.8%), intermediate at middle elevations (25.7%) and lowest at the high elevations (11.6%). The differences are highly significant ($P<0.0001$). The ratio-cline parallels clines found in the European studies. An interesting point about the system in western North America is that both white clover and the putative herbivores (primarily slugs) are comparatively recent introductions, and that the sorting out of cyanogenesis in these systems has occurred in less than a century.

There has been a good deal of discussion concerning the nature of the selection factors involved in this system. An early observation by Daday (1954a) involved the correlation between frequency of cyanogenesis in Europe and temperature. Although herbivory involving slugs and snails has been discussed as a major factor leading to the production of cyanogenic glycosides as defensive substances, the picture is apparently more complex. Jones (1972) discussed early efforts to unravel this system. Ganders (1990) pointed out that three factors—herbivores, temperature, and moisture stress—are likely involved, and that it is not possible to explain the clinal patterns on the basis of any one of these factors alone. He concluded by suggesting the need for more complex theoretical models. Although the mechanism of how this system originates is far from completely understood, the cyanogenic polymorphism remains one of the most thoroughly studied cases in the literature.

Reports continue to appear dealing with the antiherbivore properties of cyanogenic plants. For example, Saucy et al. (1999) demonstrated, using laboratory as well as outdoor enclosure experiments, that voles (*Arvicola terrestris*) showed clear preference for the acyanic form of white clover. It is interesting to note that in the outdoor experiments lasting for 2–3 weeks, animals adjusted their dietary intake of cyanogenic clover relative to the acyanic form to a sustainable level. The authors also reported feeding preferences for the acyanic form by two species of slugs, *Arion ater* and *A. subfuscus*.

Additional information relating to cyanogenesis polymorphisms can be found in a recent paper by Schappert and Shore (1999) who studied the phenomenon in *Turnera ulmifolia* L. (Turneraceae), which is used by *Euptoieta hegesia* (Lepidoptera: Nymphalidae) as its primary host plant.

A further note pertaining to cyanogenic taxa, although not concerned with geographic differences per se, could have relevance in discussions of relationships of species or populations based upon cyanogenesis. It has generally been held that a positive reaction implies the functioning of the two genes, *Ac* and *Li*, required for synthesis (one that directs making the glycoside and one that codes for the hydrolyase), and a negative reaction implies the lack of one or both. Kakes and Chardonnens (2000), working with mixed populations of the related species *Trifolium repens* and *T. occidentale* Coombe in Bretagne, demonstrated that the hydrolyase linamarase was absent from all individual plants examined (763). The apparent lack of linamarase from *T. occidentale* had been reported some years earlier (Gibson et al., 1972), but only ten plants were used in that work. The more recent results strongly suggest that cyanogenic frequencies in these two closely related species are controlled by different mechanisms. In other words, cyanogenesis does not have the same genetic meaning in these two species.

It is useful to include a recently described example of cyanogenic polymorphism here, although the species involved is neither related to the clovers nor does it occur in Europe. The work in question comes from a study of variation in cyanogenesis in two populations of *Eucalyptus polyanthemos* Schauer subsp. *vestita* L. Johnson and K. Hill from the Brisbane Ranges National Park in Victoria, Australia (Goodger et al., 2002). Two sets of trees were sampled, 100 from a population at Stony Creek, and 201 from a population at Sutherland Creek. All trees (100/100) from the Stony Creek set were cyanogenic with concentrations in the range 0.17–1.98 mg cyanide per gram dry weight leaves. Six acyanogenic trees were detected within the Sutherland Creek set (195/201) with maximum concentration of cyanide in the others reaching 2.07 mg cyanide per gram dry weight leaves. The authors, assuming no sampling error, speculated that the absence of acyanogenic trees in the Stony Creek population might be the result of selection against some localized herbivore. Again, follow-up studies of this system would be welcomed.

2.3.3 Lotus corniculatus *(Fabaceae)*

A few paragraphs above, we saw the part played by *Lotus corniculatus* in the cyanogenic glycoside story. We now turn our attention to another aspect of this species that has attracted a considerable amount of attention and has, coincidentally, involved probably the largest sampling program to be met in this review (and likely any where else!). It has been known for some time that European populations of *L. corniculatus* are usually polymorphic for keel petal color, which can be either "light" referring to yellow pigmentation, or "dark" involving deposition of anthocyanin glycosides toward the tip of the petal. The compounds involved have been identified as cyanidin and delphinidin [personal communication from P. Kakes to T. J. Crawford and Jones (1988, p. 175)]. Previous studies revealed that pigmentation is determined by a pair of alleles at a single locus, and that presence of the pigments represents the dominant situation (Hart and Wilsie, 1959).

The presence of the petal pigment polymorphism prompted Jones and Crawford (1977) to undertake a more systematic examination of the phenomenon. Although no clear-cut pattern emerged from the study, dark-keeled plants seemed to occur more commonly near the Atlantic and North Sea coasts of Europe. Limited sampling in the same study in Great Britain revealed that the darker morph was more common in plants found growing in northern England and Wales than in more southern areas. Additional information on pigment variation in Scotland came from a study by Abbott (1981), who found the color morph to occur more frequently in eastern populations than in western ones. Crawford and Jones (1986) found that the cline reported in Scotland also was evident in northern England, namely, an increase in the frequency of the colored keel feature in populations in the east. This study also recorded extremes in the appearance of the trait with the virtual absence of pigmented keels in collections made in the vicinity of Morecambe Bay (map reference: Lancaster, 54°03′N, 2°48′W) in the west to frequencies nearing 100% in the vicinity of Flamborough Head (Fig. 2.15). The authors noted that interpretation of these results was complicated by the fact that there was a steep increase in frequency of the trait over the initial 30 km of the transect, followed by a slower increase in frequency as samples were taken farther along the transect to the east. It was not possible to determine whether this reflected the "coastal" effect, as noted

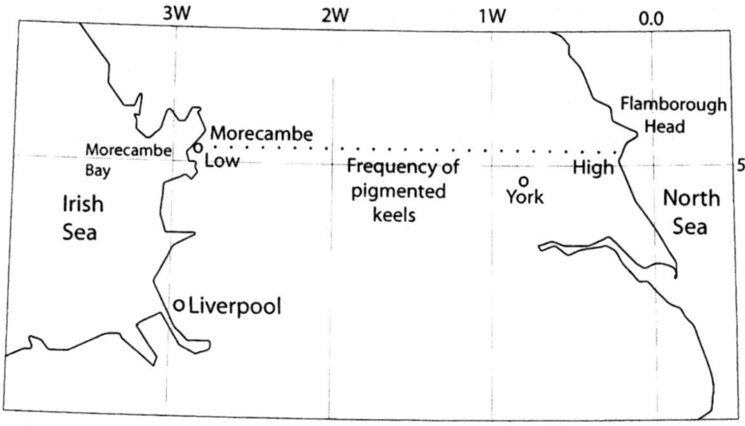

Fig. 2.15 Morecambe Bay to Flamborough Head transect in study of keel color morph in *Lotus corniculatus*

earlier in the European survey, or was simply a function of longitude. The next question addressed dealt with the effect of elevation on the expression of pigmentation, which Baker and Jones (1986) answered with transectional studies in the Alps, the Auvergne, and in the Pyrenees. Again, a trend was observed with higher frequencies of the pigmented keel observed in plants at higher elevations. We now come to the "grand survey" in which Crawford and Jones (1988), with the help of numerous volunteers, scored no less than 125,503 plants from 1348 sites representing all of the 100 km^2 in the United Kingdom National Grid. This very large sampling confirmed and extended the earlier studies, and put the west-to-east frequency trend on much firmer ground. The comparatively lower frequency of pigmented keel plants in southern England was also confirmed.

No satisfactory explanation for this color polymorphism in terms of its selective advantage over the whole range of the species has been forthcoming. Some local effects were noted, however. For example, Jones et al. (1986) observed that light-keeled plants at two sites in Yorkshire flowered earlier than their dark-keeled neighbors. Similarly, at six sites near York, the frequency of dark-keeled plants increased as the season progressed. Little advantage seems to accrue to the color morphs, however. The authors (Crawford and Jones, 1988) also remarked that attempts to correlate large-scale clines of this sort with climatic conditions is tempting, but no evidence exists to support such a relationship in this species. In concluding, they stated that, "We find it hard to accept that the distribution of the keel-color morphs could have arisen by chance. We believe, therefore, that the determination of the selective forces involved in this polymorphism provides a major challenge to ecological genetics." This conclusion could serve equally well for the discussion of Woodson's study of flower color morphs in *Asclepias tuberosa*, which will be met a little later.

Although the author has not had the opportunity to read either of the following two papers, it seems appropriate to identify another color morph described in them, this time involving *Arum maculatum* L. (Araceae). Both Burbidge (1903) and Colgan (1903) reported that leaf markings in this species appeared to be more prevalent in populations visited in England than those observed in Ireland. Without chemical data, it is not possible to equate these markings to the chevrons commonly seen in clover species, but the patterning in some leaves is caused by deposition of anthocyanin derivatives in the upper epidermis of leaves, resulting in a purple coloration. This has been shown to be the case with *Collinsia parviflora* Dougl. ex Lindl. (Scrophulariaceae), locally known as blue-eyed Mary, a plant common in southwestern British Columbia. Griffiths and Ganders (1983, pp. 93–97), studied the occurrence of this leaf marking in populations of blue-eyed Mary on Vancouver Island, islands in the Straits of Juan de Fuca and Georgia, and a few sites on the mainland. Populations were scored as unspotted, lightly spotted, or heavily spotted. Populations characterized by heavy spotting were usually found in cooler, north-facing slopes rather than on the warmer southern sides of hills and headlands. Since this species flowers in early spring, the authors suggested that the darker pigmented leaves absorb more sunlight, and thus can overcome the differential in air temperature between cooler and warmer microhabitats. Growth of plants in controlled conditions revealed that pigment deposition (presumably synthesis) does not occur if the temperature is maintained above 20°C.

2.3.4 Silene latifolia (= alba, = pratensis) *(Caryophyllaceae)*

This discussion features an analysis of the flavonoid genetics of *Silene latifolia* and how those data, among others, provide insights into the development of agriculture in Europe. Before looking at the flavonoid chemistry of the group, it is necessary to deal with some nomenclatural matters. In the earlier literature, this taxon was referred to as *Melandrium album.* Referral to *Silene* required the taxon to be named *S. pratensis* (Rafn.) Godron and Gren., since *S. alba* (Miller) E. H. L. Krause had been shown to be illegitimate by McNeill and Prentice (1981). A further change was occasioned by the discovery of a specimen under the name *S. latifolia* Poiret that carried an earlier publication date (see comments in Mastenbroek and van Brederode, 1986, p. 166). *Silene latifolia* is closely related to *S. dioica* (L.) Clairv., and is also widely distributed in Europe. They differ in flower color; *S. dioica* has red flowers, whereas *S. latifolia*, as its earliest name recognizes, has white flowers.

Before embarking on a discussion of *Silene*, a few comments on the significance of the work are in order. Biochemical systematics or chemotaxonomy as the subject was more often called in the 1960s and early 1970s, consisted in the early days essentially of comparing chromatographic patterns and developing a table of presence/absence data for the taxa under consideration. This was the "spot counting" phase of the subject. Occasionally, statistical treatments of presence/absence data were employed to establish levels of significance of the arrays of chemical data, which, incidentally, was an important contribution made by another fledgling discipline of the time, namely, numerical taxonomy. As the subject matured, more and more attention was given to determination of exact structures and, logically, how these structures were constructed in the cell. The establishment of biosynthetic pathways led to the study of the enzymes responsible for individual steps and, ultimately, to the question of the genetic control of these reactions. Flavonoids provided particularly good test subjects for genetic studies owing to the ease of scoring the products of controlled crosses, that is, the accumulation of intermediates when an enzyme was absent, or, more readily, by correlating flower color with genotype.

The subject more recently has progressed to the stage of probing of the genome using allozyme/isozyme analyses, endonuclease restriction fragment analysis (RFLP), or sequence analysis of a variety of genes themselves, although very few of these focuses on enzymes or genes coding for flavonoid biosynthetic reactions. Ready availability of these techniques has tended to overshadow use of secondary metabolites (not just flavonoids) in systematic studies. The current situation can be appreciated by quoting from a work by Crawford et al. (1992b) that addressed the comparative value of the two types of data: "Flavonoid chemistry offers few advantages over morphology because it is difficult, if not impossible, to infer genetic divergence from the arrays of flavonoid compounds sequestered by two species." The group at the University of Utrecht, whose work is examined next, has clearly demonstrated that this is not necessarily true. It has been the combination of flavonoid profile analysis and a thorough understanding of the genetic background of each profile, when translated into a measure of divergence that has led to the level of understanding of the system that we now enjoy. To be certain, the Dutch workers have studied many aspects of the evolutionary biology of *Silene*, and in that regard answer the challenge of Crawford et al. (1992b) that studies of disjunctions—and other systems to be sure—should include morphological, cytological, ecological, and geological aspects of the systems under scrutiny. The information to follow should suggest that study of secondary metabolites, and their genetic control, might well be added to that list!

The evolutionary history of *Silene* attracted the attention of members of the Department of Population and Evolutionary Biology at the University of Utrecht, whose work dominates the bulk of the following discussion. Pioneering studies of flavonoid formation in *Silene latifolia* and *S. dioica* had shown a considerable degree of allelic variation in Europe (Kamsteeg et al., 1978) with at least six loci involved in the control of glycosylation in the former. Subsequent studies (van Brederode and van Nigtevecht, 1975; Heimsbroek et al., 1980; van Brederode and Kamps-Heimsbroek, 1981) provided a detailed view of the reactions involved. The fundamental flavonoids present in these plants are isovitexin [67] and its positional isomer vitexin [68] (see Fig. 2.16 for structures 67–70), 6-*C*-glucosyl- and 8-*C*-glucosylapigenin, respectively. Three of the loci, *g*, *gl*, and *fg*, control glycosylation of isovitexin. A fourth locus, indicated as *V*, controls transfer of glucose (dominant allele is *Vg*) and xylose (dominant allele is *Vx*) to the 2″-hydroxyl group of vitexin. Three alleles have been found at the *g* locus: *g*, g^G, and g^X. The g^G and g^X alleles control the transfer of glucose and xylose, respectively, to the 7-OH group of isovitexin. The alleles at the *gl* locus are designated as *gl*, gl^R and gl^A. The latter two control the transfer of rhamnose and arabinose, respectively, to the 2″-hydroxyl group of the C-bound sugar of isovitexin. The dominant form of *fg* controls the transfer of glucose to the 2″-hydroxyl group of isovitexin. Two other loci, *P* and *Me*, are responsible, respectively, for the 3′-hydroxylation of isovitexin to isoörientin [69], and 3′-O-methylation of isoörientin to isoscoparin [70]. The biosynthetic relationships among these compounds are shown in Fig. 2.17.

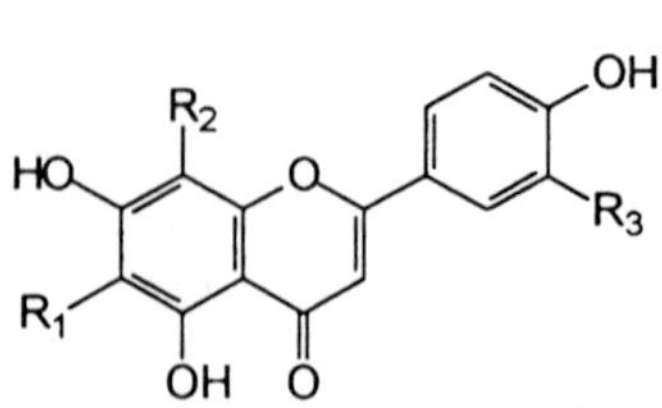

[67] R_1 = Glucose, $R_2 = R_3$ = H
[68] $R_1 = R_3$ = H, R_2 = Glucose
[69] R_1 = Glucose, R_2 = H, R_3 = OH
[70] R_1 = Glucose, R_2 = H, R_3 = OCH_3

Fig. 2.16 Flavone glycosides identified from *Silene*

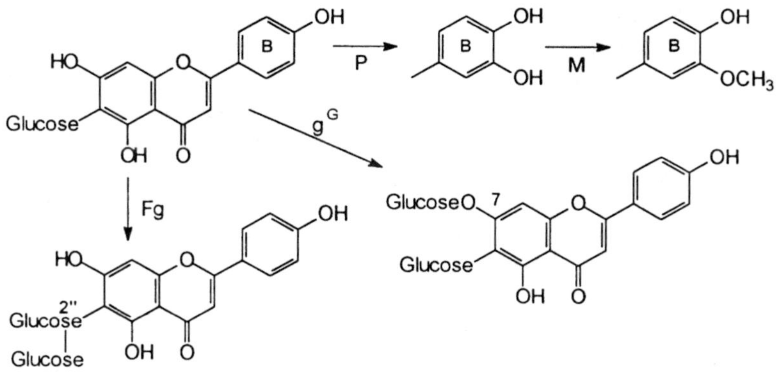

Fig. 2.17 Biosynthesis of flavone glycosides in *Silene* with glycosylation genes identified

Whereas g^X and gl^A are rare in *S. latifolia*, there is a clear-cut geographic patterning of the other genes. In one of the earlier studies, van Nigtevecht and van Brederode (1972) mapped the frequency of *g*, g^G, *gl*, and gl^R in 54 populations. The initial mapping resulted in the recognition of two races in Europe, an eastern one consisting of a high frequency of isovitexin 2″-*O*-rhamnoside along with isovitexin, and a western one consisting of isovitexin 7-*O*-glucoside and isovitexin. In an extension of this study, Mastenbroek et al. (1982) examined 285 populations, allowing a more detailed mapping of flavonoid variation patterns. Three pigment races became apparent. The first of these, which occurs in western and southern Europe and has g^G, *gl*, and *fg* present at a frequency of 1.0, is equivalent to the Western race observed in the first survey. Referring to the table, we see that this arrangement is manifested in a pigment profile consisting of isovitexin and its 2″-*O*-rhamnoside. The second race, which occurs in central Europe reaching northward into southern Norway, is characterized by a high frequency of *g* and gl^R, and a low to intermediate frequency of *Fg*. This translates into a pigment profile dominated by isovitexin and its 2″-*O*-rhamnoside with isovitexin 2″-*O*-glucoside present in some populations. The third race occurs in eastern Europe and consists of high frequencies of g^G, gl^R, and *fg*, which yields a pigment profile consisting of isovitexin, and its 2″-*O*-rhamnoside and 2″-*O*-glucoside.

The next step in fine tuning the distribution map involved collection of specimens from areas not well represented in the earlier studies, and included areas where the flavonoid races came into contact (Mastenbroek et al., 1983a; Mastenbroek and van Brederode, 1986). A further set of populations, bringing the total to 358, provided even better resolution and established the existence of transition zones between adjacent chemical races (see appendix in Mastenbroek, 1983). Analysis of the larger data set, using principal-components analysis and cluster analysis, established the existence of eight zones, the main western European race (Scotland to Spain and to the Balkans), the second main race (Central Europe: Hungary, Czechoslovakia, and Germany), the third main race (Russia, Poland, and Finland), and five intermediate zones that were hypothesized to have formed as a result of secondary intergradation of the main groups. The zones of intergradation exhibit different degrees of sharpness with regard to their boundaries. For example, there is a steep cline for the occurrence of gl^R in a narrow zone, consisting of The Netherlands, Belgium, and parts of Western Germany. On the basis of available data, it was not possible to establish whether this was the result of comparatively recent contact between the two main races or whether it was the result of selection. By contrast, no sharp transition zones were observed in Eastern Europe, which was taken, tentatively at least, to indicate that the main races had been in contact in this area for a comparatively longer time during which there was sufficient opportunity for gene exchange. An effort to represent the three major zones can be seen in Fig. 2.18.

With the establishment of clear-cut flavone races, the next obvious question was how the pattern became established. One possibility discussed by Mastenbroek (1983) involved selection for flavone profiles in response to some environmental factor or factors, climatological or edaphic being the most obvious ones. Alternatively, the present-day distribution pattern may be the result of historical factors: (1) migration involving long- or intermediate-distance dispersal; (2) range extension

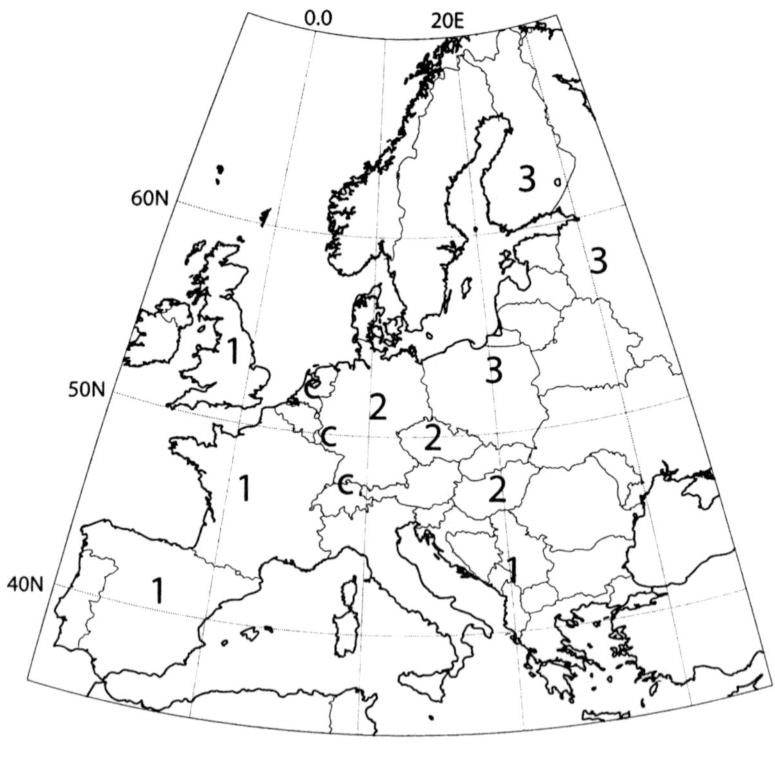

Fig. 2.18 Distribution of *Silene* glycosylation types in Europe. The numbers refer to the three major types. The zone of contact between types **1** and **2** is marked by a line of **C**'s

by means of small-scale, random dispersal; or (3) gene flow involving hybridization and introgression, involving members of the same species, or, as is possible in *Silene*, other closely related species. (Any combination of these is also possible.) Selection by environmental factors, for the most part at least, does not appear to be a major factor. Mastenbroek (1983) attempted to correlate flavonoid pattern with climatological and edaphic information without success. This is in general agreement with observations described by Niemann et al. (1980) that *S. latifolia* flavone profiles were not altered when the plants were grown under different moisture and temperature regimes. An unusual situation, which may involve selection, does exist for *S. latifolia*, however. When plants that possess only the gl^R gene are found in extreme wet and cold environments they exhibit "retarded development" (Mastenbroek, 1983). This may account for the comparative scarceness of *S. latifolia* in the northern part of its range (Fennoscandia and Russia).

The next step involved a determination of other features of *S. latifolia* that might reflect the flavone profile pattern. If other features were distributed in a manner similar to the flavone profiles, it would be reasonable to assume that the suite of features had moved as a set. This was tested by examination of seed, flower, capsule, and pollen morphology and electrophoretic behavior of several enzymes. Prentice (1979) had shown, using nonmetric, multidimensional scaling, that differences in seed morphology exist between western and eastern populations. A reexamination of Prentice's data set using a variety of multivariate techniques (Mastenbroek, 1983; Mastenbroek et al., 1983b) indicated that the pattern of seed differences was similar to that of the flavone pattern. Although differences in floral morphology (Prentice's data) also showed geographic patterning, different results were evidently obtained depending upon the numerical method used in the analysis (Mastenbroek, 1983). Mastenbroek (1983) did not elaborate upon the floral differences. In the case of capsular differences, however, clear-cut eastern and western races were shown to exist. Moreover, the pattern coincided with that observed with the seed patterns and with differences reported for pollen (McNeill and Crompton, 1978). Allozyme analysis proved the least effective in the search for patterns. Two problems manifested themselves, variation of the pattern during development (Mastenbroek et al., 1981) and a sensitivity of the pattern to environmental conditions (Mastenbroek et al., 1984). Nonetheless, a weak north-south patterning was evident, but there was no indication of any east-west differentiation. Although a one-to-one concordance among the different data sets did not emerge, it is clear that patterns in the occurrence of each feature analyzed exist. On the bases of these results, Mastenbroek (1983) concluded that it is unlikely that selection has played a major role in establishing the present-day geographic pattern. Additional evidence that selection has not played a significant part in contributing to the present levels of genetic variation in *S. latifolia* came from a study by Vellekoop et al. (1996), in which they studied random amplified polymorphic DNAs (RAPDs) from specimens representing 16 populations representing much of the species' range in Europe. The RAPD characters were readily assorted into western and eastern groups in parallel with the morphological and chemical features described in earlier papers. Attention was then turned to the question of how this species came to occupy its present range.

Mastenbroek (1983) posed two questions: (1) Did *S. latifolia* follow the spread of agriculture in Europe? (2) What was the impact of glaciation on the history of *S. latifolia*? Because *S. latifolia* requires open ground for successful colonization, it could not have advanced into central and northern Europe until appropriate habitats became available. This was accomplished through the deforestation that accompanied preparation of the land for planting. Mastenbroek (1983) suggested that *S. latifolia* existed in its natural state in nonarable sites in southern Europe and the Middle East at the time that human migration northward began. With the availability of newly cleared sites, establishment of weedy forms likely occurred. These could then have spread by different routes, northward and northwestward from the Balkans, and westward toward the Iberian Peninsula. Differentiation into proto-western and proto-eastern forms (races) may have occurred at that time. Owing to difficulties in identifying seeds and pollen as coming specifically from *S. latifolia*, it is impossible to state with certainty that the species per se was the progenitor of the modern races, or whether it evolved from some other species near the time of expansion. At any rate, expansion of the weedy forms continued as more and more sites became available leading to eventual contacts from which the present-day intermediate forms arose.

In addressing the issue of postglacial history of *S. latifolia* (or its progenitor), it is necessary to consider where it existed during the time when ice covered most of northern Europe. The species is not well adapted to survive in cold conditions (Thompson, 1973), which Mastenbroek (1983) pointed out, likely accounts for the absence of this species in the more northerly parts of Europe. He went on to say that even southern Europe may not have provided the necessary conditions for growth, and that the species may have occupied refugia in Northern Africa during maximum ice cover. This issue cannot be resolved on the basis of the data at hand.

2.3.5 **Achillea** *(Asteraceae)*

As part of an extensive series of studies on flavonoids of members of Anthemideae, Valant-Vetschera and Wollenweber (1988) examined the distribution of aglycones in leaf exudates of members of the *Achillea millefolium* L. group of species. Both qualitative and quantitative differences were noted for several species. An interesting example involves differences that were noted among specimens of *A. aspleniifolia* Vent. collected in Austria, Hungary, Romania, and Yugoslavia. Their observations involved three *O*-methylated quercetagetin derivatives, the 3,6,4′-trimethyl ether, centaureidin [71], the 3,6,7,4′-tetramethyl ether, casticin [72], and the 3,6,7,3′,4′-pentamethyl ether, artemetin [73] (see Fig. 2.19 for structures 71–73). The results of their study are summarized in Table 2.4. This system might prove to be useful in studying the action and control of *O*-methylation as has been done in *Chrysosplenium americanum* by Prof. Ragai Ibrahim and his colleagues at Concordia University in Montreal (e.g., De Luca and Ibrahim, 1985a, b).

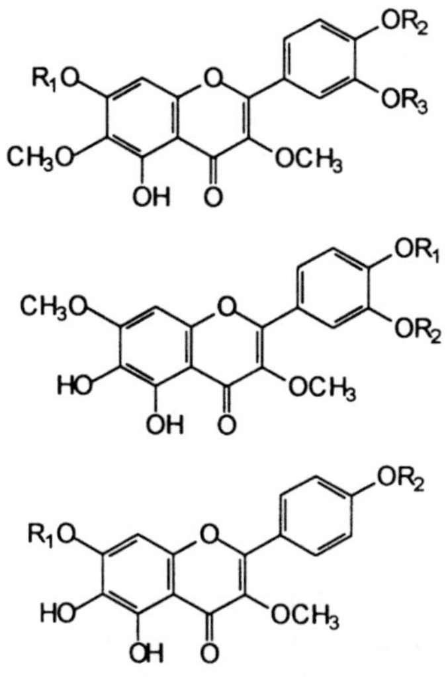

[71] $R_1 = R_3 = H, R_2 = CH_3$
[72] $R_1 = R_2 = CH_3, R_3 = H$
[73] $R_1 = R_2 = R_3 = CH_3$

[74] $R_1 = R_2 = H$
[77] $R_1 = H, R_2 = CH_3$
[78] $R_1 = R_2 = CH_3$

[75] $R_1 = H, R_2 = CH_3$
[76] $R_1 = CH_3, R_2 = H$

Fig. 2.19 Compounds 71–73, flavonoids identified from *Achillea millefolium*. Compounds 74–78, flavonoids identified from *Pulicaria dysenterica*

Table 2.4 Flavonoid variation in leaf exudates of *Achillea aspleniifolia* (from Valant-Vetschera and Wollenweber, 1988)

Country	Quercetagetin methyl ethers[a]		
	3,6,4′	3,6,7,4′	3,6,7,3′,4′
Yugoslavia		+[b]	
Austria-1			tr
Austria-2		tr	+
Hungary-1	tr	+	++
Hungary-2	tr	+	+
Romania	tr	+	+

[a] Common names: centaureidin, casticin, and artemetin, respectively.
[b] Relative amounts; tr = trace only.

Continuing their study of flavonoids of *Achillea* (Asteraceae), Valant-Vetschera and Wollenweber (2001) described exudate aglycones in *A. moschata* (Mill.) W. D. J. Koch and related alpine species of *Achillea* sect. *Ptarmica.* In addition to summarizing the exudate chemistry of all sections of the genus—a valuable contribution in its own right—those authors pointed out the predominance of 6-hydroxyflavonol 3,6,4′-trimethyl ethers in taxa native to the Balkans and southeastern Europe, in contrast to the predominance of 6-hydroxyflavone methyl ethers in taxa native to Turkey. Flavonoid aglycone diversification appears to confirm the origin of the genus in the eastern Mediterranean region. The reader might also examine the companion study of *Achillea* sect. *Filipendulinae* (Valant-Vetschera and Wollenweber, 1996).

It is convenient to mention at this point a recent study on chemical variation involving the essential oils of *Achillea millefolium* L. subsp. *millefolium* growing naturally in Lithuania. Although differentiation of chemotypes was indicated by the data, the relevant observation noted by the authors (Mockute and Judzentiene, 2003) is the existence of different chemical profiles, Portugal, Norway, and a population from Estonia in one, another population from Estonia comprising a second, and populations from Canada comprising yet another. Important leads to the literature can be found in the Lithuanian work.

2.3.6 **Pulicaria dysenterica** *(Asteraceae)*

Recent work on flavonoids of *Pulicaria dysenterica* (L.) Bernh. fits in well at this point since it too exhibits variation in the expression of *O*-methylated flavonols (Williams et al., 2000). In addition to the use of this plant in folk medicine formalized in its specific epithet, its common name in English, "fleabane," gives recognition to its value as an agent to discourage fleas and other insects from taking up residence in human quarters. Interest in the flavonoid chemistry of fleabane led to the discovery, as by the cited workers, of several chemotypes involving plants collected from different places in England and continental Europe. Whereas the vacuolar flavonoid profile, consisting of just quercetin 3-*O*-glucuronide, was observed in all specimens, the leaf exudate fractions showed differences in major components. Four chemical variants were observed based upon the major (or in some cases sole) flavonol derivative(s): (1) quercetagetin 3,7-dimethyl ether [74]; (2) 6-hydroxykaempferol-3,4′-dimethyl ether [75]; (3) a mixture of 6-hydroxykaempferol-3,7-dimethyl ether [76] and quercetagetin 3,7,3′-trimethyl ether [77]; and (4) a mixture of 6-hydroxykaempferol-3,7-dimethyl ether, quercetagetin 3,7,3′-trimethyl ether, and quercetagetin 3,7,3′,4′-tetramethyl ether [78] (see Fig. 2.19 for structures 74–78). It is difficult to assess the significance of these chemical variants owing to limited sampling and the possibility that they are responding to some environmental factor or factors. Common garden studies coupled with more extensive collections are necessary to put the existence of these forms on firm ground. Additional variation was noted with reference to the work of Pares et al. (1981) who reported quite a different flavonoid profile from material collected in Turkey.

2.3.7 **Thalictrum minus** *(Ranunculaceae)*

This example demonstrates how widely chemical profiles can differ between two populations separated by only a matter of kilometers; in this case, populations of *halictrum minus* L. in the Vojvodina area of Serbia (Popovic et al., 1992) (Fig. 2.20). Plants collected at 500 m in the Fruska Gora Mountains (Novi Sad) afforded a comparatively complex mixture of benzylisoquinoline alkaloids that consisted of

Fig. 2.20 Map of the area around Vojvodina, Serbia

the bisbenzyldihydroisoquinolines thalmethine [79], *O*-methylthalmethine [80], thalicberine [81], and *O*-methylthalicberine [82], and three monomeric compounds thaliglucine [83], thaliprophine [84], and berberine [85] (see Fig. 2.21 for structures 79–87). A population collected in nearby Beocin (45°12′N, 19°41′E) afforded a significantly reduced yield of alkaloids, as well as a simpler array of compounds consisting of the monomeric isoquinoline derivatives thalactamine [86], thalflavine [87], and berberine. It is also useful to note that plants from both of those areas were shown to be hexaploid, refuting an earlier suggestion that alkaloid composition correlated with ploidy number (Dutschewska and Kuzmanov, 1982; Kuzmanov and Dutschewska, 1982).

As noted, the alkaloid yield from the Beocin plants was low, which the authors suggested might be caused by the poor soil in which the plants were growing (Popovic et al., 1992). One could ask whether the soil conditions to which they refer might be influential in the overall alkaloid biosynthetic processes in this species. It would be of interest to see experimental studies aimed at determining the effect of soil components on these processes. In the present case, it may be a lack of, or reduction in the activity of, the oxidase(s) necessary for the dimerization process (required to form the bibenzyldihydroisoquinolines) to occur. It is also possible that the lack of dimeric alkaloids may simply reflect a concentration effect caused by the edaphic conditions. These questions should be accessible to experiment.

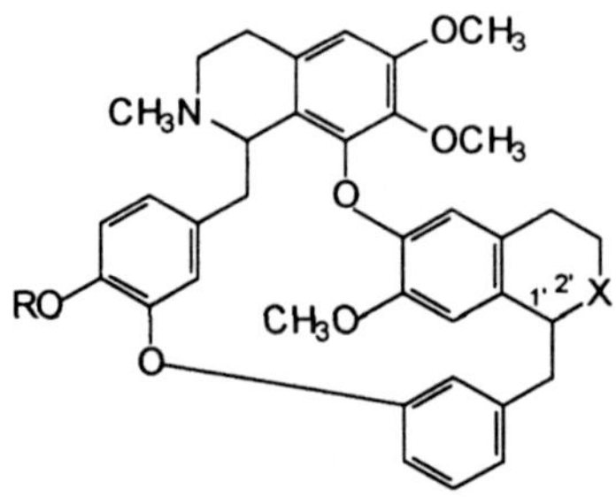

[79] R = H, X = N, 1',2'-unsat'd.
[80] R = CH$_3$, X = N, 1',2'-unsat'd.
[81] R = H, X = NCH$_3$
[82] R = CH$_3$, X = NCH$_3$

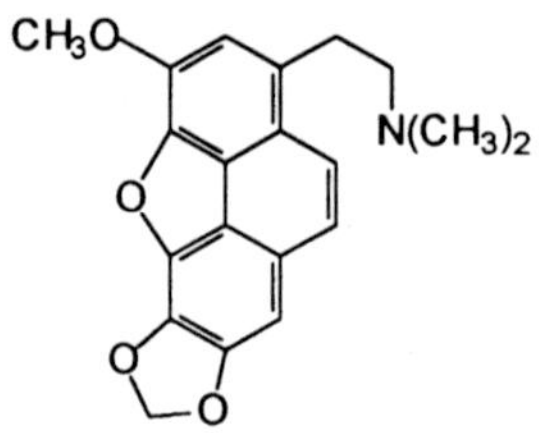

[83]

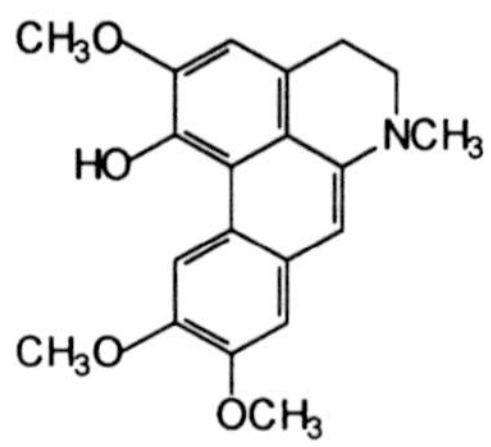

[84]

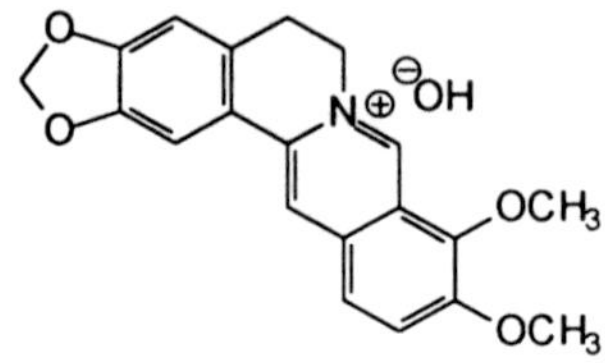

[85]

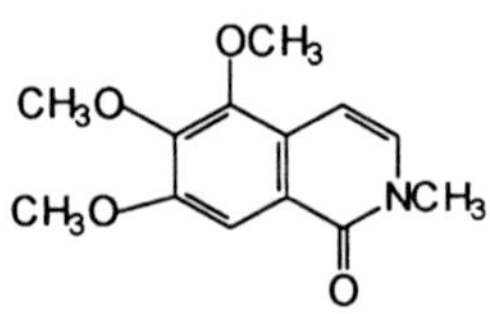

[86]

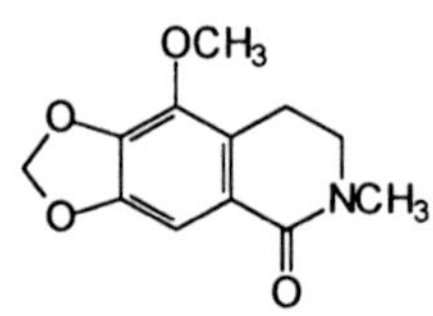

[87]

Fig. 2.21 Compounds 79–87, alkaloids from *Thalictrum minus* collected near Vojvodina, Serbia

2.3.8 Valeriana officinalis *(Valerianaceae)*

This example, which involves terpene variation in *Valeriana officinalis* L. subsp. *collina* (Wallr.) Nyman, features an impressive range of quantitative differences between two regions, the Apuan Alps in northwestern Tuscany and the Karst Plateau of northeastern Italy, and the northwestern corner of (former) Yugoslavia (Fig. 2.22). Two populations were sampled from the former, one from Pian della Fioba (840 m), referred to as Alps A in Table 2.5, and one from La Tecchia (900 m) referred to as Alps B. The sample from the Karst Plateau was collected at Sgonico (210 m), near Trieste, Italy (Coassini Lokar and Moneghini, 1989). Major quantitative differences between the two areas are summarized in Table 2.5 where values for percentage composition of common leaf terpenes are given. Statistical analysis clearly supported the existence of two chemical varieties.

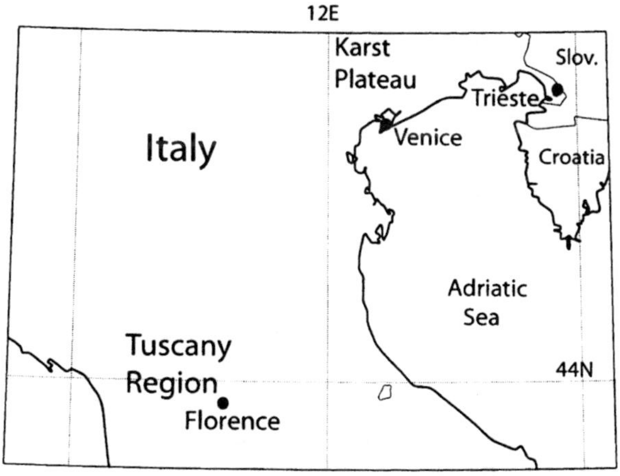

Fig. 2.22 Map of northeastern Italy and adjacent regions from which specimens of *Valeriana officialis* were obtained

Table 2.5 Selected terpenes from three populations of *Valeriana officinalis* (from Coassini Lokar and Moneghini, 1989)

Compound	Alps A[a]	Alps B	Karst
α-Pinene	4.59[b]	23.91	0.10
Limonene	0.44	0.23	2.55
n-Butyl valerate	0.19	0.28	2.46
1,8-Cineole	4.29	3.77	6.51
α-Terpineol	0.95	3.54	7.54
Bornyl acetate	21.59	33.39	0.40
Borneol + isoborneol	30.73	22.90	3.64
Valeranal	10.15	6.52	5.60
Isobornyl isovalerate	13.07	11.25	47.80
Valeranone	3.04	5.28	6.75

[a] Alps A and B are from the Apuan Alps.
[b] Percentage composition of leaf terpenes.

2.3.9 Cistus *(Cistaceae)*

As part of a study of the secondary chemistry of members of *Cistus* (the rock-rose) in France, Robles and Garzino (1998) examined the essential oil of *C. albidus* L. Plants were sampled from two areas in Provence characterized by different soil types, calcareous sites west of Marseille, and siliceous sites near Pierrefeu-du-Var and Bormes les Mimosas (PF and BM, respectively, in Fig. 2.23), which lie about 60 km and 80 km to the east, respectively, in the Massif les Maures. Regardless of the soil type, α-zingiberene [88] (Fig. 2.24) was the dominant component. Concentrations of other major components of the plants varied between the two soil types, as summarized in Table 2.6. Many other compounds were present in lesser amounts, but varied little between the two areas. A more recent paper by the same workers (Robles and Garzino, 2000) described an analysis of *C. monspeliensis* L. leaf oils, the results of which are summarized in Table 2.7.

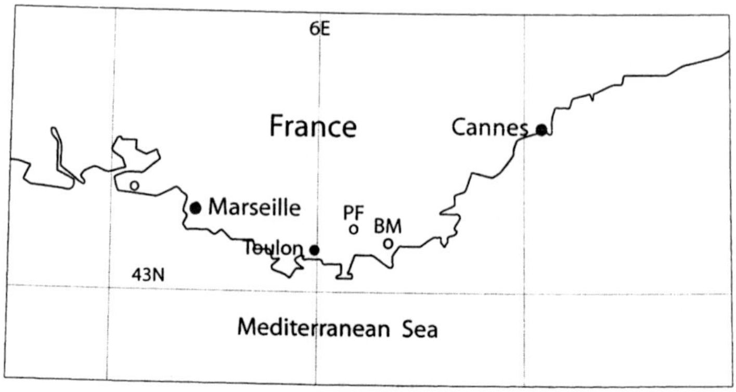

Fig. 2.23 Map of southern coastal France in the region from which samples of *Cistus albidus* were obtained

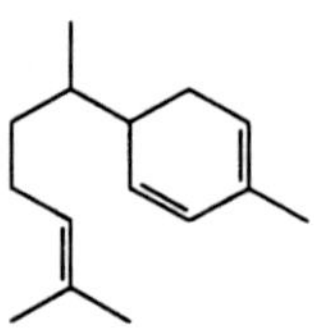
[88]

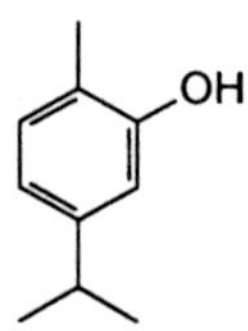

[89]

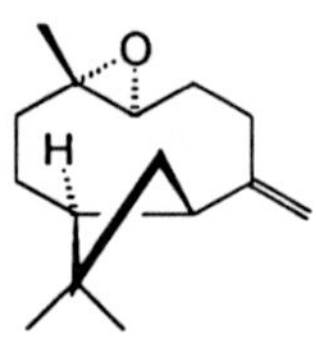

[90]

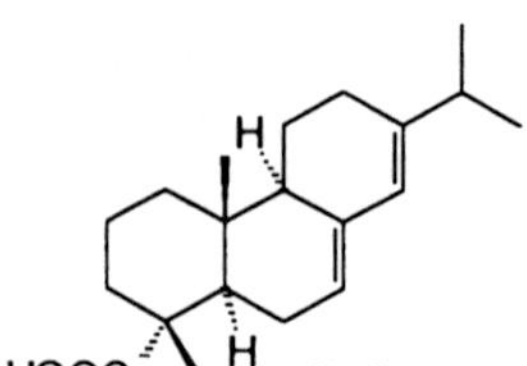

[91]

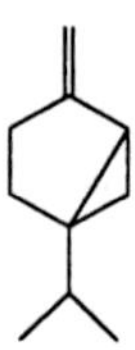
[92]

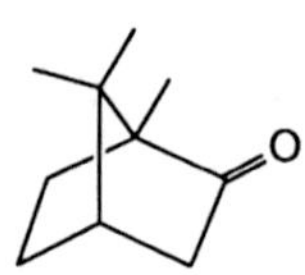

[93]

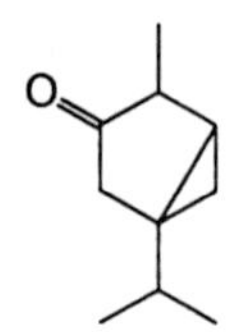

[94]

Fig. 2.24 Compound 88 from *Cistus albidus*, Compounds 89–94 from *Cistus monspeliensis*

Table 2.6 Essential-oil composition of *Cistus albidus* from different substrates (from Robles and Garzino, 1998)

Compound	Substrate	
	Calcareous	Siliceous
β-Bourbonene	10.8 ± 1.3[a]	15.4 ± 6.9
β-Caryophyllene	31.9 ± 10.9	20.0 ± 5.6
allo-Aromadendrene	32.2 ± 6.9	21.2 ± 9.0
α-Curcurmene	31.3 ± 14.9	41.1 ± 15.3
α-Zingiberene	77.9 ± 17.0	76.4 ± 31.6
Unknown	17.5 ± 0.7	24.1 ± 12.1

[a] Concentration expressed in μg/μl sample injected (GLC).

Table 2.7 Essential-oil composition of *Cistus monospeliensis* from different substrates (from Robles and Garzino, 2000)

Compound	Substrate	
	Calcareous	Siliceous
α-Cadinol	3.17 ± 1.43[a]	Not detected
Unknown (peak N)	37.62 ± 16.58	15.16 ± 9.79
α-Bisabolol	34.96 ± 32.49	1.60 ± 0.67
Unknown (peak R)	12.54 ± 6.81	30.70 ± 14.10
Diterpene (peak T)	45.03 ± 19.26	84.75 ± 40.50

[a] Concentration expressed in μg/μl sample injected (GLC).

Additional information on essential oil variation in *C. monspeliensis* comes from a study by Angelopoulou et al. (2002) who were interested in diurnal and seasonal patterns. Significant differences were noted during both time periods. It would be of interest to learn what effects cultivation in a common garden environment would have on oil expression in these species.

A second study featuring *Cistus* involves an analysis of the essential oils of plants collected from nine sites on the island of Crete, three from the western end of the island and six from the eastern end (Angelopoulou et al., 2001). In the study, 114 compounds were identified, although many of them were present in very small amounts and only in some populations. Few compounds could be described as dominant with the possible exception of carvacrol [89] (see Fig. 2.24 for structures 89–91), which was present at the level of 20.23% in one of the eastern populations, 11.13% in another, and was absent from only one of the nine populations. Caryophyllene oxide [90] reached 11.24% in one population and had one of the more consistent ranges, with the lowest percentage being 2.86%. Other compounds of note are dehydroabietol [91], which reached 10.01% in one population, and hexadecanoic acid (C-16 acid; 16:0) whose contributions ranged from 4.24% to 20.63% with the highest concentrations recorded for two of the western populations. Numerical analyses of the total data set revealed considerable structure in the data set with clear-cut distinction between an eastern and western chemotypes. Those authors failed to find any variation in morphological features throughout the range of this taxon on Crete. Comparisons with other species of *Cistus* in Greece—numbering seven—is not possible owing to the lack of comparable studies on the other taxa.

While referring to the island of Crete, it is convenient to look briefly at a recent study of monoterpenoid diversity of *Origanum microphyllum* (Lamiaceae) on the island. Extremes in monoterpene composition ranged from specimens rich in sabinene [92] (see Fig. 2.24) and its *trans*-hydrate derivative to specimens rich in *cis*-sabinene hydrate (Gotsiou et al., 2002). No geographic patterning was observed among the populations whose chemistries were not altered by growth in common garden experiments. The paper by those workers offers excellent leading references to patterns of variation in other members of Lamiaceae.

2.3.10 Three Examples from Finland (Fig. 2.25)

The first example involves a study of triacylglycerols accumulated by members of two genera, *Rubus* (Rosaceae) and *Empetrum* (Empetraceae) in Finland. *Rubus chamaemorus* L., the "cloudberry," and *Empetrum nigrum* L., the "crowberry," were collected throughout Finland by Johansson et al. (1997) and examined for their constituent seed oils by GLC. [Both *E. nigrum*, the "northern crowberry," and *E. hermaphroditum* Hagerup, the "southern crowberry," were collected without distinction for this study.] Seventeen sites were sampled, representing the latitudinal and longitudinal ranges 60.5–69.5°N and 22–29°E, respectively.

Seeds of *Rubus chamaemorus* from southern Finland were heavier than the seeds from the north, but had lower oil content; there was no difference in M_r distribution of the

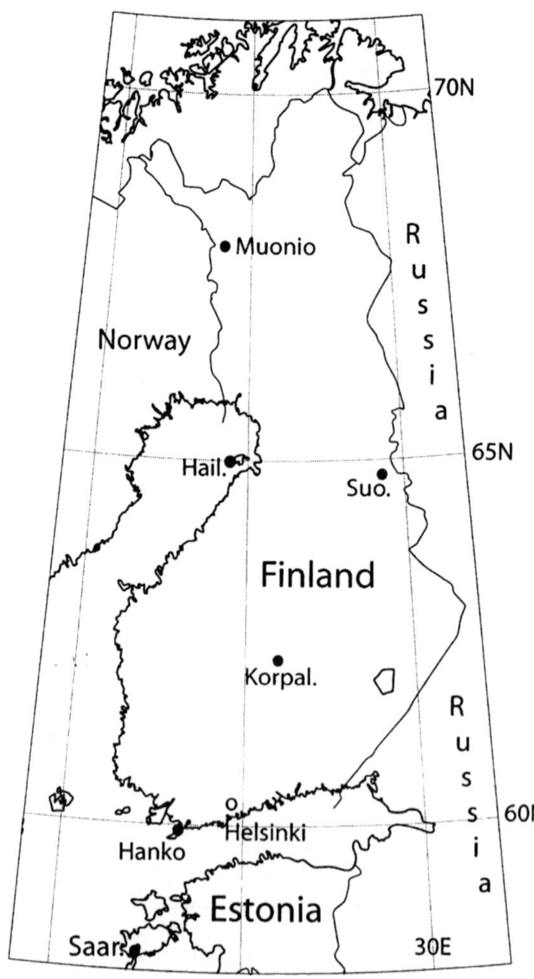

Fig. 2.25 Map of Finland and Estonia for examples including *Rubus, Empetrum* and *Pinus sylvestris*

triacylglycerols from the two areas, however. The esters from both species were based mainly (>95%) on four acids, palmitic (16:0), oleic (18:1Δ^9), linoleic (18:2$\Delta^{9,12}$), and α-linoleic(18:3$\Delta^{9,12,15}$), but compositions differed in seeds from northern *versus* southern sites. Thus, the proportion of linoleic acid in seeds of *R. chamaemorus* from northern Finland (Lapland) was highest ($P<0.05$) and that of α-linoleic lowest ($P<0.01$). In seeds of *Empetrum* from southern Finland, α-linoleic was present in the highest proportion ($P<0.001$) with linoleic present in the lowest proportion ($P<0.001$).

The second example from Finland involves a study of variation in volatile compounds in tansy (*Tanacetum vulgare* L.) by Keskitalo et al. (2001). In the introduction to their paper, those authors reviewed the numerous uses to which tansy has been put: flavoring agent for foods, herbal remedy, anthelmintic, anti-inflammatory agent, as an agent against microorganisms, and as an insect repellent. They also commented on the existence of several chemotypes throughout the plant's wide distribution in the Northern Hemisphere. The primary goal of their study was to select lines of tansy that exhibit enhanced insecticidal properties to which end a detailed study of Finnish populations was undertaken. An earlier study (Keskitalo et al., 1998), using random amplified polymorphic DNA (RAPDs), had documented the level of genetic variation among 20 accessions. The two main groups that emerged from that study were further distinguished by floral differences. The study of volatile oil composition was undertaken to see if there was any correlation between the microchemical data and the genetic patterns observed in the earlier study.

Twenty populations were sampled with plants from Hailuoto, at 65°00 N, representing the northernmost site and material from Hanko, at 59°49 N, representing the southernmost collection site (and, incidentally, the southernmost point of land in the country). Five samples represented central Finland with the remainder originating from the southern part. Fifty-five compounds were detected by GC-MS analysis, 53 of which were identified. The data obtained were subjected to complete linkage analysis, which differentiated several clusters that corresponded moderately well with geography. Genetic distance values derived from the RAPD data correlated well with chemical distance values determined from the terpene data ($r=0.41$, $P<0.0001$).

The most commonly observed terpene was camphor [93] (see Fig. 2.24 for structures 93 and 94), whose concentration exceeded 18.5% in 13 of the genotypes, but was present in less than 7.2% in the other seven. The highest levels approached 70% in some groups with a low of 0.06% in one southern sample. The majority of high-camphor genotypes occurred in central Finland, whereas those with little or low amounts were observed in specimens from southern and southwestern Finland. This finding is in agreement with an earlier study by Sorsa et al. (1968), who reported higher levels of camphor in northern specimens as opposed to southern specimens, where thujone [94] was the more prominent compound. In the present study, however, thujone was not seen as a major component in southern specimens (it was a major component in a single collection from central Finland).

An interesting dispersal scenario was described by the authors on the basis of the similarity of southern Finnish chemotypes to chemotypes found in the Netherlands, and other parts of Northern Europe. They suggest that tansy seeds, known to be common components of shipping ballast (Jutila, 1996), may have been transported to southern Finnish ports in that manner. [Other examples of seed transport via ballast will be met later in this review in a discussion of beach rocket (*Cakile*).] The more fundamental question—how to explain the chemical variation in tansy—however, remains, to use the authors' word, "elusive." It seems reasonable to suggest that tansy, over a considerable period of time, has been subjected to human intervention with an eye to improving one or another of its properties, and it is the seeds of those improved lines that have found their way to northern Europe and subsequently into Scandinavia. Environmental factors may also have played an important role (Sorsa et al., 1968). As is the case with most of these highly variable systems, carefully controlled studies focusing on likely environmental factors, coupled with a thorough understanding of the biosynthetic processes involved, should lead to a better appreciation of why such complex systems occur in nature.

The final Finnish example here deals with the question of susceptibility of Scots pine (*Pinus sylvestris* L.) to insect herbivory. Manninen et al. (1998) studied the terpene and resin acids and growth characteristics of trees from four provenances that ranged over a distance of about 1200 km: Muonio (67°56′N), Suomussalmi (65°10′N), and Korpilahti (62°0′N) in Finland, and Saaremaa (58°22′N) in Estonia. Experiments were run either in petri plates or in pots under common garden conditions. Host plant selection was tested using two species of aphids, the gray pine aphid (*Schizolachnus pineti*) and the spotted pine aphid (*Eulachnus agilis*), while measures of oviposition behavior employed the tarnished plant bug (*Lygus rugulipennis*). The terpenes, separated by HPLC, were identified as bornyl acetate [95] (see Fig. 2.26 for structures 95–103, others above), camphene [96], 3-carene [97], caryophyllene [98], copaene [99],

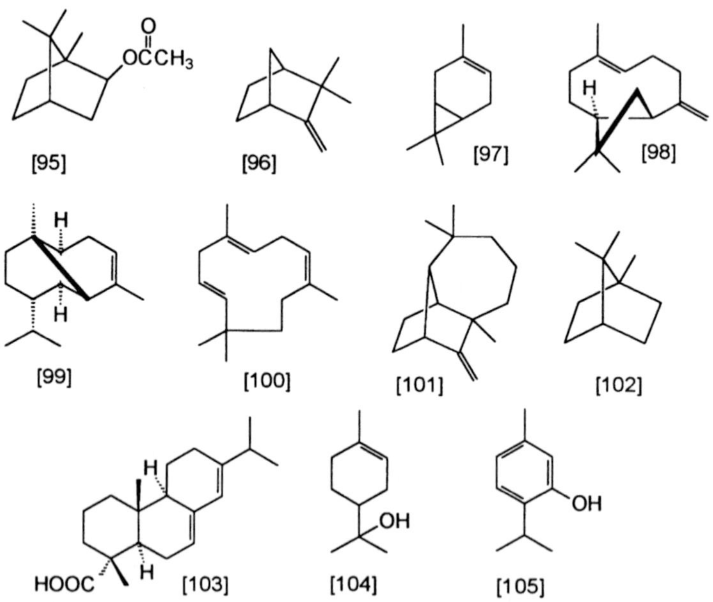

Fig. 2.26 Compounds 95–105 from *Pinus sylvestris* in Finland and Estonia

humulene [100], limonene [23], longifolene [101], myrcene [39], α-pinene [21], ß-pinene [22], sabinene [92], and tricyclene [102]. Resin acids were determined to be abietic [103] and other related diterpene acids.

Resin acids have been shown to be deterrents of insect herbivory in other species [see Manninen et al. (1998) for leading references]. In the present study, activity was highest in plant material from the southernmost provenance (Saaremaa) and decreased toward the north. Absolute concentrations of two resin acids, sandaracopimaric and dehydroabietic, correlated negatively with aphid performance, suggesting that they may act as deterrents, whereas the other acids showed positive correlation suggesting that they may in fact act as stimulants. Total terpene concentration increased toward the north. Some interaction with insects was evident; the low terpene concentration in seedlings from the southernmost provenance may be responsible for the lower preference for these plants by aphids. In general, however, chemical factors appeared to be much less significant than other factors. Seedling size in particular is a better estimator of susceptibility of Scots pine seedlings to both specialist and generalist insects.

2.3.11 Thymus *(Lamiaceae)*

Thymus caespitosus Brot. occurs on the northwestern Iberian Peninsula and on the Madeiran and Azorean archipelagos. One of the Azores group is Sao Jorge, a small island, located at ca. 38°40′N, 28°03′W, with an area of about 246 km² and a width that does not exceed 10 km at any point (see Fig. 2.27 for location of Azores). The island possesses a uniform climate and offers little in the way of environmental extremes. It is thus of some interest that significant quantitative chemical variation

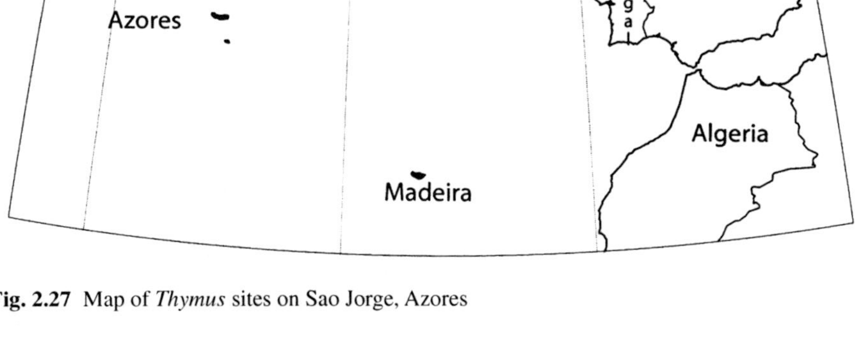

Fig. 2.27 Map of *Thymus* sites on Sao Jorge, Azores

was observed in a recent study of the essential oils of this plant (Pereira et al., 2000). Two levels of chemical variation were recorded for this species, quantitative differences of four compounds, and, perhaps more interestingly, differences in the proportions of enantiomers of two of them, sabinene [92] (Fig. 2.24) and α-terpineol [104] (see Fig. 2.26 for structures 104 and 105). Compositions of the four major compounds are given in Table 2.8. The collection sites have been reordered to reflect their geographic locations on the island (they are also identified using the authors' site numbers). UPGMA clustering emphasized the degree of non-connectedness of these populations, the "center" group (populations **1**, **2**, **3**, and **4**) for example, having widely different concentrations of sabinene, thymol [105], and carvacrol [89]. As the authors pointed out, morphological similarity and the lack of climatic and elevational effects suggest that the differences are based upon either genetic or edaphic differences. Until common garden studies are done, the effect of edaphic differences cannot be ruled out. Considering the small size of the island, it seems as likely that populations could have been readily established by medium-distance seed dispersal.

Of added interest in this study was the finding that the enantiomeric ratios of α-terpineol also differ widely among populations. In southern populations and the population from the northwestern tip of the island, the amount of (+)-α-terpineol was shown to range from 92.3 to 97.0%, while this enantiomer made up 85.6% in the upper-middle population, but only 66.0% in the population from the southeastern tip of the island. It would be of interest to see if the enzymes responsible for the biosynthesis of α-terpineol have different stereochemical requirements in these populations, or whether some isomerization has occurred in the formation, preparation, or analyses of these oil samples.

The next example involves a recent study of essential oil polymorphism in *Thymus praecox* Opiz subsp. *arcticus* (E. Durand) Jalas (syn. *T. drucei* Ronn.) on the British Isles (Schmidt et al., 2004). More than 700 specimens of the plant were collected in Ireland, Scotland, and southern England and subjected to gas chromatographic analysis (coupled with mass spectroscopy). Sixty-nine constituents were identified, the majority of which were mono- and sesquiterpenoids. Analysis of the data revealed a highly polymorphic assemblage with 13 chemotypes in Scotland, 11 in Ireland, and 17 in the south of England. Polymorphism seems to

Table 2.8 Percentage composition of essential oils of *Thymus caespitosus* on Sao Jorge, Azores (from Pereira et al., 2000)

Terpene	NW[a]	Center				Mid	South			SE
Sabinene	0.1	0.1	2.3	40.9	4.2	0.1	2.1	0.6	0.9	0.4
α-Terpineol	10.4	t	1.4	0.2	0.2	6.4	43.0	46.6	68.3	15.3
Thymol	44.4	57.9	2.5	4.3	44.5	19.8	7.0	8.4	3.0	1.4
Carvacrol	2.9	3.5	52.3	17.2	2.5	31.5	1.5	2.3	0.5	35.9

[a] Locations on Sao Jorge: NW = northwestern tip; Center = center of the island; Mid = northwest of the central site; South = southern populations; SE = southeastern tip of the island.

be more prevalent in southern areas than in more northerly sites; thus, Greenland, Iceland, and Norway had only two, five, and one chemotype, respectively.

The north–south gradient was also evident in the concentration of the most abundant chemotype, which was shown to consist of linaloöl [19] and its acetate. Most frequent in Greenland, Iceland, and Norway (90–100%), it was present in only 40% of plants collected in Scotland, and in only ca. 5% of plants from southern sites. This finding paralleled the observations of Stahl-Biskup and Laakso (1990) that a similar north–south trend exists in the oils of *Thymus serpyllum* subspecies *serpyllum* and *tanaensis* in Finland. The opposite trend was observed with γ-terpinene, which was present in southern populations but absent from those from the northern sites.

Perhaps the most interesting aspect of this set of studies is the question posed in the recent paper by Schmidt et al. (2004) and deals with the reality of the patterns they observed. Is the polymorphism observed a result of the calculation methods used in the study, neural network (NN), and multivariate statistical analysis (MVA)? Would increased sampling result in a greater number of chemotypes? It is entirely possible, of course, that the numbers obtained in this study are a true reflection of the biosynthetic capacities of the plants studied. The authors concluded—and this is a point made elsewhere in this review—that "...for a correct interpretation a good knowledge of the biosynthetic background of the components is needed."

2.3.12 Pinus uncinata *(Pinaceae)*

Pinus uncinata Ram. occurs discontinuously from the north of Spain to western Switzerland, including populations in the Jura, the Vosges, and the Massif Central. Although the taxonomy of this species is evidently open to some discussion, Lauranson and Lebreton (1991) chose to set differences of opinion aside in their study of the flavonoid profiles of this pine. (References to the taxonomic literature can be found in their paper.) For their flavonoid analysis, twigs were collected from at least 23 individuals at each of five locations, two in the Pyrenees, two in the Alps, and one in the Jura (Fig. 2.28). Site information, and relevant results of their analyses, appears in Table 2.9. Procyanidin and prodelphinidin (measured as the corresponding anthocyanidins, cyanidin [106], and delphinidin [107], respectively) were found to occur in the same ratio in each population. Following hydrolysis of the glycosidic fraction, three flavonols were identified, kaempferol [108], quercetin [109], and isorhamnetin [110] (See Fig. 2.29 for structures 106–110). Statistical analysis revealed that the five populations could be discriminated by their respective quercetin contents. The individuals from each population were sorted into three categories: (1) less than 10% quercetin; (2) 10–16% quercetin; and (3) more than 16% quercetin. Based upon the assumption that the quercetin concentration was under one locus-two allele (*A* and *a*) genetic control and that the three genotypes, *AA*, *Aa*, and *aa*, represented high,

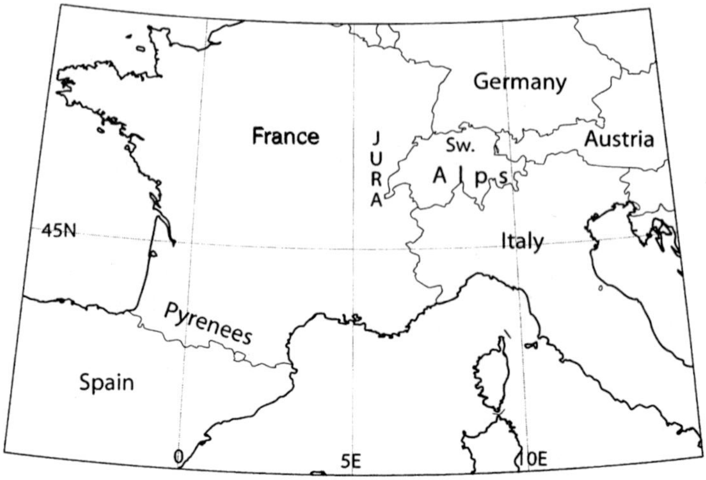

Fig. 2.28 Partial map of Western Europe showing location of the Pyrenees, the Jura, and the Alps from which *Pinus uncinata* populations were sampled

Table 2.9 Quercetin levels in *Pinus uncinatus* (from Lauranson and Lebreton, 1991)

Location	Mtns./Country[a]	Coordinates	Elev (m)	Q%V[b]	[Q][c]	p(a)[d]
Osseja	Pyrenees (Fr)	42°25′N, 1°58′E	1960–2080	30	9.0 ± 1.1	0.87
Gavarnie	Pyrenees (Fr)	42°44′N, 0°1′W	1600–1750	35	11.5 ± 1.6	0.65
Vanoise	Alps (Fr)	45°17′N, 6°54′E	1720	17	16.8 ± 1.1	0.25
Valais	Alps (Sw)	46°18′N, 7°8′E	1525	27	12.3 ± 1.3	0.55
Le Creux du Van	Jura (Sw)	46°57′N, 6°45′E	981	33	8.7 ± 1.2	0.89

[a] Fr = France, Sw = Switzerland.
[b] Average percentage of variation contributed by quercetin in each population.
[c] Average concentration of quercetin per population as % of total flavonoids.
[d] Frequency of the *a* allele in each population.

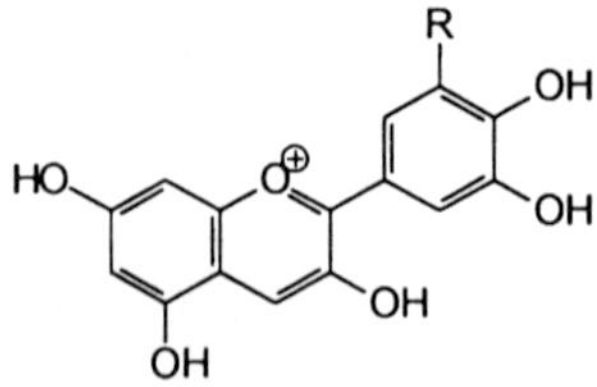

[106] R = H
[107] R = OH

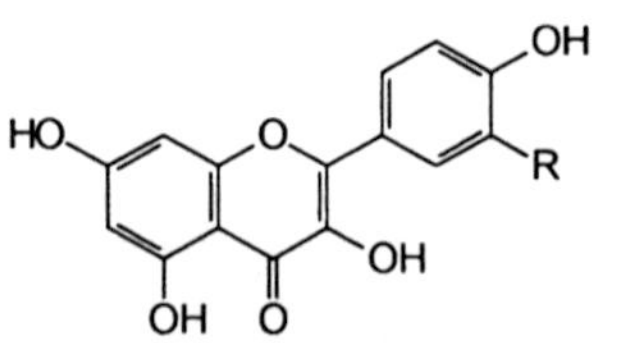

[108] R = H
[109] R = OH
[110] R = OCH_3

Fig. 2.29 Compounds 106–110, flavonoids from *Pinus uncinata* from Sao Jorge, Azores

intermediate, and low levels of quercetin, respectively, the authors calculated values for *p*, the frequency of the *a* allele. Their results also appear in the table. This method of analysis, which has also been used quite successfully in the discussion of monoterpene differences, which will appear below, provides a means to convert descriptive secondary metabolite data into a potentially more meaningful format. In this instance, it clearly demonstrates differences in the genetic composition of the respective populations of *P. uncinata*, although no easily interpreted pattern has emerged.

One point should be raised that was not dealt with in the Lauranson and Lebreton (1991) paper. It would appear that the authors examined the concentration values of the individual flavonols at face value without considering likely or possible biosynthetic relationships. In particular, isorhamnetin is produced by O-methylation of quercetin. Total quercetin produced would then be the sum of quercetin per se plus the amount of isorhamnetin present. This value, reflecting total 3′,4′-dioxygenated flavonols, would be the more accurate measure of allele frequency.

2.4 The Mediterranean Basin

2.4.1 Withania somnifera *(Solanaceae)*

Several literature reports describe efforts to identify the component(s) of *Withania somnifera* (L.) Dun. responsible for the sedative, hypnotic, and antiseptic properties of the plant. Those studies included material collected in South Africa and in different parts of India. Although none of the earlier studies resulted in the establishment of a detailed structure, it did become evident that plants from different areas exhibited different chemistries. Examination of the nonalkaloidal components of *W. somnifera* from different regions of Israel also revealed chemical heterogeneity, leading to a more detailed study of this species (Abraham et al., 1968). The study, which involved 24 populations, showed that three well-defined chemotypes exist in Israel (Fig. 2.30). Chemotype **1**, the most widespread of the three, consists of three compounds [111], known as "withaferin-A", its dihydro derivative [112], and the hydroxyl derivative [113]. Chemotype **2**, confined to northern Israel, consists of the single compound [114]. Chemotype **3**, found in five sites along the southern coastal plain, consists of the most complex array of compounds (at least seven) of the sort represented by structures [115 and 116] (see Fig. 2.31 for structures 111–116). Examination of eight individual plants of chemotype **III** showed identical profiles. Common garden studies revealed that the withanolide profiles of plants grown from seed were identical to those of plants in the wild and cultivated plants of all three types sampled at intervals throughout their growth showed only quantitative changes in withanolide content. The authors did not speculate as to how these populations had become differentiated.

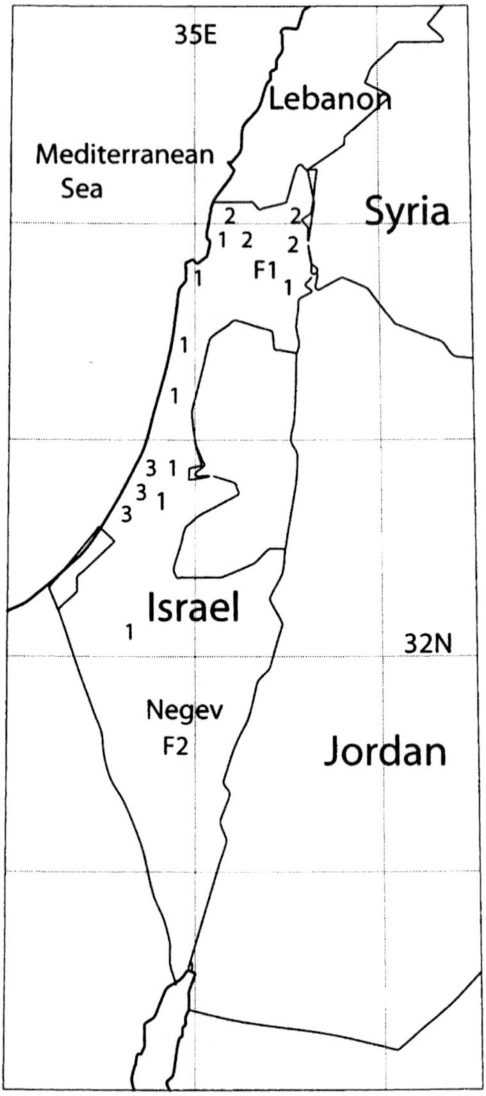

Fig. 2.30 Map of Israel with *Withania somnifera* sites marked with numerals, and *Foeniculum vulgare* sites marked as F1 and F2

2.4.2 Foeniculum vulgare *(Apiaceae)*

Barazani et al. (2002) investigated the volatile oil fraction of *Foeniculum vulgare* L. var. *vulgare* collected from seven populations in Israel. Plants from four of the populations exhibited quite different chemical profiles when individual plants from nature were compared with individual plants maintained in a common garden. In contrast, profiles of three populations were very similar to profiles of individual plants of those populations grown in a garden indicating a high level of heritability of the compounds. Two chemotypes were observed, a *trans*-anethole type [117]

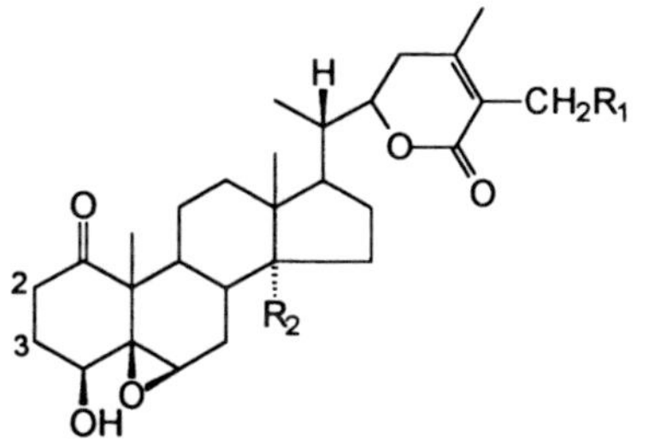

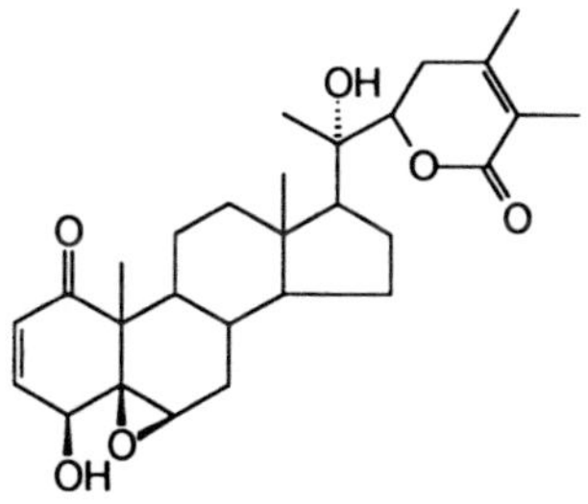

[111] R_1 = OH, R_2 = H, C2-C3 unsat'd
[112] R_1 = OH, R_2 = H, C2-C3 sat'd
[113] R_1 = H, R_2 = OH, C2=C3 unsat'd

[114]

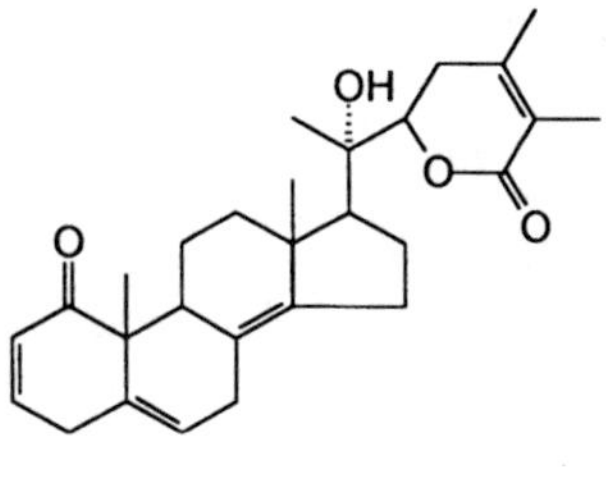

[115]

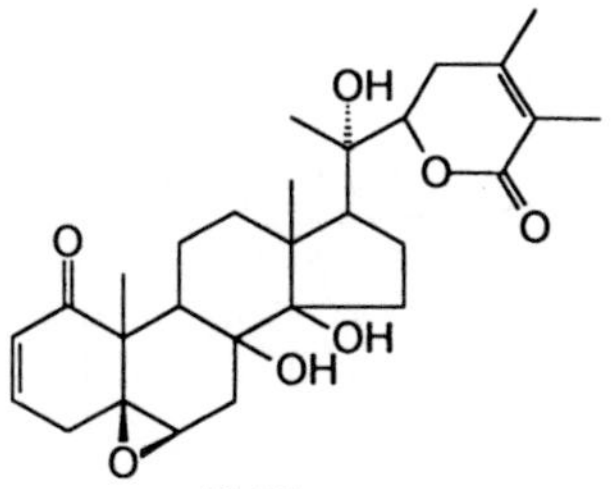

[116]

Fig. 2.31 Compounds 111–116, withanolides from *Withania somnifera*

Fig. 2.32 Compounds 117 and 118 from *Foeniculum vulgare*

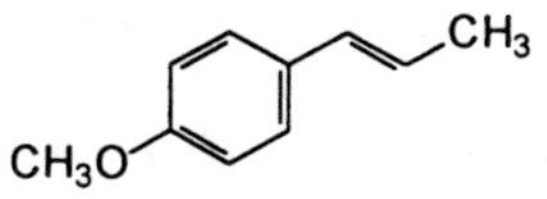

[117]

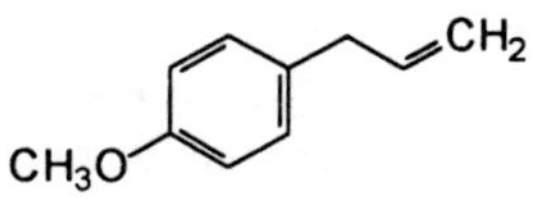

[118]

characteristic of plants from the Negev Desert and northern coastal plain, and an estragole type [118] (see Fig. 2.32 for structures 117 and 118) found in populations on Mt. Dov in the northeastern part of the country (Fig. 2.30). The authors suggest that the two consistent chemotypes remain isolated because of the strong northwesterly winds during the flowering season that prevent pollinators from carrying from pollen the northeastern populations to the coast. The two signal compounds differ only in the position of an unsaturation on the side chain, a difference whose biosynthetic background would be of interest to determine, particularly in view of their apparent genetic stability.

2.4.3 Genista *(Fabaceae)*

This example concerns the distribution of alkaloids in two species from *Genista* sect. *Erinacoides* (Fabaceae) (Gibbs, 1966), namely *G. lobelii* DC. and *G. salzmannii* DC.,

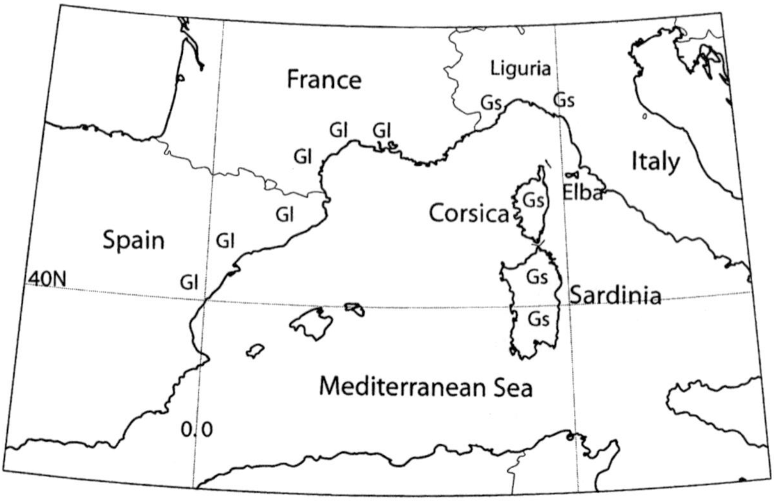

Fig. 2.33 Map of *Genista* sites in northwestern Italy, Elba, Corsica, Sardenia, and Provence. Gs = *Genista salzmannii*, Sl = *Genista lobelii*

although taxonomic opinion differs as to their distinctiveness. *Genista lobelii* occurs in southeastern Spain and southern France, while *G. salzmannii* occurs in Sardinia, Elba, Corsica, and in the Ligurian region of western Italy (Fig. 2.33). In an alternative view (Gamisans, 1973), *G. lobelii* was submerged into *G. salzmannii* with the further recognition of three varieties, var. *lobelii* in Provence, var. *lobelioides* in Corsica above 1200 m elevation, and var. *salzmannii* in Corsica below 1200 m. Because morphological variation within these taxa makes identification, and thus the assessment of relationships, difficult, additional sources of potentially useful taxonomic characters were sought (Kirch et al., 1995). These latter workers chose to examine the quinolizidine alkaloids, a class of compounds that had been shown to be useful sources of information in other legume genera, for example, *Pearsonia* (van Wyk and Verdoorn, 1991) and *Virgilia* (Greinwald et al., 1989).

Kirch et al. (1995) examined individual plants collected in Corsica, Elba, Sardinia, Liguria, and Provence for alkaloids and observed four groups, one characterized by sparteine [119] (see Fig. 2.34 for structures 119–124), one characterized by lupanine-based alkaloids [120 and 121], one that had a very low level of alkaloid production, and one that lacked sparteine and lupanine-based compounds, but did accumulate other alkaloids such as anagyrine [122], ammodendrine [123], and compounds based on cytisine [124], their "outlier" group. The distribution of these four chemotypes is presented in Table 2.10.

These workers concluded that the chemical differences among populations from the various sites were too small to support recognition of two species, pointing to the lack of correlation between morphological features and alkaloid chemistry. They noted as well that the "outlier" chemistry could be the result of developmental,

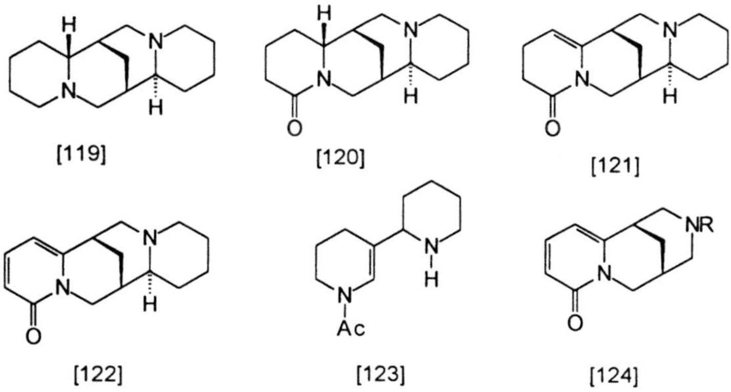

Fig. 2.34 Compounds 119–124, alkaloids from *Genista*

Table 2.10 Pattern of occurrence of alkaloids in Southern European *Genista* (after Kirch et al., 1995)

Location	No.[b]	Alkaloids[a]			
		Spar	Lupa	Lows	Others
Corsica	24	15[c]	0	3	6
Elba	14	11	0	1	2
Sardinia	6	6	0	0	0
Liguria	8	0	4	1	3
Provence	9	0	0	7	2

[a] Spar = Sparteine type; Lupa = Lupanine type; Lows = Low alkaloid content; Others = Other alkaloid types.
[b] No. =number of plants examined.
[c] Number of plants of each alkaloid type.

environmental, caryological, or lifestyle (reproductive mode) differences, indicating the need for additional study. The significance of results of this sort cannot be assessed fully without information on those factors, as well as some indication of evolutionary directions among the populations. The latter would require application of molecular genetic techniques. Were such information available, it might be possible to discuss the changes in alkaloid patterns in terms of differences in the well-known biosynthetic steps by which these compounds are formed.

2.4.4 **Pistacia lentiscus** *(Anacardiaceae)*

The following examples, starting with the pistachio, are concerned primarily with variation in profiles of common volatile components of a number of aromatic plant species native to the Mediterranean Basin. For the most part, the examples involve well-known species, most of which are of commercial importance. The literature dealing with chemical constituents of aromatic plants of the Mediterranean region and the Middle East is extensive. The examples below are only a sampling, but clearly show the existence of significant differences among populations, and species, of many plants.

Pistacia lentiscus L. (Anacardiaceae) (mastic; dessert and confectionary seeds, pistachio, come from *P. vera* L.) grows widely in countries bordering the Mediterranean Sea. In addition to its natural range, the species is much cultivated in Corsica. Castola et al. (2000) analyzed the essential oil from leaves of 105 individual plants, representing much of the natural range on the island. Three main chemotypes were identified, with one of them further divisible into two subgroups: (**1a**) α-pinene (22) and terpinen-4-ol [42] with terpinen-4-ol predominating; (**1b**) α-pinene and terpinen-4-ol with α-pinene predominating; (**2**) limonene [24] plus terpinen-4-ol type; and (**3**) myrcene [40]. No correlation between geographic region and essential oil composition was found in this study (Cassanova, personal communication).

Geographical differences do exist, however, within this species. Studies from other parts of the Mediterranean Basin had shown that *Pistacia* from Sardinia (Picci et al., 1987) and southern France (Buil et al., 1975) had profiles similar to Corsican types **1a** and **1b**, respectively. Spanish and Sicilian oils had much lower myrcene concentrations than seen in Corsican type **3** (Calabro and Curro, 1974; Boelens and Jimenez, 1991). Leaf oil from Moroccan plants had α-pinene, myrcene, limonene, β-caryophyllene, and a cadinene isomer in roughly equivalent amounts (Guenet and Aubanel, 1991). Oil from plants growing in Egypt was unique in accumulating large amounts of 3-carene (δ-carene) (DePooter et al., 1991). Differences in fruit oils from different geographic origins were also noted: Spanish material gave values for myrcene, α-pinene, and limonene of 72%, 10%, and 7%, respectively (Boelens and Jimenez, 1991), while the same compounds in Australian material were present to the levels of 39%, 28%, and 11% (Wylie et al., 1990).

From the data available, it is difficult to explain the apparent plasticity of volatile compounds produced by plants from these different areas. Subtle environmental factors may be at work. Alternatively, selection for oil yield, particular oil composition, or for some other feature, over the years may have resulted in the variation now seen. The variation seen in Corsican plants may have arisen through the arrival on the island of propagules originating in the other growing areas. In areas so long occupied by humankind, it is often difficult to sort out purely natural driving forces from the effects of cultivation.

2.4.5 Thymus *(Lamiaceae)*

Studies of the essential oils of *Thymus* species have documented the existence of several chemotypes with greater or lesser geographical distinctions among them. In an examination of the essential oils of *T. vulgaris* L. native to France, Granger and Passet (1973) described six chemical phenotypes based upon the occurrence of a number

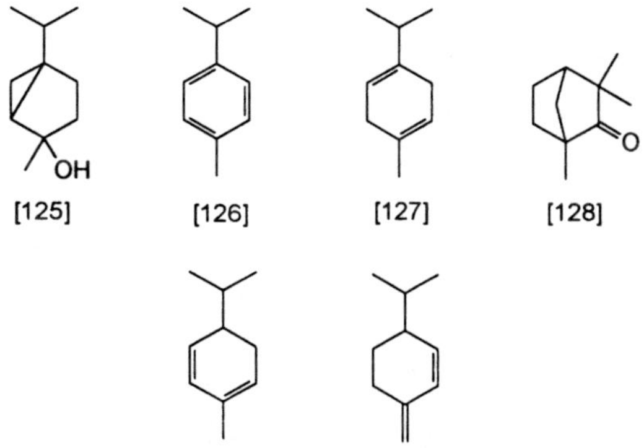

[125] [126] [127] [128]

[129] [130]

Fig. 2.35 Compounds 125–130, terpenes from *Thymus* species

of common monoterpene derivatives, namely, geraniol, linaloöl, carvacrol, and thymol. Although population differences were minimal, there was some differentiation, particularly with regard to populations rich in thujanol-4 [125], which tended to lay roughly west of Marseilles (see Fig. 2.35 for structures 125–130). Cañigueral et al. (1994) studied *T. moroderi* Pau ex Martinez and *T. antoninae* Rouy & Coincy collected in Spain. These workers found little chemical polymorphism in *T. antoninae*, but clear-cut differences in the sesquiterpene components of *T. moroderi* were evident. Thus, plants from near Alicante (38°21′N, 0°29′W) were distinguished on the basis of sesquiterpene composition from those collected around Murcia (37°59′N, 1°8′W). Plants from both areas had very similar monoterpene profiles. Sáez (1999) reported wide chemical variation, with no apparent geographical races in another Spanish species, *T. baeticus* Boiss ex Lacaita. Studies of a third Spanish species, *T. piperella* L. revealed a degree of geographic patterning with regard to one of three chemotypes (Blanquer et al., 1998) involving p-cymene [126], γ-terpinene [127], carvacrol [89], and thymol [105]. The three chemotypes defined were: (**1**) *p*-cymene-carvacrol-γ-terpinene (11 populations); (**2**) *p*-cymene-thymol (5 populations); and (**3**) *p*-cymene-carvacrol (15 populations). The distribution of chemotypes **1** and **3** can best be described as showing tendencies rather than clear-cut differences: chemotype **1** was found most often in northern populations (west of Valencia), while chemotype **3** was found most often in the southern part of the study range (north of Alicante). Chemotype **2** was observed in five populations clustered more or less in the center of the study range (about half way between Alicante and Valencia).

The finding of Mártonfi et al. (1994) that chemotype patterns in *T. pulegioides* L. depend to a significant degree upon soil chemistry may prove of wider significance in trying to establish the driving forces behind the level of variation seen in oil composition. Although many of the studies of variation involve plants that are of some commercial importance, with variation playing a significant part in quality of the product, purely biological considerations are also interesting. For example, Beker et al. (1989) studied the volatile components of two chemotypes of *Marjorana syriaca* L. as to the effect that the different chemistries (thymol/carvacrol ratio) had on honeybees. These studies indicated that bees can discriminate among the four "odor identities:"

thymolic inflorescences, thymolic leaves, carvacrolic inflorescences, and carvacrolic leaves, and that they can learn to select each of the identities separately.

2.4.6 Lavandula stoechas *(Lamiaceae)*

A study of *Lavandula stoechas* L. subsp. *stoechas* along an east–west transect in Crete demonstrated that significant variation in essential oil composition can exist on an even smaller scale than those described above (Skoula et al., 1996). Four wild populations were sampled over a distance of about 150 km: Fodele at 24°58′E, Akrotiri at 24°07′E, Lakki at 23°57′E, and Sfinari at 23°34′E (latitude was the same throughout). The main compounds identified, in order of decreasing abundance, were fenchone [128], 1,8-cineole [41], camphor [93], and α-pinene [22]. For example, fenchone concentrations ranged from 10.4% to 56.3% in inflorescence oils and from 18.7 to 48.9% in leaf oil; α-pinene from 1.8% to 14.2% of inflorescence oil and from 0.5 to 3.4% in leaf oil. α-Cadinol, which was only seen in the two centrally located populations, reached a concentration of 9.2% in inflorescence oil and 7.7% in leaf oil. Discriminant function analysis showed that the Fodele population, in which 1,8-cineole predominates with fenchone and camphor as lesser components, is clearly distinguished from the other populations. The Fodele population was thus recognized as constituting a second chemotype characterized by the 1,8-cineole/fenchone combination, as opposed to the other populations recognized as the fenchone/camphor chemotype. The numerical analysis showed a great deal of overlap among the three westernmost populations. The discovery of a chemotype dominated by 1,8-cineole takes on added significance when one appreciates that *L. stoechas* subsp. *stoechas* populations sampled from Greece (Kokkalou, 1988), Cypress (Valentini et al., 1993), Morocco (in Valentini et al., 1993), and Spain (Garcia-Vallejo et al., 1990; Negueruela, 1990) all had fenchone as the predominant essential oil component.

For additional studies of geographic (and seasonal) variation of essential oils in other members of Lamiaceae, interested readers should consult a study of *Thymus caespitosus* Brot. in Portugal (Salgueiro et al., 1997), *Origanum vulgare* L. subsp. *hirtum* (Link) Letswaart in Greece (Vokou et al., 1993; Kokkini et al., 1994, 1997), and *Origanum* X *intercedens* (a hybrid between *O. onites* L. and *O. vulgare* L. subsp. *hirtum* (Link) Letswaart) (Kokkini and Vokou, 1993). The last mentioned work involved a study of the hybrid on the Greek island of Nisyros and its resemblance to *O. vulgare* subsp. from the neighboring island of Kos.

2.4.7 Mentha suaveolens *(Lamiaceae)*

This example involves a study of free flavonoids (aglycones) present on the leaf surface of *Mentha suaveolens* Ehrh. (Lamiaceae) growing in Spain and in Algeria

(Zaidi et al., 1998). Quite significant differences inn numbers of compounds and relative degree of complexity were noted by those authors: only two compounds were found in Spanish plants, whereas plants from the Algerian mountains afforded a richer array of 6- and 8- substituted flavones, along with an assortment of simpler derivatives of apigenin and luteolin. The authors suggested that the differences could be accounted for by "geographical and climatic differences combined with efficient genetic isolation." A "high altitude" topodeme, *M. suaveolens* subsp. *timija* (Cosson ex Briq.) Harley from Morocco has been recognized. The authors were not unaware of the possibility that the chemical differences simply reflected the existence of "true chemotypes" with no correlated morphological features.

2.4.8 Teucrium polium *(Lamiaceae)*

Going slightly farther afield in this next example, we look at some of the differences that Kamal and Sandra (1994) noted in their examination of two varieties of *Teucrium polium* L. var. *album* (Mill.) Fiori occurs along the Mediterranean coast of Egypt and in Sinai, whereas var. *pilosum* Decne [= *T. pilosum* (Decne) Asch. & Schweinf] is native to Qatar. Earlier papers had described iridoids, flavonoids (Rizk et al., 1986), and volatile components (Wassel and Ahmed, 1974; Hassan et al., 1979; Vokou and Bessiere, 1985) of *T. polium* from several geographic regions, that is, Egypt, Qatar, Saudi Arabia, and Greece. Kamal and Sandra (1994) examined the volatile components of both varieties using GC-MS analysis, which allowed them to identify 70 of the 93 compounds (peaks) observed. Significant quantitative as well as qualitative differences were observed a sample of which are shown in Table 2.11. These data add significantly to the list of differences between the two varieties; they might be used as well in support of recognition of separate species status for the two, as has been suggested in the past.

Table 2.11 Variation within *Teucrium polinum* (after Kamel and Sandra, 1994)

Compound	Varieties	
	album	*pilosum*
α-Pinene	4.05[a]	nd[b]
β-Pinene	4.90	0.28
Myrcene	1.59	nd
4-Terpinyl acetate	1.03	nd
Humulene	0.28	2.28
Germacrene-D	1.44	4.42
γ-Cadinene	2.60	18.04
δ-Cadinene	4.05	8.30
Spathulenol	nd	11.59
β-Eudesmol	3.31	19.14
Patchouly alcohol	33.34	2.77

[a] Percentage of component in oil.
[b] nd = Not detected.

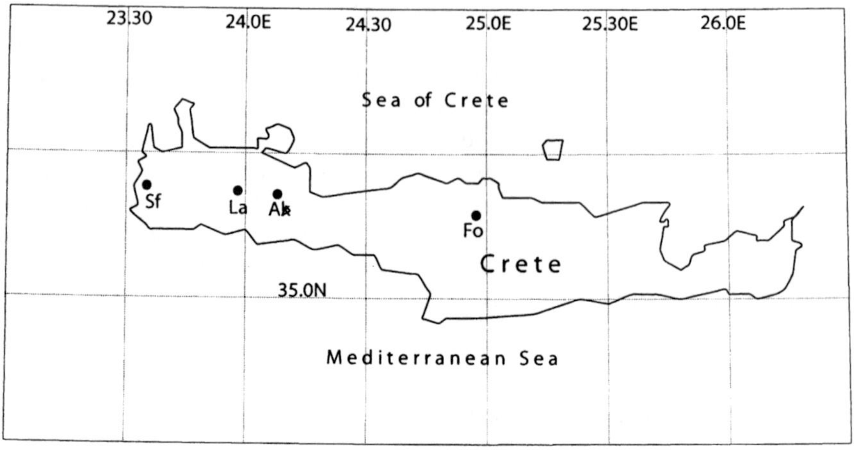

Fig. 2.36 Map of *Lavandula stoechas* sites on Crete

2.4.9 Juniperus phoenicea *(Cupressaceae)*

Rezzi et al. (2001) described an examination of the essential oils of *Juniperus phoenicea* L. subsp. *turbinata* (Guss.) Parl. Nyman (syn. *J. phoenicea* subsp. *lycia* Auct. = *J. turbinata* Guss.) growing on Corsica (Fig. 2.37). Individual plants were collected from six sites representing typical habitats and were analyzed for their terpene fraction; the major components of which were identified as α-pinene, α- and β-phellandrenes [129, 130, respectively], α-terpineol acetate, 3-carene [97], and myrcene [40]. Cluster and discriminant analyses revealed two groups of populations that could be distinguished based on contents of α-pinene, β-phellandrene, and α-terpinyl acetate with α-pinene as the major discriminating factor. Plants exhibiting the two chemotypes, identified as Group **I** with high α-pinene and low β-phellandrene, and Group **II** with lower α-pinene and higher β-phellandrene, grew intermixed but in different ratios at different sites. For example, Group **I** and Group **II** plants were found to occur in a ratio of 7:1 at a site near Porto-Vecchio (southeastern coast), whereas a ratio of 1:9 was observed at two sites, one near Borcaggio (northern peninsula) and one near Ajaccio (southwestern coast). The population at Bonifaccio (southern tip of the island) exhibited a 1:1 ratio of the two types. Whereas there seem to be no relationship of these two types with geography of the island, or with habitat differences, relationships with populations on the mainland were noted. Thus, Group **I** plants bear closest similarity to α-pinene-rich oil from *J. phoenicea* s. st. collected in Greece and Spain, whereas Group **II** plants have an α-pinene-poor oil that bears more similarity to the oil from plants growing in Portugal and also in some parts of Spain. There clearly seems to be some difficulty with taxonomic details in this system, as the authors noted in their discussion.

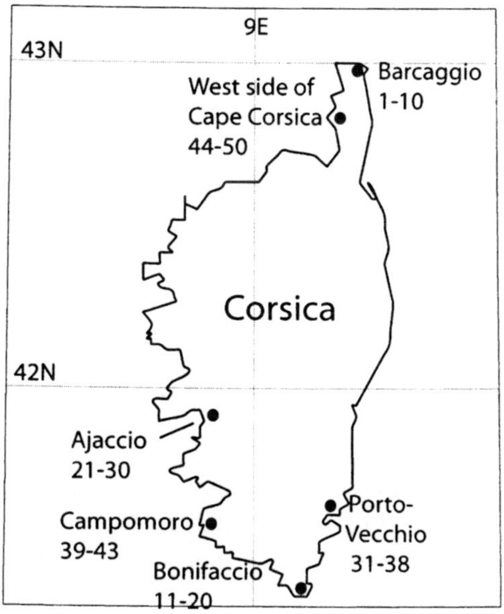

Fig. 2.37 Map of *Juniperus phoenicea* sites on Corsica

2.4.10 Phlomis lychnitys *(Lamiaceae)*

Phlomis consists of about 100 species, a dozen of which occur in Mediterranean Europe (Mabberley, 1997, p. 549). The study of interest here involves a study of the flavonoids of *P. lychnitys* L., a small plant native to Mediterranean Spain (Tomás et al., 1986). Those workers identified the common flavones apigenin, luteolin, and luteolin 3′-methyl ether (chrysoeriol) 7-*O*-glucosides and their respective *p*-coumaroyl derivatives. A brief review of the literature revealed that Mediterranean species of *Phlomis* are characterized by the presence of the flavone methyl ether, whereas continental species appear to lack *O*-methylated flavones. Species from India have been reported to lack flavones but accumulate flavonols. The suggestion was made that accumulation of flavonols represents an ancestral feature of the genus.

2.5 Asia

2.5.1 Datura *and* Berberis — *Alkaloids and Elevation*

The earlier alkaloid literature contains numerous reports of studies aimed at discovering the existence of chemotypes involving these pharmacologically important compounds. An example of this sort of study comes from the work of Karnick and Saxena (1970) who were interested in the effect of elevation on total alkaloid content of *Datura metel* L. (Solanaceae), an important source of hyoscyamine [26]

Table 2.12 Alkaloid variation with elevation in *Datura metel* (from Karnick and Saxena, 1970)

Source (Elevation)	Tissue Analyzed					
	Root	Stem	Leaf	Floral	Fruit	Seed
Bombay (sealevel)	0.27[a]	0.19	0.25	0.69	0.060	0.09
Poona (563 m)	0.52	0.29	0.32	0.86	0.079	0.10
Pachmari (716 m)	0.71	0.43	0.54	0.95	0.089	0.14
Darjeeling (2166 m)	0.89	0.46	0.58	0.99	0.097	0.19

[a] Total alkaloid expressed as percent hyoscyamine.

and related compounds. Specimens were collected from the field at four different elevations ranging from sea level (Bombay) to 2166 m (Darjeeling). Plants were separated into their constituent organs and analyzed for total alkaloid content. The results are displayed in Table 2.12. Although the report contained no statistical information, the trend of increasing alkaloid content with increasing elevation is clear. It is interesting to note that the concentration of compounds increases rather than decreases with elevation, opposite to the trend seen with the cyanogenic compounds from legumes (see above). It would be of interest to learn if herbivore pressure is higher at the upper elevation sites, or whether there is some other factor influencing alkaloid synthesis in this species.

In contrast to the increasing alkaloid concentration with elevation in the above case, Chandra and Purohit (1980) demonstrated the reverse trend in a study of *Berberis lycium* Royle (Berberidaceae) from different elevations in Garhwal, India. Roots, stems, and leaves were collected at sites ranging from 750 m to 3700 m. A clear-cut trend was noted in all three tissue types with regard to berberine [85] content with lower levels present at the higher sites. For example, roots of *B. lycium* collected at 750 m, 1400 m, 1800 m, and 2500 m gave values of 18.60%, 17.18%, 13.08%, and 11.05% dry weight berberine, respectively. Study of *Berberis asiatica* Griff. revealed a decrease in root alkaloid content, from 19.25% to 16.40%, in material collected at 1400 m and 1800 m, respectively, but an increase in the contents in both stems and leaves. Collections of *B. jaeschkeana* Schneider at 3200 m had nearly three times the alkaloid content in roots as did material from 3700 m. Stems of this species showed a slight reduction at the higher elevation, but alkaloid content of leaves was higher at the higher elevation. The overall conclusion reached by these workers was that toxic alkaloid production is reduced at the higher elevations, but no reason for this phenomenon was offered.

A recent study of secondary metabolites in *Swertia franchetiana* H. Smith (Gentianaceae) involved collections of plant material over a distance of approximately 1100 km from the area of Xining, China (36.35 N, 101.55 E) at 2200 m southwestward to the area near Lhasa, Tibet (29.41 N, 91.10 E) at an elevation of 3960 m (H. Yang et al., 2004). They reported a lack of "altitudinal trends in phytochemical constituents" of the taxon having identified the bitter principle swertiamarin, the triterpene oleanolic acid, the C-glucosyl xanthone mangiferin, and several other xanthone derivatives (see Fig. 2.38 for structures). The only compound for which altitudinal variation was detected in this work was 1,8- dihydroxy-3,7-dimethoxyxanthone (positional isomer of the compound shown in the Figure), which showed a negative correlation with

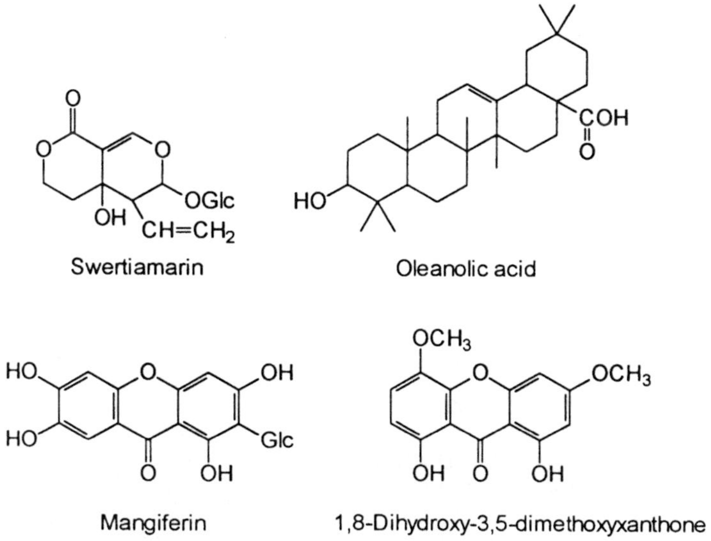

Fig. 2.38 Compounds from *Swertia franchetiana*

altitude. There was a statistically significant relationship between 1,8- dihydroxy-3,7-dimethoxyxanthone and mangiferin and latitude and longitude, however.

The authors suggest that altitudinal variation in "phytochemical constituents remains unknown." That this is factually incorrect is revealed by the well-known work on the effect of elevation on cyanogenic glycosides, as shown in the case of the clovers above, as well as in the case of alkaloid variation in *Berberis* just described.

2.5.2 Cymbopogon distans *(Poaceae)*

Cymbopogon is a genus of aromatic grasses native to tropical and warm regions of the Old World. Lemongrass, a common ingredient in southeastern Asian cooking, is *C. citratus* (DC.) Stapf. The species of interest to us here, *C. distans* (Stend.) Wats., occurs in the northwestern region of Uttar Pradesh, India, northeast of Delhi. Mathela et al. (1988) described the existence of four chemotypes of this taxon based upon different combinations of mono- and sesquiterpene derivatives. Chemotype **1**, the α-oxobisabolene [131] type (see Fig. 2.39 for structures 131–138), was characterized by high percentages of the title compound, often reaching levels as high as 68%. Geranial [132], geraniol [133], geranyl acetate [134], neral [135], and piperitone [136], all of which play major roles in the other chemotypes, were not observed. This population was collected near Hanumangarh (Naintal, 1900 m). Chemotype **2**, referred to as the "lemon-grass oil chemotype," was obtained from Munsiyari (Pithoragarh, 2200 m). This form had significant amounts of geraniol, geranyl acetate, and the two aldehydes geranial and neral. Chemotype **3**, the piperitone type, was collected on Naina Peak at 2600 m (Nainital). Its volatile oils consisted of up

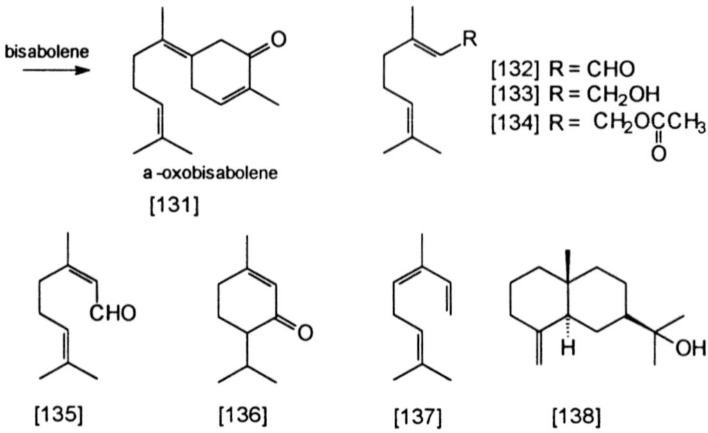

Fig. 2.39 Compounds 131–138 from *Cymbopogon distans*

to 60% piperitone, with lesser amounts of β-ocimene [137]. The fourth chemotype, found at Almora at 1800 m elevation (Loharkhet), was characterized by the presence of 48% sesquiterpene alcohols, for example, derivatives of eudesmol [138], and other related compounds. The authors pointed out that chemotypes **1** and **3** were separated by only five miles, chemotypes **2** and **3** by about 80 miles, and are over 100 miles from the populations that afforded chemotypes **1** and **2**. They also noted that, whereas chemotypes **1, 3**, and **4** have been observed at other sites in this region, chemotype **2** seems to be restricted to the region around Munsiyari. These profiles appear to be independent of environmental factors, since plants grown in a common garden produced oils of the same composition as those collected from the field. No immediate explanation for this level of variation comes to mind, especially in view of the comparative constancy of oil composition in other species.

2.5.3 Cupressus *or* Chamaecyparis? *(Cupressaceae)*

Gadek and Quinn (1987) demonstrated the usefulness of flavonoids, in particular biflavonoids, in addressing the question of generic placement of *Cupressus funebris* Endl. Geography became a factor in this otherwise straightforward chemotaxonomic exercise because earlier work on wood tropolones (Zavarin et al., 1967b; more on these compounds below) was done on material obtained from northern India, which lies outside the accepted central Chinese range of *C. funebris*. Furthermore, certain morphological features of *C. funebris*, particularly dimorphic leaves, small cone size, and low seed number, suggested that *Chamaecyparis* might be the better taxonomic home for this taxon. Based upon the knowledge that biflavonyl profiles of *Cupressus* and *Chamaecyparis* are different, Gadek and Quinn (1987) examined a specimen of *C. funebris* for these compounds. The outcome of the analysis was

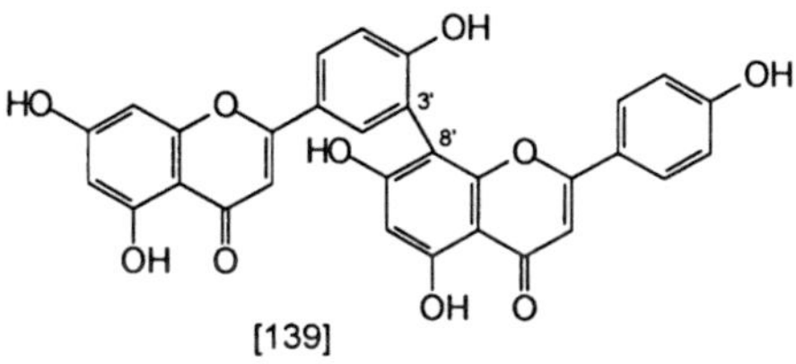

[139]

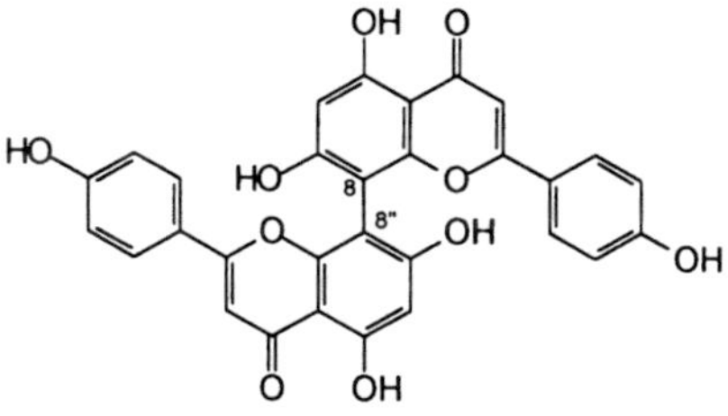

[140]

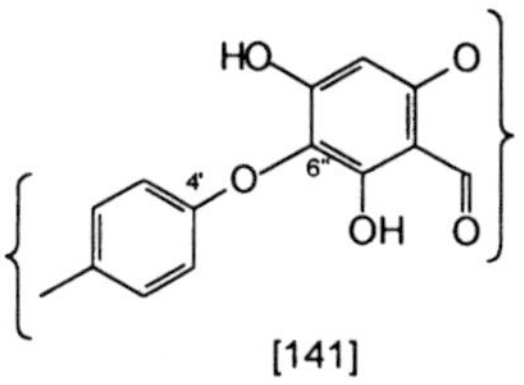

[141]

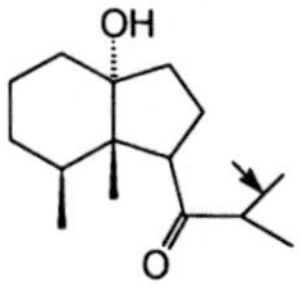

[142] Single bond (arrow)
[143] Double bond arrow)

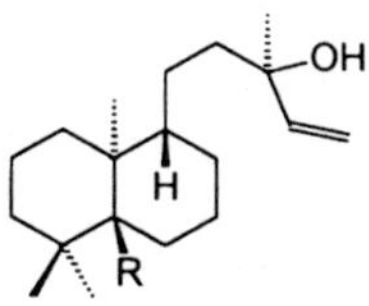

[144] R = OH
[145] R = H

Fig. 2.40 Compounds 139–141 from the *Cupressus* versus. *Chamaecyparis* study. Compounds 142–145 from *Jungermannia vulcanicola*

clear-cut: *C. funebris* yielded major amounts of amentoflavone [139] and cupressuflavone [140], with smaller amounts of hinokiflavone [141, partial structure only, the 7″-methyl ether of hinokiflavone, and traces of mono-*O*-methyl derivatives of amentoflavone and cupressuflavone (see Fig. 2.40 for structures 139–141). This profile is characteristic of *Cupressus* species, whereas profiles dominated by di- and tri-*O*-methylbiflavonoids characterize *Chamaecyparis* species.

2.5.4 Jungermannia vulcanicola *(Jungermanniaceae)*

The liverwort, *Jungermannia vulcanicola* (Schiffn.) Steph., was collected from several sites in Japan by Nagashima et al. (1996) who reported the new chiloscyphane-type sesquiterpenoid [142] from sites in the prefectures of Okayama and Shiga in southwestern Honshu. Further, the known chiloscyphane [143] was isolated from material collected at Okayama and the new labdane diterpenoid [144], and the known *ent*-13-epi-manoöl [145] from material collected at Shiga (see Fig. 2.40 for structures 142–145). These observations suggested the existence of at least three chemotypes of this liverwort in Japan, although the structural differences between the pairs of compounds involve single biosynthetic steps. It would be of value in understanding the relationships among these chemotypes were the high quality of the chemical analytical studies of this plant complemented by a thorough survey of this organism throughout its range in Japan.

2.5.5 Biebersteinia *(Geraniaceae)*

This next example involves a small genus native to central and western Asia, with one species disjunct in the eastern Mediterranean Basin, where it occurs in Greece and Anatolian Turkey. Traditionally, *Biebersteinia* has been allied with Geraniaceae (e.g., Cronquist, 1981), despite earlier chemotaxonomic investigations by Bate-Smith (1968, 1973), which had revealed patterns of polyphenolic compounds in *B. multifida* DC. and *B. odora* Stephan that differed significantly from patterns seen in members of Geraniaceae. Omurkamizinova et al. (1991) reported luteolin 7-O-glucoside [146] (see Fig. 2.41 for structures 146–159) and 7-O-rutinoside [147] from *B. multifida*, while Zhang et al. (1995) identified the rare flavone 5,7,3′-trihydroxy-8,4′,5′-trimethoxyflavone (8-hydroxytricetin 8,4′,5′-trimethyl

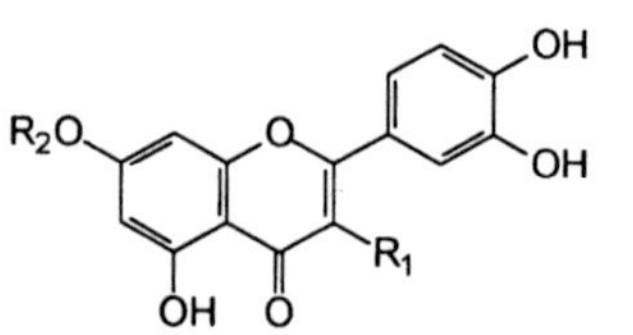

[146] R_1 = H, R_2 = Glucose
[147] R_1 = H, R_2 = Rutinose
[149] R_1 = OH, R_2 = Glucose

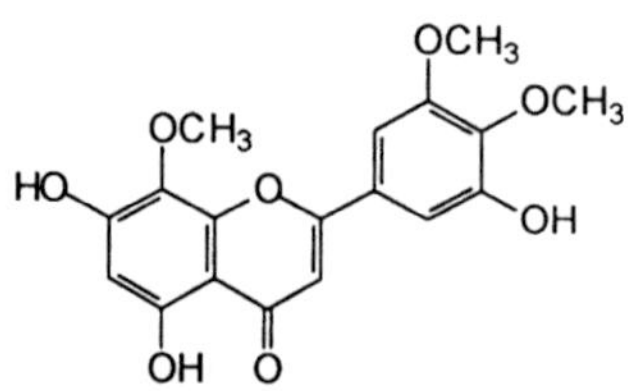

[148]

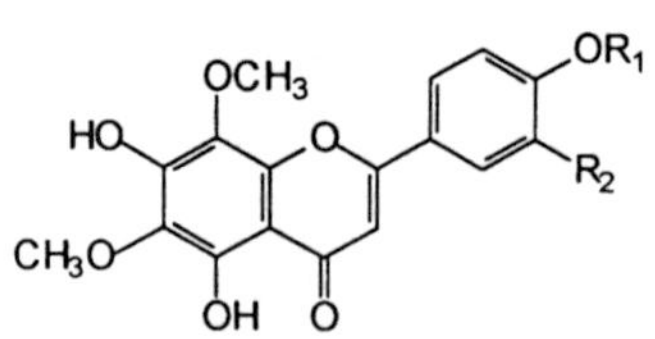

[150] R_1 = R_2 = H
[151] R_1 = CH_3, R_2 = H
[152] R_1 = CH_3, R_2 = OH
[153] R_1 = H, R_2 = OCH_3
[154] R_1 = CH_3, R_2 = OCH_3

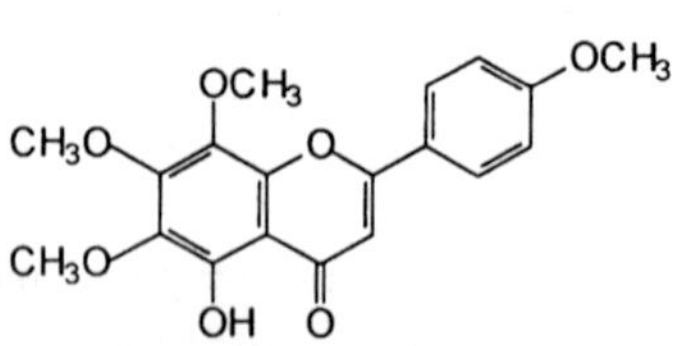

[155]

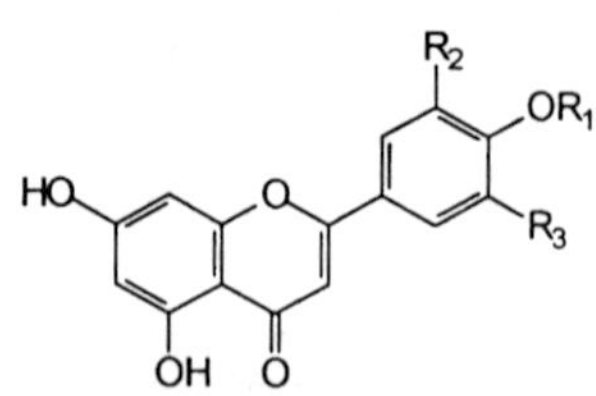

[156] R_1 = R_2 = R_3 = H
[157] R_1 = CH_3, R_2 = R_3 = H
[158] R_1 = H, R_2 = R_3 = OH

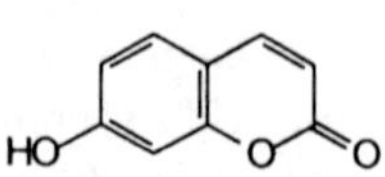

[159]

Fig. 2.41 Compounds 146–159, flavonoids and a coumarin, from *Biebersteinia*

ether) [148], a *C*-glucosylluteolin derivative, and quercetin 7-O-glucoside [149] from the Tibetan species *B. heterostemon* Maxim. A recent study of *B. ophanidis* Boiss by Greenham et al. (2001). revealed a very different picture again, particularly with regard to flavone aglycones obtained from the leaf surface. Six of the identified compounds are based on the tetrasubstituted A-ring: 5,7,4′-trihydroxy-6,8-dimethoxyflavone [150]; 5,7-dihydroxy-6,8,4′-trimethoxyflavone [151] (nevadensin); 5,7,3′-trihydroxy-6,8,4′-trimethoxyflavone [152] (acerosin); 5,7,4′-dihydroxy-6,8,3′-trimethoxyflavone [153] (sudachitin); 5,7-dihydroxy-6,8,3′,4′-dimethoxyflavone [154] (hymenoxin); and 5-hydroxy-6,7,8,4′-tetramethoxyflavone [155] (gardenin-B). Other compounds identified in the aglycone fraction included apigenin [156], apigenin 4′-methyl ether (acacetin) [157], luteolin, and the coumarin umbelliferone [159]. Vacuolar flavonoids identified were the 7-O-rutinosides of apigenin and luteolin, and the 7-O-glucosides of apigenin, luteolin, and tricetin [158]. Greenham et al. (2001) did not speculate on the evolutionary significance of these flavonoid differences; detailed sequence studies would likely be needed before relationships among these species could be discussed.

Returning to the chemotaxonomic aspect of the study, however, it is interesting to note that recent work by Bakker et al. (1998) on *rbc*L and *atp*B sequences of *Biebersteinia* place the genus close to Rutaceae. This position is fully supported by the flavonoids reported by Greenham et al. (2001), which are very similar to *O*-methylated flavones known from well-studied members of Rutaceae, such as *Citrus*.

2.5.6 Rhodiola crenulata *(Crassulaceae)*

Rhodiola crenulata (Hook. F. et Thomas) S. H. Fu occurs in southwestern China (Yunnan and Sichuan provinces) and in the Tibetan Autonomous Region. Y. Li et al. (2004) examined rhizomes of plants from these areas for their essential oils. Material from the two areas (Yunnan and Sichuan taken as one source area) shared a suite of ten compounds but differed in the relative amounts of certain individual components. The most striking differences were recorded for *n*-octanol (eight-carbon straight chain alcohol), whose content ranged from 13.4% to 21.0% of total oils for the three Tibetan populations, and from 29.6% to 33.6% for the four populations sampled in China. Levels of geraniol [18] ranged between 45.5% and 55.1% for Tibetan material, and between 14.8% and 26.9% for plants from the Chinese sites. No discussion of the significance of these results was forthcoming.

2.5.7 Hippophae rhamnoides *(Elaeagnaceae)*

Hippophae rhamnoides L., known as buckthorn or sallowthorn, occurs from Europe to northern China (Mabberley, 1997, p. 342). Of the nine subspecies recognized within this species, two occur only in China, subsp. *yunnanensis* Rousi and subsp. *sinensis* Rousi., subsp. *yunnanensis* occurs more to the south and west in China with populations sampled for chemical work from Tibet (centered at ca. 28°N, 94°50′E) and Yunnan Province (centered at ca. 27°N, 90°40′E), and samples of subsp. *sinensis* coming from Gansu Province (centered at ca. 35°40′N, 104°E) and Liaoning Province (centered at ca. 40°N, 120°E). A study of long-chain hydrocarbon derivatives obtained from these two taxa revealed clear-cut differences between the two (Tian et al., 2004). Tetracosane was present in both taxa but in significantly higher concentration in subsp. *sinensis*. The reverse was true for eicosanol where much larger amounts were seen in subsp. *yunnanensis*. A further distinction was observed with hexadecanoic acid, which was a major component in subsp. *yunnanensis*, but not detected in subsp. *sinensis*. Further distinctions could be made between populations from Tibet and Yunnan Province, and between populations from Gansu and Liaoning Provinces based upon other constituents. These authors' results were in substantial agreement with the geographic patterning of features of growth and plant hardiness reported by Yao and Tigerstedt (1995) and Yao et al. (1992).

2.5.8 Thujopsis dolabrata *(Cupressaceae)*

Thujopsis dolabrata Sieb. et Zucc., hiba, the sole species of the genus, is endemic to Japan where it occurs on all major islands. Earlier reports, cited by Takahashi et al. (2003), recorded the diterpene chemistry of leaves of *T. dolabrata* (structures not shown here), and a level of variation that led to the present study which was based on a much larger sampling of the species. Thus, over 220 individuals from 34 sites were analyzed for their diterpene and diterpene acid content by GLC methods and the data subjected to statistical analysis. Three types of *T. dolabrata* emerged from the analysis which occupy more or less definable areas: Type I area is split between Tohoku and Hokkaido (northern Japan), and the northern part of Chugoku; Type II area lies in the central part of Japan; and the Type III area includes southeastern and southern Japan. Figure 2.41.2 shows these areas.

2.6 South America

2.6.1 Vanillosmopsis erythropappa *(Asteraceae)*

Vanillosmopsis erythropappa Schultz-Bip. is a component of the Atlantic coastal forest in Brazil, where it is exploited as a source of an essential oil used in the pharmaceutical industry. A survey of specimens from four locations showed markedly different compositions of their essential oil fraction (Lopes et al., 1991). The compounds identified were bisabolol [160], costunolide [161], eremanthine [162], and five related compounds, six compounds based upon the cyclocostunolide skeleton [163], and the two esters 15-deoxygoyazenolide [164] and lychnopholide [165] (see Fig. 2.42 for structures 160–165). The distribution of these compounds is presented in Table 2.13. The amount of variation within this sample is quite impressive, considering the comparatively small distances that separate the sites; in particular, note the differences between the sites identified in the original as Rio de Janeiro (RJ Forest) and Furnas (MG Savannah). The authors interpreted these differences as a case of micromolecular diversity of sympatric species, which holds that plants acquire more complex chemistries as they migrate away from their centers of origin and meet new challenges in the newly occupied environments. This complex idea is discussed in greater detail by Gottlieb and Kubitzki (1983) and by Gottlieb (1986). The examples discussed by Gottlieb in that book do not lend themselves to easy summarization; hence, the reader should consult that source for more detailed discussions of the concept involving several different secondary metabolites.

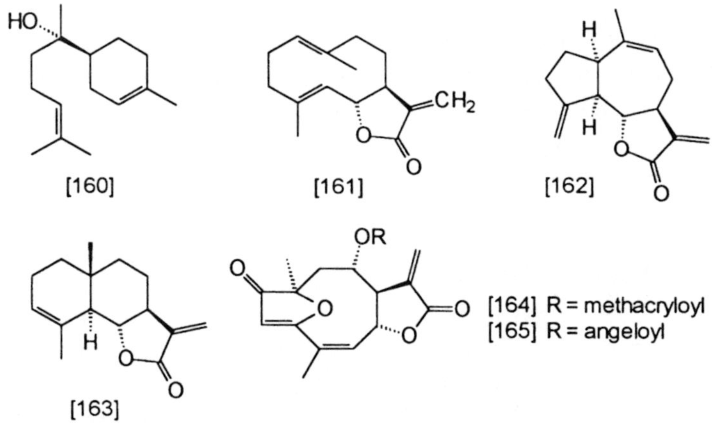

Fig. 2.42 Compounds 160–165, sesquiterpenes from *Vanillosmopsis erythropappa*

Table 2.13 Sesquiterpene variation in *Vanillosmopsis erythropappa* (after Lopes et al., 1991)

Collection Site	Compound[a]					
	[160]	[161]	[162]	[163]	[164]	[165]
Rio de Janeiro	+	+	+	–	–	–
Itabira	+	+	*[b]	*	–	–
Caraguatatuba	trace	–	–	–	+	–
Fumas	trace	–	–	–	–	+

[a] For structures see Fig. 2.41.

[b] Asterisk (*) = Present and accompanied by other compounds having the same carbon skeleton.

2.6.2 Cunila galioides *(Lamiaceae)*

According to a recent paper by Echeverrigaray et al. (2003), the genus *Cunila* consists of 22 species with two centers of diversity, Mexico with 10 species, and southern South America with the remaining 12. *Cunila galioides* plants were collected from 20 sites, 15 in Rio Grande do Sul State and five in Santa Catarina State, and subjected to analysis of their essential oils. Of the 40 compounds identified, 14 were present in levels above 10% (of the total oil yield) and were used for statistical analysis. Three chemotypes emerged from the analysis: group **1** (the citral group, C_{10} acyclic aldehyde); group **2** (the ocimene group, C_{10} acyclic alkene); and group **3** (the menthene group, C_{10} cyclic derivatives). [Note: each group consisted primarily of compounds related to the group type.] Geographic differentiation was clearly indicated by the various treatments with group **1** occurring in the northeast plateau of Rio Grande do Sul, group **2** occurring in grasslands at higher elevations, and group **3** occurring in the transitional area between the other two. Mean annual temperature, rainfall, and soil type differed between the two principal groups. That the chemical profiles were under genetic rather than environmental control was determined by micropropagation and field growth. In all cases, plants grown under controlled conditions exhibited chemical profiles identical to those of plants collected from nature. The authors concluded that chemical differences among the populations was caused by localized inbreeding, low level of recessive gene expression, and selection of particular chemical types in response to differing herbivore pressure in the different environments.

2.6.3 Cinchona *(Rubiaceae)*

One of the major diseases of humankind has been and continues to be malaria. The most effective agent used to battle malaria entered Western medicine, ironically, through the agency of the Spanish conquest of South America. One of the features of some of the native peoples of South America was a sophisticated pharmacopoeia that included the bark of a tree from Peru used to combat fever. The bark of the "fever tree," as the early explorers named it, appeared in Europe in the first half of the seventeenth century (Agosta, 1996). The plant was brought to the attention of Linnaeus who provided the generic name by which it remains known, *Cinchona*. [Agosta (1996) tells the story of how this particular name, actually misspelled, came to be chosen by Linnaeus!] Eventually, studies of the active principles of the plant resulted in the discovery of quinine [166] and a number of related compounds, such as quinidine [167], cinchonine [168], and cinchonidine [169] (see Fig. 2.43 for structures 166–169). *Cinchona*, with some 40 species, is a moderately large genus of shrubs and trees in Rubiaceae, most of which occur in the Andes but which is also represented as far north as Costa Rica (Mabberley, 1997, p. 158). The interested reader might wish to consult Steere (1945a, b) and Rainey (1946), as well as the recent description by Agosta (1996), for additional background information.

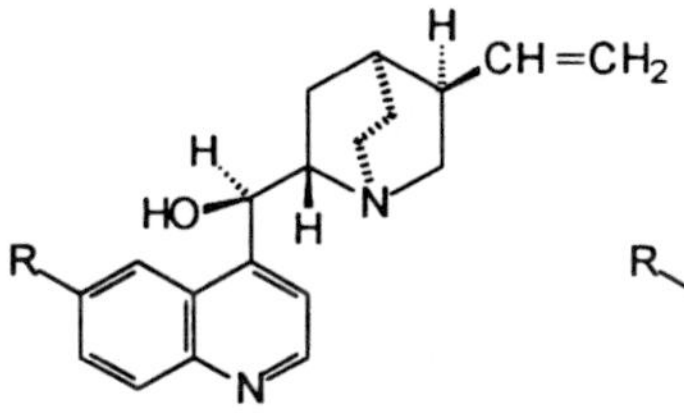

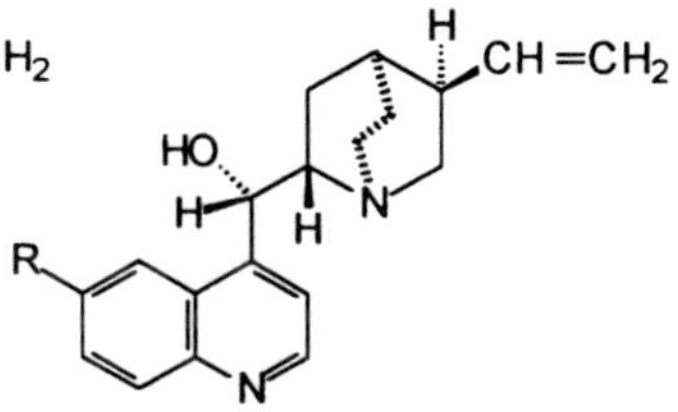

[166] R = OCH_3
[168] R = H

[167] R = OCH_3
[169] R = H

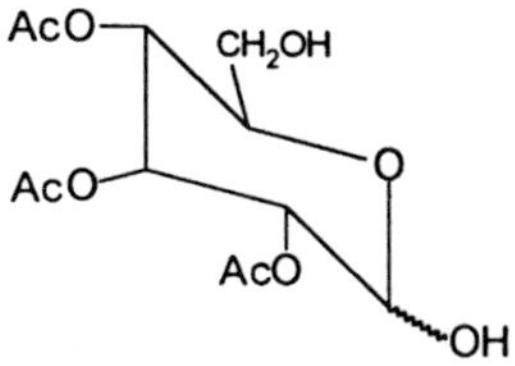

[170]

Fig. 2.43 Compounds 166–169, quinine relatives from *Cinchona*. Compound 170, a generalized sugar ester from *Lycopersicon pennellii*

One of the remarkable things about *Cinchona* and its contained alkaloids is the incredible concentrations to which some of the compounds accumulate in the bark. In some specimens, the concentration is so high that it is possible to pick individual crystals of alkaloid out of the powdered bark preparations! With concentrations of this magnitude, it is no surprise that a good deal of effort was expended in attempting to locate the highest yielding trees. The description of the quest for high-yielding trees and the results of his own study of alkaloid variation in *Cinchona* can be found in Camp's (1949) masterful treatment of the subject. His remarkable account describes analysis of trees collected along a transect running essentially along the frontal escarpment of the western Cordillera of the Andes from about 3°N latitude in Colombia through Ecuador to about 5°S latitude in northern Peru (Fig. 2.44). Three zones were defined by the study: (**1**) a low-elevation zone characterized by *C.* "*succirubra*", which ranged from 3°S latitude southward; (**2**) the mid-elevation zone characterized by *C. pubescens* s. str., which lies roughly between 0° and 2°30′S; and (**3**) the high-elevation zone characterized by *C. pitayensis* from 3°N to about 0°. Concentrations of "total crystallizable alkaloids" ranged from very low levels (<0.25%) in the low-elevation zone, to intermediate levels (ca. 3–4%) in the mid-elevation zone, to the highest values (ca. 5%) in the high-elevation zone (numbers quoted are average values). Because contact with other *Cinchona* "entities" was likely, and hybridization between species is known to occur, the entire range of trees involved in the study may be considered members of a single complex. Whatever the situation may be with regard to the origin of genomes of the various forms analyzed, the work represents one of the most involved and historically interesting studies in the literature.

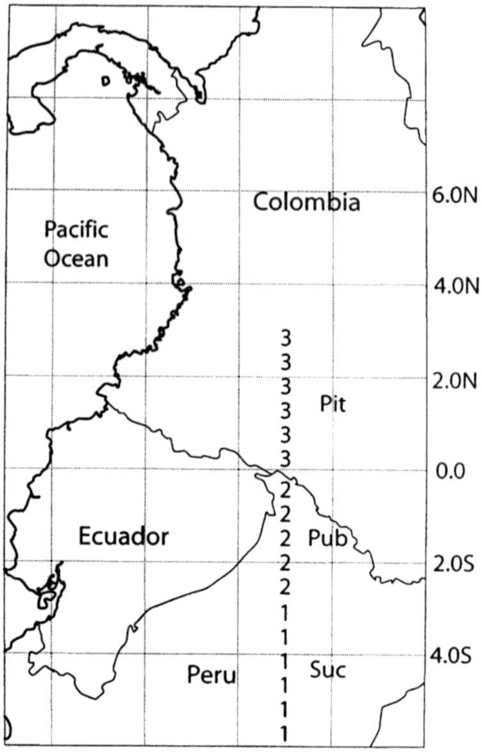

Fig. 2.44 Map of *Cinchona* sites in South America showing the three taxa involved in the complex

2.6.4 Lycopersicon pennellii *(Solanaceae)*

Lycopersicon pennellii (Correll) D'Arcy occurs in arid valleys in coastal Peru. Two varieties exist, var. *pennellii* ranging from ca. 5°S to ca. 16°10′S, and var. *puberulum* (Correll) D'Arcy, which inhabits a small area centered at ca. 14°30′S in the vicinity of Ica (Fig. 2.45). The typical variety is characterized by the production of an exudate rich in acylated sugar derivatives based upon glucose and sucrose and a variety of aliphatic acids (Shapiro et al., 1994). The powerful antiherbivore properties of these compounds are of particular interest to plant breeders and others interested in improving commercial tomatoes. (See, for example, Rodriguez et al., 1993, on the action of acylsugars on green peach aphids.) The two varieties differ significantly in their capacity to manufacture these compounds, var. *puberulum* producing only a small fraction of the quantity and variety of structures found on var. *pennellii* (Shapiro et al., 1994).

Chemical variety in the acylsugars known from the family is based upon the capacity of glucose to accommodate up to five acylating acids, while sucrose can accommodate up to six. Detailed chemical information on these compounds from various members of the family can be found in papers by King et al. (1988, 1990) and Matsuzaki et al. (1989). In the case of compounds from *L. pennellii*, the acylating acids

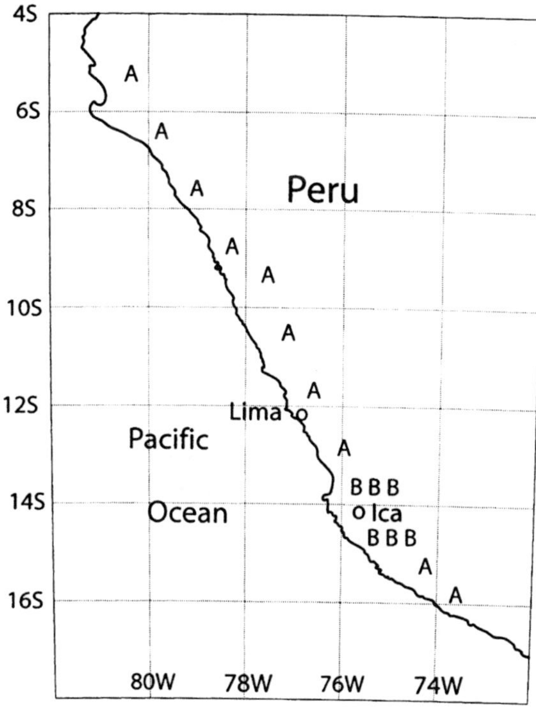

Fig. 2.45 Map of coastal Peru showing *Lycopersicon pennellii* sites

have been identified as 2-methylpropanoate (syn. = isobutyrate), 2-methylbutanoate, 3-methylbutanoate, 8-methylnonanoate, *n*-decanoate, 9-methyldecanoate, 10-methylundecanoate, and *n*-dodecanoate (Burke et al., 1987). A generalized structure of a tetra-acylated sugar is shown as [170] (see Fig. 2.43).

Shapiro et al. (1994) observed several significant differences in the biosynthetic capacities of the varieties and, in particular, members of var. *pennellii.* The most obvious difference reported by those workers is the apparent inability of some individuals of var. *puberulum* to make any acylsugar derivatives at all. In those cases where individuals do accumulate acylsugars, their capacity to do so appear limited to derivatives based on glucose. While individuals belonging to var. *pennellii* do possess the capacity to make acyl derivatives of both sugars, the total amount of acylated glucose present always exceeds the amount of acylated sucrose present. Significantly different ester chemistries distinguish northern and southern populations of var. *pennellii.* Table 2.14 lists occurrence data for the 2-methylpropionate and 3-methylbutanoate esters, along with percent acylglucoside in the acylsugar fraction, along with collection number and latitude of population. As can be seen, the 2-methylpropionate derivative predominates in the southern populations (positive correlation with latitude, $r=0.54$, $p=0.02$) while the 2-methylbutanoate derivative is the major compound in the northern populations (negative correlation with latitude, $r=-0.70$, $p=0.001$). There also appears to be a significant break in the values between samples taken at Rio Chillón, Sisicaya at 12°0′S, and Rio Cañete,

Table 2.14 Variation in selected acyl sugar concentrations as a function of latitude in *Lycopersicon pennelli* (from Shapiro et al., 1994)

Collection	LAT S	Concentrations[a]		
		2-MP	2-MB	PAG
1809[b]	5°0′	2.3	62.7	56
2560	8°45′	trace	60.7	52
1657A	9°20′	4.9	49.2	67
1376	11°0′	trace	53.9	60
1272	11°30′	trace	53.3	54
1367	11°50′	4.2	54.5	60
1282	12°0′	trace	54.5	54
1340	12°40′	43.8	2.7	77
1674	13°10′	59.1	6.9	93
1732	13°10′	54.9	4.8	94
1656	13°20′	49.5	11.0	87
1302A	13°40′	50.1	15.5	70
1946	15°50′	41.6	13.0	86
1941	15°55′	42.2	8.5	88
716	16°10′	41.8	4.0	84

[a] 2-MP = Percentage of total peak area for glucose 2-methylpropionate; 2-MB = ditto for 2-methylbutanoate; PAG = percentage acylglucoses among total acylsugars.
[b] Collection numbers from source paper.

Capillucas at 12°40′S. The authors also considered the possibility that acylsugar concentration may be affected by elevation, with the longer-chain derivatives being the more viscous of the esters, and thus providing an advantage to plants growing in the lower, hotter sites. The only relationship of note, however, involved the *n*-decanoate derivative, which showed a positive correlation with latitude ($r=0.62$, $p=0.006$). No adaptive significance was attributed to the differences in acylsugar composition, which the authors felt might simply reflect fixation of the two forms after differentiation and dispersal. In most cases, individuals within a sampled population exhibited very similar acylsugar compositions.

In an earlier study, Rick and Tansley (1981) studied this species (called *Solanum pennellii* in their paper) in order to determine the level of genetic variation exhibited throughout its range. Genetic variability was measured in terms of mean number of alleles per locus, proportion of polymorphic loci, polymorphic index, and mean actual heterozygosity. Values for these features were highest in the central northern part of the region, with the lowest values recorded for plants at the northern and southern extremities of the range.

2.6.5 C_3 and C_4 Grasses in Argentina

The example from South America ranks among the most impressive searches for variation in the literature of chemical geography. Although it does not fit

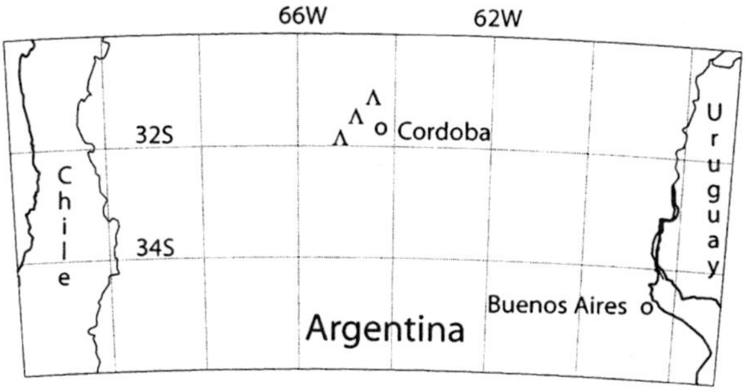

Fig. 2.46 Map of Argentina showing *Tagetes minuta* sites

comfortably within a collection of papers involving secondary metabolites, it is sufficiently interesting that I will break the rules! The work is that of Cabido et al. (1997) and concerns the distribution of C_3 and C_4 grasses along an altitudinal gradient in central Argentina. It is well known that C_3 grasses tend to occur in cooler habitats, whereas C_4 grasses tend to occur in warmer habitats. In the present study, 139 species were scored as a function of relative temperature at eight sites along an elevation gradient that extended from 350 m to ca. 2100 m in the Córdoba Mountains (32°60′S, 65°50′W) of north-central Argentina (Fig. 2.46). Fifty-nine of the species encountered were shown to be C_3, with the remaining 80 established as C_4 by the presence of Kranz anatomy. The C_3 species belonged to Aveneae, Bromeae, Meliceae, Poeae, Stipeae, and Triciceae. Grass species exhibiting the C_4 syndrome belonged to Andropogoneae, Aristideae, Cynodonteae, Eragrostideae, and Pappophoreae. Members of Paniceae, which consists of both C_3 and C_4 species, were observed at all sites except the highest one (ca. 2100 m). Although C_3 species were found at the 1400 m, 1600 m, 1800 m, and 1900 m sites, C_4 species outnumbered them at all sites except at the 1800 m level, where each was represented by two species. The three lowest sites, at 300 m, 650 m, and 1000 m, afforded only C_4 species.

The strong correlation between type of photosynthesis and environment is in accord with other studies of the distribution of C_3 and C_4 species. For example, an earlier study in a more arid region of Argentina (Cavagnaro, 1988), involving 31 grass genera, showed that the C_4 grasses, representing 19 genera, occur at lower elevations with the remaining 12 genera restricted to higher sites. The division line was located roughly between 1100 m and 1600 m. Rundel (1980) observed a similar sensitivity to the environment, in terms of elevation, in a study of 65 species of native and introduced grasses in the Hawaiian Islands, where C_3 species predominate (40 species) above 1400 m, with the C_4 species inhabiting sites at lower elevations. Other examples of photosynthetic pathway variation as a function of environment can be found in the work of Chazdon (1978) in Costa Rica, Hattersley (1983) in Australia, Wentworth (1983) in southeastern Arizona, and Schwartz and Redman (1988) in the boreal forests of northwestern Canada.

2.6.6 Tagetes minuta *(Asteraceae)*

Specimens of *Tagetes minuta* L. collected from three widely separated locations in Argentina (Fig. 2.46) exhibited identical arrays of polyacetylene derivatives in their roots, but had significantly different relative amounts of these compounds (Gil et al., 2002). Specimens were collected on the east coast near Buenos Aires (Pergamino at 33°56′S, 60°33′W), in the west near Mendoza (La Consulta at 33°44′S, 69°07′W), and in the north near Salta (Cerrillos at 24°54′S, 65°29′W). The major compounds were α-terthienyl (α-T) [171], 5-(3-buten-1-ynyl)-2,2′-bithienyl (BBT) [172] (see Fig. 2.47 for structures), and BBTOH, an hydroxyl derivative of BBT. The Mendoza and Salta populations exhibited low α-T and BBT concentrations and proportionately higher levels of the BBTOH derivative. It is interesting to note that only in the case of the material from Mendoza was there a positive relationship between total thiophenes in the roots and total aboveground biomass. How significant this relationship would likely be over a longer period of time might be open to question, considering that plants at this site exhibited the largest differences between years. Large quantitative differences in otherwise identical qualitative profiles might suggest that some sort of control system may be at work, but local environmental conditions would have to be ruled out first.

2.6.7 Hymenaea *(Fabaceae)*

The legume genus *Hymenaea* (including *Trachylobium*) comprises 16 species, all but one of which occurs in the New World (Mabberley, 1997, p. 354). The exceptional species, *H. verrucosa* Gaertner, which was formerly considered to constitute the genus *Trachylobium*, is native to coastal East Africa including the eastern coast of Madagascar. The East African species is the source of Madagascar copal. Among the New World species, *H. courbaril* L., known locally as West Indies locust, Brazilian copal, or anami gum (Mabberley, 1997, p. 354), is an important source of resins used for the manufacture of varnishes. The species occurs in Mexico, Central America, including Caribbean Islands, and in South America as far as southern Brazil, which approximates the range of the genus in the New World.

Martin et al. (1974) collected specimens from 22 locations throughout the range of *H. courbaril* and identified a number of sesquiterpenes including caryophyllene [173], humulene [174], selinene isomers [175 is β-selinene], and other related compounds (see Fig. 2.48 for compounds 173–175). Caryophyllene and two selinene

Fig. 2.47 Compounds 171 and 172, thiophene derivatives from *Tagetes minuta*

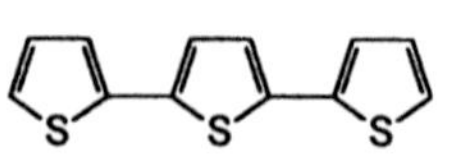

[171]

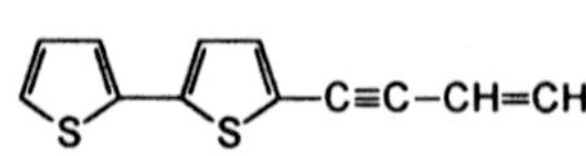

[172]

Fig. 2.48 Compounds 173–175, sesquiterpenes from *Hymenaea verrucosa*. Compounds 176 and 177, terpenes from *Minthostachys verticillata*

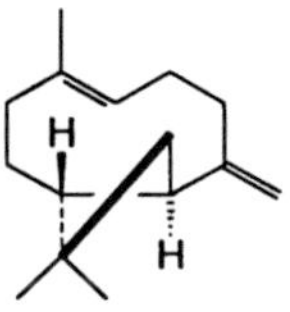

[173]

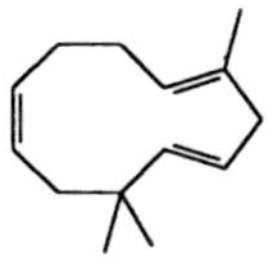

[174]

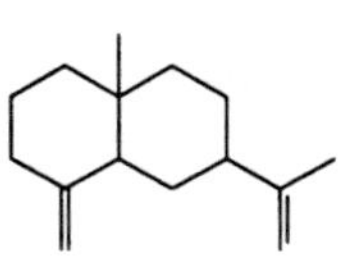

[175]

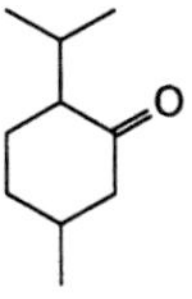

[176]

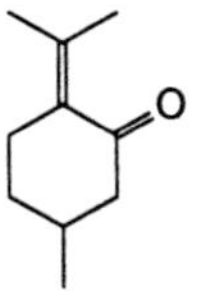

[177]

isomers accounted for about 60% of the resin sesquiterpenes in most of the specimens examined. Detailed statistical analysis revealed that South American populations often exhibited complex patterns of variation, in line with the recognition of several varieties in the region. Much lower levels of variation, by contrast, characterized populations from Mexico and Central America. The extent of sesquiterpene variation observed in *H. courbaril* is similar to patterns of variation observed in other members of the genus.

2.6.8 Minthostachys verticillata *(Lamiaceae)*

Minthostachys verticillata (Griseb.) Epl. is a labiate native to the central and northwestern parts of Argentina. Known locally as "peperina," leaves of this species are and have been used as flavoring agent for beverages, notably *maté.* It is unlikely, however, that this species has been subjected to the level of selection experienced by Mediterranean plants that we discussed above, such as thyme and oregano. A recent study of *M. verticillata* collected from the field documented the existence of chemically unique geographical races (Zygadlo et al., 1996). Ten sites were sampled from four provinces representing much of the range of this species. Most plants had at least traces of most compounds but some notable differences were observed, for example, thymol [105] and carvacrol [89] were major constituents in the oil from two sites in Catamarca Province (Potero and Sebastian) but absent from another in that province (Balcosna) and from all other sites visited. Another variant, although less dramatic, was β-pinene [23], which was present to the level of 5.1% and 6.1% in both sites in Tucumán Province but made a much smaller contribution to the oils from all other sites. Similarly, menthone [176] and pulegone [177] (see Fig. 2.48) were present in widely differing amounts from different sites both within and between provinces. These results and selected others are summarized in Table 2.15.

Table 2.15 Selected values for oil composition of Argentinean *Minthostachys verticillata* (after Zygadlo et al., 1996)

Site	Compound[a]						
	β-Pine	Limon	Menth	Puleg	Carvo	Thym	Carva
Catamarca Prov.							
Potrero	tr[b]	8.9	8.8	5.1	tr	31.8	34.0
Balcosna	tr	21.1	12.9	33.3	0.2	nd	nd
Sebastien	0.9	10.1	8.1	5.6	0.5	10.5	25.3
Córdoba Prov.							
Candonga	tr	1.4	29.2	47.0	tr	nd	nd
Cuesta Blanca	0.1	10.1	29.6	39.0	1.1	nd	nd
Paradones	0.4	25.0	15.8	31.2	1.0	nd	nd
San Luis Prov.							
Merio	0.4	13.1	27.0	39.6	0.3	nd	nd
Baños Pasos Malos	0.4	17.6	12.8	45.1	0.1	nd	nd
Tucumán Prov.							
Rancho de la Cascada	5.1	17.1	5.5	1.2	35.2	nd	nd
Potreros de las Tablas	6.1	11.9	28.3	21.1	tr	nd	nd

[a] β-Pine = β-Pinene; Limon = limonene; Menth = menthone; Puleg = pulegone; Carvo = carvone; Thym = thymol; Carva = carvacrol.

[b] tr = Trace amount; numbers are percentages of total oil; nd = not detected.

An example closely related to the previous case involves oil composition of *Minthostachys andina* (Brett) Epling collected in central Bolivia. The study, by Muñoz-Collazos et al. (1993), revealed that plants with the chemotype characterized by high amounts of pulegone and menthone (at least 65%) tended to occur at higher elevations, providing that sampling "...did not involve the transition zone between the dry highlands and humid and warmer Amazonian region."

2.7 North and Central America

2.7.1 Notholaena standleyi *(Adiantaceae)*

Notholaena is a genus of ferns with close relationships to *Cheilanthes* and *Pellaea* with which they may intergrade. The genus is represented in tropical to warm parts of the New World and includes a number of xerophytes. *Notholaena standleyi* Maxon, the subject of this study, fits well in this category owing to its occurrence in the dry regions of the United States and adjacent Mexico. It has been known for many years that members of this species exhibit different colored exudates on the undersides of their leaves. Three color forms are known, a gold-colored race, a yellow-colored race, and a yellow–green race. The gold-colored race occurs in Arizona and far northwestern Mexico (including Baja California) where it occurs mainly on rocks of igneous origin. The yellow-colored race occurs more widely having been recorded

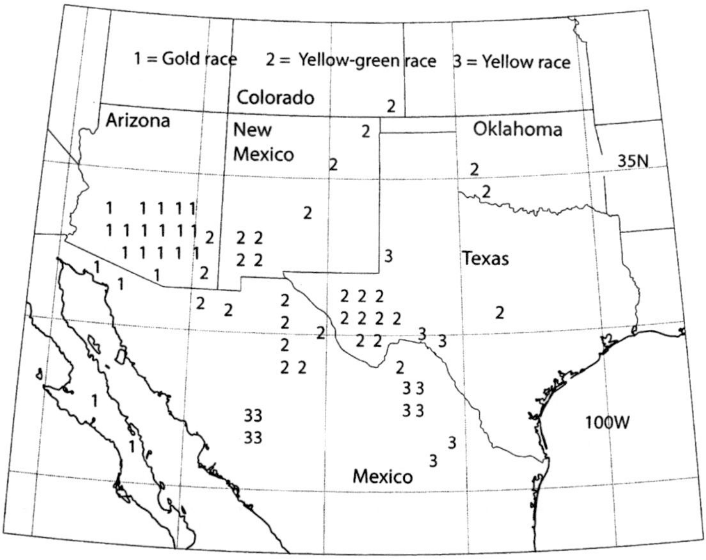

Fig. 2.49 Map of *Notholaena standleyi* sites, showing gold-race, yellow–green-race, and yellow-race populations

from sites in southeastern Arizona, New Mexico, extreme southeastern Colorado, Oklahoma, Texas, and north-central Mexico. This color form also occurs on igneous rock, but can also be found growing on sandstone. The yellow–green race is known from north-central Mexico and adjacent Texas and occurs on limestone (Fig. 2.49).

Despite the difficulty of distinguishing the three color forms by means of clear-cut morphological traits (yellow and yellow–green plants tend to be larger on the average than gold-colored plants), their flavonoid pigment profiles allow straightforward separation. Seigler and Wollenweber (1983) examined the exudate chemistry of 59 specimens representing the entire range of the species. Qualitative and quantitative differences were observed among color forms and little variation was observed within populations. No intermediate flavonoid profiles were noted. Pigment profiles were based upon the existence of various combinations of *O*-methyl ethers of kaempferol and herbacetin (8-hydroxykaempferol). The most obvious difference is the absence of herbacetin derivatives from yellow race plants. Herbacetin 7-methyl ether [178] and 7,4′-dimethyl ether [179] were observed consistently but in smaller amounts in gold race plants and both compounds were seen scattered in yellow–green race plants. Yellow race plants exhibited kaempferol [180], kaempferol 3-methyl ether [181], and kaempferol 4′-methyl ether [182] as major components. Kaempferol 3-methyl ether was not seen in gold race plants and was observed only sporadically in yellow–green plants. Structures are given in Fig. 2.50. Table 2.16 summarizes the occurrence data.

There are two obvious differences in pigment occurrence in this taxon. The first is the absence of herbacetin derivatives in the yellow race, which suggests the apparent absence of an enzyme capable of placing oxygen at the C-8 position of the

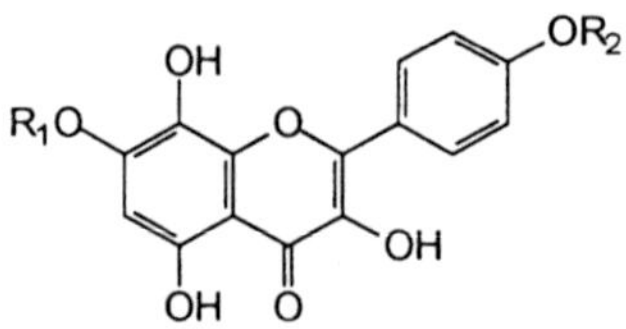

[178] $R_1 = CH_3$, $R_2 = H$
[179] $R_1 = R_2 = CH_3$

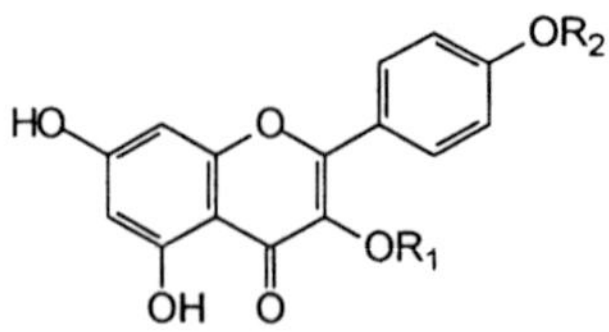

[180] R_1 $R_2 = H$
[181] $R_1 = CH_3$, $R_2 = H$
[182] $R_1 = H$, $R_2 = CH_3$

Fig. 2.50 Compounds 178–182, flavonoids from *Notholaena standleyi*

Table 2.16 Occurrence of flavonols in leaf exudates of *Notholaena standleyi* (after Seigler and Wollenweber, 1983)

Compound/ether	Frond Color Race		
	Yellow	Gold	Yellow-Green
Kaempferol	++[a]	trace	+
Kaempferol 3-methyl	++	–	var
Kaempferol 4′-methyl	++	++	++
Kaempferol 7-methyl	+	++	++
Kaempferol 3,7-dimethyl	+	–	var
Kaempferol 3,4′-dimethyl	+	–	–
Kaempferol 7,4′-dimethyl	–	trace	+
Herbacetin 7-methyl	–	+	var
Herbacetin 7,4′-dimethyl	–	+	var

[a] ++ = major component; + = minor component; var = variable within the race.

flavonoid nucleus. The second difference involves the absence of flavonol 3-methyl ethers in the gold race, which points to differences in the presence or absence of *O*-methyltransferase enzymes, and the amount of substrate that they process in the respective chemotypes. It would be of interest to examine the nature of the *O*-methyltransferases in this complex as has been done successfully with others, particularly the stepwise O-methylations that occur in *Chrysosplenium americanum* (Seguin et al., 1998, and citations therein).

Notholaena standleyi presents certain problems that, despite the clear-cut differences among the three chemotypes, make resolution of taxonomic relationships difficult. It appears that *N. standleyi* reproduces without undergoing meiosis and that there is little gene flow among individuals or populations. The flora in this part of southwestern North America and northern Mexico is thought to be an ancient one that has developed in response to increasing aridity (Tryon, 1962; Tryon and Tryon, 1982). The consistent flavonoid profiles, the differences among the three chemotypes, and the seemingly low gene flow combine to suggest that the chemotypes may be relictual forms that have become ecologically isolated and are surviving in a stable environment. Additional information on genetic variation in the system, perhaps electrophoretic studies, would be welcome.

2.7.2 Chenopodium fremontii *(Chenopodiaceae)*

Several species of weedy *Chenopodium* occur widely in western North America. In an effort to gain a better understanding of some of these taxa, D. J. Crawford and associates undertook an extensive biosystematic examination of the group. The study represents an excellent example of the advantages to be gained by the application of different techniques, including macro- and micromolecular methods, to a complex system. The primary focus of this discussion is *C. fremontii* S. Watson, which occurs over wide tracts in the southwestern states and shows wide ecological amplitude occupying habitats that range from desert to montane sites. Extensive morphological examination, including greenhouse studies, had established the highly plastic nature of the species. These workers then turned to other methods in order to estimate the level of genetic variation within the taxon. Their first analysis involved an examination of seed protein profiles of more than 210 individual plants representing 33 populations; five states were involved: Arizona, California, New Mexico, Utah, and Wyoming (Crawford, 1976) (Fig. 2.51). Preliminary studies had demonstrated that variation within populations was minimal and that plants grown from seed from the same population exhibited invariant protein profiles; morphological variation was maintained in cultivation, however. Fifteen of the 25 populations, in which plants were examined individually, showed only one protein profile; eight populations exhibited two protein profiles and two exhibited three profiles. Numerical analysis of the data showed the existence of seven clusters with identical protein profiles. The lowest level of similarity observed between any two populations was 46%. It is useful to point out that *C. fremontii* is unusual in its variable protein profiles compared to other related species whose profiles were shown to be largely invariant: *C. atrovirens* Rydberg, *C. desiccatum* A. Nelson, *C. hians* Standley, *C. leptophyllum* Nutt. ex Moq., and *C. pratericola* Rydberg (Crawford and Julian, 1976).

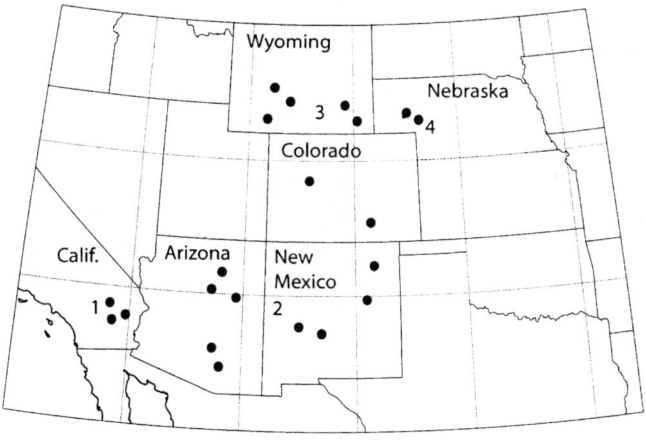

Fig. 2.51 Map of *Chenopodium fremontii* sites

Most of the protein profile groups clustered as geographical units, which can be represented broadly as occurring in the following areas: California, southern Arizona, northern Arizona, Utah, eastern New Mexico, northeastern New Mexico, and Wyoming. In general, populations from nearby localities exhibited the highest levels of similarity. Two exceptions to this generalization require comment. A population in northeastern New Mexico seems to be out of place since it was shown to link most closely with two populations from southern Wyoming. No explanation was offered for this unexpected pairing since there is little similarity between the habitats in the two areas. The second exception involves a population from northeastern New Mexico, which was found to be most closely related to several populations from eastern New Mexico. In this situation, all populations were growing in pinyon pine-juniper woodlands. It seems reasonable to suggest that the New Mexico population arose from the chance combination of a propagule and an appropriate niche. Overall, it was suggested that the occurrence of *C. fremontii* likely became successful owing to the evolution, and successful establishment, of several local races throughout its range.

Additional evidence suggesting that *C. fremontii* is genetically more homogeneous than its morphology would suggest came from a study of allozyme profiles. Collections involving over 1100 plants from 40 populations were assayed for glutamate-oxaloacetate transaminase (GOT) and leucine aminopeptidase (LAP), while over 600 plants from 22 populations were assayed for phosphoglucoisomerase (Crawford and Wilson, 1977). Sampling was expanded to include populations from Colorado and Nebraska in addition to the states listed above. These collections include a wider range of the species than was studied in the seed protein study, which presents some problems in making direct comparisons of specific sites. Nonetheless, the overall geographic patterns that emerged from the allozyme study agreed with observations drawn from the earlier data set. The most obvious result is the observation that the Californian populations were fixed for an allele of GOT not seen in any other population. This is in line with the unique seed-storage protein profile observed and, as we shall see below, with the existence of a flavonoid race restricted to the Californian populations.

Other than the Californian situation with regard to GOT alleles, the distribution pattern can be reckoned as consisting of, to a reasonable degree at least, a northern "race" found in populations from Nebraska, northern and western Colorado, Utah, and Wyoming. The southern "race" occurs in Arizona, central and southern Colorado, and New Mexico. Three populations from northeastern New Mexico, however, exhibited the highest level of variation seen in the study with 11 of the 13 heterozygous plants observed being from this region. The authors offered two possible explanations for the observed allelic arrays: (1) this area represents a zone of gene exchange; or (2) this area may represent a center of diversity. Efforts to study the possibility of hybridization between the two GOT races failed owing to technical problems with pollination.

The third paper in the *Chenopodium fremontii* study involved an investigation of the species' flavonoids (Crawford and Mabry, 1978). Twenty-two populations were sampled representing much of the range of the species. Four flavonoid races were identified: **1** California; **2** Arizona, New Mexico, and southern Colorado; **3** Wyoming and western Colorado; and **4** Nebraska. Differences among the populations were based upon both the nature of the aglycones and the glycosidic forms in which they occur. Profiles within each pigment race were qualitatively invariant. The results are summarized in Table 2.17 where data are listed for presence of the parent aglycones, the entire array of glycosides, and the summed glycosides, for example, total arabinosides. Several distinguishing features can be pointed out in the array. Races **3** and **4** can be recognized by the presence of kaempferol [183] and quercetin 7-methyl ethers [184] (See Fig. 2.52 for structures 183–189). The capacity for *O*-methylation is not unique to those populations, however, since all four races have the capacity to make isorhamnetin (quercetin 3′-methyl ether) [185].

Table 2.17 Flavonoids of *Chenopodium fremontii* (from Crawford and Mabry, 1978)

Compound(s)	Populations[a]			
	1	2	3	4
Kaempferol (K)	+	+	+	+
Quercetin (Q)	+	+	+	+
Isorhamnetin (IR)	+	+	+	+
Kaempferol 7-methyl (KM)	–	–	+	+
Quercetin 7-methyl (QM)	–	–	+	+
K Arabinoside	+	–	–	–
K Galactoside	+	+	+	+
K Glucoside	+	+	+	+
K Robinobioside	+	+	+	–
K Sophoroside	–	–	+	–
K Gal-Rhm	+	+	–	–
KM Galactoside	–	–	+	+
KM Glucoside	–	–	+	+
QM Galactoside	–	–	+	+
Q Arabinoside	+	–	–	–
Q Galactoside	+	+	+	+
Q Glucoside	+	+	+	+
Q Rutinoside	–	+	–	–
Q Robinobioside	–	+	–	–
Q Sophoroside	+	–	–	+
IR Arabinoside	+	–	–	–
IR Galactoside	+	+	+	+
IR Glucoside	+	+	+	+
IR Rutinoside	–	+	–	–
IR Robinobioside	+	+	–	–
Arabinosides	+	–	–	–
Galactosides	+	+	+	+
Glucosides	+	+	+	+
Robinobiosides	+	+	+	–
Rutinosides	–	+	–	–
Gal-Rhm	+	+	–	–

[a] Populations: 1 = California; 2 = Arizona, New Mexico, and southern Colorado; 3 = Wyoming and western Colorado; 4 = Nebraska.

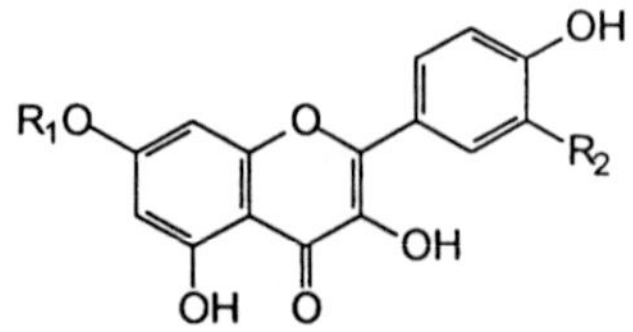

[183] $R_1 = CH_3$, $R_2 = H$
[184] $R_1 = CH_3$, $R_2 = OH$
[185] $R_1 = H$, $R_2 = OCH_3$

[186] Rutinose = Glucose ←6— Rhamnose

[187] Robinobiose = Galactose ←6— Rhamnose

[188] 2-Rhamnosylgalactoside = Galactose ←2— Rhamnose

[189] Sophorose = Glucose ←2— Glucose

Fig. 2.52 Compounds 183–189, flavonoids, from *Chenopodium fremontii*

However, the position of *O*-methylation, the hydroxyl group at C-7 and C-3′ in the latter, clearly sets these races apart. Detailed studies of flavonoid *O*-methylation by Ragai Ibrahim and his colleagues at Concordia University in Montreal (see Seguin et al., 1998 for leading references) have shown convincingly that these enzymes have high levels of position specificity. In the present case, then, it can be argued that the *O*-methyltransferases that catalyzes methylation of the 3′-hydroxyl group of quercetin (to form isorhamnetin) is different from the one that catalyzes methylation of the 7-hydroxyl group. Thus, in the case of the differences among the flavonoid races of *C. fremontii*, it is possible to infer specific enzyme, and therefore genetic, differences among the respective populations. Other differences can be seen: arabinosides are restricted to plants from populations in California (Race **1**); rutinosides [186] were observed only in plants from Arizona, New Mexico, and southern Colorado (Race **2**); robinobiosides [187] were not observed in plants from Nebraska (Race **4**); and kaempferol 3-*O*-(2-*O*-α-L-rhamnosyl)-D-galactoside [188] was seen only in plants from the southern populations (Races **1** and **2**). It is also useful noting that in only two cases did plants from both southern and northern races share any of the less common compounds: (1) kaempferol robinobioside was detected in Races **1**, **2**, and **3**, whereas it was not seen in plants from Nebraska; and (2) quercetin sophoroside [189] was seen in plants from California (Race **1**) and Nebraska (Race **4**).

Flavonoid variation in *C. fremontii* is based upon several definable enzymatic steps involving O-methylation, O-monoglycosylation with different sugars, and elaboration of a series of *O*-diglycosides involving different (outer) sugars as well as different positions of substitution on the "inner" sugar (C-2″ vs. C-6″). That these chemical differences reflect real genetic differences of the species seems an entirely reasonable position. What is needed to put this on more solid ground, of course, is a detailed analysis of the enzymology of the reactions involved and, following that, a study of the genes controlling the various reactions. Very little work has been done in the field of flavonoid biochemistry in following this line of inquiry. This would seem to be a logical next step in understanding the source(s) of variation in flavonoid profiles, and *C. fremontii* might offer a good system to study. Its flavonoid chemistry is particularly well documented, the enzymology and genetics of the flavonoid pathway are well understood, and the molecular biological tools are available.

2.7.3 Asclepias tuberosa *(Asclepiadaceae)*

Asclepias tuberosa L., butterfly weed) consists of two subspecies, based on leaf characteristics, native to North America, subsp. *tuberosa* and subsp. *interior* Woodson. The latter named taxon enjoys a range that encompasses most of the central part of the continent, from Ontario west to Minnesota and southwestward to western Utah and Arizona. Much of what is known concerning the biology of North American members of this genus has come from the work of R. E. Woodson, Jr. (1947, 1953, 1954, and 1962). The particular study of interest here deals with his study of the geography of flower color variation in subsp. *interior* (Woodson, 1964, with summarizing paragraphs by Sewell Wright).

Flower color of butterfly weed ranges from yellow to reddish orange with a definite geographic component to the different hues. The yellow color derives from a background of carotenoids with the range of yellow–orange through orange to reddish-orange being due to increasing amounts of anthocyanin overlying the yellow base. Woodson (1964) reported that a crude extract of the anthocyanin in 0.1 N HCl exhibited an absorption maximum at 515 mμ, which lies within the range of anthocyanin glycosides. Although no further comment was made concerning the identity of the pigment, it seems safe to suggest, on the basis of published maxima for anthocyanins (Harborne, 1967b, p. 17), that the pigment could be a pelargonidin 5-glycoside (λ_{max} = 513 mμ for pelargonidin 5-glucoside).

Extensive sampling of butterfly weed was conducted along transects designed to follow the trends already observed with leaf characteristics (Woodson, 1947, 1953). The starting point for the transects was a point in Camden County, Missouri (his C1 site), with a series of transects radiating at intervals of about 120 miles northward into Minnesota, northeastward into Ontario, and westward to western Colorado and Utah. A fourth transect was followed from C1 eastward to the Atlantic coast through an area occupied by subsp. *tuberosa* (Fig. 2.53). Samples collected along the transects within each of the areas defined by the radiating transects were scored for color using a standardized color system. Carotenoid concentration appeared not to vary, whereas the anthocyanin contribution varied significantly with the highest concentrations per plant and largest frequencies of red-pigmented plants seen in plants from Missouri, Illinois, adjacent Kansas, and southern Iowa. With a few exceptions, anthocyanin level dropped along the radiating transects, most precipitously along the northern track with nearly pure yellow-flowered plants observed in central Minnesota. Pure yellow plants were observed in the westernmost sites as well. Reduction of anthocyanin concentration was not uniform, however, as witnessed by the occurrence of populations in western Texas, New Mexico, and Colorado that featured significant numbers

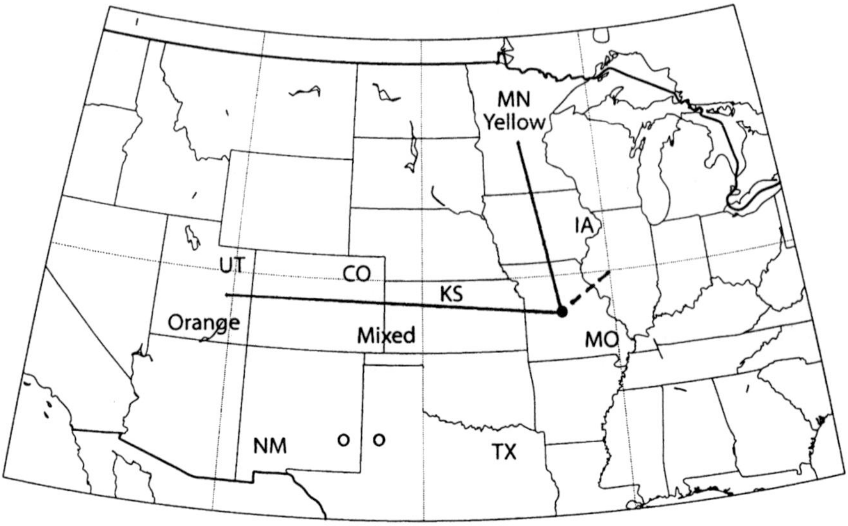

Fig. 2.53 Map showing transects for *Asclepias tuberosa* flower color study

of plants with modest to higher levels of anthocyanin. The pattern of occurrence of the color morphs was interpreted as arising from incomplete diffusion of an adaptively superior complex of features, marked by the accumulation of anthocyanins, outward from the assumed center of origin of the species, which was taken as western Missouri and adjacent areas as listed above. The concentric pattern has been broken from place to place, presumably by the lack of continuous suitable habitats, resulting in widespread colonies that are subject to extreme genetic bottlenecks. [Note that the conclusions in this paper were prepared by Sewell Wright, to whom a copy of the manuscript had been sent by Woodson for comment. Woodson never saw Wright's comments, owing to the latter's sudden death. See Editor's comments.] So far, as can be determined from Woodson's work, and the comments by Wright, no answer has been found to the question of what selective advantage anthocyanin pigmentation might provide, although it is altogether possible that this is only one part of a more complex set of features that direct selection.

2.7.4 Phlox carolina *(Polemoniaceae)*

Phlox carolina L., a tall, perennial phlox native to southeastern United States, was shown to exhibit a substantial degree of flavonoid variation by Levy and Levin (1975). In a subsequent paper, Levy and Fujii (1978) described attempts to establish geographical patterns in the occurrence of leaf flavonoids. Seventy-three collections were made representing populations in Georgia, Alabama (a total of 64 populations),

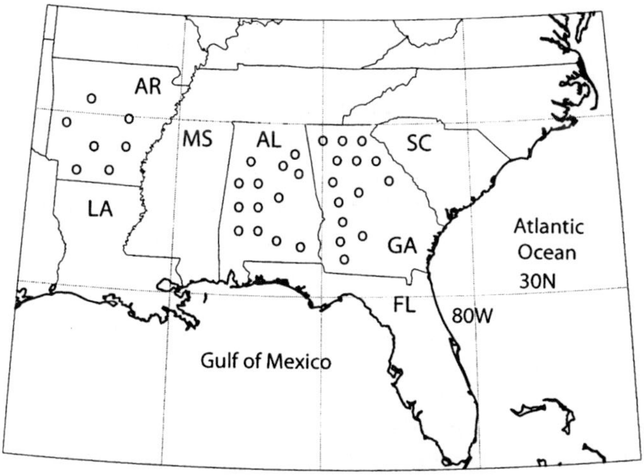

Fig. 2.54 Map showing collection sites for *Phlox carolina* study

and Arkansas (9 populations) (Fig. 2.54). Apparently, the taxon is very scarce in Mississippi and absent from Louisiana; thus the Arkansas material represents a set of populations disjunct from those in Georgia and Alabama.

From the 73 populations studied, no less than 71 unique flavonoid profiles were observed based upon two well-known flavone hydroxylation patterns, the apigenin type and the luteolin type, and two further compounds described simply as "deoxygenated variants of apigenin" and called "flavone-1" and "flavone-2." Most of the compounds identified were C-glycosyl derivatives of the base molecules, nine each of the apigenin and luteolin types, the remaining four involving flavones-1 and -2. Structural variation arose from the nature and number of C-bound sugars, the presence of O-glycosylated derivatives of some of the C-glycosyl compounds, and likely the position on the parent flavonoid where the *O*-linked sugar was attached.

A casual study of this system might well have resulted in three conclusions: (1) that variation in pigment profiles was extremely complicated (no doubt true!); (2) that there has been comparatively little flavonoid divergence between the two disjunct types; and (3) that adjacent populations may or may not be more similar to each other than they are to populations at some distance. In short, no major geographic patterns seem to exist. However, in the hands of Levy and Fujii (1978) this was anything but a casual study! Rather than simply presenting a matrix of sites studied and compounds identified, these authors discussed their chemical data in terms of the minimum biosynthetic step distance (MBSD), a measure of how closely two flavonoids are related in terms of the number of biosynthetic steps that separate them from their common precursor (see Levy, 1977, for the development of this index). The authors determined that 38 biosynthetic steps are involved in arriving at the observed flavone array for the *entire set of populations* (emphasis mine).

The overall mean number of steps per population is 19.53 and the overall mean MBSD is 14.12. Comparative mean values for the separate ranges of *P. carolina* are: continuous range = 13.14 ± 4.95 (N = 2016), disjunct range = 15.42 ± 6.40 (N = 36), and continuous versus disjunct populations = 17.47 ± 3.90 (N = 576). These differences are significant ($P < 0.001$). These values indicate that the separate ranges of the species are distinctive, but only weakly divergent with regard to their flavone profiles. Although there are no discrete patterns of flavonoid profiles, some areas in the continuous range show higher profile uniformity than others. Of particular note are populations in western Georgia, where similarities in the range 85–100% were observed. However, even in that area, adjacent populations may not be the most closely related, and those that are closely related may be separated by some distance (ca. 50 miles). Although differences in pigment profiles were found within this species, no clear-cut pigment races exist.

This finding is in general accord with aspects of the biology of the species. Levy and Fujii (1978) pointed out that restricted gene flow is characteristic of *Phlox*, which lacks the capacity for long-distance dispersal. They pointed out that the irregular occurrence of most of the compounds suggested that the gradients observed represent the results of "range dissection and contraction rather than stepping-stone dispersal from central or local centers of flavone diversity." The result would be a collection of local "islands" of flavone diversity, which, through local expansion, might come into contact with nearby populations to produce the complex array of variants seen.

This study of *Phlox carolina* represents one of the best examples in the flavonoid chemosystematic literature where workers combined thorough sampling, detailed statistical analysis, and an intimate knowledge of the biology of the system under scrutiny to produce a convincing picture of natural variation. As mentioned at the beginning of this discussion, a more casual approach would have undoubtedly overlooked the subtle differences that characterize this system.

2.7.5 Lasthenia *(Asteraceae)*

Lasthenia Cass. is mostly a Californian genus of about 17 species; the single species that occurs in Chile will be dealt with in the section on north–south disjuncts below. An examination of flavonoid pigments of all members of the genus (Bohm et al., 1974) had shown that sections, defined on the basis of morphology, cytology, and breeding behavior (Ornduff, 1966), could also be distinguished on the basis of the presence or absence of a variety of flavonoids (Bohm et al., 1974). For example, sections *Baeria* (both species), *Burrielia* (all three species), and *Hologymne* (two of three species) were characterized by an array of chalcone and aurone derivatives (collectively referred to as anthochlors), in addition to a number of common flavones and flavonols, whereas members of the other three sections exhibited only flavonols. The initial studies suggested that each section was characterized by a unique combination of pigments. Some indication of variation did emerge from those studies however, which prompted an examination of a larger number of individuals from as many of the species' respective ranges as possible.

The presence of variable pigment profiles in several other species was noted in a survey of populations (Saleh et al., 1971; Ornduff et al., 1974) and on a study of flavonoids of artificial hybrids in the genus (Ornduff et al., 1973a). In an effort to arrive at a better appreciation of how extensive flavonoid pigment variation is in the genus, the most widely ranging species, *L. californica* DC. ex Lindley [= *L. chrysostoma* (Fisch. and Mey.) Greene, in the earlier literature] was subjected to a more detailed examination. *Lasthenia californica* is not only geographically the most widely distributed taxon in the genus, but it is also known to occur on a variety of substrates, which suggested that this system might provide some insight into how the species responds to different edaphic situations. The site chosen for detailed study was the Jasper Ridge Biological Preserve of Stanford University (referred to simply as Jasper Ridge below), which lies south of San Francisco in San Mateo County in the Santa Cruz Mountains (ca. 37°25′N, 122°2.5′W). One of the features of Jasper Ridge is the presence of a serpentine outcrop that harbors an extensive population of *L. californica.* Initial sampling was done along a transect approximately 1 km in length running along the spine of the serpentine ridge. In addition to the anthochlors that characterize this species, which were observed in all individuals, differences in the nonanthochlor flavonoid profiles were observed. Race **A** was distinguished by the presence of flavonol diglycoside sulfates and the flavanone eriodictyol 7-*O*-glucoside (eriodictyol is the flavanone equivalent to luteolin), whereas race **C** exhibited only the basic array of anthochlor pigments. A third variant, called race **B** in earlier papers, was observed in a few plants but is now included within race **C**. Race **B** is characterized by the presence of luteolin 7-*O*-glucoside, in addition to the basic array of anthochlors. Plants were also collected along several other transects, each between 50 m and 65 m long, that lie at right angles to the ridge. Since no additional pigment variants were observed in these shorter transects, no further collections were made along the original long transect. We focus our attention, at least for the moment, on a single transect.

Individual plants were collected at meter intervals, starting at a fire-access road near the crest of the ridge progressing downslope for a distance of 65 m. Elevation change over this distance is approximately 5 m, which defines the drainage pattern on this part of Jasper Ridge. Analysis of the individual plants showed that race **C** plants occurred in the range 0–45 m after which only race **A** plants were observed (Fig. 2.55). This pattern was closely approximated in plants collected along other parallel transects on either side of the main transect. The site was visited each year for a period of 6 years during which the distribution pattern of race **A** and race **C** plants remained essentially unchanged (a race **B** plant was observed in the upper portion of the transect during one season). These observations have been described in a paper (Bohm et al., 1989) where it was pointed out that this was, at least to the best of the authors' knowledge, the only report of a flavonoid distribution pattern remaining constant over a period of years. Subsequent collections along the same transect have shown the pattern to have remained constant for a period of 15 years (N. Rajakaruna & B. A. Bohm, unpublished observations). Greenhouse studies

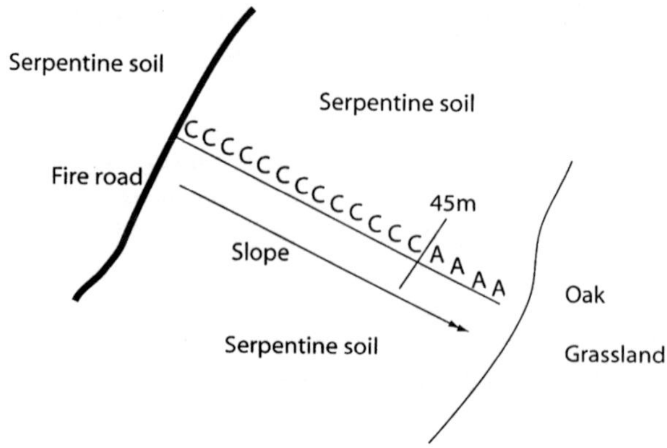

Fig. 2.55 Diagram of *Lasthenia californica* transect study

using potting soil showed that achenes from race **A** plants always yielded race **A** progeny, and that achenes from race **C** plants afforded mostly race **C** plants with an occasional race **B** individual. These observations suggested that the flavonoid profiles of *L. californica* are genetically controlled, and do not require serpentine soil for flowering or flavonoid production.

Figure 2.56 illustrates the results of an expansion of the flavonoid survey to the entire range of the species. Populations having either totally or predominantly the

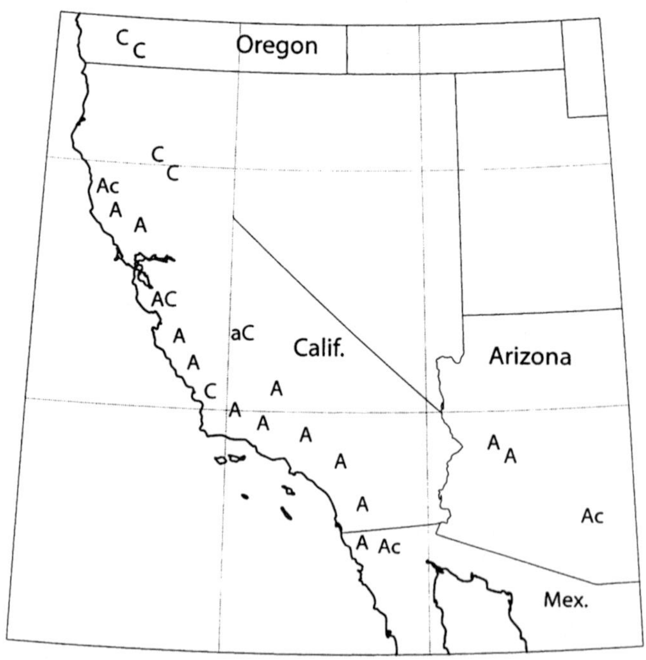

Fig. 2.56 Map of *Lasthenia californica* populations sampled for flavonoid variation study

C-type pigmentation were observed in the northern part of the range of *L. californica*. Populations exhibiting (predominantly) A-type flavonoid chemistry occurred in Arizona and central and southern California. The study area at Jasper Ridge belongs to the latter group.

Other differences between the two flavonoid races emerged from a biosystematic study of *L. californica* collected from its entire range. Differences in allozyme banding patterns and pappus structure correlated extremely well with the flavonoid types described above (Desrochers, 1992; Desrochers and Bohm, 1995). A more recent examination of achenes from the Jasper Ridge transect has shown that the two pigment races can be differentiated on the basis of achene dimensions and weight as well (Rajakaruna et al., 2003 and citations therein). A detailed study of soil chemistry has shown a series of trends along the original short transects involving pH, soil moisture, clay content, cation exchange capacity, soil conductivity, and concentrations of several elements (Rajakaruna et al., 2003). There was no abrupt change in any of those factors at the 45 m point, where we have seen the transition between race **C** plants (upper part of slope) and race **A** plants (lower part of slope). That there is a difference between the soils above and below the transition point is clear from the results, but what factor or combination of factors is/are responsible for the change in race within the transition zone is not known. A series of experiments involving germination of achenes, growth of seedlings, and maturation of plants (seed set) add to the view that strong selection is operating in this system. Achenes from both races germinated and grew in soil from either the upper part of the transect (type **C** soil) or from the lower part (type **A** soil), although not equally well. (Both races performed better in potting soil.) The most striking finding of this study was that race **C** plants never reached maturity—they set no seed—when grown in soil from the lower part of the slope. A parallel set of experiments was run using an aqueous extract of upper and lower soils to irrigate achenes of both races; the results were identical to those from the soil-growth experiments.

Two additional sets of observations provided further evidence that two unique biological "entities" (thus avoiding any taxonomic decision, at least at this time!) along the transect at Jasper Ridge. Field observations suggested that the flowering times of race **A** and race **C** plants differ by as much as 10 days, race **C** being the earlier to reach reproductive maturity. Greenhouse studies supported this with the difference in flowering times ranging from 7 to 10 days. Earlier attempts to determine the breeding behavior of these races were inconclusive owing to small sample numbers, but recent work (Rajakaruna and Whitton, 2004) has shown that **A**×**A** and **C**×**C** crosses (all done reciprocally) produced approximately 80% seed set, whereas **A**×**C** crosses (reciprocal) gave lower percentages of seed set (ca. 20%) with the lowest level seen in crosses of **A** and **C** plants originating from Jasper Ridge (ca. 5%).

An interesting case of variation among populations over comparably short distances emerged from studies of flavonoids in *L. burkei* (Greene) Greene (Bohm and Banek, 1987). This species occurs in a comparatively small area in the north coast ranges of California, with populations in Lake, Mendocino, and Sonoma counties. Five populations were sampled, populations **A** and **B** from Sonoma County, and populations **C**, **D1**, and **D2** from Lake County. Populations **A** and **B** were located between Santa Rosa and Healdsburg and are separated by about 10 km. Population **C** was found in

Lake County just south of Lower Lake, and lies at a distance of about 40 km from population **B**, the northernmost of the Sonoma County sites. Populations **D1** and **D2** lie about 9 km west of population **C** and are separated from each other by about 300 m. It is very likely that these two populations were at one time a single large population, but construction of California Rt. 29, with associated roadside maintenance, has effectively separated them (refer to any California atlas for details). Other sites reported in the literature, or represented by voucher specimens, could not be located, owing to the heavy impact of commercial development in this area of California.

The pigment profile of *L. burkei* collected from these sites consists of three quercetin glycosides and eight patuletin glycosides (patuletin is 6-methoxyquercetin) plus traces of two unidentified flavonol glycosides. Only quercetin and patuletin 3-*O*-glucosides were present in all members of all populations. Clear-cut differences among the populations were observed with glycosides of both flavonols. Some populational variation was also observed at all five sites. These results are summarized in Table 2.18. Populations **D1** and **D2** are very similar to each other, which is not surprising in view of their close proximity; in fact, they may at one time have been part of a much larger population that was disrupted by road construction. Populations **A** and **B**, the two populations from Sonoma County, are also very similar to each other and are markedly different from the **D** populations. Despite the fact that population **C** lies geographically much closer to the **D** populations than it does to populations **A** and **B**, it is the pigment profile of these latter two that it most closely resembles, although it clearly exhibits its own characteristic pattern of compounds. Examination of the table reveals that all differences, both within and among populations, are based upon the nature of the glycosides accumulated. Without additional information on the history of the species, the relative ages of the populations, and the genetic control of the glycosylation reactions, it is not possible to draw any conclusions regarding relationships among these five groups of plants. This would seem to be an excellent example of a system for which a population phylogeny would likely be very revealing. With the ever-increasing pressure of development, however, systems of this sort are disappearing and taking their genetic histories with them.

Table 2.18 Flavonoid variation in a small area, the case of *Lasthenia burkei* (after Bohm and Banek, 1986)

Population	No.[b]	Flavonoid Derivative[a]												
		PT	PD	PI	PX	QD	PG	PX	?G	?X	PJ	QG	PG	QN
Sonoma Co. (A)	24	nd[c]	24[d]	24	24	nd	24	24	17	17	3	24	2	2
Sonoma Co. (B)	24	1	24	24	24	nd	24	24	19	19	4	24	8	8
Lake Co. (C)	24	1	24	24	24	9	24	24	21	21	nd	24	nd	nd
Lake Co. (D1)	18	4	nd	1	nd	18	18	nd	18	nd	nd	18	18	18
Lake Co. (D2)	18	2	nd	4	nd	18	18	nd	18	nd	nd	18	18	18

[a] P = Patuletin; Q = quercetin; ? = unidentified flavonol; T = triglycosides; D = diglycoside; X = xyloside; G = glucoside; I and J = isomeric glucosides; N = glucuronide.
[b] No. = Number of plants sampled in each population.
[c] nd = Not detected.
[d] Number of plants exhibiting the compound.

2.7.6 Brickellia cylindracea *(Asteraceae)*

From the southern part of the United States comes an example where flavonoid profiles from disjunct populations indicate a closer relationship than might otherwise have been recognized. *Brickellia cylindracea* Gray & Englem. occurs over an area including southern Texas and displays a level of morphological variation that suggested to at least one worker that a population in the Big Bend area of western Texas (Fig. 2.57) might be sufficiently different from populations in central Texas to warrant its recognition as a formal taxon (A. M. Powell, personal communication to Timmermann and Mabry, 1983). As part of a large-scale study of flavonoids of *Brickellia*, Timmermann and Mabry (1983) investigated the pigments in collections from both areas. Identical arrays of flavonols, including both aglycones and glycosidic forms, were observed for each. The compounds identified were derivatives of quercetagetin, including the 3,6,4′-trimethyl [190], 3,6,3′,4′-tetramethyl-[191], and 3,6,7,3′,4′-pentamethyl ethers [192] (Fig. 2.58). These data were taken as evidence

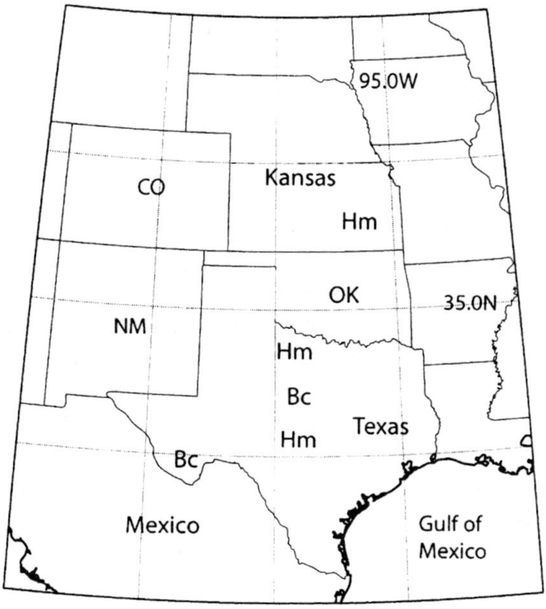

Fig. 2.57 Map showing sites for *Brickellia cylindrica* (Bc) and *Helianthus maximiliani* (Hm) studies

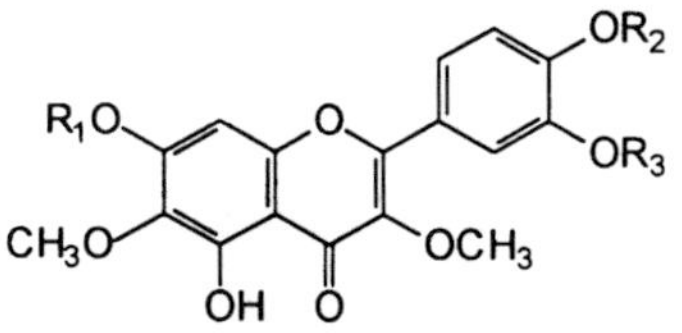

[190] $R_1 = R_2 = H$, $R_3 = CH_3$
[191] $R_1 = H$, $R_2 = R_3 = CH_3$
[192] $R_1 = R_2 = R_3 = CH_3$

Fig. 2.58 Compounds 190–192, flavonoids from *Brickellia cylindrica*

that the two populations clearly belong to the same species. Additional weight to the argument came from observations that *Flyriella parryi* (A. Gray) R. M. King & H. Robinson, a segregate from *Brickellia* generally accepted by systematists, presents a flavonoid profile based upon quercetin, quite unlike the profiles that characterize *Brickellia*. Outgroup situations of this sort are not commonly encountered in systematic applications of secondary metabolites; their value is highly significant.

The studies of variation patterns in *Lasthenia* and *Brickellia* hardly break the surface of a vast and complex literature on chemical variation within Asteraceae. The subject has been discussed in detail, with reviews focusing on polyacetylenes (Bohlmann et al., 1973), sesquiterpene derivatives (Seaman, 1982), and flavonoids (Bohm and Stuessy, 2001). The next examples come from the sesquiterpene lactone literature and, again, represent only a sample of the applications that have been made using those data. Examples covering the taxonomic hierarchy within Asteraceae up to the early 1980s can be found in the monumental review of the family prepared by Seaman (1982).

2.7.7 **Helianthus maximiliani** *(Asteraceae)*

The first example involves the sesquiterpene lactones from *Helianthus maximiliani* Schrader initially described by Herz and Kumar (1981). Working with plant material collected in Coffey County, Kansas (see Fig. 2.57), those workers isolated and identified several closely related heliangolides exemplified by compounds [193, with a second set having a double bond at the starred position] and [194] (see Fig. 2.59 for structures 193–205). Subsequently, Gershenzon and Mabry (1984) examined a population of this species from Travis County in south-central Texas. The chemistry reported by these latter workers was completely different from the array of compounds obtained from the Kansas collection consisting of five guaianolides, represented by [195], where the R group represents a series of aliphatic acids, the germacranolides [196 and 197], and the labdane derivative [198]. A further collection of this species from a north central Texas population showed an array of compounds consisting of [197] and its (2′*S*,3′*R*)-epoxyangelate isomer, but no heliangolides or guaianolides. This infraspecific lactone variation was not considered unusual by those workers who pointed out that chemical races have been reported in *Ambrosia*, *Artemisia*, and *Iva*, among other genera, referring to Seaman's (1982) review.

2.7.8 **Ambrosia** *(Asteraceae)*

The next examples concern species of *Ambrosia* L. that are characterized, in the first, by a north–south variation, and in the second, an east–west one, with the added feature of different ploidy levels. The work is that of Seaman and Mabry (1979a, b). In the first example, we will look at *A. ambrosioides* (Cav.) Payne sampled from southern

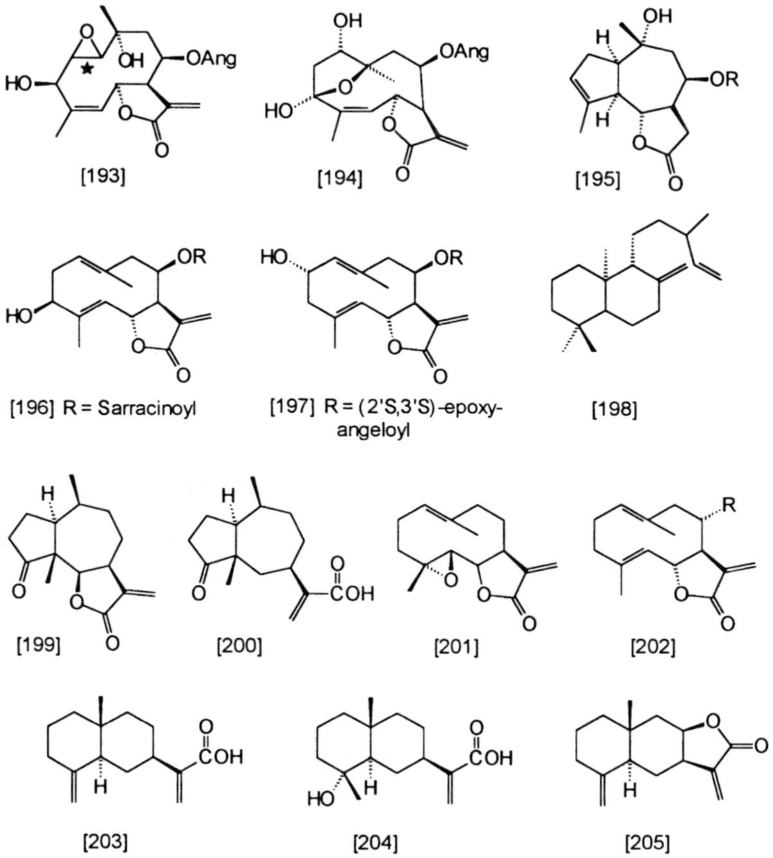

Fig. 2.59 Compounds 193–198 from *Helianthus maximiliani*; 199–201 from *Ambrosia ambrosioides*; 202–205 from *Ambrosia camphorata*

Arizona, northwestern Mexico, and Baja California (Fig. 2.60). The species is morphologically uniform throughout its range and is a diploid ($n = 18$). Earlier chemical workers had observed some variation in sesquiterpene lactone chemistry with different components identified in material collected in Arizona and in Sinaloa, Mexico. Seaman and Mabry collected material from 13 sites, ranging from south central Arizona to the southern tip of Baja California, plus a site on the mainland somewhat farther south. A definite difference between northern and southern populations was observed with damsin [199] identified as the major component of northern specimens with some specimens also exhibiting damsinic acid [200]. Damsinic acid, however, was the major component of five of the six populations from southern Baja California and the more southerly mainland population. A peculiar observation, for which no explanation was offered, concerned two populations from mainland Mexico (numbers 42 and 43 in table). Plants from neither of these sites contained damsin or damsinic acid, with the total lactone fraction composed of parthenolide [201]. Parthenolide was

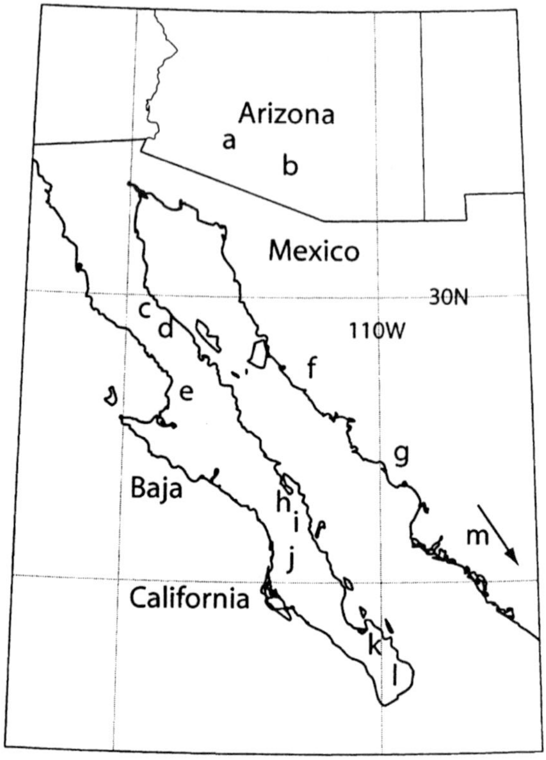

Fig. 2.60 Map of sites for the *Ambrosia ambrosioides* study

also present in two populations from Baja California along with hydrodamsinic acid otherwise not seen in this taxon. These data are summarized in Table 2.19.

Ambrosia camphorata (Greene) Payne is a shrubby native of the Sonoran Desert that has been variously described as consisting of two varieties, var. *camphorata* and var. *leptophylla* Gray, or two species (Rydberg, 1922). A reinvestigation of the complex led Payne (1964) to conclude that the problem could not be resolved on the basis of the available information. In an effort to overcome this state of affairs, Seaman and Mabry (1979a, b) examined the sesquiterpene lactone chemistry of individuals collected from 14 sites in Baja California and one from the Mexican mainland near Puerto Libertad (see Fig. 2.61). Chromosome number was established for individuals from nine of the Baja Californian sites; four were found to be diploids ($n=18$) and five tetraploids ($n=36$). Two of the diploids (Seaman and Mabry populations No. 17 and 217) occurred near the northern extent of the collections and two from much farther south (No. 174 and 175). Four of the tetraploids were also found in this area, with the fifth from near the tip of the peninsula. Also noteworthy is the apparent heterogeneity of floral structure. Two types of head structure have been observed, one with small heads with few spines that characterize plants throughout the length of Baja Californica, except for an area that lies roughly between 24°30′N

Table 2.19 Distribution of sesquiterpene lactones in *Ambrosia ambrosioides* (from Seaman and Mabry, 1979b)

Sample[b]	State	Lat.	Sesquiterpene Lactone[a]			
			DA	HAD	PAR	DAM
ROD	a[c] ARIZONA	33°N	–	–	–	100
63	b ARIZONA	32°30′	35	–	–	65
1766	c BAJA	30°	–	–	–	100
90	d BAJA	29°	65	–	–	35
94	e BAJA	28°	55	–	–	45
42	f SONORA	28°	–	–	100	–
43	g SONORA	27°30′	–	–	100	–
100	h BAJA	26°30′	100	–	–	–
1838	i BAJA	26°	100	–	–	–
103	j BAJA	25°30′	50	32	10	–
115	k BAJA	24°	25	55	20	–
116	l BAJA	23°	100	–	–	–
120	m SINALOA?	<23°	100	–	–	–

[a] DA = Damsinic acid; HAD = hydrodamsinic acid; PAR = parthenolide; DAM = damsin. Values are relative percent.
[b] Collection numbers in source paper.
[c] Locator on map in Fig. 2.60.

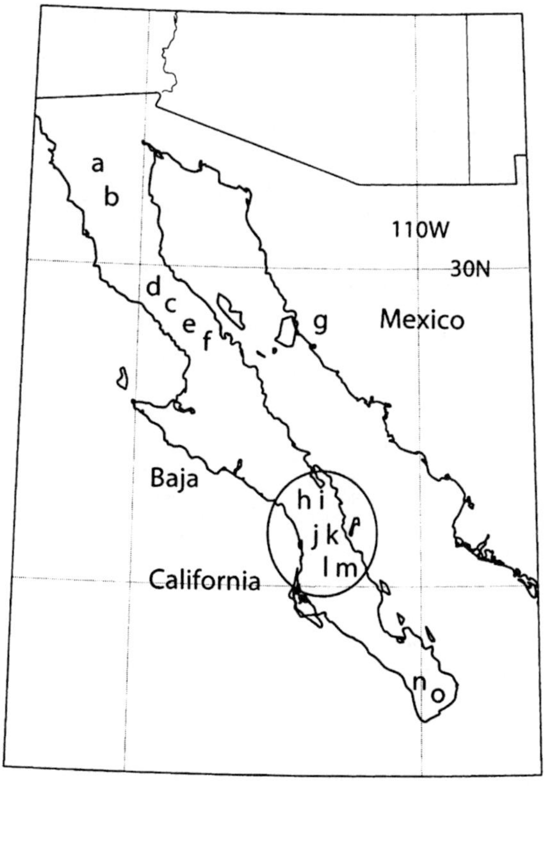

Fig. 2.61 Map of sites for the *Ambrosia camphorata* study

and 26°30′N, and the area within that latitudinal range where the heads are large and many spined. Some individuals within this range are diploid; some are tetraploid.

Seaman and Mabry (1979a, b) identified five sesquiterpene lactones, the germacranolides costunolide [202, R=H] and tulipinolide [202, R=acetoxy], and the eudesmanolides costic acid [203], ilicic acid [204], and isoalantolactone [205]. The distribution of these compounds, along with chromosome number, head size, and range, is summarized in Table 2.20. Although there is no correlation between sesquiterpene lactone chemistry, head morphology or location of the specimens within the sampled range, there are some interesting connections between chromosome number and sesquiterpenes. Tulipinolide, the acetylated derivative of costunolide, appears restricted to tetraploids (4 of 5). This can be rationalized in terms of two processes, one that establishes the hydroxyl function at the particular position, and one that involves acetylation, clear-cut differences between the two ploidy levels. Costunolide itself occurs in all five tetraploids, in two of the four diploids counted, and in several additional individuals whose chromosome numbers were not determined. The tetraploids, all five in this instance, lacked isoalantolactone and costic acid, compounds consistently present in diploids. The only compound present in all individuals was ilicic acid. A specimen of *A. camphorata* collected from a disjunct

Table 2.20 Sesquiterpenes, chromosome numbers, and flower head features for *Ambrosia camphorata* (after Seaman and Mabry, 1979a)

Sample[b]	Site[c]		n[d]	Head[e]	Sesquiterpene[a]				
					ISOL	COSA	ILLA	COST	TULP
17	a	N	18	small	50	10	30	10	–
217	b	N	18	small	10	21	57	12	–
14A	c	N	nd	small	22	–	45	33	–
15G	d	N	nd	small	–	15	60	25	–
1770	e	N	nd	small	–	60	10	30	–
1804	f	N	nd	small	–	60	10	30	–
66	g	N	nd	small	–	26	74	–	–
175	h	S	18	large	72	16	12	–	–
174	i	S	18	large	51	31	18	–	–
173	j	S	36	large	–	–	38	14	48
106	k	S	36	large	–	–	5	50	45
107	l	S	36	large	–	–	5	50	45
171	m	S	36	large	–	–	39	7	54
165	n	S	36	small	–	–	56	44	–
118	o	S	nd	small	–	–	14	–	86

[a] ISOL = isoalantolactone; COSA = costic acid; ILLA = ilicic acid; COST = costunolide; TULP = tulipinoside; Values are relative percentage of each compound based on total sesquiterpenes.

[b] Collection numbers from source paper. Small Arabic letters refer to sites on map in Fig. 2.61.

[c] N = Northern group, ca. 29°–30°30′N; S = Southern group, ca. 23°30′–26°30′N.

[d] Meiotic chromosome number.

[e] Small heads with few spines; Large heads with many spines.

population near San Luis Potosi (22°10′N, 101°W) was shown to accumulate ilicic and costic acids (Higo et al., 1971) and have $n=36$ (Payne, 1964).

2.7.9 Gaillardia pulchella *(Asteraceae)*

Gaillardia pulchella Foug. is a morphologically and cytologically diverse taxon that enjoys a wide distributional range, from the southeastern United States westward as far as Arizona (Fig. 2.62). Its chemistry has been the subject of numerous studies, revealing the existence of several chemotypes. The coastal chemotype, ranging from the southeastern seaboard through Florida to an area near Houston, Texas, is characterized by a set of alkaloidal derivatives exemplified by pulchellidine [206] (see Fig. 2.63 for structures 206–209). The Rio Grande collection afforded the ambrosanolide spathulin [207]. Collections from Live Oak County, Texas revealed the guaianolide derivative gaillardin [208]. Plants collected in New Mexico and

Fig. 2.62 Map of sites for the *Gaillardia pulchella* study

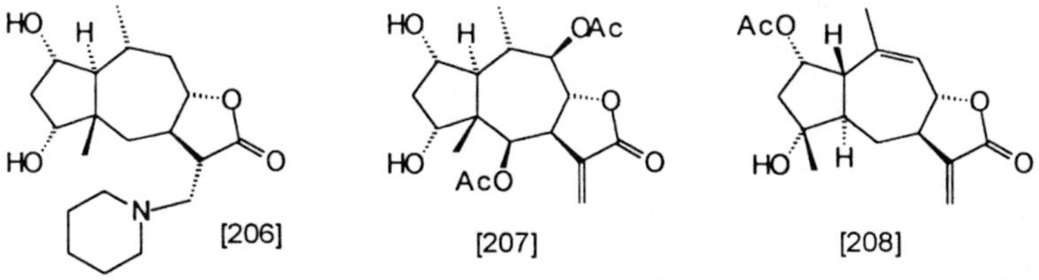

[206] [207] [208]

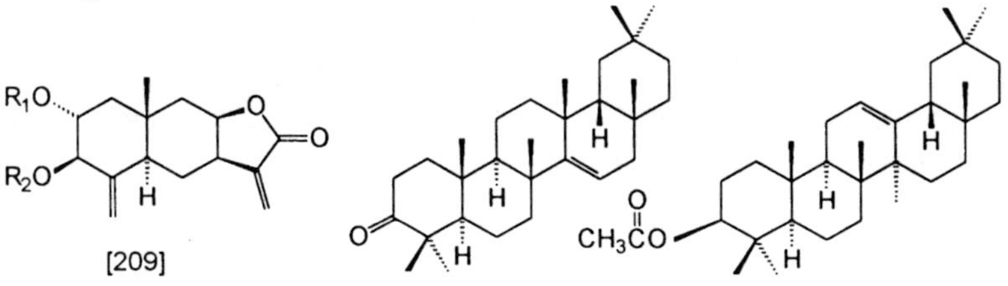

[209] [210] [211]

Fig. 2.63 Compounds 206, an alkaloid, and 207–209, sesquiterpenes from *Gaillardia pulchella*. Compounds 210 and 211, triterpenes from *Dudleya*

Arizona yielded a series of pulchellins [209], with various arrangements of substituents on the hydroxyl groups (Herz and Roy, 1969; Yoshioka et al., 1970). These chemical data provide some support for taxonomic distinctions that have been made within *G. puclhella* over the years. Thus, Turner and Whalen (1975) recognized the Gulf Coast populations as a distinct taxon, *G. pulchella* var. *picta* (Sweet) A. Gray. The population that occurs in Val Verde County (referred to as the Rio Grande Collection), and is characterized by the presence of spathulin, falls within the area occupied by *G. pulchella* var. *pulchella.* (But very close to an area in which var. *australis* B. Turner and Whalen occurs.) Stoutamire (1977) found that the western populations, those from New Mexico and Arizona, were easily distinguished from their eastern counterparts by means of chromosome behavior and pollen stainability. He considered that a distinct boundary exists between the New Mexico and Texas populations. The finding of a series of eudesmanolides, the pulchellins, uniquely in the western populations lends support to this view (Stoutamire, 1977). The westernmost populations of var. *pulchella* have been recognized at one time or another as a distinct species, *G. neomexicana* (Biddulph, 1944) or as a subspecies of *G. pulchella* (Stoutamire, 1954). A recent examination of allozyme variation in *G. pulchella* (Heywood and Levin, 1984) revealed little variation among the varieties. Neither morphological nor chromosomal races were revealed by clustering of population genetic distances. Genetic distances were significantly correlated with geographical distances "... suggesting that gene flow may be an important deterrent to differentiation."

2.7.10 Dudleya *(Crassulaceae)*

Dudleya consists of about 40 species that occur in California, Arizona, and Baja California. A comparatively high degree of morphological variation along with a tendency to hybridize in nature combines to make taxonomy difficult in the genus. In an effort to find additional characters by which species and species groups might be distinguished, Manheim et al. (1979) examined some of the components of the epicuticular wax of 19 species of *Dudleya* subgen. *Dudleya* (eight from Baja California; 11 from California). Thin layer chromatography revealed two different profiles, one representative of more northerly and/or montane sites, the other centered in Baja California. The northern/montane species have taraxerone [210] as the major triterpenoid, while plants from the southern sites are characterized by ß-amyrin acetate [211] (see Fig. 2.63). Those workers reported a correlation between triterpene type and corolla tube length where species with taraxerone have shorter tubes (petals united for less than one-third of their length), while those with ß-amyrin have corolla tubes fused for more than one-third of their length. They concluded that the correlation between two seemingly unrelated features likely has phylogenetic significance.

By way of contrast, a study of the leaf cuticle alkanes of *Sedum lanceolatum* Torrey (Crassulaceae) from 44 populations showed no correlation with elevation.

Plants were collected along an east-west transect lying at approximately 40°40 N in the southern Rocky Mountains of Colorado, and represented sites that ranged from 1598 to 3568 m (Bowman, 1983). The compound with 33 carbons was the major alkane in all samples with the C_{31} compound next in abundance. Clustering of the *Sedum* populations on the basis of alkane proportions showed, for example, that the nearest neighbors based upon total alkane variability were two populations that occupied elevational extremes. In other instances, nearest neighbors along the transect were nearly identical, which was taken to indicate the possibility of "clonal propagation or localized biotypes."

2.7.11 Delphinium variegatum *(Ranunculaceae)*

Another example of variation on a local scale in southern California comes from the work of Dodd and Helenurm (2000) on *Delphinium variegatum* Torrey and Gray (Ranunculaceae), an attractive larkspur native to grasslands and open woodlands throughout its range. This species is thought to consist of three subspecies, subsp. *Variegatum*, which occurs exclusively in central California from the coast to the foothills of the Sierra Nevada, and two that are endemic to San Clemente Island, subsp. *kinkiense* (Munz) M. J. Warnock and subsp. *thornei* (Munz) M. J. Warnock. Three features distinguish the two insular taxa, lateral sepal length, lower petal blade length, and sepal color, which range from white through pale to dark blue. So far, no chemical study has been done on the pigment(s) of this species of larkspur, but it seems a fair guess that glycosides of delphinidin are involved. Depth of color may simply be the result of greater concentrations of pigment in certain lines, or there could be a more complicated explanation based upon the known capacity of blue pigmentation to involve intricate complexation of the fundamental anthocyanin with metallic ions. White, of course, can be taken to indicate the complete absence of part, or all, of the anthocyanin biosynthetic pathway.

Neither metric character sepal length nor petal length can be used to distinguish between the two subspecies on San Clemente Island. Sepal color provided somewhat better separation, but the distribution of pigmented individuals was not clear-cut. Seven of the eight northernmost populations had white or pale blue flowers; the eighth had a small fraction of darker blue flowers. Populations in the central part of the island had varying fractions of pale, intermediate, and dark flowers, whereas as the majority of populations in the southern part of the island have predominantly dark flowers, but almost always with some flowers of lighter hue. The southernmost population, with primarily white or pale flowers, resembles populations from mid-island. Dodd and Helenurm (2000) speculated that the "problem" may reside in sampling and in the nature of the specimens studied, primarily herbarium material in the taxonomic study (Warnock, 1990a, b) compared to the extensive population sampling of the present work. In short, the more one looks, the more one may find. The authors went on to suggest that subspecific recognition may not be justified, suggesting rather that varietal status may be more appropriate, but added that further study may, in fact, reveal that taxonomic dissection may not be justified at all. The problem may be purely academic, however, as both forms are considered at risk.

A more recent report from these workers (Dodd and Helenurm, 2002) described an electrophoretic study of seven populations of the mainland subspecies and all known populations (24 in all) of the two island subspecies. Information was obtained representing 19 loci. Although the mainland populations have higher levels of polymorphic loci (33.6% vs. 24.5%) and greater number of alleles per locus (2.61 vs. 2.15) than the island populations, observed heterozygosities are lower for the mainland populations (0.041 vs. 0.071), which was attributed to lower levels of outcrossing for the mainland populations compared to island populations. These values fall within the limits usually associated with comparisons of insular and mainland subspecies (Hamrick and Godt, 1989). The island subspecies are nearly identical genetically (mean $I=0.997$). The island populations were very closely related with no taxonomic distinction possible. Values for mainland populations ranged from 0.752 to 1.0 (mean 0.929)

2.7.12 Isomeris arborea *(Capparaceae)*

Glucosinolates, the naturally occurring glucosides from which the so-called mustard oils arise, have been studied in considerable detail with regard to their putative role as defensive chemicals in a variety of brassicaceous taxa, for example, *Cardamine cordifolia* A. Gray (Louda and Rodman, 1983a, b). Blua et al. (1988) have also investigated ecological factors that may influence the formation of the glucosinolate produced by *Isomeris arborea* Nutt. (Capparaceae) in Southern California. This species offers the advantage of possessing a single glucosinolate, methyl glucosinolate, also known as glucocapparin [213]. The presence of only one compound offers the combined advantages of ease of quantitative analysis and absence of interactional effects of a more complex array of compounds. Plants were collected from four sites in southern California, allowing a comparison between desert and nondesert environments to be made. The desert sites sampled were the low desert in Anza-Borrego State Park (San Diego County) and a high desert site in the Mojave Desert (Kern County). A valley grassland site near Bakersfield (Kern County) and a coastal sage scrub site in Torrey Pines State Preserve (San Diego County) provided material from nondesert environments. Glucocapparin was quantified from five tissues: immature leaves, mature leaves, flower buds, capsule walls, and seeds. Seeds were eliminated from further comparison owing to the lack of any statistically significant difference among their glucocapparin concentrations. Significant differences were observed

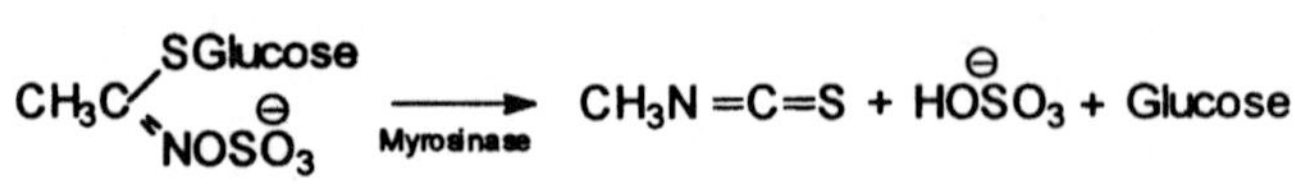

[212]

Fig. 2.64 Breakdown of the glucosinolate glucocapparin, compound 212

between desert and nondesert plants, however. With only one exception (flower buds from Anza-Borrego plants), tissues from desert plants had higher concentrations of the glucosinolate than those from more moderate environments. Experiments with *Isomeris* grown under laboratory conditions revealed that glucocapparin content varied inversely with soil nitrogen, suggesting that the availability of this element in nature strongly influences the formation of the defense compound, which is in accord with the views of Coley et al. (1985) who hold that species that grow slowly owing to limitations of resources evolve "immobile defenses" (phenolic compounds or fiber) whereas "mobile defenses" are produced by plants that are capable of rapid growth in the presence of ample resources. *Isomeris arborea*, which is capable of rapid growth (Blua et al., 1988), falls within this latter category.

2.7.13 Soils and Asters

One of the earliest studies of variation involving flavonoids was that of Abrahamson and Solbrig (1970) who examined these compounds, including an analysis of total anthocyanins, in several species of *Aster* along a transect running from extreme northwestern Pennsylvania to eastern Massachusetts. Soil and plant samples were taken from 19 sites. Soil samples were examined for a variety of features: nitrate content, moisture, color, and clay content. Plants were taken to the University of Michigan Botanical Gardens where they were maintained in a uniform garden environment. Extracts of the cultivated plants were analyzed by paper chromatography and total anthocyanin (determined to be cyanin) content was determined colorimetrically. The species were identified as *A. ciliolatus* Lindl. (two sites), *A. cordifolius* L. (12 sites), *A. lowreianus* Porter (one site), and *A. undulatus* L. (one site). Considerable variation was seen in the chromatographic profiles, with six compounds (spots, actually) common to all 83 plants, ten present in over 60 plants, and nine variable within populations. No compound (spot) was species specific. Anthocyanin concentration also varied without evident pattern. Likewise, there appeared to be no relationship between flavonoid profiles and soil. Significantly different chromatographic profiles were observed at different times throughout the growing season, which led the authors to caution future workers to be wary of randomly collected samples and to make use of common garden studies whenever possible. Their advice has not always been followed.

2.7.14 **Lupinus sericeus** *(Fabaceae)*

In this example, individual plants of *Lupinus sericeus* Pursh were collected over a distance measuring approximately 1500 km, including most of the range of the species. In practice, this amounted to a north–south transect running from the Alberta-British Columbia border to northern Arizona. *Lupinus sericeus* occurs from southeastern British Columbia and southwestern Alberta to the Kaibab Plateau in northern Arizona, covering extensive areas in Alberta, British Columbia, eastern Washington, Idaho, western Montana, northeastern Oregon, Utah, and northern Arizona, although the range is interrupted in places. Taxonomic difficulties with this species can be appreciated by reflecting on the fact that no fewer than 23 synonyms for *L. sericeus* are recorded in *Vascular Plants of the Pacific Northwest* (Hitchcock et al., 1961), and that Fleak (1971) studied 61 taxa thought by some workers to belong to the *L. sericeus* complex.

Following an extensive study of lupines in western North America (Nicholls and Bohm, 1982a), a detailed study of *L. sericeus* was undertaken (Nicholls and Bohm, 1982b). Visual comparisons of two-dimensional thin layer chromatograms representing 181 individual plants from 32 populations covering the entire range of the species (Fig. 2.65) showed major differences in size of the spot corresponding to orientin (8-C-glucosylluteolin) [213, R_1 = hydrogen, R_2 = glucose] (see Fig. 2.66 for structures 213 and 214). High-performance liquid chromatography (HPLC)

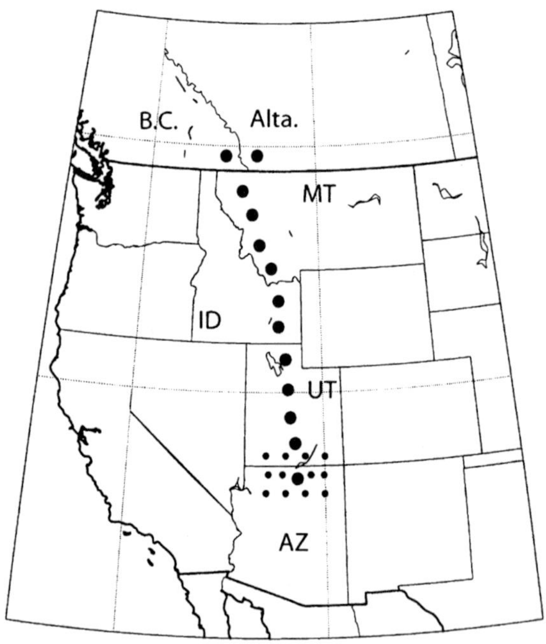

Fig. 2.65 Map showing the collection sites for *Lupinus sericeus* study

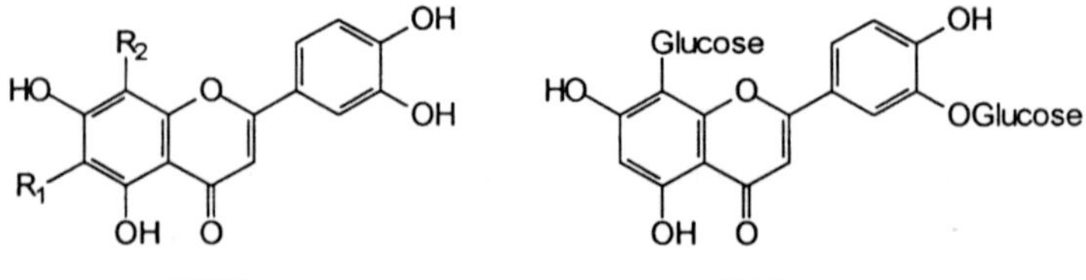

[213] [214]

Fig. 2.66 Compounds 213 and 214, key flavonoids from *Lupinus sericeus* study

confirmed the visual comparisons. The mass of information generated by the HPLC analyses was subjected to multivariate statistical analysis. Principal-components analysis revealed a large degree of dispersion in the data, with orientin dominating axis I (axis I accounted for 72% of total variation), isoörientin [213, R_1 = glucose, R_2 = hydrogen] dominating axis II (12%), and orientin 3′-*O*-glucoside [214] dominating axis III (10%). Despite the large contribution from orientin, no clear-cut grouping of populations appeared to exist in most of the northern and central parts of the range. The southernmost populations are more discernible than those throughout the rest of the range, but the major outcome of these analyses was recognition of clinal variation in a north–south axis, with the high-orientin plants originating from the southernmost populations. This agrees well with an observation made by Fleak (1971) on plants from the southern populations, which consistently display heavy anthocyanin pigmentation in the lower part of the stems. Stem pigmentation was sufficiently variable in plants from the rest of the range of the species that Fleak and Dunn (1971) recognized the southern plants as *L. sericeus* subsp. *huffmanii*. Accumulation of high concentrations of orientin in these individuals provided support for their view. Flavonoid data provide no support, however, for three other subspecies defined by Fleak (1971).

A resampling of populations, again representing the entire range of the species, provided seed so that plants could be grown under controlled conditions of day length and temperature. The flavonoid fraction from each individual was then subjected to HPLC analysis as before. Under uniform conditions, all quantitative differences in the flavonoid profiles disappeared! Further studies to determine what environmental factor or factors is/are responsible for the quantitative differences in orientin concentration in nature have not been undertaken. It is interesting to note in passing that plants from the southern populations retained their red pigmentation under the controlled conditions of the growth experiment, thus providing support for the use of anthocyanic pigmentation as a character in suggesting subspecific status for those individuals.

2.7.15 Satureja douglasii *(Lamiaceae)*

Terpene pattern variation has provided a useful means of investigating evolutionary and ecological situations in a variety of plant groups. The next example, a study of the monoterpenes of *Satureja douglasii* (Benth.) Briq. by Lincoln and Langenheim (1976), is an excellent example of an investigation aimed at establishing possible factors responsible for chemical variation exhibited by a species throughout its range. *Satureja douglasii* occurs in a narrow band paralleling the coast from southern British Columbia to southern California and disjunctly in a narrow band in the western half of the Idaho panhandle and adjacent British Columbia. Sampling for this study involved the coastal and near-coastal sites and constituted a north–south transect between approximately 50°N and 34°N latitude (Fig. 2.67). (A specimen from Idaho was analyzed but does not figure in the study.)

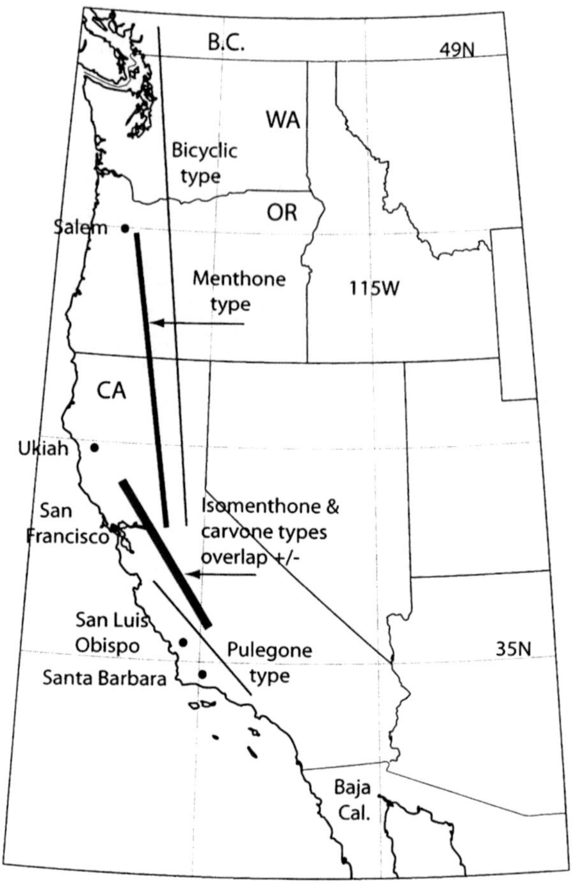

Fig. 2.67 Map showing range of occurrence of *Satureja douglasii* chemotypes

As is the case with many members of Lamiaceae, *Satureja douglasii* produces abundant essential oil from glandular trichomes on the leaves. Gas chromatographic analysis of the leaf oils from specimens collected throughout the species' range revealed the presence of some dozen and a half well-known compounds. The major compounds identified were camphene [215], camphor [216], which, taken together, were considered to comprise the "bicyclic" type, carvone [217], pulegone [218], menthone [219], and isomenthone [220] (see Fig. 2.68 for structures 215–220). The predominance of each of these major components defined a terpene "type." (All compounds were observed in each of the terpene types, most in comparatively small amounts, some only as traces.)

The major structural difference between monocyclic and bicyclic compounds should be obvious; camphene and camphor exhibit bridged structures. It should be noted, however, that the monocyclic compounds can be further distinguished from each other based on the position of the oxygen atom, located at C-2 in carvone, and at C-3 in pulegone, and in the menthones. Work by others had shown that the

Fig. 2.68 Compounds 215–220, terpenes from *Satureja douglasii*

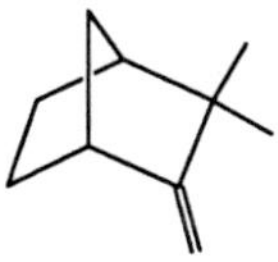

[215]

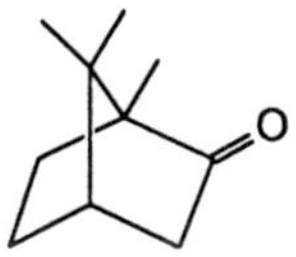

[216]

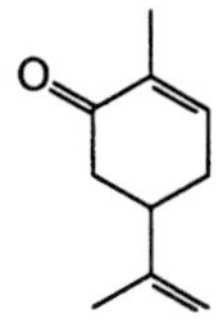

[217]

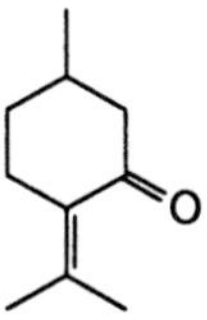

[218]

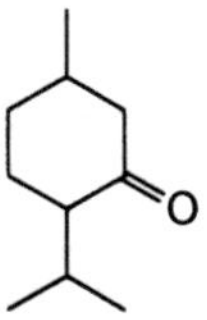

[219]

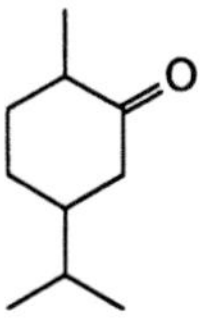

[220]

position of oxygenation on the terpene ring is under genetic control (see Lincoln and Langenheim, 1976, for leading references). The general assumption is made that partitioning of carbon into pathways leading to monocyclic versus bicyclic terpenes is also under genetic control. Plants transplanted to a common garden from their respective locations along the transect remained true to their field-observed terpene type, although the relative amounts of terpenes varied somewhat; these differences did not affect the overall pattern.

The most geographically localized terpene phenotype, referred to as the pulegone type, occurs in the extreme southern portion of the range of the species, roughly from the Santa Barbara area (33°29′N, 119°01′W) north to the Monterey Peninsula (36°35′N, 121°55′W). This type consists of ca. 40% pulegone, with lesser amounts of camphor and camphene [amounts estimated from Fig. 2.3 of the Lincoln and Langenheim (1976) paper]. The isomenthone type overlaps the pulegone type very slightly in the south, and extends north to the vicinity of Ukiah (Mendocino County) (39°09′N, 123°12′W). This type consists of roughly equal amounts of isomenthone, camphene, and camphor. The carvone type overlaps the prior two types ranging, roughly, from the San Luis Obispo area (35°16′N, 120°40′W) to somewhat south of Ukiah. Carvone is the prominent component in this type and occurs with camphene and camphor. The menthone type, which consists of camphor, camphene, and menthone—concentrations decreasing in that order—extends roughly from the San Francisco Bay area north to near Salem, Oregon (44°57′N, 123°01′W). The most widespread is the bicyclic type, which consists of camphor and camphene with the former contributing the larger share. This type extends from the San Francisco Bay area north through Oregon, Washington, and into British Columbia. By plotting the concentrations of terpenes as a function of latitude, the investigators noted discontinuities with both the bicyclic and *p*-menthane derivatives (refers to cumulative monocyclic compounds). *p*-Menthane derivatives were highly correlated with latitude from 34°N to 41°N (r=0.–801, P<0.01). However, concentrations of *p*-menthane derivatives from more northerly sites (41–50°N) did not correlate with latitude (r=–0.067). The total concentration of bicyclic compounds also correlated with latitude in the southern part of the transect ($r = 0.740$, $P < 0.01$), whereas in the northern part of the range there was no significant correlation ($r = 0.264$).

Subsequent papers on *Satureja douglasii* terpenes (Lincoln and Langenheim, 1979, 1981) addressed the question of what effect environmental factors might have on the expression of the chemical forms, and what the genetic basis of the different chemical forms is, respectively. In the first of these studies, the effects of light intensity and degree of herbivory were assessed. It was shown that high-yielding genotypes tend to occur under low light-high herbivore pressure, whereas low-yielding genotypes occurred under high light-low herbivore pressure. Attempts to establish the genetic basis for monoterpenoid composition were not conclusive, but did suggest that tight genetic control does function in this system. Despite the limited conclusions concerning the genetics of the system, this study of *Satureja douglasii* represents one of the better-documented examinations of secondary metabolite profile variation available in the literature.

2.7.16 Pentagramma (Pityrogramma) triangularis *(Adiantaceae)*

Pentagramma triangularis (Kaulf.) Yatskievych, Windham, and Wollenweber, the most recent taxonomic reincarnation of *Gymnogramma triangularis* Kaulf. (see discussion below), is a member of the fern family, Adiantaceae. This fern is characterized by copious amounts of solid exudate (often called a farina) on the undersides of its leaves, which has given rise to the common names of silverback or goldback fern, depending upon the color of the material present. The taxon occurs in rock crevices and open rocky slopes from southern British Columbia (Vancouver Island, Gulf Islands, and only very rarely on the mainland; author's comments) to northern Baja California, east into Arizona, southern Nevada, and in extreme southwestern Utah (Hitchcock et al., 1969). Since the range of occurrence of this fern, at least along the coast (refer to Fig. 2.67 for general area map), is very similar to that of *Satureja douglasii*, it seems appropriate to discuss differentiation of its chemotypes at this point.

Before getting to the chemical and related studies of this system, a note on recent taxonomic work is in order. This species was first described as a member of *Gymnogramma* in 1824, but was placed in *Pityrogramma* in 1913 (see Yatskievych et al., 1990 for generic taxonomic details; D. M. Smith et al., 1971 for varietal/subspecific details). Questions have been raised, however, as to the inclusion of *P. triangularis* (Kaulf.) Maxon in the genus. It was pointed out by Tryon (1962) that this taxon does not fit comfortably with the "central group" of species in the genus, and suggested that it might well stand on its own, but did not make any formal change to its status. This position was echoed by Tryon and Tryon (1982), but again no formal changes were made owing to lack of sufficient information. The problem was addressed by Yatskievych et al. (1990) who assembled information on rhizome scale features, stipe structure, laminar shape, shape of pinnae, venation, spore structure, and base chromosome number in support of the recognition of the taxon as *Pentagramma triangularis* consisting of four subspecies: *triangularis*, *viscosa*, *semipallida*, and *maxonii*.

(See D. M. Smith, 1980, for a study of flavonoid profiles of the varieties.) The overall flavonoid profile of *P. triangularis* is fully in accord with the unique status of the species. A detailed discussion of the chemistry of this system, which is beyond the scope of the present treatment, can be found in a paper by Wollenweber and Dietz (1980). An example of the complexity of flavonoid biosynthesis in this species can be found in a description of biflavonoids present in the farinose exudate (Iinuma et al., 1994).

An early hint of the complexity of chemical variation lurking within this species came from the work of Dale Smith and his colleagues at the University of California at Santa Barbara (D. M. Smith et al., 1971). The group studied chromosome numbers, spore size, and chromatographic profiles of plants obtained from three populations in Santa Barbara County: Hoffman Hill, Painted Cave, and Refugio Pass. Chemical differences among these populations were evident in the different colors of the leaf exudates that characterize this species, yellow, white, or green. Cytological study revealed diploids, triploids, and tetraploids based upon $x = 30$. One-dimensional paper chromatograms sprayed with ferric chloride solution revealed several spots, one of which corresponded with ceroptene, a compound whose structure had been determined some years earlier by Nilsson (1959). Other compounds were tentatively identified as kaempferol methyl ethers by comparison of spot color with colors generated with known flavonols, but structural work was not taken further. Four pigment patterns emerged, referred to as **I**, **II**, **III**, and **IV** by those workers. Plants from the Hoffman Hill site were either yellow or green backed. The yellow-backed plants exhibited pigment pattern **I** and chromosome counts revealed them to be tetraploids. The two green-backed plants were triploids with pigment pattern **IV**. The Painted Cave population consisted of yellow-backed and green-backed plants. The yellow-backed plants exhibited two chromatographic patterns, **I** and **II**. All plants of both colors were shown to be tetraploids. The three green-backed plants examined exhibited chromatographic pattern **IV**; two were triploid, and one was tetraploid. The collections at Refugio Pass yielded green-, yellow-, and white-backed individual plants. Five yellow-backed plants had chromatographic pattern **I**; three of the plants were diploid, one was a tetraploid, and the fifth was a triploid. Yellow-backed plants from this population with chromatographic pattern **II** were either diploid or triploid. White-backed plants with pattern **III** were tetraploids, and so on. In a summary statement, the authors pointed out that 10 of the 12 possible combinations of ploidy level and chromatographic pattern have been observed. It was clear that pigment type and ploidy level were not correlated, with the possible exception that all white-backed plants in the Refugio Pass sample were tetraploids. The complexity of the data precluded any definitive statement as to whether the observed combinations originated via auto- or alloploidy.

Additional information on chemical variation within this species came from studies of exudate and internal (vacuolar) pigments based on broader collections of the fern (Star et al., 1975a, b). Exudate flavonoids, identified from specimens representing the range of the species, were identified as ceroptene [221], the C-methylflavonol pityrogrammin [222], and two derivatives of kaempferol, the 4′-methyl ether [223] and the 7,4′-dimethyl ether [224] (see Fig. 2.69 for structures 221–224).

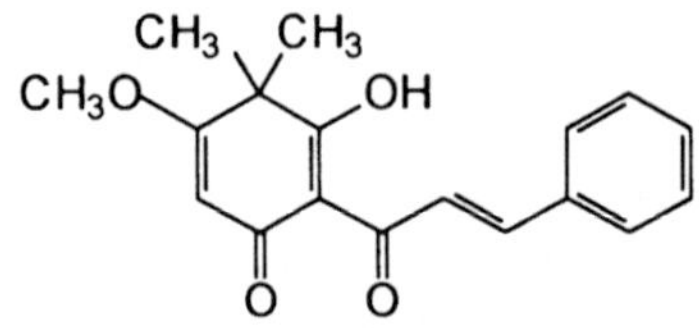

[221]

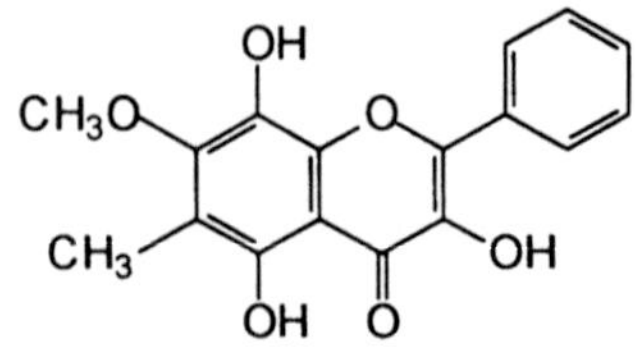

[222]

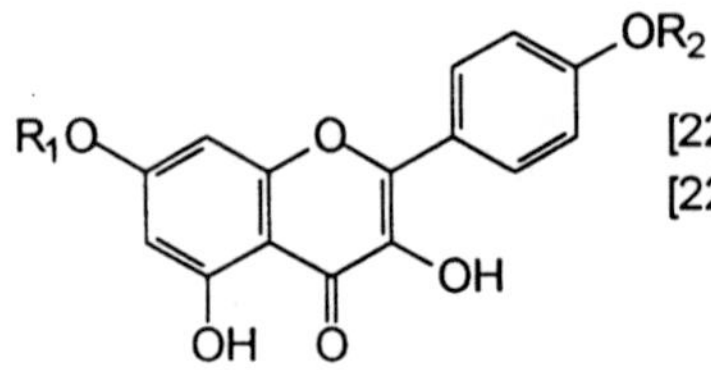

[223] $R_1 = H, R_2 = CH_3$
[224] $R_1 = R_2 = CH_3$

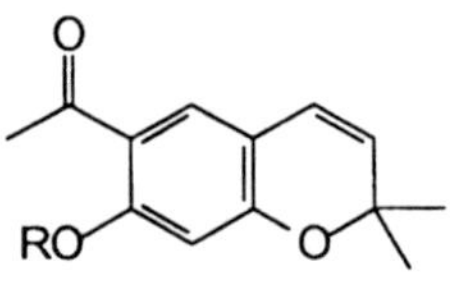

[225] $R = CH_3$
[228] R = H

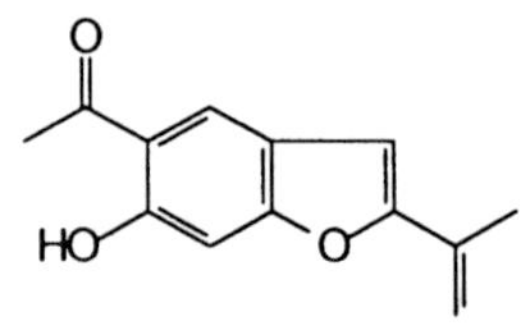

[226]

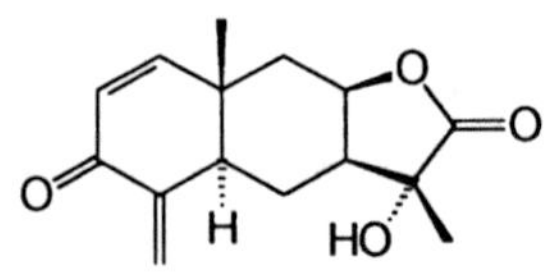

[227]

Fig. 2.69 Compounds 221–224, flavonoids of *Pentagramma* (*Pityrogramma*) *triangularis*. Compounds 225–228 from *Encelia farinosa*

Kaempferol methyl ether-bearing plants occur throughout the fern's range, whereas the ceroptene-pityrogrammin chemotype occurs from, roughly, central California to the southern extent of the species in northwestern Baja California. Each of these chemotypes consists of diploids and tetraploids that cannot be distinguished on the basis of their exudate chemistry. An examination of the vacuolar components, however, revealed the existence of flavonol glycosides the presence of which distinguished ploidy level within each exudate chemotype with only minor variation. Thus, tetraploids of the ceroptene-pityrogrammin chemotype consistently exhibited kaempferol 7-methyl ether 3-*O*-glucoside and 3-*O*-rhamnoglucoside as major components. Diploid plants from two sites had the diglycoside in trace amounts. Neither of these compounds was seen in plants belonging to the kaempferol methyl ether exudate chemotype. A parallel situation exists with regard to the kaempferol methyl ether exudate chemotype, wherein one sees quercetin 3-*O*-glucoside and 3-*O*-rhamnoglucoside in members of the tetraploid race but not in the diploid race.

One final note on this topic involves an examination of diploid, triploid, and tetraploid plants for their leaf exudate hydrocarbons (Seigler et al., 1975). *n*-Alkanes having odd numbered chains from C_{25} to C_{33} were recorded for almost all plants collected from several sites with much more limited distribution of hydrocarbons based on C_{35} and C_{28}. The hydrocarbon data do not assist in distinguishing groups within this species.

2.7.17 Encelia farinosa *(Asteraceae)*

Taxonomic troubles with *Encelia*, a genus of perennial shrubs found primarily in the Sonoran and Mojave Deserts of southwestern North America (two species occur in South America and one occurs on the Galapagos Islands), include questions of species number, relationships with other genera within its tribe, and, in fact, its proper tribal affiliation. In addition to traditional taxonomic studies, examination of several classes of secondary metabolites has revealed quite complex patterns of variation. For example, Wisdom and Rodriguez (1982) analyzed plants from 16 populations of *E. farinosa* A. Gray in Torr. for the benzopyran encecalin [225], the benzofuran euparin [226], and the sesquiterpene lactone farinosin [227] (see Fig. 2.69 for structures 225–227). The three populations from California, along with three populations from adjacent Baja California, were characterized by high percentages (>80%) of encecalin with about equal amounts of the other two compounds making up the balance. One population from extreme northwestern Mexico and nine populations from Arizona were characterized by concentrations of encecalin that varied from about 60% to zero. The proportions of euparin and farinosin varied widely in these populations. Compounding these profiles was the existence of both seasonal and ontogenetic variation, which, though clearly evident, did not detract from the existence of a western and an eastern chemotype.

Some years later, Proksch and Clark (1987) examined 18 species (19 taxa) for the presence of encecalin, 7-demethylencecalin [228] (Fig. 2.69), and euparin. These compounds allowed a sorting of the taxa into several clear-cut groups. The issue that concerns us in the present context, however, is those workers' examination of variation within *E. farinosa*, which was done in an effort to assess the degree of stability of species profiles. In most cases differences between different populations were quantitative. The species was sampled at ten sites along a transect running from Palm Springs, California (33°49′N, 116°34′W) to Tempe, Arizona (33°25′N, 111°55′W), a distance of about 500 km (Fig. 2.70). Highly significant differences were observed

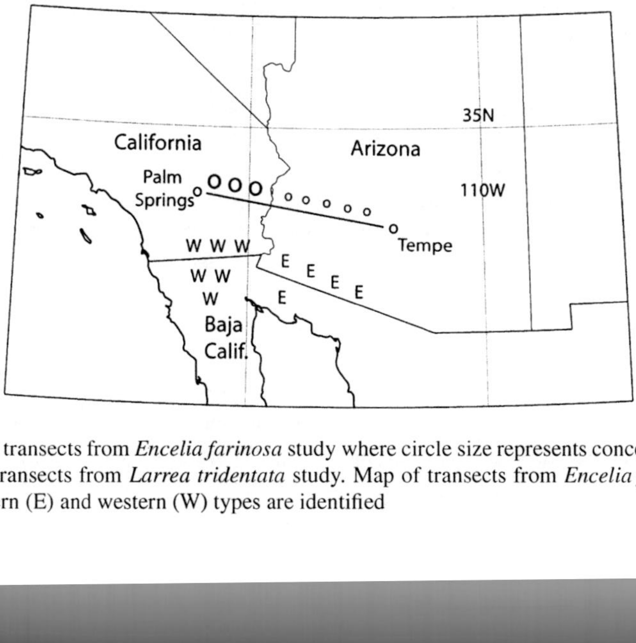

Fig. 2.70 Map of transects from *Encelia farinosa* study where circle size represents concentration differences; and transects from *Larrea tridentata* study. Map of transects from *Encelia farinosa* study where eastern (E) and western (W) types are identified

along the transect with the four Californian populations clearly distinguishable from the Arizona populations. The total combined chromene-benzofuran fraction of the California populations was several times larger than the corresponding fraction in the Arizona populations. With regard to individual compounds, encecalin was the predominant compound in the four California populations, with amounts ranging from about 90 to nearly 150 μmol g^{-1} dry weight. The Palm Springs population differed from all the rest with a concentration of euparin of about 75 μmol/g dry weight, while in all of the other nine populations, concentrations of this compound did not exceed about 25 μmol g^{-1} dry weight. No explanation for these striking differences was offered.

An earlier study of *E. farinosa* by Kyhos (1967) suggested that floral color variation within the species might be linked to environmental factors. *Encelia farinosa* comprises two varieties, *E. farinosa* var. *farinosa* and *E. farinosa* var. *phenicodonta* (S. F. Blake) I. M. Johnston. (A third variety, *E. f.* var. *radicans*, now considered *E. radicans* Brandegee, occurs mainly in the Cape region of Baja California and does not concern us here.) Variety *phenicodonta* occurs in most of Baja California extending around the northern shore of the Gulf of California, in the Colorado River Valley, and in scattered sites at higher elevations in southern California. Variety *farinosa* occurs in much of the northern Sonora Desert. The two varieties are most readily distinguishable on the basis of floral pigmentation: var. *farinosa* is characterized by yellow ray and disk florets, whereas var. *phenicodonta* has brownish-purple disk florets (a color usually associated with anthocyanins). Furthermore, the color differences appear to be inherited as a single genetic dominant. Kyhos (1967) observed differences in disk floret color as he followed a transect from Rice, California, southeastward toward Blythe, California, and then across the Colorado River and east into Arizona. In the vicinity of Rice, only var. *farinosa* was observed. About 26 miles northwest of Blythe, individuals of var. *phenicodonta* began to appear. Over the next mile, their proportion rose to 5% and then to 38% over the next 6 miles. Within about 10 miles of Blythe the frequency of var. *phenicodonta* had risen to 70%. The census could not be sustained in the immediate vicinity of Blythe owing to extensive agricultural development, but just east of the Colorado River the frequency of var. *phenicodonta* was 100%. There then followed a steady decrease so that at the 13–14 mile mark (along US Rt. 60) the frequency had dropped to 7%. Further east only var. *farinosa* was seen. Very similar counts were recorded on other transects (e.g., in the vicinity of Yuma, Arizona). The only apparent exception to the appearance of var. *phenicodonta* along watercourses is its occurrence at several sites at higher elevations where frequencies approaching 20% were recorded. The predominance of var. *phenicodonta* near the coast, along watercourses, and at higher, moister sites in otherwise desert locales strongly suggested a higher moisture requirement for this variety. Kyhos (1967) reasoned that the correlation between higher moisture and frequency of occurrence of var. *phenicodonta* might indicate a sensitivity of this variety to higher temperatures, but no correlation with maximum, minimum, or mean temperatures was evident from a study of local meteorological records. The possibility that different floral colors might be the product of selection for pollinator preference was also tested, but the major pollinator, the beetle *Tanaops abdominalis*, visited flower heads indiscriminately in mixed populations. Thus, the involvement of floral color in the establishment and/or maintenance of the geographic partitioning of the two forms of *Encelia farinosa* remains a mystery.

Interested readers can find information on geographic variation in aboveground architecture and leaf characters of *Encelia farinosa* in response to temperature and available water in Housman et al. (2002), and a discussion of chemical variation and defense against predation in a report by Wisdom (1985).

2.7.18 **Larrea tridentata** *(Zygophyllaceae)*

Later in this review we examine the genus *Larrea* with reference to the origin of the North America–South American disjunction of the genus. Here, we look at variation within one of the components of that system, *Larrea tridentata* (Seese & Moc. ex DC.) Coville, the creosote bush, which occurs widely in the deserts of the southwestern United States, Baja California, and northwestern Mexico in the State of Sonora. The chemistry and biology of "creosote bush" have been reviewed in detail in a volume edited by Mabry et al. (1977). The name derives from the copious amounts of resin that accumulate on aerial parts of the plant, material that is thought to serve in several capacities, including waterproofing, as an ultraviolet shield, and as a feeding deterrent. Downum et al. (1988) undertook a detailed study of this species in order to gain a better idea of the composition of and seasonal dynamics of the exudate. To this end, they collected material over the entire range of the species, including many sites in Sonora and along two transects: (1) one that ran the length of Baja California; and (2) one that ran across southern California and Arizona (Fig. 2.70). Analyses were performed to determine the amount of exudate produced and variation, if any, of the major component, nordihydroguaiaretic acid [229] (NDGA) (see Fig. 2.71). Mean percentages of hexane-soluble and methanol-soluble components varied, but the variation did not correlate with either latitude or longitude. In contrast, the concentration of NDGA correlated reasonably well ($r^2=0.59$) with latitude, with a low mean value of 6.4 mg g^{-1} (dry weight basis) near the southern boundary of the species' range (southern Baja California, 24°N) and a high of 60.2 mg g^{-1} for plants near the northern extent of the range (32°N). Seven collections in a roughly east to west transect across southern Arizona (111°W) and California (115° 35′W) showed a relation between NDGA concentration and longitude, but the r^2 value was much lower (0.22).

2.7.19 **Chrysosplenium glechomaefolium** *(Saxifragaceae)*

Worldwide, *Chrysosplenium* consists of at least 55 species; most occur in the Northern Hemisphere, but the genus is also represented in southern South America where two species are known (Hara, 1957). *Chrysosplenium glechomaefolium* Nutt. occurs

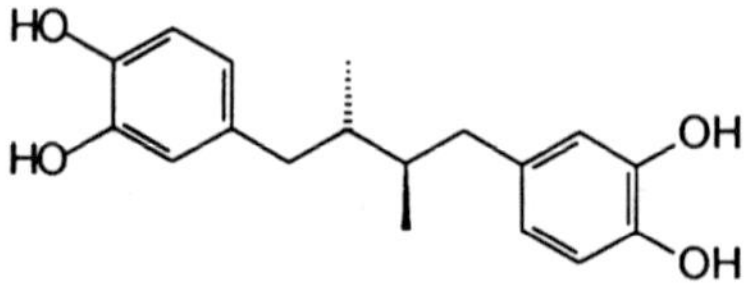

[229]

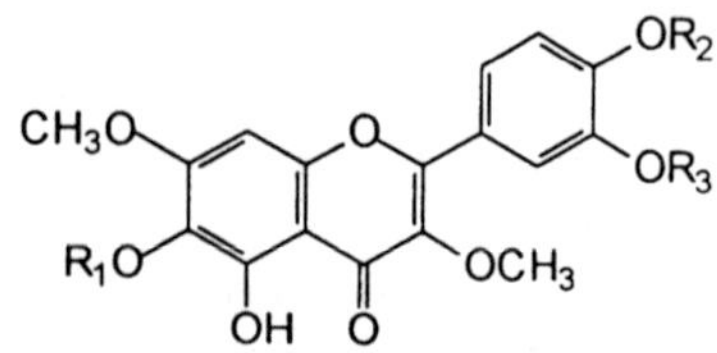

[230] $R_1 = R_2 = R_3 = H$
[231] $R_1 = CH_3$, $R_2 = R_3 = H$
[232] $R_1 = R_3 = H$, $R_2 = CH_3$

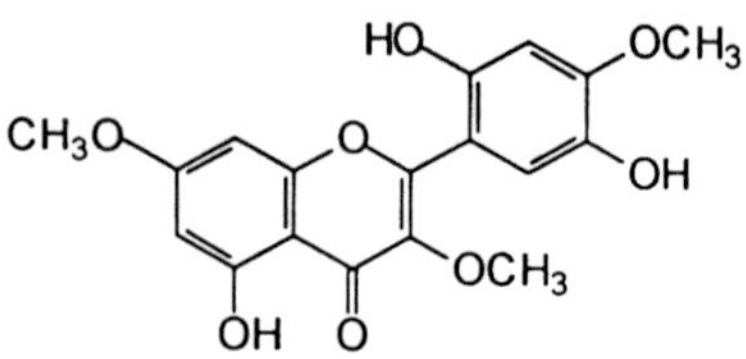

[233]

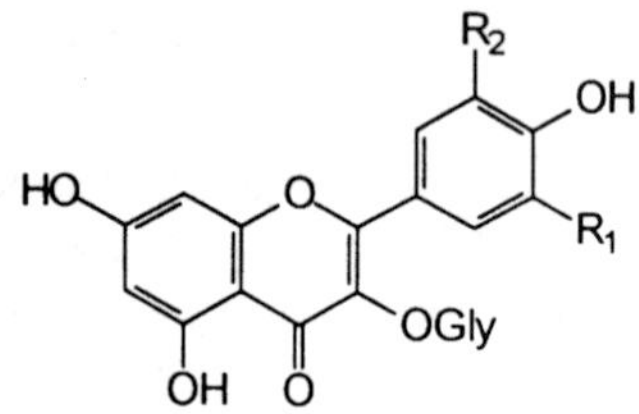

[234] $R_1 = R_2 = H$
[235] $R_1 = OH$, $R_2 = H$
[236] $R_1 = R_2 = OH$

Fig. 2.71 Compound 229 from *Larrea tridentata*. Compounds 230–233, flavonoids from *Chrysosplenium glechomaefolium*. Compounds 234–236, flavonoids from *Quercus rubra*

along the western coast of North America from the Olympic Peninsula of Washington south through Oregon and into northern California. (It was not possible to verify these reports of its occurrence in British Columbia.) It generally occurs not only in wet habitats west of coastal mountain ranges throughout its range, but it is also known from higher elevations in mountains in central Oregon (Mary's Peak, Benton/Lincoln Counties). An examination of flavonoids of specimens representing the range of the species (Bohm and Collins, 1979) revealed several *O*-methylated flavonols the aglycones of which were identified as 5,6,3′,4′-tetrahydroxy-3,7-dimethoxyflavone [230], 5,3′,4′-trihydroxy-3,6,7-trimethoxyflavone [231], 5,6,3′-trihydroxy-3,7,4′-trimethoxyflavone [232], and 5,2′,5′-trihydroxy-3,7,4′-trimethoxyflavone [233] (see Fig. 2.71 for structures 230–233). All of these compounds occurred as glucosides, with the sugar attached at the 2′-position in the last named compound and at 6′ and/or 4′-positions in the others. The observation of note is the presence of the 2′-hydroxy compound in plants from only six of the 13 sites from which collections were made. Plants possessing this flavonoid occurred in the more northerly sites—those from Washington and northern Oregon—whereas, the plants lacking this compound occurred in California and southern Oregon (Fig. 2.72). The two pigment types overlapped in central coastal Oregon, although the populations sampled had either one or the other type. No population was found that had both flavonoid types although the sampling was not extensive enough to place this conclusion on strong ground. All populations sampled had

Fig. 2.72 Map of *Chrysosplenium glechomaefolium* populations. "Yes" populations have the 2-hydroxyflavonol, "no" populations do not. "Yes/No" represents a general area where both types occur (but not mixed)

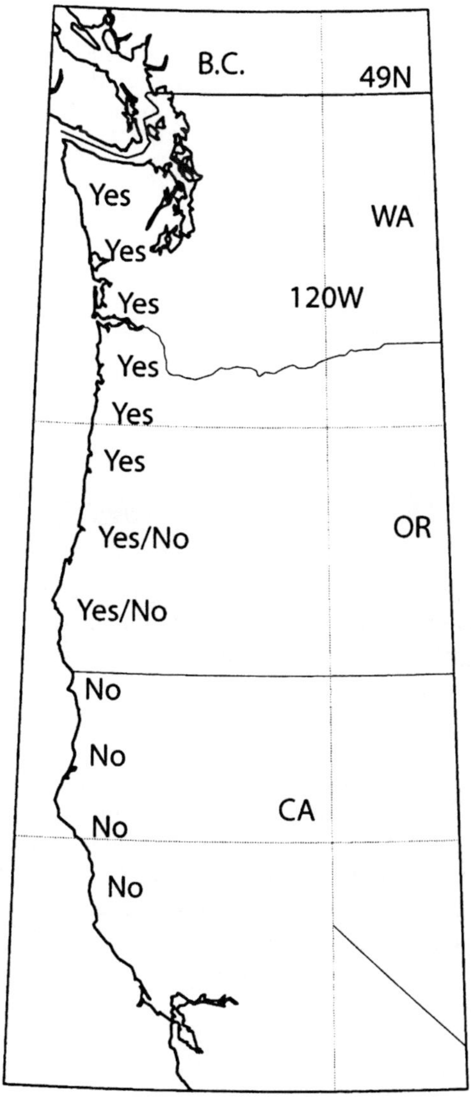

the other three flavones, although the nature of glycosylation varied among them. Other than to point out the obvious correlation with latitude, with the obvious exception of the apparently overlapping populations in the center of the species' range, the authors offered no suggestion as to why this polymorphism might exist. So far we are unaware of any other differences among populations of *C. glechomaefolium* have been reported. It is interesting to speculate, however, that the north–south differentiation parallels that seen in other taxa whose different biochemical profiles have been discussed in terms of population retreat and advance in response to Pleistocene glaciation (Soltis et al., 1997).

Chrysosplenium fits into several phytogeographic categories. In addition to the local infraspecific differentiation just described, it also falls into the category of genera with an eastern Asia–eastern North America–western North America distribution. For example, several members of the opposite-leafed species of *Chrysosplenium* occur in Japan with the group represented in North America by two species, *C. americanum* Schwein. ex Hook., which occurs along the eastern seaboard of North America, and *C. glechomaefolium* the western North American representative. Evolutionary relationships between these two North American species were discussed by Soltis et al. (1993) as part of a study of the genus using DNA sequence data. Included in that study were two species native to extreme southern South America, which will be discussed below under the category of polar disjunctions. Eastern Asian–eastern North American disjunct taxa will also be discussed later in this review.

2.7.20 Quercus rubra *(Fagaceae)*

Quercus rubra L., red oak, is a major component of the deciduous forests of eastern North America, with a range that extends from Ontario south to the Gulf States (Sargent, 1965). Part of the variability of the species has been formalized by the recognition of var. *borealis* (Michaux f.) Farwell to account for trees in the northern part of the species' range, and at higher elevations in the southern Appalachians. The two varieties, *rubra* and *borealis*, are thought to intergrade between 1200 m and 1350 m in the southern Appalachians.

An examination of the flavonoids of red oak by McDougal and Parks (1984) revealed altitudinal differences in the flavonol profiles. One chemotype, characterized by the presence of kaempferol [234] and quercetin 3-*O*-glycosides [235], occurred in trees from lower elevations, whereas in trees from higher elevations myricetin 3-*O*-glycosides [236] (see Fig. 2.71 for structures 234–236) were added to the array. Mixed populations were observed between 600 m and 1050 m, somewhat lower than the elevational range of overlap of the morphological varieties known to grow between 1200 m and 1350 m (Sargent, 1965). The influence of environmental and genetic factors on the expression of flavonols in red oak was the subject of a second paper by McDougal and Parks (1986). Acorns were collected from eight sites representing an elevational gradient of 1455 m, divided equally, and planted in nursery plots at 75–1140 m elevation. After 5 years, leaf samples were analyzed for flavonoids qualitatively and quantitatively. Regardless of the elevation of the nursery, seedlings originating from high-elevation myricetin-positive trees retained their capacity to synthesize myricetin, while seedlings from low-elevation nonmyricetin stock continued to produce only kaempferol and quercetin derivatives. Quantitative analyses indicated that the total amounts of flavonoids varied somewhat with elevation, but the array of compounds produced was not affected. Although the flavonoid chemotypes correspond to the two varieties of red oak in the southern Appalachians, the zone of overlap of chemical and morphological characters does not agree.

The authors suggested that extrapolation beyond the specific area studied was not warranted. It would be of interest to see what DNA sequence data might have to say about red oak throughout its range.

2.7.21 Cnidoscolus spinosus *(Euphorbiaceae)*

Cnidoscolus is a genus of about 75 New World species, some of which have useful properties such as *C. chayamansa* McVaugh, eaten as a spinach-like vegetable (chaya), and *C. elastica* Lundell, which is true to its name in yielding a latex rich in rubber (Mabberley, 1997, p. 166). The species of interest to us in this chapter is *C. spinosus* Lundell, native to the western coast of Mexico. Kolterman et al. (1984) examined the flavonoid chemistry of this and related species from Mexico and Central America. Eight populations were sampled, three from the northwestern part of the range (ca. 22°N, vicinity of Tuxpan, State of Nayarit), two from the southeastern part of the range (ca. 17°N, vicinity of Acapulco, State of Guerrero), and three from roughly midway between the other two (ca. 19°N, vicinity of Manzanillo, State of Colima) (Fig. 2.73). The flavonoid profiles consisted of 3-*O*-glycosides of kaempferol and quercetin and, interestingly from the chemosystematic point of view, lacked the C-glycosylflavones reported from the other two taxa. Significant differences were observed among the three populations as summarized in Table 2.21. The populations

Fig. 2.73 Map of *Cnidosculus* sites

Table 2.21 Distribution of kaempferol and quercetin glycosides in *Cnidoscolus spinosus* (after Kolterman et al., 1970)

Population	No.[b]	Kaempferol glycosides					Q-Glys[a]	
		GAL	GLC	RHM	RGR	RUT	RHM	RUT
Northwestern	3	0[c]	0	3	3	3	1	0
Central	3	2	2	1	2	2	1	1
Southeastern	2	0	0	0	2	0	1	2

[a] Q = quercetin; Glys = glycosides; GAL = galactoside; GLC = glucoside; RHM = rhamnoside; RGR = 3-*O*-rhamnosylgalactoside; RUT = rutinoside.
[b] Number of populations sampled.
[c] Number of populations exhibiting the compound.

in the middle of the range possessed all of the compounds observed in the species, with reduction in numbers in both the southeastern and northwestern populations. The authors suggested that range expansion in both directions from ancestral populations in the middle of the present distribution could account for the observed profiles. Sample size is a problem in this study, as noted by the authors.

2.7.22 Cupressus *(Cupressaceae)*

Zavarin et al. (1967b) described the occurrence of tropolone derivatives, seven-membered ring compounds, in species representing both the Old World and New World members of *Cupressus*, the true cypresses. Owing to the difficulty of acquiring samples of these heartwood constituents, coupled with difficulties in purification and analysis, sampling was very limited. Nonetheless, sufficient differences between populations were noted to suggest that a wider sampling using more conservative sampling methods, allied with more sensitive analytical techniques, might provide useful insights into relationships within the genus. Two examples serve to illustrate the variety of structures involved, and hence biosynthetic step differences between populations. *Cupressus sargentii* Jepson occurs from Santa Barbara County in the south to Mendocino County in the north (Fig. 2.74) and is felt by some workers to exist in two forms, one coastal and one from inland. The inland form was shown to accumulate four compounds, nootkatin [237] as a major component (see Fig. 2.75 for structures 237–241), β-thujaplicin [238] and β-thujaplicinol [239] as lesser components, and β-dolabrin [240] as a trace component. The coastal form exhibited the simpler profile, in that it lacks β-thujaplicinol and β-dolabrin. The two forms share the capacity to make both the basic terpene-derived tropolones, as well as the sesquiterpene-derived tropolone nootkatin, but differ in the oxidation level of the tropolone ring.

Two collections of *C. arizonica* Greene also exhibited differences involving the same set of compounds: a sample from Mexico afforded β-thujaplicin as a minor component and nootkatin, whereas a specimen from the Chiricahua Mountains in

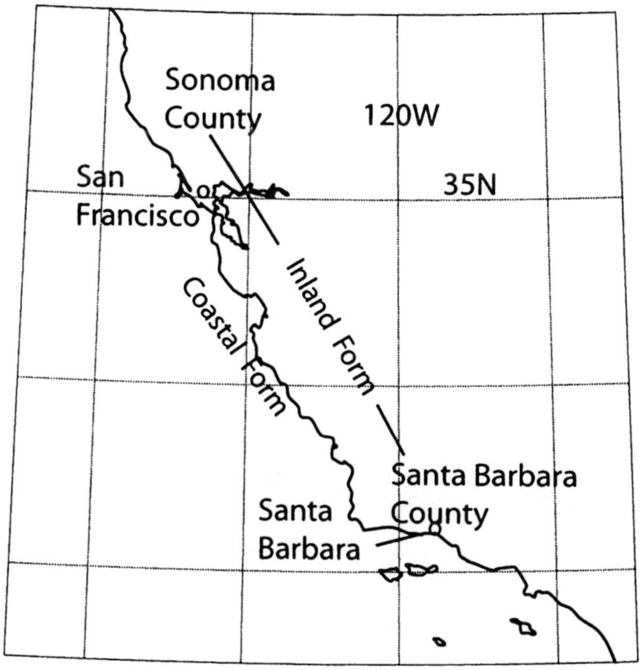

Fig. 2.74 Map of *Cupressus* sites

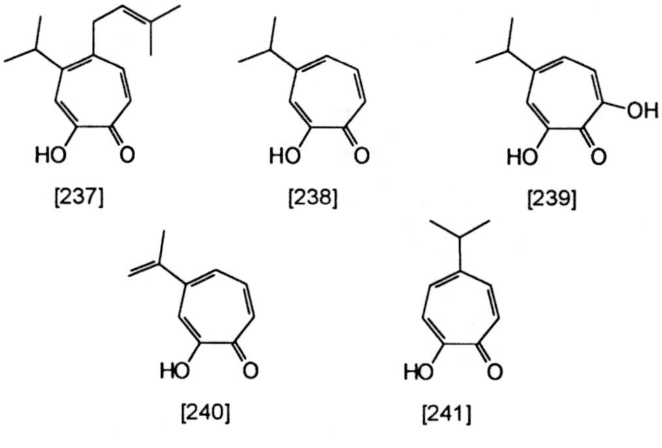

Fig. 2.75 Compounds 237–241, tropolone derivatives from *Cupressus*

Arizona yielded β-thujaplicin and nootkatin as major components along with lesser amounts of γ-thujaplicin [241] and β-thujaplicinol. At a higher taxonomic level, two subspecies of *C. bakeri* Jepson (subsp. *bakeri* and subsp. *mattewsii* Wolf), gave identical profiles consisting of β-thujaplicin and nootkatin.

2.7.23 Mimulus aurantiacus *(Scrophulariaceae)*

Mimulus aurantiacus Curtis (syn. *Diplacus aurantiacus* Jepson), sticky monkey-flower, is one of a group of woody *Mimulus* species native to southern California characterized by the presence of copious amounts of sticky exudate on leaf surfaces (<30% dry weight). Among the constituents of the exudate are several C-geranyl flavonoids, for example, [242 and 243], and the unusual geranyl pyrone [244] (Lincoln, 1980; Wollenweber et al., 1989; Phillips et al., 1996) (see Fig. 2.76 for structures 242–244). Recently, Hare (2002) tested the hypotheses that populations in more arid settings have relatively higher concentrations of the more highly O-methylated flavonoids, and that populations attacked by the lepidopteran *Euphydras chalcedona* should have lower levels of total exudate and lower concentrations of *ortho*-dihydroxy flavonoids. Although common garden studies revealed that chemical variation is under genetic control, both hypotheses were rejected. Populations nearer the coast were shown to have higher levels of exudate than inland populations from the arid sites.

2.7.24 Cladonia cervicornis *(Cladoniaceae)*

A study of *Cladonia cervicornis* (Achar.) Flotow from two widely disjunct populations in California, one from coastal Mendocino County and one from the foothills of the Sierra Nevada in Amador County, revealed a difference in the frequency of occurrence of the depside atranorin [245] (see Fig. 2.76) (Hammer and Ahti, 1990). The compound, considered to be a primitive feature of this lichen species, was present in 80% of the

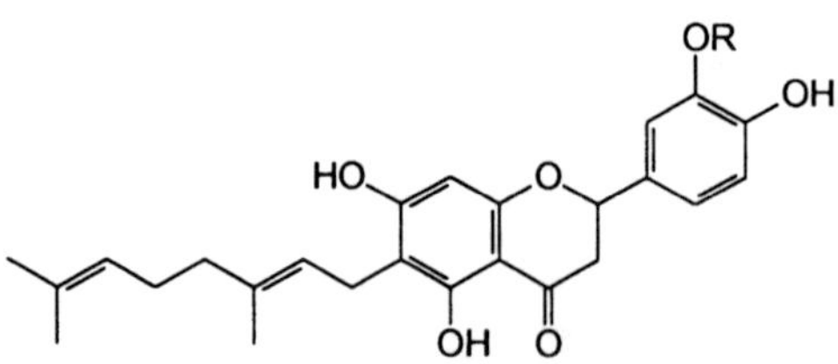

[242] R = H
[243] R = CH_3

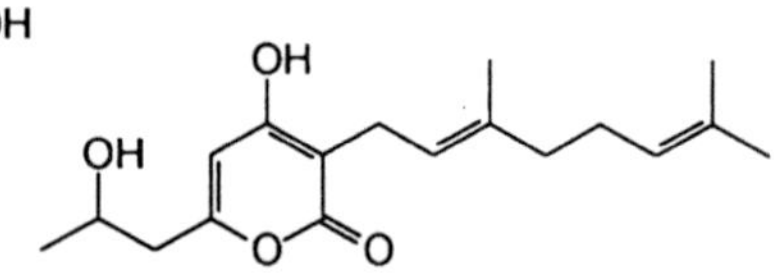

[244]

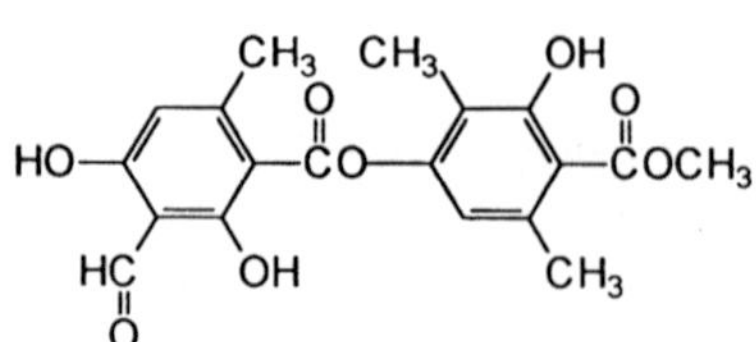

[245]

Fig. 2.76 Compounds 242–244, exudate constituents from *Mimulus* (*Diplacus*) *aurantiacus*. Compound 245, a depside from *Cladonia cervicornis*

specimens obtained from the Amador County site and only about 10% of those collected in Mendocino County. Populations at these sites are considered to be relictual and are thought to have been separated from one another for a considerable period of time.

2.7.25 Echinacea angustifolia *(Asteraceae)*

Echinacea angustifolia DC., one of the most sought after medicinal herbs in North America, occurs in a variety of habitats ranging across much of the Great Planes of the United States north to the Canadian prairies, east to the Appalachian uplands and to the southeastern coastal plains. The species grows in a variety of habitats and exhibits a range of morphological forms. A qualitative and quantitative study of secondary products, with a view to evaluating the phytochemical diversity, was undertaken by Binns et al. (2002) who sampled nine populations, five in Oklahoma, two in Kansas, and one each in Nebraska and Iowa, with an overall latitudinal range of 34.368–42.917N. Between 8 and 17 individual plants were taken from each population. Roots of *Echinacea* are a rich source of caffeic-acid derivatives, long-chain unsaturated amides, and long-chain unsaturated ketones. Representative structures of some of the 28 compounds identified in this study are illustrated below (Fig. 2.77). The family of amides are here represented by structure [246],

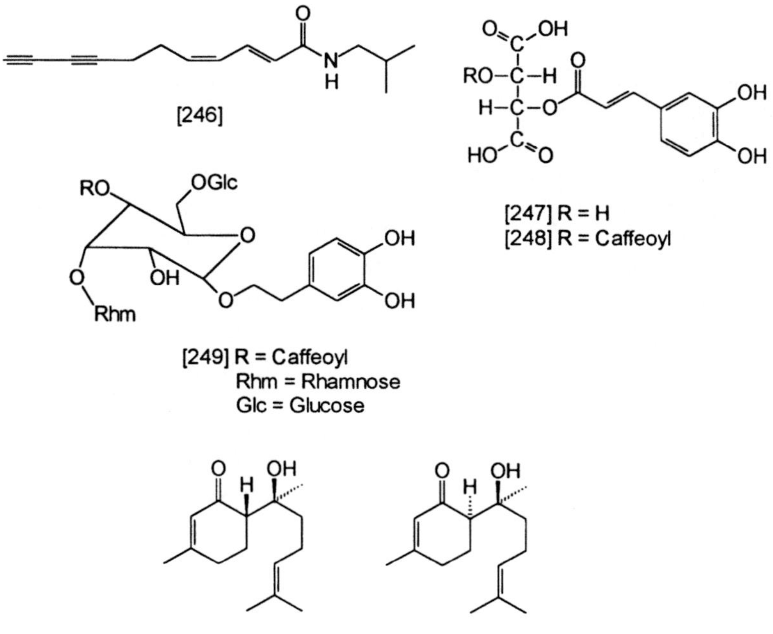

[246]

[247] R = H
[248] R = Caffeoyl

[249] R = Caffeoyl
Rhm = Rhamnose
Glc = Glucose

[250] [251]

Fig. 2.77 Compounds 246–249, caffeic-acid derivatives from *Echinacea angustifolia*. Compounds 250 and 251, sweet principals from *Lippia dulcis*

undeca-2*E*,4*Z*-diene-8,10-diynoic acid isobutylamide. Other members of this group differ in length of chain, degree of unsaturation, and constitution of the amine function. Caffeic-acid derivatives include the widely distributed chlorogenic acid (3-caffeoylquinic acid), cynarin (dicaffeoyl quinic acid), caftaric acid [247], and cichoric acid [248], which are mono- and diesters of tartaric acid and caffeic acid, respectively, and echinacoside [249].

In addition to chemotype patterning revealed by statistical analysis of the data, latitudinal relationships emerged as well. Concentrations of cichoric acid and echinacoside and one of the amides, undeca-2*Z*,4*Z*-diene-8,10-diynoic acid 2-methylbutyl amide [246], with a larger amine-bearing group increased with latitude. The inverse relationship was observed for caftaric acid and hexadeca-2*E*,9*Z*-diene-12,14-diynoic acid isobutylamide. The latitudinal differences may be of importance in helping to identify populations for selection of seed for propagation.

2.7.26 Lippia dulcis *(Verbenaceae)*

The last example in this section features *Lippia dulcis* Trev., whose range extends from southern Mexico to Panama and Colombia and on the Caribbean islands of Puerto Rico, Cuba, and Hispaniola, among others. This example underlines a fundamental question that must always be borne in mind: is the taxonomy of the subject species correct? The plant has long been used in traditional medicine for a variety of illnesses and is characterized by having intensely sweet-tasting leaves and flowers, and a highly aromatic scent. Earlier work by Compadre et al. (1985) described identification of the bisabolene sesquiterpenoid hernandulin as the sweet principle of plants collected in Mexico (known locally as "hierba dulce"). Subsequent studies from the same laboratory (Compadre et al., 1985, 1987) resulted in contradictory findings, in that some specimens had very little hernandulin, while others, in fact, had substantial quantities of camphor (up to 53%) who's bitter taste is not at all in accord with the various names by which the plant is known, as well as accounts in older literature that the plant is "sweeter than honey" (Souto-Bachiller et al., 1997). These last named workers subjected plant material from Puerto Rico to detailed chemical analysis with particular emphasis on identifying the camphoraceous component. (+)-Hernandulcin [250] and its epimer (−)-*epi*-hernandulcin [251] (see Fig. 2.77) were found to be the major components of the herb. The level of camphor, if present, was less than 0.01%. In addition to the hernandulcin epimers, numerous other volatile components of the Puerto Rican material were detected, almost none of which were detected in extracts of the Mexican material analyzed under identical conditions. Moreover, those compounds that are held in common are present in often very different proportions. Differences in summed volatiles are also significantly different: 8.14% total monoterpenoids in Puerto Rican material opposed to 86.29% in Mexican and 79.52% versus 6.00% total sesquiterpenoids, respectively. Differences in total volatile fractions were of the same magnitude.

The authors discounted the existence of two chemotypes of *L. dulcis* on the grounds that no evidence appears to exist as to this level of plasticity in this species. Rather, they suggested that the problem likely arose in the source of the Mexican specimen, which consisted of plants collected in the vicinity of Tlayacapan (Morelos) and mixed with "hierba dulce" purchased in a market in Mexico City. Adulteration of the commercial product, not an uncommon practice, likely accounts for the striking chemical differences between the two specimens.

Chapter 3
After the Ice

According to Pielou (1991), the most significant recent event to have had major influence on North American vegetation was Pleistocene glaciation. One fears little argument in expanding this generalization to include the entire globe! Regardless of one's home continent, advance of the ice front removed all vegetation in its path. Although we tend to focus our attention on the impact of glaciation on the local biota (especially those of us who now live where the ice lay 2 km deep!) and upon likely refugia, we should not lose sight of the fact that glaciation was, to a greater or lesser extent, a global phenomenon. The eventual retreat of the ice fronts once again released the land, and into the newly available sites the plants returned. Exhaustive study of revegetation following retreat of the glaciers has resulted in a vast literature on all aspects of the subject: faunal, floral, geological, and meteorological. The global impact was enormous. In fact, it seems likely that most plant species in the world experienced some of the effects of glaciation, ranging from the severe extirpation to something as comparatively minor as a change in local climate. It is interesting to speculate on whether there were any taxa not affected by glaciation: likely very few.

In this section, we examine a selection of plants whose ancestors survived the ice in their respective refugia, subsequently produced offspring who returned to their original homeland (more or less), and established the patterns of vegetation we are familiar with today. Examples will be met in the following paragraphs that address some combination of the following questions: (1) Where did the surviving ancestors spend the glacial years? (2) How long did it take for them to establish their current ranges? and (3) What were the routes followed? We will also see examples of "nunataks," a term used to describe ice-free areas—islands in a sea of ice—that might have existed in more northerly (or southerly, for that matter) latitudes during the time when the mass of ice extended farther south (or north). [See Fernald (1925) for an historical view of ice-free areas in Boreal America.]

We are concerned here with systems that have been studied using secondary products—flavonoids and terpenoids in particular—but other information, including micro- and macrofossils, and occasionally chromosome numbers, will be included in the discussions when such information is available. The majority of current research on postglacial reestablishment of plant distribution patterns is based on DNA sequence information. In a few instances below, reference will be made to such information, but this in not the place for a detailed review of that literature.

B.A. Bohm, *The Geography of Phytochemical Races*,

3.1 North America

3.1.1 Luetkea pectinata *(Rosaceae)*

Luetkea pectinata (Pursh) Kuntze, the sole member of the genus, is a small, trailing species that grows in moist or shaded, primarily subalpine areas along the Pacific Coast of North America ranging from Alaska to northern California, in the Cascades, and in the Rocky Mountains from southern Alberta and British Columbia to eastern Idaho and western Montana. Identification of a rare tricetin derivative in a specimen of *L. pectinata* from southern British Columbia (Wells and Bohm, 1988) prompted wider sampling of the species throughout its entire range. The unusual compound, identified as tricetin 3′-*O*-glucoside [252] (see Fig. 3.1 for structures 252–255), was subsequently seen in plants from all 27 sites visited (ranging from the Kenai Peninsula in Alaska to Lassen National Forest in California). Quercetin 3-*O*-mono- and diglycosides [253] were similarly ubiquitous, as was luteolin 7-*O*-glucoside [254]. The flavanone eriodictyol 7-*O*-glucoside [255] was observed in plants from all but one of the sites. The only clear-cut difference that correlated with latitude was the increased concentration of luteolin 7-*O*-glucoside in the four most southerly populations studied (one from Washington, two from Oregon, and one from Lassen National Forest, California). An acylated derivative of quercetin 3-*O*-glucoside (acyl function not identified) occurred sporadically in the northern part of the range, but was consistently observed in specimens from southern British Columbia southward, with increased amounts in the two most southerly locations, Crater Lake (Oregon) and Lassen National Forest. An acylated derivative of luteolin 7-*O*-glucoside (acyl function not identified) was seen in scattered populations in the northern and central parts of the species' range but occurred as a major component in the two most southerly sites. In contrast, the amount of eriodictyol 7-*O*-glucoside present in the two most southerly sites was lower than seen in plants from any of the other sites. It is difficult to assign any great significance to these observations as they stand, but the trends at the northern and southern extremes do suggest that a more detailed analysis using one or more of the macromolecular techniques now available might be worth the effort.

3.1.2 Tolmiea menziesii *(Saxifragaceae)*

Tolmiea menziesii (Hook.) Torrey and A. Gray, the sole member of the genus, occurs from northwestern California through western Oregon, Washington, British Columbia, and into Alaska. The taxon gained a measure of recognition through the work of Doug Soltis and associates, which established it as one of the clearest examples of autopolyploidy in nature (Soltis, 1984; Soltis et al., 1989). The distribution of plants with different ploidy levels was also of interest in that northern populations, those from Alaska, British Columbia, Washington, and northern Oregon, were tetraploid

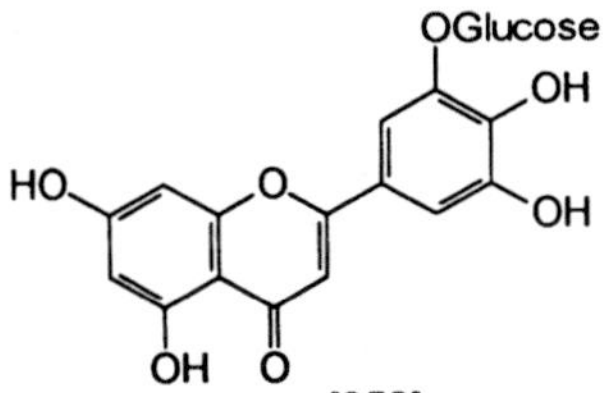

[252]

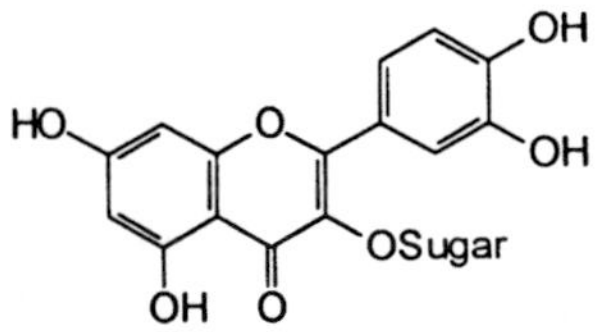

[253]

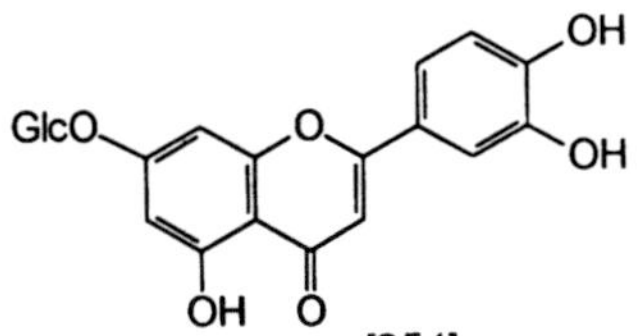

[254]

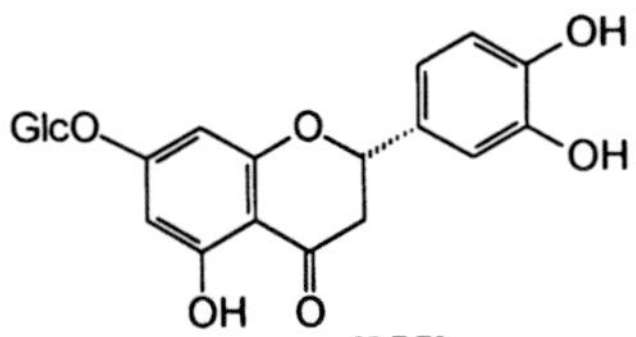

[255]

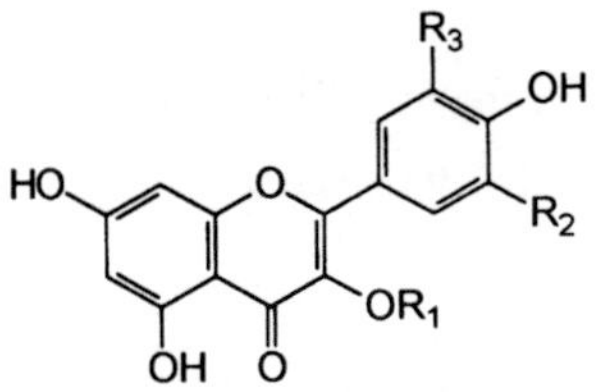

[256] R_1 = Sugar, $R_2 = R_3$ = H
[257] R_1 = Sugar, $R_2 = OCH_3$, R_3 = H
[258] R_1 = Sugar, $R_2 = R_3$ = OH

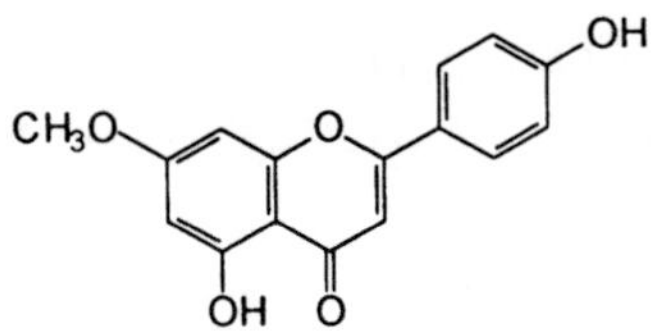

[259]

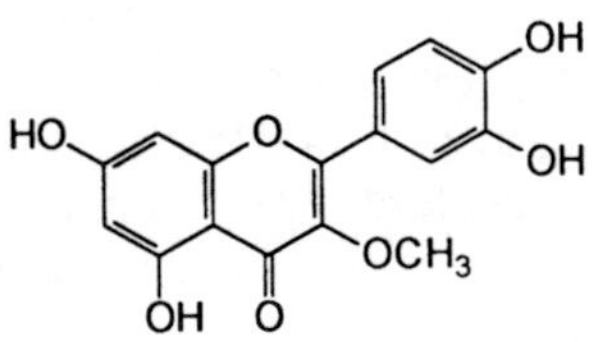

[260]

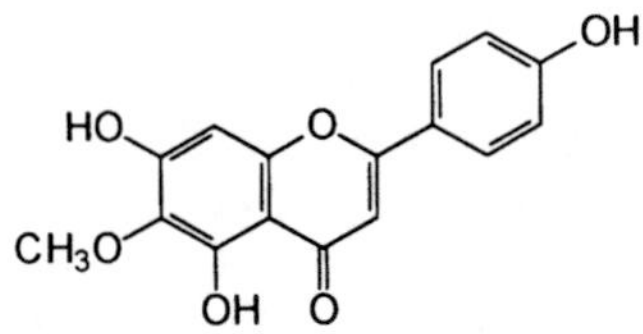

[261]

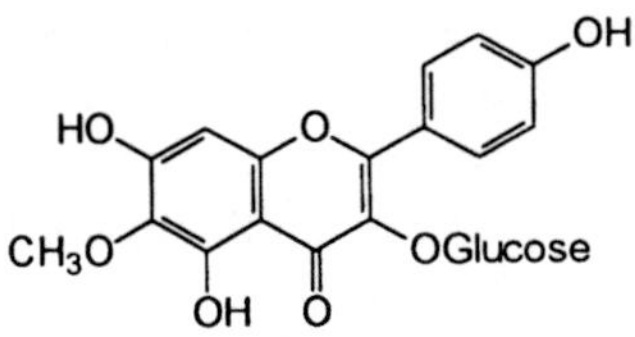

[262]

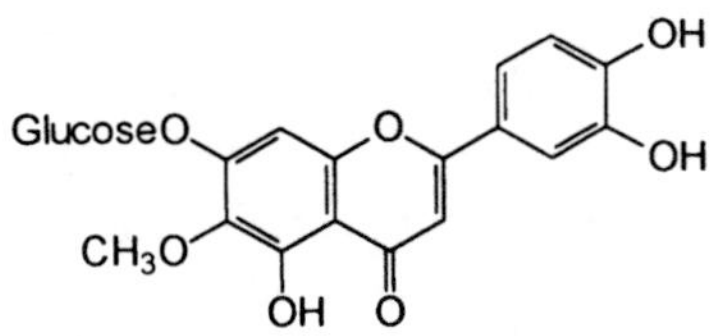

[263]

Fig. 3.1 Compounds 252–255, flavonoids from *Luetkea pectinata*. Compounds 256–258, flavonoids from *Tolmeia menziesii*. Compounds 259–263, flavonoids from *Arnica cordifolia*

(x=7, 2n=28), while those from the southern part of the range, northern California, and southern Oregon, were diploid (2n=14). The flavonoids of *T. menziesii* were first reported in 1979 (Bohm, 1979) before the existence of two cytotypes had been reported. Owing to the limited sampling in that work (six populations), and the more recent discovery of autopolyploidy, it was decided to reinvestigate the situation.

Specimens from 29 populations, representing the entire range of the species, were analyzed and found to exhibit extremely similar profiles (Soltis and Bohm, 1986). Kaempferol [256], quercetin [253], isorhamnetin [257], and myricetin [258] 3-*O*-mono- and 3-*O*-diglycosides were observed in all specimens with variation limited to 3-*O*-triglycosides (see Fig. 3.1 for structures 256–258). Since the triglycosides were present in much smaller amounts than the other glycoside classes and limited amounts of plant material were used in the analyses, it was not possible to say with certainty that triglycosides were absent (visual inspection of two-dimensional thin-layer chromatograms). With that caveat in mind, it was concluded that there is no distinction between the cytotypes based upon flavonoid profile, which is what one would anticipate in an autotetraploid system.

3.1.3 **Arnica** ***(Asteraceae), a Note on*** **Sedum** ***(Crassulaceae), and a Comment on*** **Antennaria** ***(Asteraceae)***

The first species of *Arnica* that we examine belong to subgenus *Austromontana* Maguire, a North American group that consists of nine species (Maquire, 1943). Members of the group occur over a large area in western North America, although some of its constituent species have very limited ranges. *Arnica cordifolia* Hook. and its derivative species *A. latifolia* Bong. are considered ancestral and occupy the widest ranges within the subgenus. The first detailed study of the flavonoids of the group focused on *A. cordifolia*, an apomictic polyploid complex that extends from the Yukon Territory to central California and northern New Mexico, with disjunct populations in Ontario and Michigan. Wolf and Denford (1983) sampled 33 populations representing all five known chromosome races with numbers ranging from $2n=38$ (diploids) to $2n=114$ hexaploids). Tetraploids are widespread and triploids occur along the front ranges of the Rocky Mountains, while diploids, pentaploids, and hexaploids are rare and are widely scattered. There was no correlation of flavonoid profile with chromosome number, reminiscent of the findings of Glennie et al. (1971) on the *Senecio radicans* complex in Africa and adjacent regions where eight chromosome races were involved. Similarly, diploid and tetraploid races of *Lasthenia californica* (Asteraceae) show no relationship with flavonoid profiles throughout the range of the species (Desrochers and Bohm, 1993).

Quercetin 3-*O*-gentiobioside and quercetin 3-*O*-diglucoside [233-*O*-Glycs) were, with a single exception, ubiquitous in the 33 populations of *A. cordifolia.* Apigenin 7-methyl ether (genkwanin) [259] (see Fig. 3.1 for structures 259–263) was seen in the one population collected in California, and quercetin 3-methyl ether [260] was seen in two populations, one in British Columbia and one in Oregon. 6-Methoxyapigenin (hispidulin) [261], 6-methoxykaempferol 3-*O*-glucoside [262], and the 3-*O*-glucosides of kaempferol and quercetin were all more or less randomly distributed throughout the populations. Two compounds, however, showed regionalism in their distribution, luteolin 7-*O*-glucoside [254] and 6-methoxyluteolin 7-*O*-glucoside [263]. Luteolin

7-*O*-glucoside occurred only in populations in the northern part of the species' range, that is, in populations from Alberta, British Columbia, Washington, and the Yukon Territory. An apparent anomaly was pointed out by Wolf and Denford, namely, that the population from Ontario, which would have been expected to mirror other northern populations, did not exhibit luteolin 7-*O*-glucoside, although this compound had been reported in floral tissue from plants from this general area (Borokowski et al., 1966). 6-Methoxyluteolin 7-*O*-glucoside was observed in fewer populations than the previous compound, but their ranges did overlap. This compound also was observed in one of two Montana populations sampled.

The authors suggested that the present flavonoid profiles represent, for the most part, the result of depletion of the ancestral profile through the action of founder effect, genetic drift, and a reduction of "phenotypes-chemotypes" as a result of apomictic reproductive isolation. The existence of two different arrays with respect to luteolin 7-*O*-glucoside and 6-methoxyluteolin 7-*O*-glucoside was taken to indicate the possible existence of two different refugia during Pleistocene glaciation, although the possibility of other explanations was not ruled out. Other possible effects of glaciation will be encountered below when we examine the remaining species in subgenus *Austromontana*.

Examination of all members of subgenus *Austromontana* for flavonoids (Wolf and Denford, 1984a) and morphological features, followed by multivariate statistical analyses (Wolf and Whitkus, 1987), confirmed the close relationship among the species and supported the view that *A. cordifolia* and *A. latifolia* were ancestral types. Differentiation of the remaining species seems to have occurred primarily in the Klamath region of northwestern California and southwestern Oregon, from which area dispersal has occurred (Fig. 3.2). A quick look at the range of each species will set the scene for speculation as to how the present overall pattern became established. *Arnica cordifolia* is a very wide-ranging species, as noted in the paragraphs above. *Arnica latifolia*, which is considered to have originated from *A. cordifolia,* is also widely distributed, ranging from Alaska through Colorado and northern California. It is comprised of mostly diploid populations. Chemical, cytological, and morphological information led to the conclusion that these two taxa represent a progenitor-daughter species pair. Additional support for the close relationship between them comes from the observation that they have hybridized to produce *A. gracilis* Rydb. (Wolf and Denford, 1984b).

Arnica discoidea Benth. is of more limited occurrence than either *A. cordifolia* or *A. latifolia* with a range that includes the Californian Coast Ranges, foothills of the Sierra Nevada, and in the Cascades as far north as southern Washington. Diploid, triploid, and tetraploid races have been reported with the diploids largely restricted to the Klamath area. Although the flavonoid profile of *A. discoidea* resembles that of *A. cordifolia* quite closely, some interesting geographic patterns exist. For example, luteolin 7-*O*-glucoside and its 6-methoxy derivative, known otherwise only from *A. cordifolia*, occur only in populations of *A. discoidea* in the Klamath area. Populations in the Klamath area in general exhibit greater flavonoid diversity than populations that occur farther to the east and north. In addition to simpler flavonoid profiles in the non-Klamath populations, plants in these areas tend to be

Fig. 3.2 Map of northwestern California and southwestern Oregon showing the Klamath area. Arrows represent general dispersals from the refugial area

characterized by higher ploidy levels (3*n* and 4*n*). These observations are consistent with the origin of *A. discoidea* from *A. cordifolia* in the Klamath region, with subsequent radiation accompanied by polyploidization and simplification of flavonoid profile.

Arnica spathulata Greene is a largely diploid, serpentine endemic of the Klamath region thought by Wolf and Denford (1984a) to have arisen from *A. discoidea*. The flavonoid profile of *A. spathulata*, which is a subset of the *A. discoidea* profile, is concordant with that origin. The much more uncommon tetraploid form of this species has a simpler flavonoid profile, again suggesting that radiation from the center of origin of these taxa is accompanied by an increase in chromosome number and flavonoid-profile simplification. *Arnica venosa* H. M. Hall, also a diploid, is the most restricted species in this set, with a range that includes arid sites in western Shasta County and eastern Trinity County, California. The flavonoids identified from this species represent a subset of the profile of *A. discoidea*, from which it is thought to have arisen. Possible hybrids between *A. discoidea* and *A. venosa* have been found to occur naturally suggesting a close relation between the two. A third rare, serpentine endemic in the Klamath area is the diploid *A. cernua* Howell. Similar floral and leaf morphology suggest that *A. cernua* may have been derived from *A. cordifolia.* The flavonoid profile of *A. cernua,* which consists of three compounds, kaempferol 3-*O*-glucoside, quercetin 3-*O*-diglucoside, and quercetin 3-*O*-gentiobioside, represents one of the most depauperate profiles in the subgenus. An even more reduced profile (only two compounds) has been reported from a rare diploid population of *A. cordifolia* lending support to the possibility of a close relationship between the two.

Arnica viscosa A. Gray, a diploid, is the rarest species in subgenus *Austromontana.* It is known from only seven localities and appears to be restricted to volcanic soil at high elevations. The flavonoid profile of this species is complicated by the presence of several unique compounds: 5,6,7,3′-tetrahydroxy-4′-methoxyflavone [264]; 5,4′-dihydroxy-6,7,3′-trimethoxyflavone (cirsilineol) [265]; and 3,5,7,4′-tetrahydroxy-6,3′-dimethoxyflavone (quercetagetin 6,3′-dimethyl ether, spinacetin) [266] (see Fig. 3.3 for structures 264–269). The structural features characterizing these compounds, extra hydroxylation, and a moderate level of O-methylation, along with the presence of compounds seen in several other species of subgenus *Austromontana*, makes it difficult to do more than speculate on phylogenetic relationships, although the authors (Wolf and Denford, 1984a) tend toward *A. latifolia* as the most likely candidate. They conclude that it is probably the most recently derived species in the subgenus occupying habitats that are less than 14,000 years old.

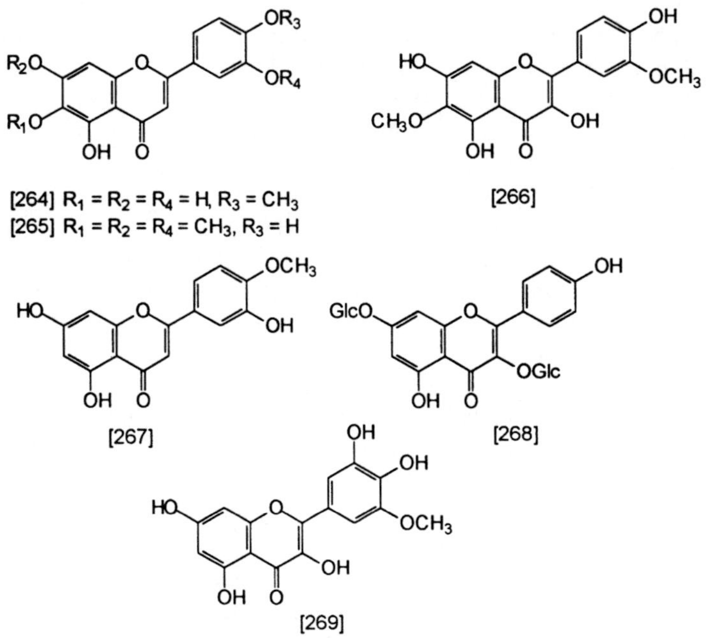

[264] $R_1 = R_2 = R_4 = H, R_3 = CH_3$
[265] $R_1 = R_2 = R_4 = CH_3, R_3 = H$
[266]
[267]
[268]
[269]

Fig. 3.3 Compounds 264–269, flavonoids from *Arnica viscosa*

The last species in this group, *Arnica nevadensis* A. Gray, is interesting in that it exhibits pappus features characteristic of other subgenera, in addition to features suggesting that it has been derived from *A. cordifolia.* Flavonoid evidence supporting its placement in subgenus *Austromontana* comes from the presence of a suite of quercetin glycosides common to most other species. Luteolin 4′-methyl ether (diosmetin) [267] and 6-methoxyluteolin (nepetin), the aglycone of [253], however, are unique to *A. nevadensis.* Sampling was not sufficient to establish whether these compounds are characteristic of the Klamath populations.

The general conclusion drawn by Wolf and Denford (1984a) concerning subgenus *Austromontana* is that *A. cordifolia*, with its more primitive morphology, cytology, wide geographical distribution, and flavonoid profile, represents the ancestral species. Differentiation into the other species is thought to have occurred primarily in the Klamath area resulting in the establishment of four local, edaphically specialized endemics (*A. spathulata, A. venosa, A. cernua,* and *A. viscosa*) and three other species whose ranges have expanded to greater or lesser degrees.

A separate example of a group of diploid species centered in the Klamath region, with apparent radiation from the center involves *Sedum* section *Gormania* (Crassulaceae). The section consists of 12 taxa that occur in western Oregon and northern California. A study of the flavonoids of all members of the section (Denton and Kerwin, 1980; see also Denton, 1979) revealed the presence of three compounds, two of which were identified as a kaempferol 3,7-di-*O*-glycoside [268] and 3,5,7,3′,4′-pentahydroxy-5′-methoxyflavone (laricytrin) [269]. The third compound was identified only as an *O*-methylated flavone derivative. All 12 taxa exhibited the first two compounds, but the presence of the flavone derivative allowed definition of two geographical groups with only slightly overlapping ranges. Six species, *S. laxum* (Britton) Berger (four subspecies), *S. moranii*, R. T. Clausen, *S. oblanceolatum* R. T. Clausen, and *S. oregonense* (Wats.) Peck identified as group **a**, occur in the Cascade Mountains of Oregon and in the Klamath Mountains of Northwestern California. *Sedum albormarginatum* R.T. Clausen and *S. obtusatum* A. Gray (four subspecies), identified as group **b**, occur primarily in the Sierra Nevada Mountains in California with only a small overlap into the Klamath area. The authors suggested that the capacity to make the flavone was lost with migration from a center of origin in the Klamath Mountains (group **a** taxa) eastward and southward. There appears to be no relationship between flavonoid chemistry and ploidy level (diploids and polyploids are known in each group).

We can return now to the observations on the last group of *Arnica* species, the edaphic specialists. Wolf and Denford's (1984a) observations are in general agreement with hypotheses concerning endemism in *Parthenium* (Mears, 1980b). The flavonoid chemistry of all North American members of *Parthenium* (Asteraceae) was studied by Mears (1980a), who recorded the occurrence of 34 compounds that were about equally divided between aglycones and glycosides. Species enjoying wide ranges of occurrence on a variety of noncalcareous substrates tend to accumulate larger arrays of compounds, owing primarily to glycoside diversity. Species limited to limestone or gypsum-based substrates, on the other hand, tend to have much simpler pigment profiles with *O*-methylated aglycones as the predominant structural type. The possible function of flavonoids as antiherbivore defense compounds was discussed, but Mears (1980b) found no correlation between complexity of flavonoid profile and latitude, as might be the case if complexity of profile increases as one goes from temperate to more tropical climates with concomitant increase in insect predators. Although the number of samples studied was not large, there was a relationship between latitude and complexity of pigment profiles in taxa restricted to calcareous substrates. No driving force for this apparent relationship is evident.

Returning to our consideration of North American *Arnica*, we look next at the flavonoid studies of subgenus *Arctica* (Downie and Denford, 1986b). The first taxa to be studied by those workers were *A. frigida* Meyer ex Iljin subsp. *frigida*, which occurs in eastern Siberia, Alaska, the Yukon, and in a few sites in northern British Columbia; *A. frigida* subsp. *griscomii* (Fernald) S. R. Downie, restricted to the Gaspé Peninsula (Quebéc) and northwestern Newfoundland; and *A. louiseana* Farr, which occurs at high elevations in the Rocky Mountains of Alberta and British Columbia (this taxon was previously considered as a subspecies of *A. frigida*; taxonomic revision by Downie and Denford, 1986a). *Arnica frigida* subsp. *frigida* consists of three chromosome races with $2n=38$, 57, and 95, while *A. frigida* subsp. *griscomii* and *A. louiseana* both have $2n=76$. The significance of these numbers can be appreciated by noting Barker's (1966) observation that no well-developed sexual species occur in glaciated areas, and that no polyploid taxa appear to occur in unglaciated areas. Downie and Denford (1986a) established that populations of *A. frigida* from unglaciated areas, in fact, reproduced sexually.

The second paper from these workers (Downie and Denford, 1986b) concerned the application of flavonoid data to relationships among the three subspecies. The pigment profiles were straightforward consisting of the 3-*O*-mono- and diglycosides of kaempferol and quercetin and luteolin 7-*O*-glucoside. Individual profiles ranged from two to six compounds. Quercetin 3-*O*-galactoside and 3-*O*-diglucoside were observed in all specimens studied; all other compounds varied to a greater or lesser degree. The key observation from this study is the high degree of similarity between western subsp. *frigida* and eastern subsp. *griscomii*: the profile of the Québec population, consisting of kaempferol 3-*O*-glucoside, quercetin 3-*O*-galactoside, and 3-*O*-diglucoside, was identical to four $2n=38$ populations from Alaska and three $2n=57$ populations from each Alaska and the Yukon. The Newfoundland profile (two populations) was identical to the profile from eight $2n=38$ populations from Alaska, as well as three $2n=57$ populations from the Yukon. The close morphological similarity between the two subspecies suggests that they may be remnants of a once transcontinental distribution that became disrupted by Pleistocene ice, and that eastern and western members survived in glacial refugia with subsequent expansion of the western component. The observation that the flavonoid profile of subsp. *griscomii* falls within the limits of the profile observed for subsp. *frigida* adds further weight to this suggestion. It was not possible to draw any firm conclusion as to whether *A. louiseana*, the Rocky Mountain member of the group, represents the product of some hybridization event or whether it, or an ancestor, also survived glaciation. The authors concluded by pointing out the need for studies of genetic variation of these three taxa. An analysis of allozyme variation, as they suggested, or other macromolecular approach, would certainly seem to be in order.

The remaining members of *Arnica* subgenus *Arctica* have been subjected to flavonoid analysis as well (Downie, 1988; Downie and Denford, 1988). Relevant to the present discussion of glacial refugia are their observations on the distribution and flavonoid chemistry of the *A. angustifolia* Vahl complex. Reappraisal of morphological data using multivariate analysis led Downie (1988) to realign previously recognized subspecies into just two: (1) *A. angustifolia* subsp. *angustifolia*, which comprises subspecies *angustifolia, attenuata, sornborgeri, intermidia, iljinii,* and *alpina* and *plantaginea* (see Downie, 1988 for taxonomic details); and (2) *A. angustifolia* subsp. *tomentosa* (J. Macoun) G. W. Dougl. & G. Ruyle-Dougl. *Arnica angustifolia* has much in common with *A. frigida* Meyer ex Iljin in terms of its breeding biology, the existence of several cytological races, and possession of a comparatively simple flavonoid profile. Although the distribution of *A. angustifolia* subsp. *angustifolia* is essentially continuous across North America—ranging from Alaska to the eastern Arctic and Greenland—the combined evidence suggests that the present range has resulted from reoccupation of these areas from refugia following retreat of the Wisconsin ice sheet.

Owing to the high level of apomixis within the taxon, few genera in Asteraceae offer as much challenge in species definition as does *Antennaria.* Anderberg (1994) summarized the situation well in describing the genus as consisting of "… ca. 70 to several hundred species depending on species concept." This situation is very well represented in the North American flora by *A. rosea* E. Greene. Morphometric analysis revealed that the main source of morphological variation in the complex comes from six sexually reproducing progenitors. This situation has been reviewed in depth by Bayer (1990, and citations therein). The bulk of Bayer's work has been directed at understanding the contributions of the various sexual species involved in the establishment of each of the agamospermous species complexes. Of concern to us in the present context is the pattern of distribution of the *A. rosea* complex in North America. *Antennaria rosea* occurs over a very large area in western North America ranging from southern California, Arizona and New Mexico north to the Arctic and east to Hudson's and James Bay. It also occurs disjunct in eastern Canada and along the shores of Lake Superior. The northern half of its range is similar to the range occupied by some of the members of *Arnica* discussed above. Bayer's work addressed questions of geography only as they pertained to ranges of occurrence of the participants in the formation of the species complexes. Since correlations of flavonoid profile complexity and ploidy level relative to the origins of present distribution patterns provided interesting results in the *Arnica* complexes, it would seen reasonable that studies of *Antennaria* might yield similarly interesting results.

3.1.4 Menziesia *(Ericaceae)*

Menziesia Smith, a genus of perhaps eight species, is one of many genera that exhibit an eastern North American, western North American, eastern Asian distribution.

Two species concern us in this discussion, *M. ferruginea* Smith from western North America and *M. pilosa* (Michx.) Juss. from eastern North America. The western taxon has been further subdivided into var. *ferruginea*, which occurs along the coast from Alaska south to northern California, and var. *glabella* (Gray) Peck, which is native to the Rocky Mountains. However, there has been some difference of opinion as to the level at which the variation in the species should be recognized. Hickman and Johnson (1969) argued for the elimination of infraspecific recognition; Peck (1961) treated them as varieties, whereas Calder and Taylor (1956, 1968) recognized them as subspecies. *Menziesia pilosa* consists of only the single taxon and is native to the Appalachian Mountains. Two more recent studies of the North American members of the genus have been described, one that focused on the flavonoids (Bohm et al., 1984), and one that concerned itself with allozyme variation (Wells and Bohm, 1994).

Extensive collections of both species (all taxa) provided an opportunity to examine the flavonoid profiles in detail. In addition to the almost ubiquitous presence of the "usual suspects," glycosides of kaempferol, quercetin, and myricetin (general structures above), the dihydroflavonol dihydromyricetin [270], the flavanone [271], and the unusual flavonol gossypetin [272] were identified (see Fig. 3.4). Dihydromyricetin was observed only in the eastern species, whereas gossypetin was seen only in the western species. Insofar as dihydroflavonols are part of the normal flavonol biosynthetic pathway, the presence of dihydromyricetin may only reflect a concentration effect (sequestration is the term usually used for this). Gossypetin, however, having an oxygen function at a position (C-8) represents a unique biochemical step.

Structural data of flavonoids from several hundred individual plants were analyzed by a variety of statistical and multivariate techniques. Whereas significant differences were noted between the eastern and western species, the level of variation within and between populations in each species was such that no pigment races could be defined. Plants from Alaskan populations of *M. ferruginea*, for example, were as likely to emerge next to populations from California or Oregon in these treatments, as they were to emerge near populations from the northern or southern Rocky Mountains. The number of combinations of flavonoid pigments observed in each species nearly equaled the number of populations examined, a level of variation that approximates that described by Levy and his collaborators for species of *Phlox*, as mentioned above. A good deal more insight into the *Menziesia* story came from electrophoretic studies.

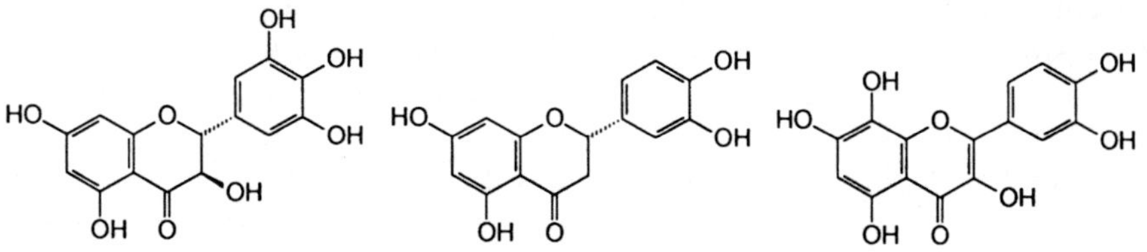

[270] [271] [272]

Fig. 3.4 Compounds 270–272, flavonoids from *Menziesia*

The two species were sampled from throughout their respective ranges and subjected to electrophoretic analysis. Twenty-eight populations of *M. ferruginea* s.l. and 15 populations of *M. pilosa* were studied for 13 enzyme systems. Populations of *M. ferruginea* subsp. *ferruginea* along the Pacific coast and Cascade Range are not distinguishable from each other based upon allozymes, and are only slightly different from populations of subsp. *glabella* in the Rocky Mountains. This lack of differentiating allozyme features corresponds well with the high degree of morphological similarity observed within this taxon (Wells, 1992), but differs from the high degree of flavonoid variation described above. What little differentiation there is likely arose as a result of the isolation of the coastal/Cascades populations from the Rocky Mountain populations as the Cascades were formed during the Pliocene. A major result of this orogeny was the development of dry intermontane valleys that lacked suitable habitats for *Menziesia* (Daubenmire, 1978). Changes in local climate during the Pleistocene provided the opportunity for the two forms to reunite in areas, such as the Columbia River Plateau, which at the present time is too dry to support the species, except in a very few scattered localities. The allozyme data suggest a divergence between the coastal/Cascades form and the Rocky Mountain form in the range of 60,000–80,000 years ago. Morphological evidence suggests an area of overlap in the vicinity of Mount Hood, Oregon (Hickman and Johnson, 1969; Wells, 1992).

The most distinctive populations of *M. ferruginea* were those collected from the northern limit of the species in Alaska. These populations exhibit the low levels of intrapopulational genetic variation characteristic of recolonization from propagule sources that existed in unglaciated refugia in central Alaska (Heusser, 1983). In contrast, populations from the Alaskan panhandle have alleles characteristic of populations located farther south along the coast and in the Cascades. Overall, however, the panhandle populations are less variable than those from the more southerly sites. It is generally held that the panhandle populations arose by recolonization from refugia on offshore islands such as the Queen Charlotte Islands (Haida Gwaii) (ca. 53°N, 132°W).

The genetic identity between the two western taxa is high ($I=0.98$) suggesting a comparatively recent divergence as commented upon above. More surprising is the high degree of genetic identity between *M. ferruginea* s.l. and *M. pilosa* ($I=0.92$). This value points to contact between the two taxa as recently as the Pleistocene, possibly 400,000–800,000 years ago, much more recent than was previously thought. It seems reasonable to suggest that the progenitor of the eastern and western species may have enjoyed a larger, and possibly continuous, distribution across what are now the northern United States and adjacent Canadian provinces, and that suitable habitats in the middle of the range were subsequently eliminated by climatic changes. Intriguingly, there have been reports of *Menziesia* growing in the vicinity of Duluth, Minnesota (Gleason, 1952; Rosendahl, 1963). However, the absence of herbarium records substantiating this claim (Rosendahl, 1963) puts this suggestion in doubt. Were this claim substantiated, however, suitable habitats for *Menziesia* could well have existed in late glacial *Picea-Larix* forests that occupied the area (Watts, 1983).

3.1.5 Thuja plicata-*Western Red Cedar (Cupressaceae)*

Thuja plicata D. Don is one of the most important forest trees in coastal, western North America, providing a weather-resistant wood much in demand as a source of shingles as well as for more decorative purposes. The species also has great significance within the First Nations people throughout its range, which includes moist to swampy sites from southern Alaska southward along the coast to Humboldt County, California, and in central British Columbia southward through Washington, eastern Oregon to northern Idaho and northwestern Montana. The coastal and inland ranges of western red cedar are separated by a system of valleys that experience much lower rainfall and thus provide few suitable habitats. Both segments of the species' range likely represent northward advances from refugia following glacial retreat. Patterns of variation have been studied using protein electrophoresis as well as gas chromatographic analysis of volatile oil components. We will look at the terpene chemistry of the species first.

Interest in volatile oil chemistry of western red cedar goes back to the early days of the twentieth century references, which can be found in von Rudloff's (1962) paper on gas–liquid chromatography (GLC) of red cedar terpenes. That work, although mainly concerned with within-tree variation, did report similar analytical data from trees sampled at some distance from each other, central versus southwestern British Columbia. The major components of the oil were thujone [273] (76–87%), isothujone [274] (7.0–9.0%), and sabinene [275] (1.0–8.0%) (see Fig. 3.5). Also observed, but at much lower concentrations, were α-pinene, camphene, car-4-ene, limonene, 1,8-cineole, terpinene, p-cymene, and terpinolene. A subsequent study of terpene variation as a function of geographic origin of the species involved material analyzed from 29 sites in British Columbia, Idaho, Montana, Oregon, and Washington (von Rudloff and Lapp, 1979). Analysis of quantitative values for 18 compounds was performed using simple linkage clustering of *F*-1 weighted similarities.

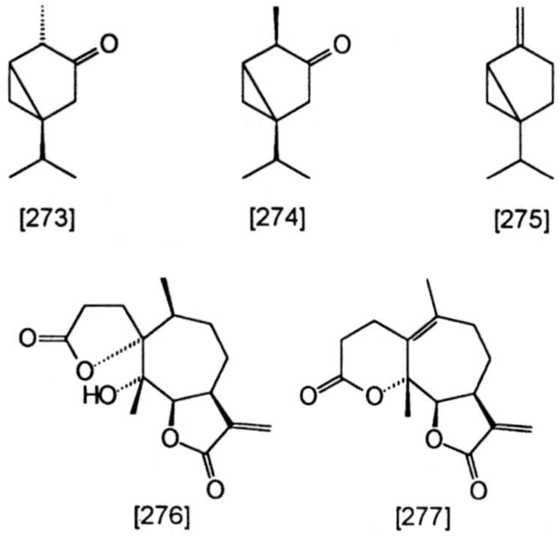

Fig. 3.5 Compounds 273–275, terpenes from *Thuja plicata*. Compounds 276 and 277, sesquiterpenes from *Ambrosia psilostachya*

No geographic structure was revealed by this analysis, with trees from Arch Cape, Oregon, Whitefish, Montana, and sites from northern British Columbia, including Queen Charlotte Islands, being closely associated. The authors remarked on the lack of differentiation between coastal and interior populations. A reinvestigation of red cedar from 55 sites (3–6 trees per site) provided a new data set that was analyzed by numerical and discriminant-function analyses (von Rudloff et al., 1988). These analyses confirmed the low intra- and interpopulational variation seen in the earlier study, but did reveal small differences between coastal and interior populations. No correlations between northern and southern populations emerged from the analyses; likewise, elevation had no effect on terpene composition.

Parallel with the study just described was an examination of genetic variation in red cedar using enzyme electrophoresis (Yeh, 1988). Useful data from 2300 megagametophytic samples from 230 trees in eight populations were obtained for 15 enzymes representing 19 loci. Sampling ranged between 48°49′ and 52°37′N latitude and 116°40′ and 124°10′W longitude. The number of alleles per locus averaged 1.17 and observed heterozygosity ranged from 0.024 to 0.055 (average 0.036). The level of variability of western red cedar approaches the lower limit for primarily outcrossing taxa (Hamrick et al., 1979) and is in accord with the very low levels of variation reported from the terpene analyses. Yeh (1988) suggested that the low levels of variation in red cedar, an outcrossing species that would be expected to be more variable, could be accounted for by comparatively rapid expansion following a bottleneck. He goes on to suggest that divergence time since the bottleneck is in the neighborhood of 10,000 years. It is known from the fossil record that red cedar did not reach the northern end of Vancouver Island until about 3000 years ago. Estimates of the timing of red cedar migration on the mainland range from 4900 to 10,000 years ago, which is consistent with the value obtained from the electrophoretic data. Further insights require information on effective population size and mutation rate per locus per generation (Yeh, 1988).

3.1.6 Liriodendron tulipifera *(Magnoliaceae)*

Liriodendron, tulip tree, consists of two species, the North American *L. tulipifera* L., which we will examine here, and the Asian *L. chinense* (Hensley) Sarg. The relationship between the two species will be discussed later in this review. The North American tulip tree occurs naturally throughout a wide range in eastern North America from 28°N to 43°N latitude, predominantly east of the Mississippi River. It is also commonly planted as a decorative species in many parts of North America (it fares well in Vancouver, B.C. at 49°13′N). In its natural setting, the tulip tree occurs in two forms, an upland form that can occur in large populations in the northerly part of its range, and an ecotype usually found in low, wet habitats along the coastal plains from North Carolina to Louisiana (Parks and Wendel, 1990, and citations therein; Parks et al., 1994). The latter report described an electrophoretic study of allozyme variation across the range of the species, from which three forms emerged based on differences in allele frequencies: (1) a form ranging from the northern extent of the range to southern parts of the Appalachian uplands, (2) an isolated Florida peninsula form, and (3) an intermediate form that occurs along the southeastern coastal plain. The peninsular and intermediate forms are similar in distribution to the southern ecotype noted above.

Essential oils are commonly occurring constituents of conifers but are less frequently encountered in hardwood species. Examples of trees that do produce these compounds include species of *Sassafras* (Lauraceae) and *Magnolia* (Magnoliaceae). To this list one can add *Liriodendron tulipifera*, which A. L. Smith et al. (1988) have shown to produce a rich array of common monoterpenes: α-pinene, β-pinene, camphene, myrcene, 3-carene, limonene, β-phellandrene, *cis*-β-ocimene, γ-terpinene, terpinolene, and bornyl acetate (see Fig. 3.7 for structures). The analyses were done on twig samples collected from 11 provenances, representing sites in Alabama, Georgia, Louisiana, Mississippi, Ohio, South Carolina (coastal plain and piedmont), Tennessee, and Virginia. With the single exception that the piedmont specimen from South Carolina lacked β-phellandrene, all 11 compounds were present in all specimens. Variation was significant in most cases with the following ranges for concentrations (expressed in % of dry $\text{wt} \times 10^{-3}$): α-pinene, 5.00–16.10; β-pinene, 5.14–24.78; camphene, 1.03–5.58; myrcene, 2.23–18.26; 3-carene, 0.51–2.57; limonene, 3.31–18.77; β-phellandrene 0.00–24.56; *cis*-β-ocimene 33.30–75.21; γ-terpinene, 0.08–2.37; terpinolene, 1.85–12.20; and bornyl acetate, 2.40–21.84. There appears to be no clear geographic pattern in the terpene concentrations.

Restriction endonuclease site analysis, including results from the earlier study (Parks and Wendel, 1990), revealed five sequence changes within populations of *L. tulipifera.* Of these, three occur exclusively in the "northern" haplotype and two only in the "southern" haplotype. Based upon estimates of sequence-divergence time (Parks and Wendel, 1990), it was suggested by Sewell et al. (1996) that the two forms diverged about 1.2 million years ago. To put this time estimate in perspective, it is necessary to refer to comments by Parks and Wendel (1990) on the fossil record for *L. tulipifera. Liriodendron* was present as an element of a temperate deciduous forest during early to mid-Miocene. Later cooling resulted in a southward migration of the temperate forest effectively isolating the Asian and North American populations. During Pleistocene glaciation, the temperate forest is thought to have survived in only a limited number of areas in the southeastern part of the continent: (1) Nonconnah Creek, Tennessee (Delcourt et al., 1980), at which fossil evidence of *L. tulipifera* was found; (2) Goshen Springs, Alabama (Delcourt, 1980); and (3) Sheelar Lake, Florida (Watts and Stuiver, 1980). Parks et al. (1994) suggested that possible refugial areas were likely to lie along the bluffs of the Apalachicola River, where the northern haplotype might have survived, and the Ocala highlands area of north-central peninsular Florida, where the southern haplotype could have survived. This latter area, which corresponds with the peninsular allozyme group, existed as an island during the Pliocene (Stanley, 1986). With the retreat of the last glaciers, populations of the two types came into contact, allowing hybridization to occur, the result of which was the intermediate type seen along the southeastern coastal plain.

3.1.7 Ambrosia psilostachya *(Asteraceae)*

Ambrosia psilostachya DC. is a ragweed that displays a high level of morphological variation throughout its range, which extends from Canada to central Mexico. It was noted by the University of Texas group, however, that populations on the Texas Gulf Coast islands were noticeably more homogeneous than the species at large, suggesting to them that the island populations had arisen from a single source. The possible usefulness of sesquiterpene lactone data in shedding some light on relationships among populations of this species emerged from observations by Miller et al. (1968) that plants from an island population exhibited the same set of sesquiterpene dilactones based on psilostachyin, for example, psilostachyin [276] and psilostachyin-B [277] (see Fig. 3.5) that they observed in plants from the mainland dunes suggesting that the latter population may have been the source of propagules that gave rise to the island populations. It was also noted that plants from the vicinity of Vera Cruz, Mexico, exhibited the same array of dilactones.

The sesquiterpene lactone chemistry was reinvestigated in greater detail by Porter and Mabry (1972), who used plant material from 20 populations of *A. psilostachya*

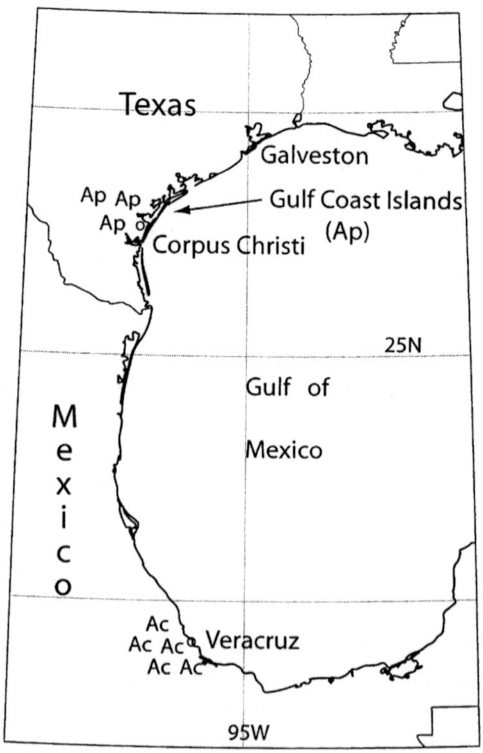

Fig. 3.6 Map of Texas coast and adjacent Mexican sites from *Ambrosia psilostachys* (Ap) study. Populations of Ambrosia cumanensis are represted by "Ac."

from the Texas mainland and Gulf Coast Islands and seven populations of the closely related *A. cumanensis* Kunth. from near Vera Cruz, Mexico, where both species grow sympatrically (Fig. 3.6). In addition to the sesquiterpene fractions, these workers also examined the volatile terpenes of these plants. Numerical analysis of the combined chemical data revealed a closer relationship of the island populations with the population from Vera Cruz than with populations on the Texas mainland. The earlier suggestion that the island was colonized from achenes originating from the mainland has to be discarded in view of the strong support for the parental populations lying farther south, possibly in the vicinity of Vera Cruz as suggested by the newer chemical data. It is thought that the islands originated within the last 5000 years as a result of post-Pleistocene sea level changes. Colonization of a newly formed island by means of a single achene, or a few achenes from a localized source, would explain the observed morphological homogeneity. The existence of populations on the mainland that exhibit the dilactone chemistry characteristic of island plants can be rationalized simply by transport of propagules to the mainland by prevailing winds during the time of year when *A. psilostachya* fruits are available.

3.2 North American Conifers

The next series of examples involve conifers whose secondary chemistry has been of major interest for many years. The resulting literature in this area, particularly with regard to the application of terpenoids to taxonomic and evolutionary problems, is a large and rich one. In the paragraphs that follow, only a representative sampling of the available examples will be described. An important early work is Mirov's (1967) book *The Genus Pinus*, in which he summarized much of the available literature on the application of turpentine composition to unraveling relationships among the pines. Von Rudloff's 1975 review is also a valuable source of information.

The use of volatile chemicals as systematic markers has the obvious advantage of lending itself to quantification through GLC. In many, if not most, of the cases discussed below, qualitative differences in monoterpene profiles would not have been sufficient to allow distinctions to be made between taxa, or even between individuals within a population. This is true because most conifers synthesize many of the same monoterpenes, although often in vastly different relative concentrations. It is these quantitative differences that have been constructively used in the following examples. Structures of the terpenes commonly studied are presented in Fig. 3.7.

3.2.1 Abies *(Pinaceae)*

Abies—the true firs—comprise a moderately large genus, 49 species (Mabberley, 1997, p. 1) that include some of the most important commercial species among the conifers, particularly in North America. Some of the more important North American

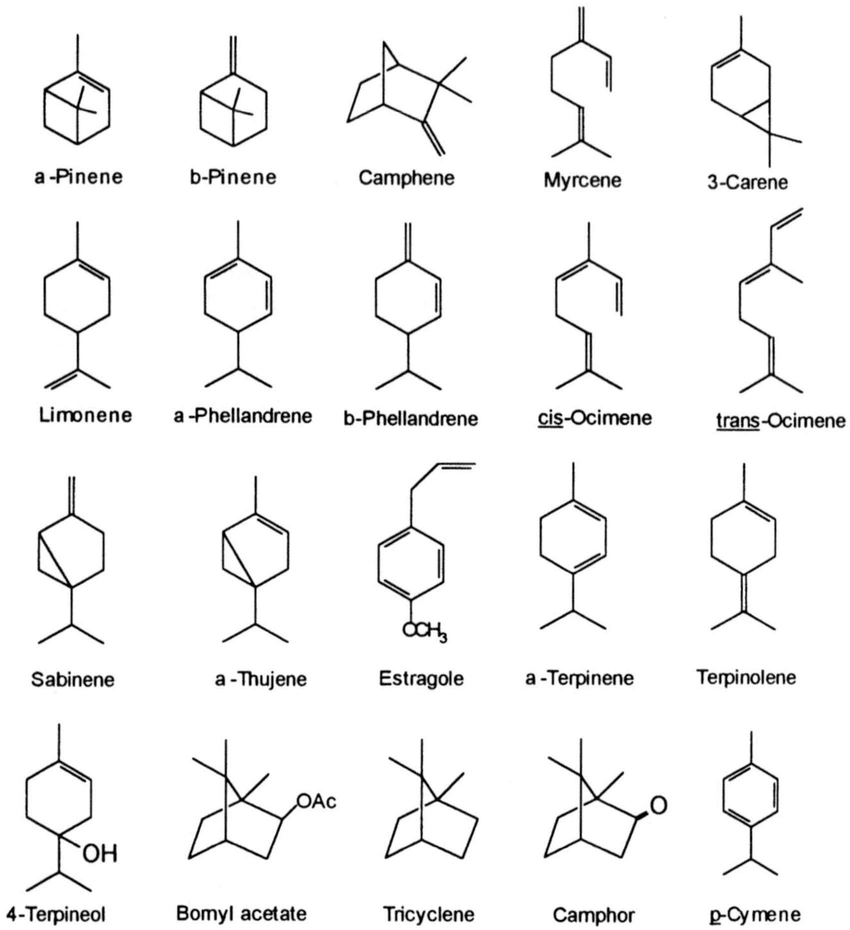

Fig. 3.7 Common terpenes seen in conifer studies

firs are *A. amabilis* (Dougl.) Forbes (Pacific silver fir), *A. balsamea* (L.) Miller (balsam fir), *A. concolor* (Gord. & Glend.) Lindl. (lowland white fir), *A. grandis* (Dougl.) Lindl. (grand fir), *A. magnifica* A. Murr. (California red fir), and *A. procera* Rehder (syn. *A. nobilis*) (noble fir).

Zavarin and his associates have described detailed analyses of the terpenoid constituents of several species of fir. We can start with their study of *A. amabilis* collected throughout the range of the species (Zavarin et al., 1973). The Pacific silver fir occurs from southeastern Alaska to northern California at mid to higher elevations with a major contribution to the forest flora in coastal British Columbia, the Olympic Mountains (Washington), and in the Cascade Mountains in Washington and Oregon. It is represented in California in only a few sites in the north. Analysis of cortical terpenoids from over 100 trees from 15 sites revealed a comparatively homogeneous profile with α- and β-pinenes, 3-carene, and β-phellandrene as major components, with lesser amounts of α-phellandrene, myrcene, limonene, and terpinolene. Statistical analysis revealed little in the way of geographic patterning, which is in accord with the view that silver fir occupied a single refugium south of the Cordilleran ice in Washington and Oregon during the most recent glacial episode (Hultén, 1937, as cited by Zavarin et al., 1973). It was noted that *A. amabilis* is a slow migrator and that it may still be in the process of northward movement.

Two closely related species of firs are *Abies grandis,* which occurs in northern California, Oregon, Washington, Idaho, Montana, and southern British Columbia at mid elevations, and *A. concolor*, which enjoys a more southerly range at somewhat higher elevations. Populations from northern and eastern California have been recognized as *A. concolor* var. *lowiana* (Gord.) Lemm., whereas those form Nevada, Utah, Colorado, New Mexico, and Arizona constitute var. *concolor.* Populations along the Pacific Coast and in Washington, British Columbia, northern Idaho, and northern Montana are considered to be typical *A. grandis*, whereas populations in northern California, central and eastern Oregon, southern Washington, and eastern Idaho exhibit features that suggest incursion of genes from *A. concolor.* Features by which the two taxa can be distinguished include stomatal placement, needle-tip shape, needle length, and the color of bark periderm.

Analysis of the cortical turpentine of *A. grandis* revealed a mixture consisting for the most part of α- and β-pinenes, camphene, β-phellandrene, and bornyl acetate. Present in lesser amounts were 3-carene, limonene, myrcene, and tricyclene (Zavarin et al., 1977). Camphene and 3-carene figure significantly in distinguishing between *A. grandis* and *A. concolor.* An earlier study (Zavarin et al., 1975) had revealed that *A. concolor* could be resolved into three distinct chemical races on the basis of relative amounts of two compounds: (1) the Rocky Mountain race with large amounts of camphene and 3-carene; (2) the Cuyamaca race from southern California and northern Baja California characterized by large amounts of 3-carene and only trace amounts of camphene; and (3) the western race with these two compounds present in only trace amounts. *Abies grandis* turpentine exhibits large amounts of camphene and has 3-carene in trace amounts only. Based on the monoterpenes of the intermediate populations, the workers concluded that introgressive hybridization had occurred and might still be occurring.

The comparatively low level of interpopulational variation in the monoterpenes of *A. grandis* resembles that for *A. amabilis* commented upon above. As noted by Zavarin et al. (1977), it is likely that *A. grandis* occupied much the same areas during interglacial periods that it occupies now. Advancement of the ice sheet forced its retreat to refugia in Washington and Oregon. Present-day populations in the northern parts of its range, therefore, represent a return to the interglacial situation. The general lack of chemical variation observed suggests a relatively slow northward progress during which time comparatively little differentiation has occurred. Halliday and Brown (1943) considered the present-day distribution in British Columbia as having resulted from postglacial migration of fir along the coast toward Vancouver Island and along the Kootenai Valley toward the moist Rocky Mountains. Pollen records in the Puget Sound area, the Cascade Range of Washington, and northern Washington support this scenario. In contrast, *A. concolor* tends to occupy drier habitats such as occur east of the Cascades.

Abies procera and *A. magnifica* and intermediate forms constitute a large complex stretching from north central Washington to the southern Sierra Nevada. *Abies magnifica* tends to grow at higher elevations with *A. procera* occupying somewhat lower sites. The question has been raised as to whether the intermediate forms are the result of introgression between two "good" but similar species or whether they are the result of recent evolutionary differentiation. Zavarin et al. (1978) addressed this issue by examining the essential oil components of 352 individuals collected at 35 sites. Four compounds were found useful in studying the transitional populations, 3-carene, limonene, β-phellandrene, and α-pinene. The chemical data allowed the specimens to be sorted into three groups: (**1**) north of 44° (*A. procera*); (**2**) between 44° and 40° (transitional); and (**3**) south of 40° (*A. magnifica*). Transition-like populations were also found in the southern Sierra Nevada and in the vicinity of Mt. Shasta. The authors also compared published values for seed weight and cotyledon number and found that those data paralleled the trends seen within the chemical transitional region. The authors pointed out that neither the chemical nor morphological observations clearly differentiate between the two hypotheses, whereas paleobotanical evidence suggests that the transitional forms are the product of recent evolutionary change.

Abies lasiocarpa (Hook.) Nutt. is a western North American taxon of considerable significance. Differences in terpene composition among extensive collections led Hunt and von Rudloff (1977) to suggest that the variation should be reflected in the recognition of three taxa, *A. lasiocarpa*, *A. balsamea* (L.) Miller, and *A. bifolia* Murr. Subsequent studies of volatile components of material from populations in California, Oregon, and Colorado (Cope, 1983) reinforce that view. The populations from Colorado differed from the others in having higher concentrations of santene, α-pinene, and camphene, and smaller amounts of β-phellandrene and limonene.

3.2.2 Cupressus bakeri *(Cupressaceae)*

The next example is the Baker cypress, *Cupressus bakeri* Jepson, which occurs in scattered populations in the Siskiyou, Cascade, and Sierra Nevada Mountain ranges of northern California and southern Oregon (Fig. 3.8). Despite the limited range of the species, there have been suggestions that the morphological variation observed is best handled by recognizing two subspecies, subsp. *typica* C. B. Wolf from the southern part of the species' range and subsp. *matthewsii* C. B. Wolf from the northern part (Wolf and Wagener, 1948). Two studies have also been directed toward documenting chemical variation within this species, one using megagametophytic fatty acids (Rafii et al., 1992b), and one using mono- and sesquiterpenes (Rafii et al., 1992a).

The fatty acid composition of the Baker cypress is comparatively complex. The major saturated acids were identified as palmitic (16:0) [chain length:number of double bonds], stearic (18:0), and arachidic (20:0) with only minor contributions from

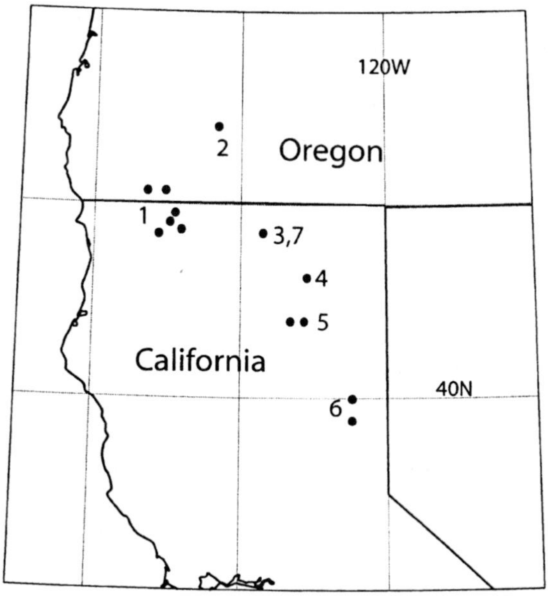

Fig. 3.8 Map of *Cupressus bakeri* sites

margaric (17:0), behenic (22:0), and lignoceric (24:0). The unsaturated acid fraction consisted of oleic (18:1), linoleic (18:2), linolenic (18:3), *cis*-11-octadecenoid (*cis*-vaccenic) (18:1), *cis*-11-eicosenoic (20:1), *cis*-5,11-eicosadienoic (20:2), *cis*-11,14-eicosadienoic (20:21), *cis*-5, 11,14-eicosatrienoic (20:3), *cis*-11,14,17-eicosatrienoic (20:3), and *cis*-5,11,14,17-eicosatetraenoic (20:4). Canonical discriminant analysis revealed statistically significant separations between most population pairs but the sensitivity of the analysis to small changes in fatty acid concentration rendered the results less meaningful. Although there was some separation of the two subspecies with some of the canonical variables, other variables showed no separation at all. The authors chose the more conservative approach and concluded that fatty acid composition provides only marginal support for recognition of the infraspecific groups.

Foliage from 63 trees representing seven populations of Baker cypress was collected and analyzed for mono- and sesquiterpenes (Rafii et al., 1992a). The array of compounds identified from this species is quite impressive; structures were established for 31 of the 54 compounds that were detected in greater than trace quantities. Stepwise discriminant analysis for the combined mono- and sesquiterpene data reduced the number of variables to a more manageable dozen: β-pinene, 3-carene, *p*-cymene, β-phellandrene, camphor, bornyl acetate, and four unknowns. Canonical discriminant analysis of the resulting variables yielded a three-dimensional array that accounted for 84% of the total variation. When the first two variables were plotted, the northernmost populations, **1** and **2**, formed a group separated from all others. Population **6**, the southernmost population, was also well separated from the others. Populations **4** and **5**, constituting the sample from the middle of the range, fell together but were not clearly separated from populations **3** and **7**. The latter two populations, representing Goose Nest Mountain in eastern Siskiyou County, were not separated from each other and were only peripherally separated from populations **4** and **5**. Separation of the populations at the northern and southern extremes of the species would seem to offer support for the recognition of the subspecies. Problems arise, however, in that the populations in the middle of the range do not fall nicely into either camp. The conclusion of the authors that the terpene data do not support recognition of infraspecific taxa seems reasonable. The unexpected similarity of populations **3** and **7** from Goosenest Mountain prompted Dodd and Rafii (1994) to examine a larger sample for terpene variation. Fifty-one trees from five populations were analyzed and found to exhibit a larger level of chemical and genetic variation than seen in any other population. This was in total agreement with earlier morphometric studies. The authors suggested that these levels of variation could be accounted for if the Goosenest Mountain area is considered as a region of transition for the species. Further, the high level of variation between adjacent populations was taken to indicate restricted gene flow within the species.

3.2.3 Picea *(Pinaceae)*

Picea—the spruces—consists of about 40 species (Mabberley, 1997, p. 555), members of which occur widespread in cooler habitats in the Northern Hemisphere. *Picea* is a major element in the forest flora of many areas of North America: *P. breweriana* S. Wats. (Brewer's spruce) of central California; *P. engelmannii* Parry (Engelmann's spruce) of western North America; *P. glauca* (Moench) Voss and *P. mariana* (Mill.) B.S.P. (white and black spruce, respectively) of boreal North America; *P. pungens* Engelm. (blue or Colorado spruce) of the Rocky Mountains; *P. rubens* Sarg. (red spruce) of eastern North America; and *P. sitchensis* (Bong.) Carr. (Sitka spruce) of the Pacific coast. All of these species have been examined chemically to a greater or lesser extent. Much of the earlier information on chemotaxonomic studies of spruce, along with other conifer genera, has been reviewed by von Rudloff (1975).

The value of spruce-oil chemistry in sorting out problems of hybridization and introgression—major factors in *Picea* taxonomy—was succinctly summarized by von Rudloff who defined three situations: (1) Terpene variation is limited such that it is not possible to use these characters in studies of introgression; this is the case in eastern North America where the ranges of black spruce and red spruce overlap. (2) Sufficient variation in terpene profiles exists for the compounds to be useful markers in systematic studies as seen in white spruce, Brewer's spruce, and Sitka spruce. (3) Tree-to-tree variation in terpene content is so variable that use in chemosystematic studies is precluded, or at least requires very large sample sizes for statistical reliability, as seen with Engelmann's spruce.

The overall usefulness of terpene data for defining geographical races is shown in the comparison of profiles of eastern and western white spruce (see von Rudloff, 1975, for specific citations). Leaf oil analysis revealed consistent differences in camphor, which occurs in higher concentrations in western trees, and bornyl acetate, which is present in larger amounts in eastern trees. This geographic differentiation parallels the findings of Wilkinson et al. (1971) who studied oleoresin composition in representatives of 16 populations maintained in a common plantation in southern Michigan. Two clusters emerged from their study, one that encompassed trees from the Great Lakes northeast to Labrador, and one that extended northwestward to Alaska. Trees from the eastern type had significantly higher amounts of β-pinene, 3-carene, and β-phellandrene, whereas limonene was present in higher concentrations in western trees.

In von Rudloff's leaf oil study, distinctly different profiles appeared in certain valleys in the Rocky Mountains and in central British Columbia where the ranges of white spruce and Engelmann's spruce overlap. Hybridization with subsequent introgression has resulted in individuals with greater or lesser amounts of marker compounds indicating several degrees of backcrossing. von Rudloff's review is extensive and should be consulted for additional examples and greater detail.

3.2.4 Pinus *(Pinaceae)*

Pinus is well represented in the North American flora with several species having been thoroughly studied with respect to terpenoid composition. A few selected examples will demonstrate the usefulness of these compounds in studying relationships and migrational history of members of this moderately large genus. Examples to be discussed range from commercially important species such as *P. contorta*, (lodgepole pine), *P. monticola* (western white pine), and *P. ponderosa* (ponderosa pine) to the bristlecone pines, noteworthy because of their exceptional age.

3.2.4.1 Lodgepole Pine

Pinus contorta Dougl. ex Loudon, the lodgepole pine, occurs in a variety of habitats over extensive ranges in western North America. The morphological variation, and to some extent, its geography, led Critchfield (1957) to recognize four subspecies: subsp. *contorta*, the coastal form that extends from Alaska to northern California; subsp. *bolanderi* (Parl.) Critchf., which occurs in an isolated pocket north of San Francisco; subsp. *murrayana* (Grev. and Balf.) Critchf., which occurs in the Cascades Range and south through the Sierra Nevada and into the San Bernardino Mountains; and subsp. *latifolia* Engelmann, which occurs from the interior of British Columbia, north to the Yukon, and east to Alberta and the Rocky Mountains. Earlier studies of terpenes of lodgepole pine, reviewed briefly by Forrest (1980a), were sufficient to reveal significant variation, but not comprehensive enough to allow any clear-cut chemical types to be defined. A study of geographic variation of lodgepole pine in relation to population history was published by Cywnar and MacDonald (1987).

Efforts to remedy this situation were made by Forrest (1980a, 1981) who investigated the cortical oleoresin constituents of trees collected from 150 sites throughout the range of the species. Three compounds were found to be ubiquitous, α-pinene, β-pinene, and β-phellandrene; three were seen consistently but in smaller amounts, camphene, α-terpinene, and terpinolene; and two, limonene and 3-carene, were highly variable, ranging from absent to moderate concentrations. Fairly clear-cut chemotype patterns emerged from GC analysis three of which were common: type **A**=β-phellandrene>>β-pinene~α-pinene; type **B**=β-phellandrene>β-pinene>α-pinene; and type **C**=β-pinene>β-phellandrene>α-pinene. Three subtypes were identified based upon presence, absence, or relative concentrations of 3-carene and limonene: **1**=3-carene and limonene either absent or present in low concentrations; **2**=3-carene present in moderate to high concentrations; and **3**=limonene present in moderate to high concentrations. Several patterns, based on relative amounts of the pinene isomers, referred to as "rare types", were identified that tended to be restricted to small areas. Finally, an "introgression" type, his type **E**, was identified that was characterized by relatively high concentrations of camphene along with high concentrations of α-pinene and/or β-pinene. Type **E** trees occurred in southwestern Northwest Territories and west central Alberta and were thought to have resulted from introgression with Jack pine, *Pinus banksiana* Lambert. There was fair agreement between chemotypes and subspecific taxonomy, except for subsp. *murrayana*, which could not be distinguished from subsp. *latifolia* on the basis of the terpene data.

The next step in the analysis was to determine the geographic distribution of the various types. Type **A1** trees (that is, type **A**-subtype **1**), for example, were found to occur in north coastal British Columbia, the Queen Charlotte Islands, and western Vancouver Island. Local, or at least comparatively short distance, differences were common in the database as one sees in the differences between western Vancouver Island trees and those on the eastern side of the island where high β-pinene and 3-carene concentrations were observed in concert with decreased amounts of β-phellandrene. Type **C1** trees were characteristic of the Puget Sound area, with **A1** and **B1** types present in populations along the coastal strip of Washington and Oregon. One feature of the distribution of chemical types in the context of refugia, and subsequent gene flow, is the homogeneous nature of populations at the southern and western peripheries of the range of lodgepole pine. Uniformity of these populations might be the result of the founder effect, or, alternatively, peripheral populations might have resulted from local selection. The most heterogeneous populations were those in central British Columbia with others extending farther to the north and northeast. In these areas, population density is high and thus open to greater pollen exchange. Forrest (1987) discussed this feature of lodgepole pine in terms of possible glacial refugia at the northern extent of the species. Despite the considerable amount of information provided by these studies, we are left with questions that cannot be answered, principally, what has been the direction of movement of genes in this system? What is needed to answer this—and the same question can be asked about virtually all studies of this sort—is a phylogenetic hypothesis that provides a framework of interpopulation relationships. We do not know, for example, whether high concentrations of any given terpene represent a derived or an ancestral trait. Until a gene sequence-based analysis of this system (and, it follows, the others of this sort) we are left with no definitive idea of the group's evolutionary history.

Pinus albicaulis Engelm., the whitebark pine, is a timberline species that ranges from central British Columbia south through the Sierra Nevada Mountains of California and from the Rocky Mountains of eastern British Columbia through Idaho and western Montana into northwestern Wyoming, with a disjunct population in the Ruby Mountains in Nevada (Elko County). There are several reports of chemical constituents of whitebark pine in the literature, but the most comprehensive examination comes from the work of Zavarin et al. (1991), who studied the wood monoterpenes from 243 trees representing 24 populations. Four groups were clearly delineated when the complete data set was subjected to discriminant-function analysis: (**1**) an eastern group encompassing the Rocky Mountains plus populations sampled in eastern Oregon and northeastern Nevada; (**2**) a northwestern group encompassing the coast ranges of British Columbia, Washington, and the northern half of the Oregon Cascades; (**3**) a small intermediate group consisting of populations in the southern half of the Oregon Cascades; and (**4**) the Californian group. The most striking difference observed, evident even in simple inspection of the data, was the very high levels of 3-carene in the Californian populations. Table 3.1 presents the mean concentration values for 3-carene and selected other monoterpenes taken from Table 3 of Zavarin et al. (1991).

The predominance of 3-carene in the Californian chemotype provides an excellent opportunity to explore the usefulness of this compound in chemosystematic and phytogeographical studies. Whereas concentrations of most monoterpenes seen in the studies described here are distributed to a greater of lesser degree around a single central value, concentrations of 3-carene frequently exhibited a trimodal distribution. This led to the suggestion that genetic control of 3-carene biosynthesis could be accounted for by a single locus two-allele model, with allele *C* coding for high concentration of 3-carene and allele *c* coding for low. The heterozygous condition, *Cc*, then would yield an intermediate amount of 3-carene. (Hanover, 1966a; see also, Hanover, 1966b, 1971).

Table 3.1 Concentrations of selected monoterpenes from populations of *Pinus albicaulis* (from Zavarin et al., 1991)

Terpene	Location (Number of populations)			
	Eastern (9)	Northwestern (4)	Intermediate (2)	Californian (9)
α-Pinene	25.1 (7.7)[a]	12.7 (4.6)	8.7/5.6	5.5 (2.3)
β-Pinene	10.0 (5.4)	14.8 (4.6)	6.2/5.6	1.9 (1.6)
3-Carene	20.2 (8.5)	26.6 (8.5)	53.5/52.7	70.0 (6.2)
Limonene	23.5 (7.3)	11.8 (1.8)	7.3/8.5	4.5 (1.9)
β-Phellandrene	10.8 (4.0)	15.5 (8.1)	1.6/1.4	1.5 (1.7)
Terpinene	0.6 (0.3)	2.2 (1.7)	2.2/1.2	3.0 (1.2)
p-Cymene	0.5 (0.3)	0.9 (0.7)	0.8/3.3	1.2 (0.7)

[a] Percentage means and standard deviations, except for intermediate populations where percentage values only are available.

Zavarin et al. (1991) assumed that the same genetic mechanism was operational in *P. albicaulis* and assigned the intervals of 0–8%, 8–56%, and 56–100% 3-carene to the genotypes *cc*, *Cc*, and *CC*, respectively. Next, they compared the individual genotype ratios with the theoretical ratio using the Hardy–Weinberg law and found the values to be in agreement. Next, they plotted the occurrence of allele *C* against latitude, which revealed that the highest values characterized the Californian populations. Populations north of 42°N had frequencies of allele *C* in the vicinity of 70%, with the value dropping to roughly 45% in populations north of ca. 45°N latitude.

Zavarin et al. (1991) also discussed the relationship between total monoterpene variability ($\sum$V) and latitude as it applied to the evolutionary history of *P. albicaulis*. The least variable populations of *P. albicaulis* occurred among the Californian populations, with increasing values of $\sum$V characterizing populations north of 42°N latitude. The greatest variation in monoterpenes was found within the northernmost populations (>48°N). The significance of these values can be appreciated as they apply to alternative explanations for the present-day distribution of the species. One suggestion is that *P. albicaulis* evolved from an ancestor of subsect. *Cembrae*, all members of which are Asian, either before the separation of North America and Eurasia at the end of the Mesozoic, or after the ancestor had crossed the Bering Land Bridge and become established in North America. This view is based upon the characteristics that the four Asian species have in common, but are otherwise rare in the genus (wingless seeds and closed cones at maturity). The alternative view is that *P. albicaulis* evolved from an ancestor in subsect. *Strobi*, and that the peculiar cone and seed features developed independently from the Asian taxa. The level of chemical variation is not consistent with a recent arrival, but rather supports the view that the species has been in western North America long enough for it to accumulate considerable variation. Adding weight to this scenario is the fact that the differences between coastal and Rocky Mountain forms of *P. albicaulis* are very similar to differences observed in other conifers, that is, *Pinus flexilis*, *Pseudotsuga menziesii*, *Abies concolor*, *A. lasiocarpa*, and the bristlecone pines, all of which have been explained by the disappearance of suitable habitats through the formation of the arid Great Basin.

An explanation was also put forward by Zavarin et al. (1991) to account for the low level of variation in the southern populations, which is contrary to what might have been expected, had this area served as a refuge during periods of glacial advance. It was suggested that the refuge existed a good deal to the north, possibly in the Klamath area (northwestern California and southwestern Oregon), to which reference has been made above for other taxa, and that populations lying to the south are of more recent origin, and thus have had much less time to diversify.

Ponderosa pine, *Pinus ponderosa* Laws., is among the most widespread of western conifers, with a range that extends from British Columbia along the coast to southern California and along the Rocky Mountains as far south as Mexico. It occurs at mid-elevations in drier habitats and often covers considerable acreage. Three varieties have been recognized (Critchfield and Little, 1966), the typical western var. *ponderosa*, the northeastern var. *scopulorum* Engelm., and the southern var. *arizonica* (Engelm.) Shaw. Not unexpectedly, not all workers agree with this dissection of the species. A brief review of studies documenting racial, ecotypic, and altitudinal variation can be found in the introduction to the chemosystematic work of von Rudloff and Lapp (1991). In addition to those sources of information, the reader may wish to consult a comprehensive listing of literature published on ponderosa pine up to 1965 compiled by Axelton (1967).

Of interest to us here is the chemical study by von Rudloff and Lapp (1991), who examined the leaf oil composition of trees from 37 northwestern sites, with eight of these representing the northeastern variety (*scopulorum*), and a single specimen representing var. *arizonica.* In addition to the dozen identified compounds, several unknowns (homogeneous by GLC analysis), two sets of isomers, and several minor components were reported. Although there was some overlap in the ranges of values for most of the compounds observed, overall differences are evident. Within the major components, β-pinene was present in 40–63% relative amounts in var. *ponderosa* as opposed to 32–45% in var. *scopulorum.* Similarly, α-pinene was present in 8–18% and 15–26% and estragole (*p*-methoxyallylbenzene, 118) was present in 1–21% and trace—6% in var. *ponderosa* and var. *scopulorum*, respectively. Several minor components, including limonene, ocimene, and camphene, tended to be present in higher concentrations in eastern trees. There was also a tendency for trees along the Continental Divide to exhibit terpene profiles of somewhat more intermediate composition, with some components present in concentrations more typical of "western" trees while others resembled "eastern" profiles. Thus, this area appears to be an area of transition between the two varieties. Although only one specimen of var. *arizonica* was studied, the differences observed suggest that major differences exist relative to the northern varieties. Thus, whereas the northern varieties featured β-pinene as the major oil component (total range 32.0–63.0%), this compound was present in only 18.4% in the specimen of var. *arizonica.* The major component of oil from var. *arizonica* was α-pinene at 37% as compared to the total range of 8.0–26% for the northern varieties. Other compounds that were present in levels higher than seen in the northern varieties include limonene and linaloöl. Given the obvious limitations placed on these conclusions because of the sampling, it would seem a reasonable hypothesis to suggest that the terpene profile of var. *arizonica* more closely approximates that of var. *scopulorum* than it does to that of var. *ponderosa.*

Pinus monticola Dougl., the western white pine, occurs in the coast ranges of British Columbia south through the Cascade Mountains of Washington and Oregon, and in the Sierra Nevada, and in a second band in the Rocky Mountains of British Columbia south through northern Idaho, and western Montana, and in the mountains of northeastern Oregon. An earlier chemical study of specimens from 20 sites in British Columbia, Washington, and northernmost Oregon and Idaho showed little geographical patterning of foliar monoterpenes (Hunt and von Rudloff, 1977). Hanover (1966b) studied the genetic control of monoterpene levels in this species of pine.

In order to expand the chemical sampling of this species to its entire range and to investigate xylem terpenoids in detail, Zavarin et al. (1990) collected wood samples from 191 trees from 20 areas, isolated the monoterpene fractions, and subjected them to GLC analysis. The major component was identified as α-pinene, with β-pinene, 3-carene, and limonene as lesser, but still important, components. Minor components (up to 1.5% of total monoterpene fraction) included camphene, sabinene, myrcene, β-phellandrene, γ-terpinene, terpinolene, and *p*-cymene. In general, levels of variation within and among populations were high with very little difference between the profiles of trees from coastal British Columbia and Washington and those from the Rocky Mountains, all of which suggested that these populations have occupied their current range for a comparatively short time. Critchfield (1984) suggested, on the basis of glacial history, fossil evidence, and dispersibility factors, that *P. monticola* reoccupied this region some 5800–7000 years ago, having survived in a refugium in southwestern Oregon.

From central Oregon south, however, the situation appears to have been quite different. Again we rely on the occurrence of 3-carene to provide insights into the events of the past. With few exceptions, most populations with the *C* allele were observed in the southeastern Sierra Nevada Mountains in California with frequencies of the allele often approaching 100%. Moreover, Zavarin et al. (1990) suggested that these populations, including some that are quite isolated, originated from a second refugium that lay east of the area glaciated during the Wisconsin period. Following retreat of the ice, *P. monticola* expanded from its southern refugium both toward the west and north. At some point, the northward expansion would have brought trees with the southern genotype into contact with trees from the northern refugium that was, in their turn, moving northward into Washington and British Columbia and southward into the northern Sierra Nevada.

A discussion of the bristlecone pines requires comment on all three members of *Pinus* sect. *Parrya* Mayr subsect. *Balfourianae* Engelm.: *P. aristata* Engelm., the Rocky Mountain bristlecone pine; *P. longaeva* D. K. Bailey, the Great Basin bristlecone pine; and *P. balfouriana* Grev. & Balf., the foxtail pine. Until 1970, the term bristlecone pine referred only to *P. aristata*, but the situation changed with the recognition of *P. longaeva* by D. K. Bailey (1970). More specifically, *P. aristata* occurs in the Rocky Mountains of Colorado south into New Mexico and in a disjunct population in the San Francisco Mountains in north central Arizona (map reference Flagstaff). *Pinus longaeva* occurs in Utah, extensively in the mountains of east central Nevada, and disjunctly in the Spring Mountains of southern Nevada, and in the White Mountains of east central California (Fig. 3.9). (The proximity of *P. longaeva* and *P. balfouriana* will be commented on below.)

Until the work of Zavarin and Snajberk (1973a), the only chemical information available on the bristlecone pines was the report by Haagen-Smit et al. (1950) that the major component of *P. longaeva* (at that time referred to as *P. aristata*) turpentine was α-pinene and, interestingly, that the optical rotation of α-pinene from *P. longaeva* and southern specimens of *P. balfouriana* was *dextro* rotatory (+26.12° and +23.72°, respectively), whereas that of northern specimens of *P. balfouriana* was *levo* rotatory (−10.23°).

The first comparative study of monoterpenes in these taxa was that of Zavarin and Snajberk (1973a) who examined wood monoterpenes from 67 trees at three sites representing *P. aristata*, and four sites representing *P. longaeva* and reported very significant differences. The major component of *P. longaeva* turpentine was, as originally reported by Haagen-Smit et al. (1950), α-pinene, which constituted

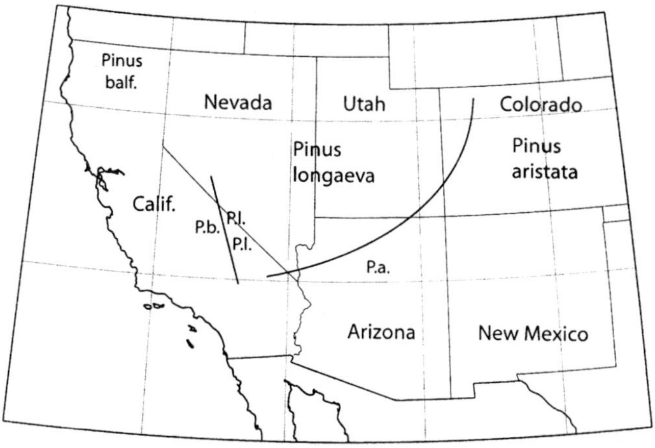

Fig. 3.9 Map for Bristlecone pine discussion: *Pinus aristata* (P.a.), *Pinus longaeva* (P.l.), and *Pinus balfouriana* (P.b.)

85.3–98.2% of the monoterpene fraction with the balance comprised of 3-carene and limonene. Turpentine of *P. aristata*, however, consisted mainly of 3-carene with concentrations in the range 62.7–92.1%. Limonene reached a value of 17.9% in one population with α-pinene ranging between 1.4 and 5.3% at the three sites. Minor components also contributed to the distinctive chemistries with sabinene, β-phellandrene, myrcene, and terpinolene regularly present, albeit in small amounts. These compounds were observed only sporadically, and then only in trace amounts, in collections of *P. longaeva.* Running the risk of overstatement, it is useful to point to the authors' finding that even small differences in the concentrations of camphene and β-pinene were statistically significant, and that such vivid chemotaxonomic differences are rarely seen.

With the establishment of clear-cut differences between the two species, the finding of trees from both that deviated widely from their respective population chemistry takes on special significance. Selected data document these differences (Table 1 in Zavarin and Snajberk, 1973a). In order to address the problem of inadequate sample size, the areas were revisited. In the second report from these workers (Zavarin et al., 1976), the monoterpenes of both wood and foliage of over 200 additional trees were analyzed (including additional *P. balfouriana* specimens). These new data fully substantiated the findings of the earlier work, as well as served to locate additional trees with deviant chemistries. The wood monoterpene profiles of the northern Arizona population agreed with the earlier findings (normal profile), but the newly analyzed foliar monoterpene profile clearly suggested an intermediate position between the two species with closer affinities to *P. aristata.* The authors suggested that selection for the present-day chemical profile has occurred, but offered no suggestion as to what the selecting factor or factors might have been that favored particular terpene compositions.

These workers also addressed the problem of abnormal production of 3-carene by certain trees in the California population of *P. longaeva*. Controlled crosses with *P. balfouriana* individuals having known levels of 3-carene production led to the conclusion that gene flow between *P. balfouriana* and *P. longaeva* could be ruled out as an explanation for the deviant profiles. The alternative explanation would involve the selective elimination of the capacity to make 3-carene. Again, no immediate selective advantage of the carene-free state comes to mind. The existence of 3-carene-producing trees in the first place could have been due to limited gene exchange between *P. aristata* (or an *aristata*-like ancestor) and *P. longaeva* (likewise) through a path south of the Grand Canyon of the Colorado River that no longer exists.

The third paper in the bristlecone pine series (Snajberk et al., 1979) addressed the issue of chemical differentiation within *P. balfouriana*. This species exhibits a disjunct distribution with populations known from the mountains of northern California (Marble Mountains, Salmon Mountains, Trinity Alps, and Yolla Bolly Mountains), and in the southern Sierra Nevada where the range lies just to the west of the range of *P. longaeva* (they are separated by 35 km across the Owens Valley). Analysis of the monoterpenes of representative specimens revealed that the northern and southern populations can be distinguished primarily on the basis of their different foliar α-pinene concentrations. It is interesting to note, recalling the significance of differences between wood and foliar monoterpenes in the Arizona populations of *P. aristata*, that monoterpenes from the wood of *P. balfouriana* did not show any differentiation. Other differences between northern and southern populations of *P. balfouriana* included higher limonene content in wood from northern populations and higher 3-carene content in foliar monoterpenes from northern populations. These chemical differences were taken as support for an earlier suggestion that the northern and southern forms of this species deserve subspecific recognition (Mastrogiuseppe, 1976).

The final paper in this series Zavarin et al. (1982) described studies of resin canal structural differences within the subsection, concluding that despite having *P. aristata*-like monoterpenes, the trees in the Panamint Mountains of California rightfully belong in *P. longaeva*. They also identified a chemical-latitudinal gradient in trees producing 3-carene. The gradient involves frequency of the I^{hc} allele (hc = high 3-carene as opposed to lc = low 3-carene), which exhibited the highest frequency (1.0) in trees in the Panamint Mountains (ca. 36°05′N), and ranged to a low value of 0.10 in the White Mountains (Esmerelda, Nevada, 37°49′N). Intermediate frequencies, 0.269 and 0.318, were observed for populations in the Inyo Mountains.

The level of genetic variation in *Pinus longaeva* in eastern Nevada and adjacent western Utah was investigated by Hiebert and Hamrick (1983). Seeds were collected from five populations, two from the Markagurt Plateau of Utah, and three from Nevada, one each representing the Snake Range, the Egan Range, and the White Pine Range. Electrophoretic analysis provided information for seven enzyme systems representing 14 loci. Variation among populations was low but statistically significant, whereas variation within populations was high. In general, the overall level of genetic variation observed in this study of bristlecone pines is in reasonable agreement with average values for 20 other conifer species (Hamrick et al., 1981): 78.6% polymorphic loci per population as opposed to 67%; 2.35 alleles per locus compared to 2.29, and 0.327 for mean individual heterozygosities, compared to 0.207, the mean conifer value. Hiebert and Hamrick (1983) suggested that the latter value was higher for the bristlecone populations primarily because of an even distribution of allele frequencies at the polymorphic loci.

There are two possible explanations for the present distribution of *P. longaeva* in the eastern Great Basin. In the first, long-distance dispersal of a small number of seeds via birds could have established the species' presence in the area. In this case, one would predict a relatively low level of intrapopulational variation, since each population would have originated from a very limited sampling of the parental genome and a relatively high level of interpopulational variation because the populations would not, in all probability, have arisen from related parents. In the second scenario, the present distribution of bristlecone pines reflects fragments of a once continuous distribution individual populations, of which became separated from others through a series of climate changes, resulting in extinction of lower-elevation stands associated with Pleistocene glacial advance and retreat. In this situation, one would expect to see higher levels of intrapopulational and lower levels of interpopulational variation. Since most of the observed genetic variation occurs within populations, the second explanation for the present distribution seems the more likely. This view is supported by paleo-ecological observations, including radiocarbon dating of bristlecone pine needles, 10,000–40,000 years old, found in pack rat middens at much lower elevations (1900 m) than those at which the trees exist at the present time. It was also noted by the authors that, since glacial periods tend to be longer than interglacials, it is likely that bristlecone pines spent a considerable amount of time in large, continuous populations. The low level of genetic variation observed by Hiebert and Hamrick (1983) suggests that bristlecone pine populations in this area have not lived in isolation from one another for a long time.

Pinus torreyana C. Parry ex Carrière, the Torrey pine, occurs in southern California and exhibits an interesting natural distribution consisting of two disjunct populations, one on Santa Rosa Island, which lies about 50 km south of Santa Barbara (Santa Barbara County), and one on the mainland immediately south of the town of Del Mar on the northern edge of San Diego (San Diego County). They are separated by about 280 km. The relationships and taxonomy of Torrey pine were discussed in detail by Haller (1966, 1986), who described the two populations as two subspecies, *P. torreyana* subsp. *torreyana* of the mainland, and *P. torreyana* subsp. *insularis* Haller from the island. Features upon which the distinction was made included crown shape, needle color, cone features, seed width, and seed color. A further distinction between the two forms came from a study of the composition of their turpentines (Zavarin et al., 1967). Table 3.2 summarizes their results. Two qualitative differences can be noted; *n*-decylaldehyde (ten-carbon hydrocarbon aldehyde) and cineole [41] were present in small amounts in the mainland material, but were not seen in material from the island. Concentration differences were observed with all components but most were not statistically significant. Two quantitative differences that were judged significant by statistical tests involved limonene [24] and

Table 3.2 Comparison of concentrations of selected components from Santa Rosa Island and mainland (Del Mar) populations of *Pinus torreyana* turpentine (from Zavarin et al., 1967)

Compound	Santa Rosa	Del Mar
Limonene	73.4 (2.1)[a]	84.2 (2.2)
β-Phellandrene	8.7 (1.2)	0.1 (0.15)
Cineole	nd	0.9 (0.5)
n-Decylaldehyde	nd	0.4 (0.2)

[a] Percentage of total terpenes (mean deviation).

ß-phellandrene [130]. The ß-phellandrene difference amounts to a factor of nearly 90, which represents a significantly different flow of carbon through that particular part of the terpene biosynthetic pathway.

Land bridges, rafting islands, and long-distance dispersal have all been suggested as means by which the present distribution became established. Geological evidence suggests that the island separated from the mainland during the Miocene, and that it has experienced rotation both northward and westward (Axelrod, 1980). Based on the amount and rate of genetic divergence between the mainland and island populations, however, Ledig and Conkle (1983) tentatively concluded that Torrey pine reached the island from the mainland more recently by "sweepstakes" dispersal. Coupled with their genetic information, they cited the absence of a Pleistocene land bridge (Junger and Johnson, 1980) and a disharmonic fauna (Wenner and Johnson, 1980) as evidence for the "sweepstakes" hypothesis. Haller (1986), however, pointed out that the flora of the larger Californian islands is neither depauperate nor disharmonic relative to areas of comparable size and habitat variation on the mainland. He also noted that *P. torreyana* is not alone as an endemic on Santa Rosa; *Quercus tomentella* Engelm., *Prunus lyonii* (Eastw.) Sargent, and *Lyonothamnus floribundus* A. Gray are also relic species that were well known on the mainland during the Tertiary, and that they have not been known on the mainland for more than 3 million years (see Haller for citations). Other possibilities include a formerly much more extensive distribution of Torrey pine, including other islands, and that both current populations are relics. After summarizing the various sources of evidence, Haller (1986) concluded that the island population can be explained perfectly well by means of either past land connections or by "normal" dispersal (Axelrod, 1952). More recently, Waters and Schaal (1991) added valuable information from their comparison of chloroplast DNA from plants collected at the two sites. The two populations were monomorphic at over 150 restriction sites (26 enzymes) and the chloroplast genomes were identical in length. Those authors suggested that *P. torreyana* has been monomorphic since before the populations became isolated. It is perhaps unnecessary to note that situations such as the Torrey pine, which appear simple, are only deceptively so. Their resolution requires all of the tools available to the biogeographer.

These examples of terpenoid analyses represent only a sampling of the information available on pines. Several other studies involving North American and Mexican pines are listed in Table 3.3.

Table 3.3 Additional studies of *Pinus* terpenes

Species	Common name	Reference
P. cembroides Zucc.	Mexican piñon pine	Zavarin and Snajberk (1985)
P. flexilis James	Limber pine	Zavarin et al. (1993)
P. monophylla Torr.	Single-leaf piñon	Snajberk et al. (1982) Zavarin et al. (1990)
P. nigra Arn.	European black pine	Rafi and Dodd (1996)
P. quadrifolia Parl.	Parry piñon	Snajberk et al. (1982)
P. radiata (Little) D. Don	Monterey pine	Cool and Zavarin (1992)
P. remota (Little) Bailey & Hawksworth		Snajberk and Zavarin (1986)

3.2.5 Pseudotsuga

Pseudotsuga is represented in North America by two species, *P. menziesii* (Mirb.) Franco, the Douglas fir, and *P. macrocarpa* (Vasey) Mayr, commonly known as big cone Douglas fir. True Douglas fir is widespread, ranging form British Columbia south through the Coast Ranges, along the Cascade Range, in the Rocky Mountains, in the Sierra Nevada, with a small population near Lompoc, California (Santa Barbara County). Big cone Douglas fir occupies a much smaller range in southern coastal California. Both species have been subjected to detailed analysis of their monoterpenes.

A study of terpene composition of Douglas fir by von Rudloff (1973) established existence of geographic pattern in the volatile components. Zavarin and Snajberk (1973b) examined *P. menziesii* using specimens collected from its entire range. Four distinct chemical races were found to exist: (1) the coastal race (Oregon, Washington, and British Columbia) characterized by high levels of sabinene and terpinolene; (2) the northern inland race (Rocky Mountains of Canada and the United States north of the Snake River Basin) characterized by high levels of α-pinene with only traces of sabinene; (3) the southern inland race, characterized by high levels of α-pinene and limonene with only traces of sabinene; and (4) the Sierra Nevada race (central Sierra Nevada Mountains) characterized by high levels of α-pinene and β-pinene and traces of sabinene. Other compounds identified, but not figuring in the definition of chemical races, were tricyclene, α-thujene, camphene, 3-carene, myrcene, β-phellandrene, γ-terpinene, and *trans*-ocimene.

The coastal race conforms to *P. menziesii* var. *menziesii*, whereas the inland races accord well with *P. menziesii* var. *glauca* (Beissn.) Franco, the form known from the Rocky Mountains. There is no established morphological equivalent to the Sierra Nevada race; trees from that area are usually referred to var. *menziesii.* Zavarin and Snajberk (1975), however, summarized differences between trees harvested in the Sierra Nevada Mountains and those harvested in northwestern California with regard to commercial applications, particularly in the manufacture of plywood or the production of composite materials (particle board). Differences in such factors as wood permeability (important for production of glued products), pH, bending strength, and brittleness, etc., are well known in the industry. Such factors, of vital importance for commercial applications, and likely based upon genetic differences to a certain extent; however, do not figure in taxonomic considerations.

Although the races described above were statistically significantly different, a degree of intergradation was evident. Thus, the northern inland race and coastal race intergrade in central British Columbia, northeastern Washington, and northern Idaho as well as in the mountains of central and east central Oregon. The coastal race intergrades with the Sierra Nevada race in coastal and northern California. A subsequent study, involving over 300 trees from 30 localities throughout California and southern Oregon, confirmed intergradation of the Sierra Nevada race with the coastal race north of about 40°N latitude. Pure, unhybridized Sierra Nevada trees occurred south of 40°N latitude (Zavarin and Snajberk, 1975). The chemical data suggested a closer alignment of the Sierra Nevada race with the inland races than with the coastal form. The present-day geographical separation of the Sierra Nevada and inland races is thought to have occurred as a result of the development of the Great Basin through increasing aridity that occurred during the Pliocene-Pleistocene epochs.

The third paper in this set Zavarin and Snajberk (1976) described their efforts to detect chemical races within big cone Douglas fir. Analysis of the cortical monoterpenoid fraction of 33 trees revealed that the major component was α-pinene, with β-pinene, 3-carene, and limonene present in lesser amounts. The monoterpene profiles of different populations varied somewhat from each other, but the overall profile of big cone Douglas fir was clearly different from that of Douglas fir. There was no evidence for gene flow between the southernmost population of Douglas fir at Lompoc and the closest population of big cone Douglas fir at Figueroa, sites separated by only 34 km.

A recent examination of conifer terpenes and their interaction with bark beetles involves *Pseudotsuga menziesii* as one of the subject taxa (Pureswaran et al. 2004). Quantitative analysis of leaf and bole terpenes was conducted on *P. menziesii*, *Pinus contorta* var. *latifolia* Engelm. (lodgepole pine), *Picea engelmannii* × *glauca* (interior spruce), and *Abies lasiocarpa* × *bifolia* (interior fir) (nomenclature from that paper) growing sympatrically at three sites in British Columbia. Collections were made along a transect running south to north over a distance of about 475 km [Princeton in the south (49°25′N, 120°35′W), 100 Mile House in the center and Prince George in the north (53°55′N, 122°49′W)]. A fourth sample of *P. menziesii* from Maple Ridge, B.C., represented the coastal form of Douglas fir.

The four species shared the same array of simple terpenes, but significant differences in amounts of individual compounds were observed. The differences were considered large enough to support the idea that species-specific bark beetles were responding to the individual mixtures of compounds produced by the trees. In the case of Douglas fir, for example, (+)-α-pinene rose from being a minor component of bole terpenes in the Maple Ridge (coastal) collection to significant contributor in trees from the Princeton site, to being the major compound in mixtures at the 100 Mile House and Prince George sites. (−)-β-Pinene followed much the same pattern.

Foliage of Douglas fir had only trace amounts of (+)-camphene in trees from the coast but major concentrations in trees from the other three sites. In contrast, terpene mixtures from lodgepole pine leaf and bole tissues had nearly the same levels of the major compound, (−)-β-phellandrene, and minor components regardless of origin. Patterns of variation in the other species were less extreme.

3.2.6 Sequoia sempervirens *(Cupressaceae)*

Sequoia sempervirens (D. Don) Endl., the coast redwood, is one of the most striking members of coastal forests of the western United States. The range of the redwoods extends from southwestern Oregon to the Santa Cruz region of central California. Although restricted to a strip of coastal North America at the present time, fossil evidence shows that *Sequoia* was well developed in the Upper Jurassic and that the genus enjoyed a wide distribution in appropriate habitats throughout the Northern Hemisphere. Despite the importance of this species, both from the commercial forestry standpoint as well as a species of major tourist attraction, comparatively little work had been done on establishing limits of variation with the apparent exception of varieties based upon leaf and branch morphology (Dallimore and Jackson, 1966).

In order to rectify this state of affairs, Hall and Langenheim (1987) undertook a study of monoterpenes of the leaves of redwood collected from sites representing essentially the entire range of the species. This was done in two ways, sampling of 13 wild populations and sampling of trees grown from seed collected at eight locations throughout the species' range and grown and maintained under cultivation for about 20 years at the Russell Reservation, Contra Costa County, California. Fifteen compounds were identified and quantified as percentages of the total monoterpene fraction. α-Pinene was the major monoterpene followed, in descending order, by the common terpenes limonene, sabinene, β-phellandrene, γ-terpinene, and myrcene. The other identified compounds, α-thujene, camphene, β-pinene, α-phellandrene, α-terpinene, the *cis*- and *trans*-isomers of ocimene, *p*-cymene, and terpinolene, were present at levels too low to be of use in numerical analysis. In general, results from analysis of wild-collected material and cultivated material were in agreement with similar levels of interpopulation variation which points to the source of variation being genetic rather than environmental.

Cluster analysis revealed the existence of three groups of populations with a strong break in the vicinity of San Francisco Bay (Fig. 3.10). Population group **3** occurred exclusively south of San Francisco Bay, whereas the other two types occur scattered throughout the northern part of the range of the redwoods. Although the three population groups exhibited all six marker terpenes, obvious distinctions can be seen, even with visual examination of the data. Thus, all northern populations, that is, population groups **1** and **2**, exhibited higher levels of α-pinene and β-phellandrene. The highest levels of limonene, sabinene, myrcene, and γ-terpinene were observed in type **1** populations, although the level of variation among populations was substantial.

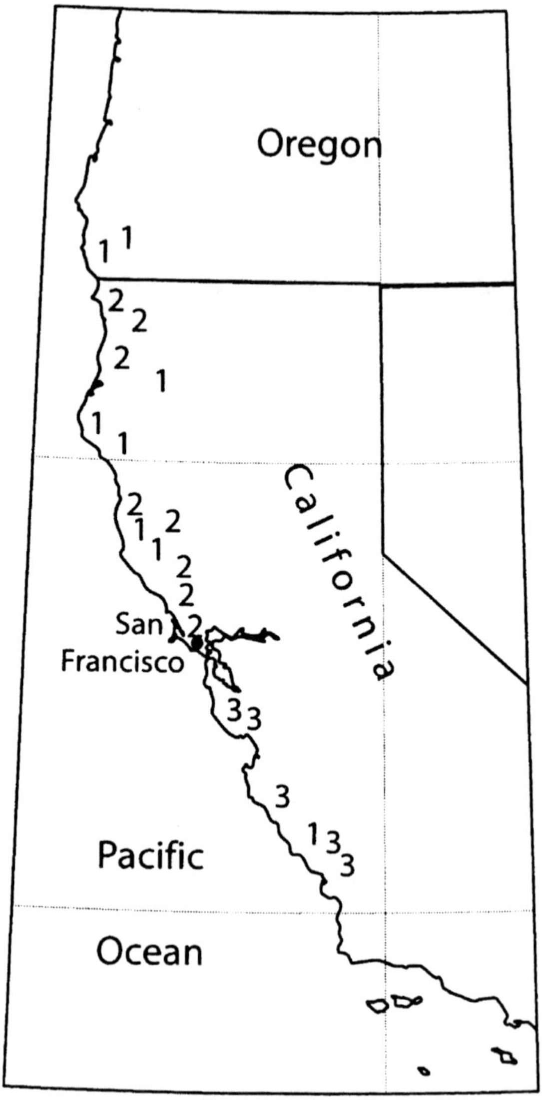

Fig. 3.10 Map of Redwood *(Sequoia)* distribution. The numbers refer to the three chemotypes

Further distinctions between the northern population groups can be made. Thus, only group **2** populations were found in Marin and Sonoma Counties. Central Mendocino County appears to be an area of transition characterized by the appearance of group **1** populations, which then continue northward through Humboldt and Del Norte Counties into Oregon. Located near the northern end of the range of the redwoods, however, were three populations, one in extreme northern Humboldt County and two in Del Norte County very near to the Oregon border, that bear more similarity to populations in the transitional area in central Mendocino County than they do to type **1** populations in Oregon. In a discussion of this phenomenon, Hall and Langenheim (1987) pointed out the possibility that the Klamath region, which is well known as a center of conifer diversity, might have served as a link between the northern populations and those populations farther south in Mendocino County. The author follows their lead in quoting Whittaker (1961) who considered it as "… a center toward which mesophytic forests of the past have shrunk, and as a center of accumulation of species of varied evolutionary history in the diverse habitats of ancient land surfaces." He went on to say that the area has served "… as a reservoir of species populations of diverse environmental adaptations and of genetic diversity within some species and species-complexes, from which populations have evolved and migrated into other areas."

3.2.7 Juniperus *(Cupressaceae), and a Comment on* Marshallia graminiifolia *(Asteraceae)*

The next example involves taxa native to the southeastern United States, *Juniperus virginiana* L. and *J. silicicola* (Small) Bailey. The third species of juniper from that corner of North America, *J. communis* L., is not involved in the following story and is not considered here. *Juniperus virginiana* is a widespread upland species of much of eastern North America, while *J. silicicola* was considered to be restricted to coastal sand dunes ranging from North Carolina southward through much of northern Florida, westward along the Gulf Coast into southeastern Louisiana, and disjunctly in southeastern Texas. This was the situation that existed at the time of the detailed examination of morphological and terpenoid features of both taxa described by Adams (1986). Data were accumulated for 15 morphological features and 10 terpenes from plants at 14 sites ranging from near Washington, D.C. to two sites in southeastern Texas. Limonene was the major compound identified in most specimens, with sabinene and 4-terpineol making significant contributions to the profile of other sites. Multivariate analysis revealed that the two taxa were more similar in both morphological and chemical features than had been thought. In fact, two characters used in diagnostic keys, female-cone size and leaf-tip shape, were not significantly different between the two taxa. That finding, coupled with the absence of any qualitative terpene differences between the two taxa, suggested that recognition of two species is not justified and that the coastal juniper should best be treated at the varietal level, *J. virginiana* var. *silicicola.* A further result of the study was that the Texas populations, originally thought to be *J. silicicola*, exhibited features, both chemical and structural, that fell within the ranges of variation established for *J. virginiana* var. *virginiana.*

Some readers may wish to look at the study of Watson et al. (1994) involving the *Marshallia graminifolia* (Walter) Small (Asteraceae) complex, which occurs along the Atlantic and Gulf coastal plains essentially paralleling the juniper example just discussed. Those workers observed no geographic patterning among populations in a study of nine enzymes encoded by 16 loci.

3.3 Europe

3.3.1 Pinus halepensis *(Pinaceae)*

The three examples from Europe all involve species of pine, the first of which is the Aleppo pine, *Pinus halepensis* Mill. The Aleppo pine enjoys a natural range that extends from Spain and Morocco in the west to Jordan in the east and from Israel in the south to the Rhone Valley in France (Nahal, 1962; Jalas and Suominen, 1973). Variation in this taxon has been studied using a variety of approaches, which provides an opportunity to examine its past from several points of view. An early division of the species into three geographical groups, "Oriental," "Occidental," and "North African", was based upon palynological data (Nahal, 1962). Differences in needle anatomy and morphology were shown by Calamassi (1986) to be correlated with latitude, longitude, and altitude. Schiller et al. (1986) found that isozyme data led to the division of the species into two major groups, eastern and western Mediterranean. Further resolution of the main western group into four races was based upon allele frequencies and genetic distances. A study of cortical monoterpenes led Shiller and Grundwald (1987b) to distinguish three groups, Greek, western European, and North African. Baradat et al. (1995) studied the terpene chemistry of more extensive collections and concluded that the variation observed did not conform to any straightforward pattern of migration and could well reflect a degree of human intervention.

A recent study of chemical variation in Aleppo pine involves the work of Kaundun et al. (1998a, b) on needle flavonoids. In this study, 215 trees from nine populations were sampled. Six of these originated from experimental gardens; three were collected from nature (two from France, one from Algeria). Countries of origin were: Morocco, Tunisia, Algeria, Spain, France, Italy, and Greece. Plant material was extracted and the extracts subjected to HPLC analysis which allowed identification of procyanidin, identified as cyanidin [278], and prodelphinidin, identified as delphinidin [279], kaempferol [280], quercetin [281], isorhamnetin [282], myricetin [283], larycitrin [284], and syringetin [285] (see Fig. 3.11 for structures). Differences between trees sampled in nature and those grown in experimental gardens were significant, but flavonoid differences between trees grown in the gardens were minimal. Of the eight compounds identified in this study, only myricetin showed significant quantitative differences among the populations. Cluster analysis pointed to the existence of three major geographical groups: (1) Greece; (2) North African/Western European (populations from Tunisia, Morocco and Spain); and (3) Southern European (populations from France and Italy). This grouping reflected a decrease in the myricetin level as one went from highest

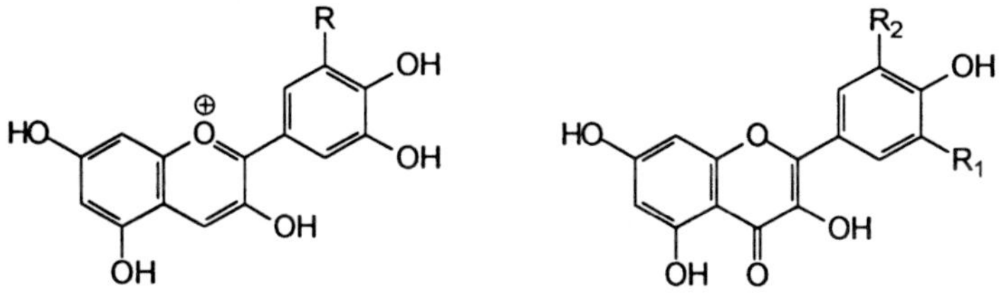

[278] R = H
[279] R = OH

[280] $R_1 = R_2 = H$
[281] $R_1 = OH, R_2 = H$
[282] $R_1 = OCH_3, R_2 = H$
[283] $R_1 = R_2 = OH$
[284] $R_1 = OCH_3, R_2 = OH$
[285] $R_1 = R_2 = OCH_3$

Fig. 3.11 Compounds 278–285, flavonoids from *Pinus halepensis*

amounts, ca. 16.6% in Greek populations, to the lowest amounts, ca. 10.2%, in the northernmost populations. The authors reasoned that since loss of trihydroxylated B-rings is considered an advanced feature of flavonoids (Harborne, 1967; Swain, 1975) the Greek population, with the highest concentration of myricetin, must represent the "archetype population" of the species. With the lowest concentrations of myricetin, the Italian and French populations would, therefore, represent the more advanced state. This trend in flavonoid oxidation level was taken to indicate that Aleppo pine occupied regions in the Middle East during the height of glaciation and that migration, after retreat of the ice, followed a pathway westward across North Africa to western Europe. Migration into southern Europe was a later event, although Kaundun et al. (1998a, b) offered no estimate of the time involved. This conclusion is in line with that of Shiller et al. (1986) based on terpenoid data, but is contradicted by the conclusion of Nahal (1962) and Panestos (1975) who suggested, on the basis of palynological and morphological evidence, respectively, that Aleppo pine originated in southern and central Europe. Macromolecular data would likely resolve this issue.

In a study of isozymes of *P. halepensis*, Shiller et al. (1987a, b) found alleles characteristic of *P. brutia* in trees from the Greek province of Chalkidiki. Other features suggested that these two species had undergone hybridization in this area [see Kaundun et al. (1998a) for further references]. The flavonoid profile of trees from that area (Kaundun et al., 1998a, b) was typical of *P. halepensis*, however, which was unexpected in light of their earlier work documenting differences between the flavonoid profiles of the two species (Kaundun et al., 1997). Two explanations were offered: (1) hybridization does occur in the area, but offspring exhibit only the *P. halepensis* profile; or (2) hybridization does not occur. The author would like to suggest that a third explanation might be equally possible. The differences between the two species lie in relative quantities of quercetin and myricetin: quercetin is a major component in *P. brutia*, with myricetin present in much smaller amounts, whereas in *P. halepensis* the situation is reversed. It is well known (Levy, 1976) that hybrids often exhibit flavonoid profiles that differ from both parents, a phenomenon that might result from disruption of control systems caused by mixing of genomes.

3.3.2 Pinus brutia *var.* brutia *(Pinaceae)*

Pinus brutia Ten. is a complex assemblage of subspecies that occurs on islands in the eastern Mediterranean and extends as far east as Iraq. Four subspecies are known but only the most widely distributed one, *P. brutia* subsp. *brutia*, will be dealt with in detail here. This species was first encountered above in a brief statement concerning the similarity of its flavonoid components to those of *P. halepensis.* Our interest at this point is a report describing flavonoid profiles observed in six populations, one from Crete, one from Greece, and four from Turkey (Kaundun et al., 1998b). Using HPLC, 150 trees were sampled and their flavonoid profiles were determined. Procyanidin, prodelphinidin, kaempferol, quercetin, isorhamnetin, myricetin, larycitrin, and syringetin were present in all individuals, but total proanthocyanidins and concentrations of isorhamnetin varied significantly, allowing recognition of eastern and western forms. This observation is in accord with both terpene data (Shiller and Grundwald, 1987a) and information from isozymes (Conkle et al., 1988). Eastern populations of *P. brutia* subsp. *brutia* approximate *P. brutia* subsp. *eldarica*, which Kaundun et al. (1998b) suggest originated from the former following glaciation.

3.3.3 Pinus sylvestris *(Pinaceae)*

Pinus sylvestris L. (Scotch pine) is a Eurasian species of very considerable economic importance used for both construction purposes and for pulp. Scotch pine has been introduced to North America where it grows well and is also of commercial value, including, among other things, Christmas tree production. The natural range of Scotch pine involves extensive tracts at low and mid elevations in northern Europe, whereas in southern Europe it is confined to much smaller patches at higher elevations. The considerable morphological variation characterizing the species has been recognized by definition of several varieties (*altaica, acquitana, armena, borussica, haguensis, hercynica, iberica, illyrica, lapponica, mongolica, pannonica, polonica, rhodopaea, rigensis, scotia, septentrionalis,* and *uralensis*). In an effort to gather additional information that might illuminate the evolutionary history of Scotch pine, Tobolski and Hanover (1971) undertook an examination of the monoterpenes of the cortical oleoresin of the species. One hundred eight trees representing the listed varieties, plus some local variants, were grown in a common garden in Michigan. GLC analysis of cortical monoterpene preparations revealed the presence of the common terpenes α- and β-pinenes, camphene, 3-carene, cymene, limonene, myrcene, β-phellandrene, α- and γ-terpinenes, and terpinolene. Numerical analysis revealed that α-pinene and 3-carene were the most variable of the compounds identified, with 3-carene present in highest concentrations in most specimens from northeastern Europe and Siberia. By contrast, 3-carene was present in much lower levels in most populations from southern Europe. The very low levels of 3-carene in the southern populations was taken as indication that those populations have not been in (reproductive) connection with the northern European ones for a very long time, possibly since Tertiary times. These data are in substantial agreement with several other studies that pointed to the existence of southern refugia. Further, the high concentrations of 3-carene in trees from southern Scandinavia and from northern Europe was taken to suggest that these were at one time part of a continuous population, that is, when land bridged the two areas. One should note that Scotch pine was much used by early humans and that some of the variation observed in this study may have been the result of early reforestation since it is known that extensive, and often severe, harvesting of this species was practiced, at least in England and Denmark (Godwin, 1975).

The range of *P. sylvestris* includes an area of highland Scotland, some 500 km from the nearest populations in Europe. This disjunct population has, of course, drawn the attention of several botanists interested in its origin, or origins as it turns out. Multiple colonizations of the Scottish Highlands from France were suggested by Birks (1989) on the basis of pollen data, whereas Bennett (1995) hypothesized two routes, one northward from France and a second along a more westerly route possibly involving the north of Ireland. Multiple colonizations would predictably lead to genetic heterogeneity. Evidence in the literature suggests that such is the case. Thus, Forrest (1980b) reported significantly lower levels of 3-carene from trees sampled in the Wester Ross area than from any other Scottish population. Subsequent studies of isozyme variation by Kinloch et al. (1986) revealed that a population at Shieldaig (57°31′N, 5°37′W), which lies within the Wester Ross group, was the most divergent from all other Scottish populations included in their study. They speculated that the isozyme markers were characteristic of ancestral plants, and that the markers have been retained for some 50 generations.

More recently, Sinclair et al. (1998) applied DNA sequence techniques to needle and dormant bud tissue from 466 trees representing 20 natural populations of Scots pine in Scotland. They employed a mitochondrial marker known to be maternally inherited in order to avoid genetic noise associated with extensive pollen flow among populations. Using probes for the *cox*1 gene, and following the usual procedures, these workers observed three restriction fragment patterns, two that were referred to as mitotypes **a** and **b**, and a third in a single individual that did not figure further in their analysis. Seventeen of the populations exhibited only mitotype **a**, with mitotype **b** present in three populations, Shieldaig (coordinates above), Glen Loy (56°5′N, 5°08′W), and Doire Darach (56°32′N, 4°47′W). Mitotype **b** was represented in these three populations at frequencies of 0.16, 0.59, and 0.08, respectively. It was also reported in that paper that mitotype **b** was not detected, using the *cox*1 probes, in 85 trees from northern Europe (northern France, Germany, Poland, southern Sweden, and Russia). That observation and the presence of mitotype **b** only in western Scottish populations were taken as clear support for a second route of colonization, as had been postulated earlier. The hypothesis that the western route involved populations of *P. sylvestris* in Ireland cannot be tested owing to the extinction of natural populations in that country at least 1000 years ago (Birks, 1989). The trail goes cold at this point; the postglacial source of the "western" Scots pine must remain a matter of speculation, at least until additional information becomes available.

Although the existence of a western pathway to account for the western Scottish strain of Scots pine has not been resolved, additional information on its genetic diversity in other parts of its range has been obtained (Sinclair et al., 1998). Thirty-eight populations were analyzed: Scotland (20 from the earlier work), Russia (1), Germany (1), France (1), Poland (1), Sweden (2), Norway (1), Finland (2), Italy (2), and Spain (7). Application of the mitochondrial *cox*1-based probe revealed the existence of three mitotypes, **a**, **b**, and **d**. The populations in Spain exhibited the highest level of heterogeneity with all three major mitotypes present. Noteworthy is the finding that mitotype **d** was observed only in the southernmost population near Baza (37°30′N, 2°45′W). In this population, 15 trees exhibited the **d** mitotype with mitotype **a** seen in only one. Two other Spanish populations, at Gudar and Orihuela (both southeastern Spain), exhibited mixed mitotypes, 30 **a** and five **b** in the former, four **a** and five **b** in the latter. Four other populations were homogeneous, two having only mitotype **a**, and two with only **b**. The rest of the European populations were homogeneous. The Italian, Finnish, Norwegian, and one of the Swedish populations exhibited exclusively mitotype **b**, while the other Swedish population plus those from Poland, France, Germany, and Russia were exclusively mitotype **a**. The preponderance of mitotype **a** in Scottish populations was discussed above. Several interpretations of these data were made, the first of which held that the southern Spanish site represents the descendants of one of the glacial refugia, and that the unusual mitotype **d** has been lost as descendants migrated northward. As attractive as this hypothesis might sound, other data do not support it. Thus, in the case of flavonoids (Lebreton et al., 1990), terpenoids (Tobolski and Hanover, 1971), and isozymes (Prus-Glowacki and Stephan, 1994), considerable differentiation has occurred between Spanish and northern European populations of *P. sylvestris.* The authors suggest that a more plausible explanation is simply that these sites represent isolated populations that survived glaciation but did not contribute to recolonization of northern Europe. Long isolation has led to the accumulation of sufficient differences that taxonomists may recognize discrete taxa, *P. sylvestris* subsp. *nevadensis* (Christ.) S. Rivas-Martinez, A. Asensi, J. Molero-Mesa and F. Valle, in the case of Scots pine (Vidakovic, 1991).

Sinclair et al. (1998) also addressed the origin of the difference in distribution of mitotypes **a** and **b** in northern Europe, northern France, Germany, Poland, and southern Sweden for the former, and northern Sweden, Norway, and Finland for the latter. They prefer to view this distribution as reflecting dual colonizations, one from the south through Denmark and the other from the northeast via Finland. Available data do not allow a resolution to this problem at the moment.

3.3.4 Abies alba *(Pinaceae)*

Abies alba Mill., the silver fir, is an important forest tree in Europe. Its size—it is the tallest on the continent—and quality of wood have attracted attention since early historic times. Turpentine from silver fir has also been used medicinally for many years (Mabberley, 1997, p. 1). Lang (1994) undertook a study of twig resin in an effort to see if the widespread natural distribution of the species in southern Europe was reflected in patterns of variation in monoterpene composition.

Trees (N = 1543), ranging from 3 to 6 years old, originated from 63 provenances representing the entire range of the species. Concentrations of four compounds, α-pinene, β-pinene, limonene, and β-phellandrene, allowed a sorting into three groups of provenances: (1) the Bavarian Forest, the eastern part of the alpine foothills, and some areas in the Alps; (2) western alpine foothills, the Black Forest, France, northern Italy, the Alps, and parts of eastern Europe; and (3) central and southern Italy, central and southern Yugoslavia, Romania, and Bulgaria. These conclusions are in general agreement with groupings obtained using other data, as reviewed by Lang.

3.4 South America

3.4.1 Fitzroya cupressoides *(Cupressaceae)*

Fitzroya cupressoides (Molina) I. M. Johnston, Alerce, the sole member of the genus, has been identified as one of the longest-lived tree species, second only to the bristlecone pines that we discussed earlier. In this regard, a tree from southern Chile has been dated at 3622 years by Lara and Villalba (1993), who used the tree ring information in a study of long-term climate changes. Our interest in *Fitzroya* here, however, involves two studies, one an attempt using secondary metabolites to gain a measure of genetic variation within its range, and one that directly addresses its glacial history.

The first study was that of Cool et al. (1991) who examined the patterns of variation in foliar terpenes. Foliage was sampled from 57 trees representing six locations in Chile, ranging from La Union (40°08′S) to Rio Negro (41°55′S) (Fig. 3.12). The monoterpene composition of foliage from all sites was quite similar, with α-pinene as the major component in all (88.2–91.1%); myrcene was the second most abundant compound (3.7–5.7%), followed by limonene (2.3–3.0%). No geographic patterning was observed in this data set as judged by discriminant function analysis.

The sesquiterpene results were a bit more encouraging, with clear-cut qualitative differences among groups of trees. Although the arrays of sesquiterpene derivatives were complex, and involved several unidentified compounds, sufficient information was available to allow the workers to identify three chemotypes. The basic chemotype (Type **0**) consisted of high amounts of germacrene-D [286], with lesser but still substantial amounts of caryophyllene [287], α-copaene [288], α-humulene [289], δ-cadinene [290], 4α-hydroxygermacra-1(10), 5-dinene [291], and a number of other sesquiterpene alcohols. Also present, but at much lower concentrations, was a group of compounds related to β-funebrene [292], β-acorenol [293], and

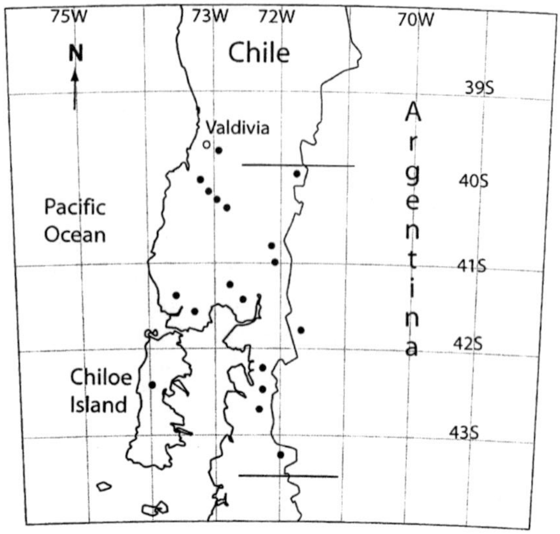

Fig. 3.12 Map of *Fitzroya cupressoides* occurrence

β-acoradiene [295]. These compounds were seen in all other trees but apparently at significantly different quantitative levels. For structures 286–298 see Fig. 3.13.

The second chemotype (their Type **1**) had, in addition to the Type **0** array, substantial amounts of α-longipinene [297] and an unidentified sesquiterpene alcohol. The third chemotype (their Type **2**) was distinguished by the presence of, among other compounds, cedrene isomers, [α-cedrene is shown as 298], and large amounts of the isomeric sesquiterpene alcohols α-acorenol [294] and its β-isomer [295]. The acoradiene isomers [295 and 296] were also identified. Some geographic patterning was observed in the Type **0** chemotype when the data were subjected to numerical analysis: a trend in the reduction of caryophyllene content was revealed in a west to east direction. The data sets for Types **1** and **2** were too small to allow for similar analysis.

The second study, which addressed glacial history directly, involved an analysis of isozymes and how patterns of variation could be used to test whether there had been a single refugium or whether there had been two, or possibly more, refugia (Premoli et al., 2000). Geomorphological and palynological evidence revealed that a refugium existed from northwestern Chiloé Island (ca. 40° S) and along the coast on the Chilean mainland to about 40° S latitude (Heusser and Flint, 1977; see also Vuilleumier, 1971). If this had been the sole refugium (their Single Refugium hypothesis), then one would predict that genetic variation of trees along the eastern slopes of the Andes would be lower than that of trees in the refugial population.

The situation becomes somewhat more complicated if one considers the effects on genetic variation if more than one refugium had existed. Premoli et al. (2000) discussed two effects: the Cordillera effect and the extent-of-the-ice effect. In the case of the Cordillera effect, the assumption was made that there was incomplete ice coverage, resulting in patches of forest remaining intact on either side of the Andes; this amounts to the existence of multiple local refugia that would supply propagules

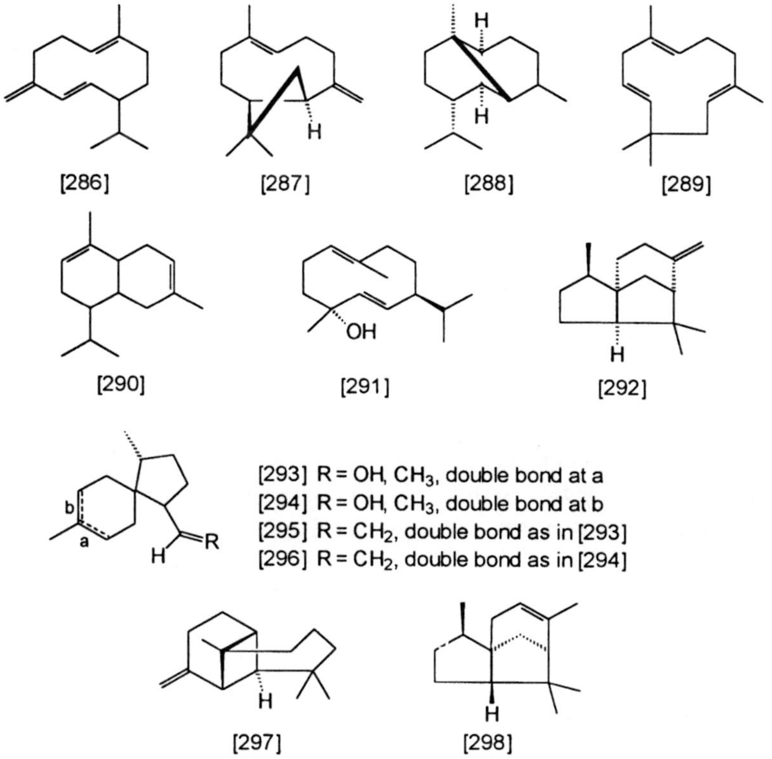

Fig. 3.13 Compounds 286–298, terpenoids from *Fitzroya cupressoides*. The horizontal lines at ca. 40°S and 43°30′S represent the range of occurrence of *Fitzroya*

to reestablish populations on their respective slopes of the mountains. The greatest genetic variation would be expected when comparing populations from eastern versus western slopes of the Andes. In this scenario, the mountain chain would have served as an effective barrier to gene exchange between trees on the opposite sides of the range.

The extent-of-the-ice effect is based on the assumption that there would be differential effects along a north south transect. This was based on the knowledge that the extent of ice cover was about twice as large at 42°30′S than it was at 39°50′S. If multiple refugia had existed, one would expect a greater degree of genetic isolation of populations near the present southern limit of the species than at its northern limit. Northern populations of *Fitzroya* could have originated from long-distance dispersal of propagules from a southern refugium, or they may have survived glaciation in patches and continued to exchange genes during the period of maximum glaciation. This scenario would lead to the prediction of greater genetic identity between northern populations and their southern source population(s) if long-distance dispersal had occurred, or between northern populations if there had been local survival.

In order to test these predictions, Premoli et al. (2000) collected plant material from 24 populations of *Fitzroya*, 12 on each side of the Andes, and subjected the foliage to appropriate isolation procedures and electrophoresis. Reliable information was obtained for 21 loci. Populations on the eastern side of the Andes were more variable than populations on the western side, as the following pairs of figures show: number of alleles per locus, 1.58 versus 1.37; total number of alleles, 33 versus 29; number of rare alleles, 5 versus 3.08; percent polymorphic loci, 39.3 versus 26.6; observed heterozygosity, 0.090 versus 0.057; and expected heterozygosity, 0.091 versus 0.063. Also, genetic identities of populations on the western slopes of the Andes were higher than those of populations from the eastern slopes. The higher levels of variation in eastern populations were in accord with preliminary studies using DNA markers (Allnutt et al., 1999).

The isozyme frequency data were then subjected to discriminant function analysis. Poor discrimination between groups was observed when the single refugium model was tested. However, when the data were analyzed using the groups suggested by the multiple refugia hypothesis, both scenarios, Cordillera effect and extent-of-the-ice effect, were supported. Although the data clearly suggest that multiple refugia existed, it is not possible on the basis of the available data to say how many were there. The data do suggest, however, that the southernmost populations in Argentina have been separated from western ones for a considerable period of time, and that a southeastern refugium may have existed. Much of the variation can be accounted for by northward flow from the southeastern refugium and eastward flow from the Chilean coastal refugium. The authors pointed out that the existence of a southeastern refugium, farther south than was expected, is another example of how patterns of glaciation differed in South America, where ice cover was often patchy, as compared to North America where ice cover was essentially total. Revegetation in South America, thus, started from several (many?) refugial centers, whereas revegetation in North America involved continent-wide movement northward.

3.4.2 Austrocedrus chilensis *(Cupressaceae)*

Although the impact of Pleistocene glaciation was not addressed in the following study, it did involve attempts to measure genetic variation in another South American tree species, *Austrocedrus chilensis* (D. Don) Pic.-Ser. and Bizz. This monotypic genus occurs in the cordillera of Chile and Argentina and disjunctly in the coastal mountains of Chile. Dodd and Rafii (1995) examined the seed megagametophyte fatty acids from 104 trees from 13 natural populations representing the range of the species, San Felipe in the north (32°39′S) to Corcovado in the south (43°32′S). Identified acids ranged from palmitic (16:0) to 5,11,14,17-eicosatetraenoic acid (20:4). The main compounds were shown to be oleic (18:1), linoleic (18:2), linolenic (18:3), and 5,11,14,17-eicosatetraenoic acids. Multivariate statistical analysis revealed a separation of the specimens into two groups, one comprising populations in the

Mediterranean sites in Chile and the steppe ecotone of Argentina, and a second group from mesic sites in the Argentinean lakes area and Chilean rain forest. The latter group was distinguished on the basis of their 20-carbon unsaturated acids.

3.4.3 Araucaria araucana *(Araucariaceae)*

Another example from South America involves *Araucaria araucana* (Molina) K. Koch. This tree occurs in the Andean cordillera between 37°27′S and 40°03′S, where it occurs on both sides of the crest of the range, and disjunctly in the Cordillera de Nahuelbuta of the Chilean coastal mountains. The species is considered a Chilean National Monument and is listed internationally as an endangered species. In an attempt to learn something about variation within this species, Rafii and Dodd (1998) examined foliar wax alkanes extracted from 40 individuals representing four natural populations. The major hydrocarbons identified were heptacosane (27 carbons), nonacosane (29 carbons), hentriacontane (31 carbons), and tritriacontane (33 carbons), with the 29-carbon compound present in highest concentrations in all populations. Numerical analysis of the data set, and construction of a dendrogram based on a cluster analysis of Nei's genetic distance measure revealed that there was a greater difference between west side Andean populations of *A. araucana* than there was between Andean populations and populations from the coastal mountains. These results are in accord with the hypothesis that plants living in more arid habitats produce higher percentages of longer-chain hydrocarbons. Chemotypes were not detected in this species.

Chapter 4
Intercontinental Disjunctions

4.1 Across the Atlantic Ocean

4.1.1 Datiscaceae

Until comparatively recently, Datiscaceae were thought by some taxonomists (e.g., Cronquist, 1981), to consist of four species in three genera, the monospecific *Octomeles* (*O. sumatrana* Miq.) and *Tetrameles* (*T. nudiflora* R. Br. ex Benn.) and *Datisca. Octomeles* and *Tetrameles* are large trees native to southeastern Asia; *Datisca* consists of *D. cannabina* L., which occurs from southwestern Asia to Crete, and *D. glomerata* (Presl) Baill., which occurs in California and northern Baja California, Mexico.

Earlier work on the flavonoid components of *D. cannabina* by Grisebach and Grambow (1968) and Pangarova and Zapesochnaya (1974) had revealed the existence of an unusual array of flavonoids that included glycosides of the rare 3,5,7-trihydroxyflavone galangin [299] and its 7-methyl ether [300], and 3,5,7,2′-tetrahydroxyflavone, datiscetin [301], and its 7-methyl ether [302] (see Fig. 4.1), along with the common flavonols: kaempferol and quercetin. Since galangin and datiscetin, and their methyl ethers, represented unusual substitution patterns among the flavonoids, it seemed reasonable to look at the other members of the putative family for their possible presence. A detailed study of *D. glomerata* showed that the unusual B-ring-substituted flavonoids were indeed present as major components, whereas the Asian taxa exhibited only common glycosides of kaempferol and quercetin (Bohm, 1988). The flavonoid profile of *D. glomerata* was homogeneous. This is not the case with isozymes or chloroplast DNA.

Despite the morphological similarity of *D. cannabina* and *D. glomerata*, and their distinctive flavonoid profiles, an electrophoretic examination revealed an entirely different picture (Liston et al., 1989). Calculation of the mean genetic identity among populations of both species, based upon variation of enzymes coded for at 21 loci, yielded a value of $I=0.142$. Intraspecific genetic identities were $I=0.649$ for *D. cannabina* and 0.847 for *D.*

The authors suggested the following sequence of events as a possible explanation for the present-day distribution. It has been thought that the DG-S plastome,

Fig. 4.1 Compounds 299–302, flavonoids from *Datisca*

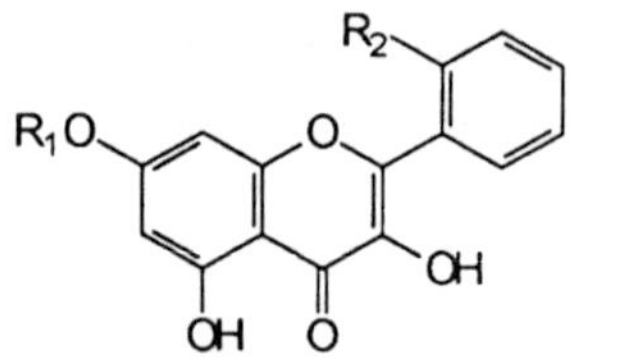

[299] $R_1 = R_2 = H$
[300] $R_1 = CH_3, R = H$
[301] $R_1 = H, R_2 = OH$
[302] $R_1 = CH_3, R_2 = OH$

considered the ancestral type from phylogenetic analyses, was widely distributed throughout the range of the species in pre-Pleistocene times. Climatic cooling would have resulted in retreat of the species south of the Transverse Ranges but with localized populations surviving north of the Transverse Ranges. The existence of relictual populations in Monterey and Santa Cruz Counties would be consistent with the existence of other relictual taxa in suitable habitats in that area (Raven and Axelrod, 1978). A northern population that survived the climatic changes could have become fixed for unique chloroplast DNA mutations and given rise to the present populations by range extension. The existence of the DG-N' population in Mendocino County attests to the potential for differentiation within the northern form. The authors also pointed out that the differentiation of these two forms may have occurred several million years ago and in fact have nothing to do with Pleistocene climate extremes.

A comparison of the cpDNA data for *D. cannabina* and *D. glomerata* revealed a sequence divergence of 0.87% ± 0.17%, which Liston et al. (1992) translated into a divergence time of 8.7 million years (± 1.7 million years) placing the divergence in the late Miocene. Sampling of *D. cannabina* was not sufficient to allow further speculation on the relationship between the two species. The authors noted in their concluding remarks that this pair of species represents another system where genetic differentiation has occurred with comparatively little accompanying morphological change.

4.1.2 Armeria maritima *(Plumbaginaceae)*

This example involves *Armeria maritima* (Mill.) Willd., which is distributed over central and coastal Europe, Great Britain, and parts of North America. Attempts to accommodate the amount of variation that characterizes this taxon are reflected in the existence of several infraspecific taxa, among which are subsp. *maritima* native to the coasts of Europe from northern Spain to Norway, subsp. *calaminaria* (Petri) Rothm. that appears principally in metalliferous surroundings, such as mine tailings; subsp. *sibirica* (Tuscz ex Boiss.) Nyman that occurs in the Arctic and in subarctic areas; and subsp. *californica* (Boiss.) Posild. Areas of occurrence of these subspecies are not necessarily contiguous as indicated by specimens of subsp.

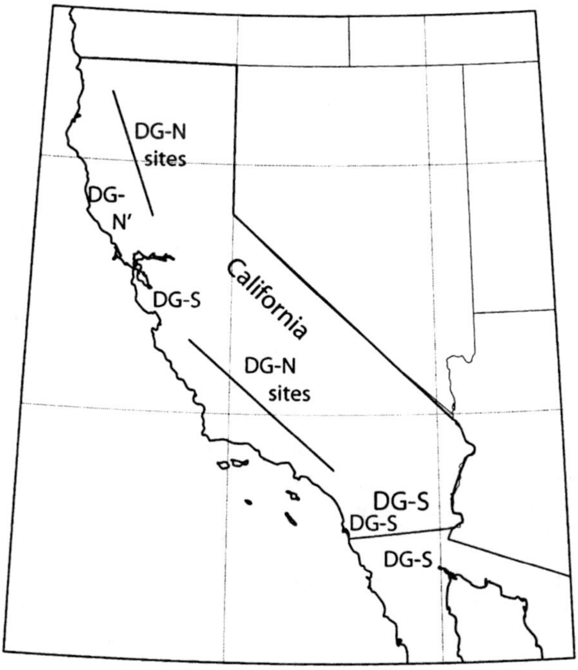

Fig. 4.2 Map of *Datisca glomerata* "plastones." Populations are identified as PG-N from the north and central parts of the state, and PG-S from southern California and northern Baja California. PG-N' represents a variant on the northern form

calaminaria from zinc-lead alluvial deposits along the River Tyne in northeastern England, which are very similar to plants growing near Aachen, Germany, the latter representing the "Continental metal" form.

Lauranson et al. (1995) investigated the flavonoid profiles of populations representing the range of the species. Compounds identified were mono- and diglycosides of the widespread flavonols: kaempferol, quercetin, isorhamnetin, and myricetin, "polyglycosides" of quercetin and myricetin, and caffeic acid derivatives (chlorogenic acids?). Seven different profiles (combinations) of these compounds were observed in plants from the five main parts of the range: maritime Europe (seven profiles), Continental metallicolous (three profiles), British metallicolous (two profiles), Arctic and subarctic (three profiles), and North American West Coast (two profiles). Earlier studies of allozyme variation revealed patterns that are in close agreement with the flavonoid results (Vekemans, 1992; Vekemans et al., 1992). Twenty alleles representing five loci in European plants contrasted with the almost total monomorphic situation observed in plants from California. Within California, differences in allozyme profiles were observed among populations, suggesting that local factors may exert significant influence on genetic structure as well. There has also been a change of breeding system in this species, from outcrossing in Europe to total or nearly total self-compatibility in populations in North America. Those workers suggested that the limited variation seen in North American populations was likely the result of the founder effect following long-distance dispersal.

4.1.3 Cakile *(Brassicaceae)*

The next example features glucosinolates in *Cakile*, the beach rocket. According to Mabberley (1997, p. 113), *Cakile* consists of seven species that occur along the strands of Atlantic and Mediterranean Europe, Arabia, Australia, and North America. Rodman (1974, 1976) has the number of species somewhat higher (14), but the subject of this section is the distribution of the genus, rather than a commentary on taxonomy. Of special interest is the use to which Rodman (1976) has put the glucosinolates (mustard oil glucosides) in sorting out origins and possible directions of migration of selected species. Glucosinolates are well-known constituents of the families comprising the order Capparales: Brassicaceae (alias Cruciferae), Capparaceae, Moringaceae, Resedaceae, Tovariaceae, and, newly added to the list, Setchellanthaceae (Iltis, 1999; Tobe et al., 1999).

Morphological variation in *Cakile maritima* Scopoli, one of the European species, is accommodated, at least in some taxonomies (see Rodman, 1974 for details), by recognition of three subspecies: subsp. *baltica* (Rouy & Roucaud) P. W. Ball, which grows on the coast of the Baltic Sea; subsp. *maritima* from the western and southern coasts of Europe, and subsp. *euxima* (Pobedimova) Nyárády from the Black Sea area. Their glucosinolate profiles exhibit some striking differences. Subspecies *baltica* is more homogeneous (among individuals) than most other taxa with a profile dominated by allyl glucosinolate [303] (see Fig. 4.3 for structures 303–311), with trace amounts of the biosynthetically related compounds: 3-butenyl [304] and 3-methylthiopropyl glucosinolates [305] (methionine pathway). *sec*-Butyl glucosinolate [306] (isoleucine pathway) is also a major component, but isopropyl glucosinolate [307] (valine pathway) was seen only in trace amounts. The two collections of subsp. *maritima* exhibited significant differences as can be seen in Table 4.1. Material from western Europe is characterized by the presence of *sec*-butyl glucosinolate as a major component and only a trace of the methionine-derived 3-methylthiopropyl glucosinolate, whereas material from the Mediterranean coast of Europe exhibited a much lower concentration of the *sec*-butyl compound and moderate amounts of three methylthioalkyl glucosinolates (propyl, butyl, and pentyl).

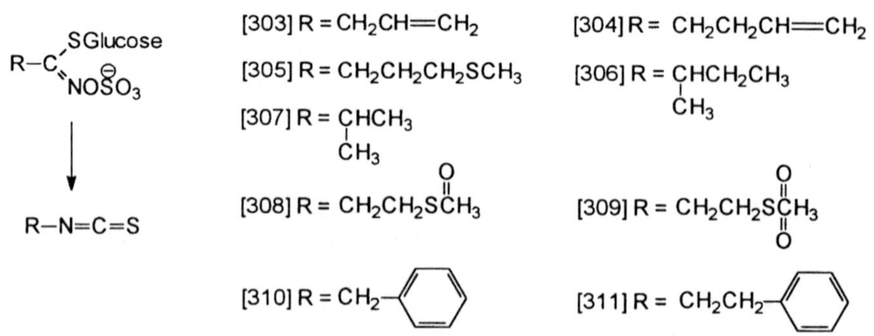

Fig. 4.3 Compounds 303–311, glucosinolates from *Cakile maritima*

Table 4.1 Glucosinolates in seeds of *Cakile maritima* and *C. arabica* (from Rodman, 1976)

Glucosinolate	*C. maritima* subsp.[a]				*C. a.*
	CMBA	CMMW	CMMM	CMEU	CARA
Isopropyl	trace	3.7	23.8	0	58.5
Allyl	80.3	12.2	9.0	99.9	23.0
sec-Butyl	16.3	84.8	34.0	0	17.0
3-Butenyl	trace	0	2.8	0	trace
3-Methylthiopropyl	1.0	trace	4.9	trace	0
4-Methylthiobutyl	0	0	12.1	0	0
5-Methylthiopentyl	0	0	10.8	0	0

[a]CMBA = subsp. *baltica*; CMMW = subsp. *maritima* from western Europe; CMMM = subsp. *maritima* from the Mediterranean; CMEU = subsp. *euxima*; C. a. = CARA = *C. arabica.*

Unique among the collections reported in this study was the chemistry of subsp. *euxima* whose glucosinolate profile consisted essentially of a single compound, allyl glucosinolate.

It is interesting, and relevant, to look at Rodman's views on the evolution of *Cakile*, as discussed in his 1976 paper. *Cakile arabica* Velen. is seen as the primitive species in the genus (or most closely resembling the primitive form) based on morphological considerations, breeding system (it is strongly self-incompatible), and the capacity to utilize the aliphatic amino acids isoleucine, methionine, and valine for the production of glucosinolates. It is also important to note that *C. arabica* is endemic to the deserts around the Persian Gulf and that the evolution of the beach lifestyle in other *Cakile* species is considered to be an adaptation to the expanding availability of sandy beach habitats associated with the development of the Tethys Sea (proto-Mediterranean Sea). Migration of *Cakile* to new areas in Western Europe was accompanied by partial relaxation of self-incompatibility, simplification of glucosinolate profiles in terms of the number of compounds, for example, *C. maritima* subspecies *euxima* and *baltica*, and in some cases, elongation of the precursor amino acids and oxidation of methylthioalkyl glucosinolates to the corresponding sulfinyl [308] and sulfonyl derivatives [309].

Colonization of the New World, thought to have occurred during late Pliocene or Pleistocene times, was accompanied by further reduction of self-incompatibility and simplification of glucosinolate profiles. A complicating factor in the overall chemical simplification trend is the capacity of the New World species to utilize aromatic amino acids as precursors in glucosinolate biosynthesis; phenylalanine is converted to benzyl glucosinolate [310] and to 2-phenylethyl glucosinolate [311] after one round of elongation. Rodman (1976) pointed out that it is not possible to distinguish between this new biosynthetic pathway having been developed de novo in New World taxa, or whether this represents a release of a repressed pathway.

It had been suggested by Rodman (1974) that the naturalized *C. maritima* along the western coast of North America originated from plants growing along the shore of the western Mediterranean. This possibility gained a good deal of support from the discovery that the glucosinolate profile of the North American plants bears close similarity to that of plants from the western Mediterranean. It also appears that the western

Mediterranean was the source of propagules of this species responsible for colonization of southern Australia. However, plants collected from the coast of South Australia exhibited a somewhat different glucosinolate profile, suggesting that a second colonization (possibly more?) may have occurred at some point. Morphological heterogeneity in the southern Australian specimens further suggested that hybridization between individuals having different chemical profiles might have occurred.

Reference to the colonization of western North American beaches by *Cakile* offers an opportunity to look at rates of dispersal of newly introduced organisms (Sims, 1968; Barbour, 1970; Barbour and Rodman, 1970). Strand plants provide a particularly convenient vehicle since their movement is restricted to two directions, essentially north and south along the coasts in North America and along the southern coast of Australia. *Cakile maritima* was first collected in North America in May, 1935, on Stinson Beach, Marin County, California (just north of San Francisco Bay), growing with the naturalized *C. edentula* (Bigel.) Hook. subsp. *californica* (Heller) Hult. (*Cakile edentula* is native to eastern North America.) Two years later, it was found growing on Salmon Beach, about 40 miles north of Stinson. It had reached Samoa Beach, Humboldt County, California, by 1940 and Sunset Bay, Coos County, Oregon, by 1942. By 1957, it had become established on the beaches in the Queen Charlotte Islands, British Columbia. Thus, in about 30 years, *C. maritima* had advanced its range northward by approximately 1000 miles, which amounts to a rate of about 33 miles per year. Its movement south appears to have resulted in a population observed on Cedros Island, Mexico, in 1963 (ca. 28°30′N latitude). A similar tale can be told about the appearance of *C. maritima* in the Southern Hemisphere. The species was first noticed in Australia in 1897, in the vicinity of the port of Fremantle (Perth), Western Australia. It had spread northward along the coast as far north as 31°S by 1963 (Sauer, 1988). After introduction in Western Australia, *C. maritima* moved steadily eastward, reaching the coast of Victoria by 1922 (Fig. 4.4).

Cakile edentula (recognized on the west coast of North America as subsp. *californica*, but not native to that area) first appeared in the floral records of western North America in the form of a specimen collected in 1882 on the shore near Berkeley, California (eastern shore of San Francisco Bay). It is thought to have traveled from its eastern North American home in wet ballast. It wasted little time in expanding its range, advancing southward to the beaches near San Diego, where it was sighted in 1906 (warmer water presumably prevents it from going further south). It is its northward journey that is most striking, however. Without detailing its intermediate stops, *C. edentula* reached Kodiak Island, Alaska, by 1931, which amounts to a rate of advance of about 50 miles per year. It is not likely that either *C. edentula* or *C. maritima* moved steadily northward (or southward), but rather reached new habitats through jumps of intermediate lengths. The involvement of the tides and ocean currents is very likely as indicated by observations on propagule viability made by Rodman (1974), who observed that seeds of *C. edentula* retained a significant degree of viability after emersion in seawater for up to 10 weeks, and that upper joints of fruits could remain afloat and viable after 11 days in seawater. This level of propagule viability in combination with tides and ocean currents created a potent force for migration.

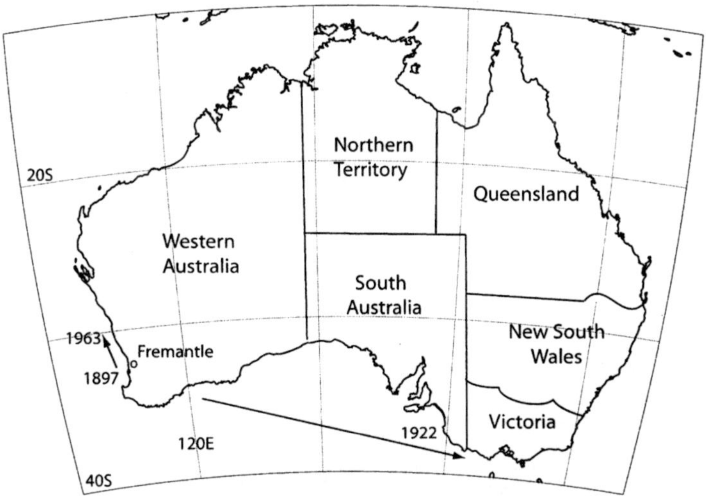

Fig. 4.4 Map showing the spread of *Cakile* in Australia, north to approximately 30°S and east to the coast of Victoria. The northward spread was likely halted by increasing ocean water temperature

Some years ago, Breckon and Barbour (1974) published a review of beach vegetation for the North American Pacific Coast that contains detailed information on a wide variety of strand taxa. A number of taxa discussed in their work have disjunct distributions, involving other continental beaches suggesting other taxa worthy of detailed phytogeographic study.

4.1.4 Cypripedium calceolus *(Orchidaceae)*

In this example, we examine differences in the floral aroma chemistry of three "species" of *Cypripedium* as determined by Bergström et al. (1992). Before the recognition of the North American species (*C. parviflorum* with two or three varieties), *C. calceolus* L. was thought to consist of three entities, whether varieties or subspecies is not relevant, the European "calceolus," and two North American entities, "parviflorum" and "pubescens."

A key factor for survival of these organisms is that they must resort to a chemical strategy to overcome the absence of nectaries in their efforts to attract pollinators. It has been observed that the most important structure for attracting bees to these flowers is the staminode, which appears to be the source of the volatile attractants. Using plants maintained in a common garden, Bergström et al. (1992) analyzed the floral-fragrance components of the three entities, selected results of which are summarized in Table 4.2. The table includes the information in two forms, concentrations of individual components in the upper panel with a notation

Table 4.2 Selected floral fragrance components of *Cypripedium* (top panel) and classes of compounds (bottom panel) (after Bergström et al., 1992)

Compound	3CAL[a]	PAR	PUB	Class[b]
Myrcene	<0.5[c]	6.0	<0.5	I
cis-β-Ocimene	trace	18.0	0.5	I
trans-β-Ocinene	trace	17.0	2.0	I
Hexyl acetate	2.5	nd	nd	F
Octyl acetate	30.0	nd	nd	F
Benzaldehyde	1.0	<0.5	6.5	P
Linaloöl	8.5	4.0	<0.5	I
Caryophylene	trace	11.0	9.5	I
Decyl acetate	28.0	nd	nd	F
α-Farnesene	4.5	13.0	5.0	I
1,4-Dimethoxybenzene	4.0	20.0	nd	P
Dihydro-β-ionone	<0.5	5.0	nd	I
Dodecyl acetate	5.5	nd	nd	F
p-Methoxybenzaldehyde	nd	1.0	4.0	P
Tetradecyl acetate	7.5	nd	nd	F
1,3,5-Trimethoxybenzene	nd	trace	69.0	P
Hexadecyl acetate	1.5	nd	nd	F
Alkyl acetates	77	–	–	
Monoterpenes	<0.5	42	3.5	
Sesquiterpenes	5	26	15	
Phenyl derivatives	7	21	80	

[a] CAL = *C. calceolus* (Europe); PAR = var. *parviflorum*; PUB = var. *pubescens*.
[b] Class of compound: I = isoprene; F = fatty acid; P = phenolic.
[c] Values are means of three analyses; nd = not detected.

of the class to which each compound belongs, and a summary of classes of volatiles in the lower panel. Differences among the three entities are striking. Perhaps the most noteworthy difference lies in the floral aroma of "calceolus" where 77% of the volatiles are alkyl acetates, a class of compound not seen in the other samples. Differences in the concentration of phenolic compounds are equally impressive, reaching a value of 80% in "pubescens." These data serve to point out just how closely each orchid species, or infraspecific taxon in this case, can be tied to its pollinator. A particularly good discussion of this phenomenon can be found in an essay written by Dodson (1970) in a volume dedicated to the general subject of coevolution. The story centers on the *Euglossini*, a group of long-tongued bees from the tropics, and the tight chemical association that they share with members of several genera of orchids.

4.1.5 Ephedra *(Ephedraceae)*

Ephedra is a genus of 40–50 species widespread in Old World and New World desert areas. Preparations of many different species have been used in folk remedies for a very long time. The physiological effects of these preparations are very likely caused by some combination of the nitrogenous components of the species, ephedrine derivatives taken as a case in point. A recent paper by Caveney et al. (2001) surveyed 26 species of *Ephedra* for nitrogenous compounds. Several interesting differences among different groups of species were observed, but the observation of interest to us here is the presence of ephedrine [(–)-1-phenyl-1-hydroxy-2-*N*-methylaminopropane] and pseudo-ephedrine [the (+)-isomer] in Eurasian species and their nearly total absence from New World species. The only exception was the detection of a trace of pseudo-ephedrine (>0.01% wet mass) in *E. californica.*

4.1.6 Cactaceae and Aizoaceae

An early review of chemistry of disjunct (allojunct in the paper) taxa was that of Turner (1972), one of the strongest proponents of the chemotaxonomic school of the 1960s and 1970s. One of the subjects he discussed was the disjunct family pair Cactaceae and Aizoaceae. His story started with a reference to a much earlier paper by Vierhapper (1919) that suggested that these two families were disjuncts. Both are succulents, with Aizoaceae predominantly South African in distribution, whereas Cactaceae is predominantly New World. As emphasized by Turner, Vierhapper's suggestion was remarkably prescient considering that plate tectonic theory lay well in the future and, in line with our concerns in this review, nothing was yet known about the unusual secondary chemistry that links these families and ties them together with others in Caryophyllalles (Centrospermae) (Wohlpart and Mabry, 1968). The chemistry to which we refer here, of course, is the presence of betacyanins (betalains is the more general term) as floral pigments in place of the more common anthocyanins. Figure 4.5 illustrates the two compound types, with betanin [312] representing the betacyanins and cyanidin [313] representing the typical anthocyanins.

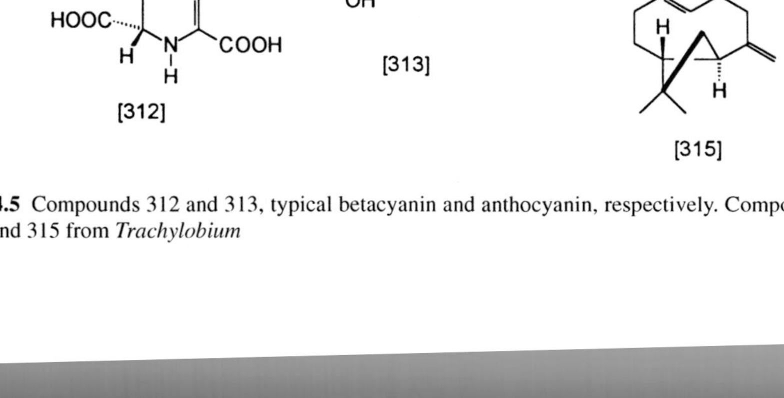

Fig. 4.5 Compounds 312 and 313, typical betacyanin and anthocyanin, respectively. Compounds 314 and 315 from *Trachylobium*

Turner speculated that the two families arose from a common ancestor that likely occupied xeric or halophytic habitats in Gondwana in an area that encompassed what were to become southwestern Africa and southeastern South America. The ancestral types do not still exist, but both families share the unusual, betanidin-based floral chemistry, and the equally unique sieve-tube plastids seen in Caryophyllales but are absent elsewhere in the plant kingdom (Behnke and Turner, 1971).

Cactaceae are certainly among the most readily recognized groups of flowering plants known. Their distinctiveness is based upon a wide variety of growth forms, including, but by no means limited to, the striking barrel and columnar cacti of the American deserts. The very extensive diversity exhibited by this group has led a number of workers over the years to conclude that the family must be an old one in order to accommodate the level of diversity now evident. Speculation put the age of the family before or at the beginning of the Tertiary, perhaps 65–90 mya (Axelrod, 1979; Gibson and Nobel, 1986; Mauseth, 1990). That amount of time could both account for the diversity of structure and coincide with the separation of South America and Africa and thus explain the absence of Cactaceae in Africa.

Recent studies employing macromolecular techniques have brought that scenario into question. A study by Hershkovitz and Zimmer (1997) employing ITS sequence divergence of nuclear ribosomal DNA revealed that the cacti nested phylogenetically within arid-adapted members of Portulacaceae. The degree of sequence divergence suggested an origin of the family in about mid-Tertiary, perhaps 30 mya with later diversification coincident with the formation of American deserts. Most recently, Nyffeler (2002) described studies aimed at determining phylogenetic relationships within the family. Although not specifically concerned with the age of the family, the recorded information on sequence divergence of the *trn*K intron, the *mat*K gene, and *trn*L-*trn*F sequences, support the time frame argued by Hershkovitz and Zimmer (1997) for the appearance and subsequent diversification of the family.

4.2 Across the Indian Ocean (Primarily)

4.2.1 Hymenaea verrucosa *(Fabaceae)*

Hymenaea verrucosa (Gaertn.) Oliv. occurs along the coast of eastern Africa (Kenya, Tanzania, and Mozambique) and on the Islands of Madagascar, Mauritius, the Seychelles, and Zanzibar (Fig. 4.6). S. S. Martin et al. (1973) examined populations from Kenya and Madagascar for their leaf-pocket resins and identified the following sesquiterpenes: α-cubebene, α-copaene, copacamphene, caryophyllene, β-humulene, γ-muurolene, α- and β-selinenes, and δ-cadinene. Profiles from the two areas were qualitatively identical, but differed significantly (1% level) in the concentration of two components, copaene [314] (11.9 and 7.6%, Kenya vs. Madagascar, respectively) and caryophyllene [315] (31.2 and 41.3%, Kenya vs. Madagascar,

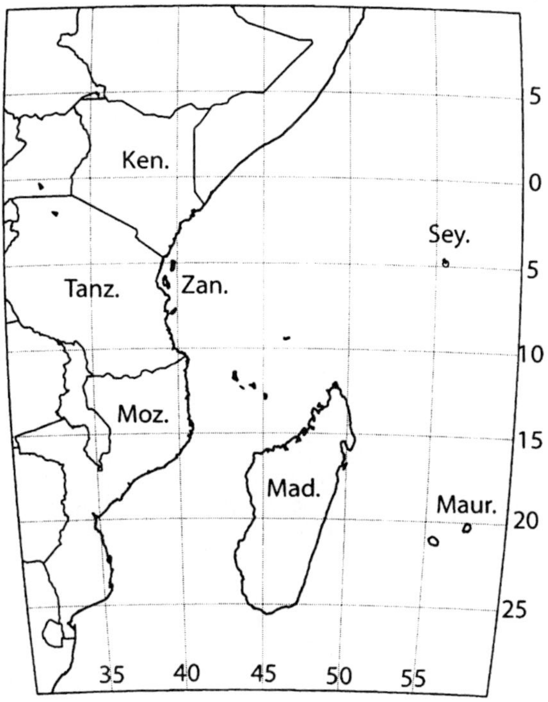

Fig. 4.6 Map of East Africa and islands in the western Indian Ocean. Abbreviations: Ken. = Kenya, Mad = Madagascar, Maur. = Mauritius, Moz. = Mozambique, Tanz. = Tanzania, Zan. = Zanzibar

respectively). See Fig. 4.5 for structures. An examination of collections from other mainland and island populations would be of interest.

4.2.2 *Restionaceae*

According to Dahlgren et al. (1985), in their treatment of the Monocotyledonae, Restionaceae consist of 40 genera with about 400 species distributed in southwestern Australia, southern Africa, New Zealand, New Guinea, southeastern China, Vietnam, Malaysia, Indonesia, and Chile (one species). Two centers of diversity exist, one in southwestern Australia and one in the Cape Province of South Africa, where, as Mabberley (1997, p. 610) pointed out, there are 10 endemic genera with about 180 species. In early taxonomic treatments, three genera were considered to be represented in Australia and South Africa, namely, *Hypolaena*, *Leptocarpus*, and *Restio* (see, e.g., Pillans, 1950). This idea was challenged on the grounds of extensive anatomical studies that showed that these genera were, in all probability, unnatural (polyphyletic) groups (Cutler, 1969, 1972).

An early study of flavonoids in *Hypolaena fastigiata* had shown the presence of novel and potentially useful compounds, including the newly reported 8-hydroxyluteolin [316] (see Fig. 4.7 for structures 316–320), which took its name, hypolaetin, from the source genus (Harborne and Clifford, 1969). A more recent survey of the family

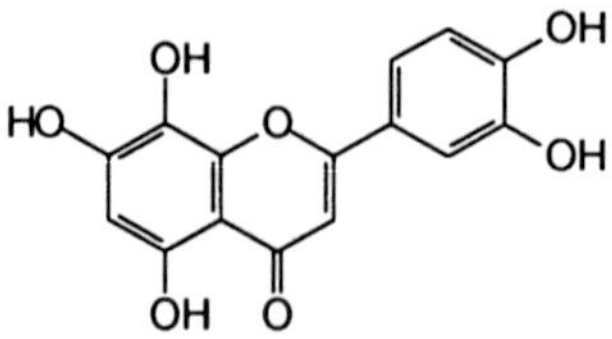

[316]

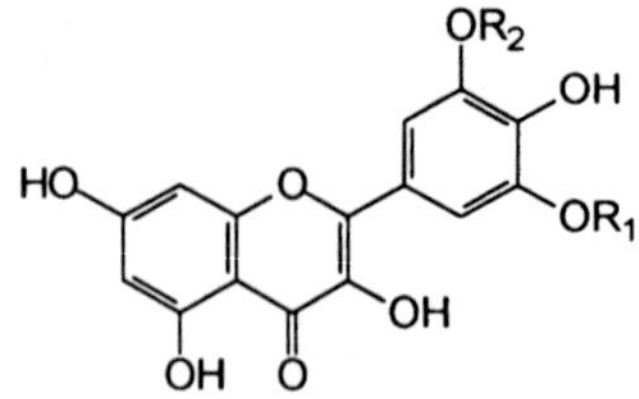

[317] $R_1 = H, R_2 = CH_3$
[318] $R_1 = R_2 = CH_3$

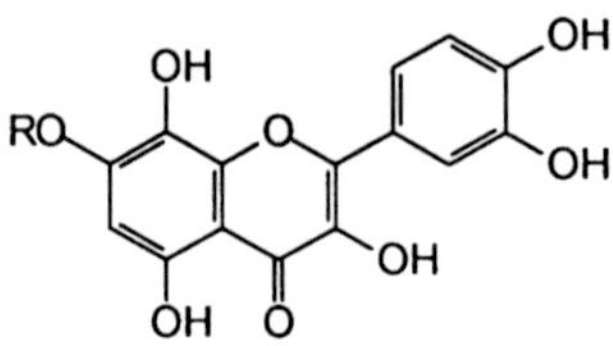

[319] R = H
[320] $R = CH_3$

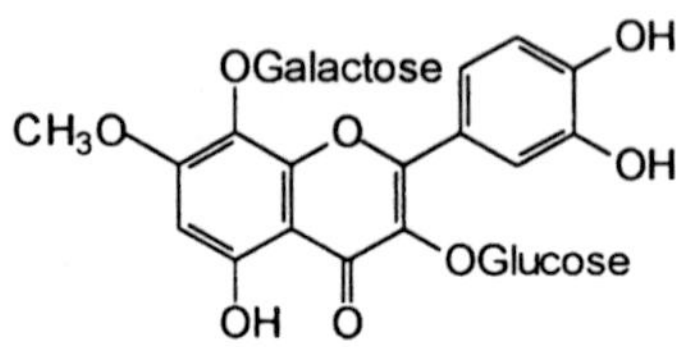

[321]

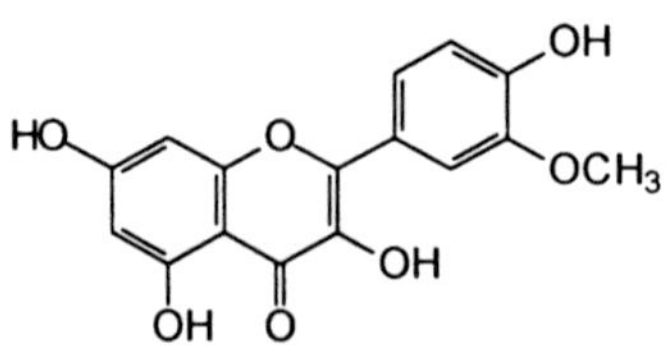

[322]

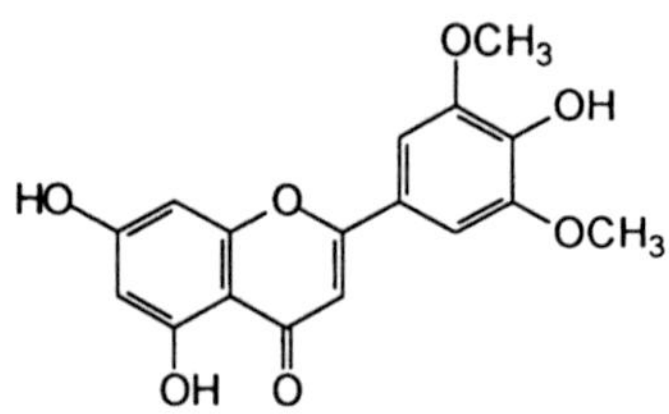

[323]

Fig. 4.7 Compounds 316–323, flavonoids from some Restionaceae

involving a much broader sampling has been described (Harborne, 1979). Fourteen Australasian species representing six genera and 33 species representing 10 African genera were subjected to chromatographic separation, and identification of their flavonoid glycosides. Flavonols identified included quercetin, myricetin, the myricetin methyl ethers larycitrin [317] and syringetin [318], gossypetin [319], and gossypetin 7-methyl ether [320]. Flavones identified were apigenin, luteolin, chrysoeriol, and hypolaetin. *C*-Glycosylflavones and proanthocyanidins were identified from several species and anthocyanins from a few.

The distribution of these compounds proved to be of considerable taxonomic value. The most interesting observation centered on the distribution of flavonoids in members of the three disjunct genera. The pigment profiles of Australasian and African members of all three correlated with their geographic origins, and not with other members of their respective genera. A solution to this problem was soon to appear in the form of a total revision of the family (e.g., Briggs and Johnson, 1998a, b). As mentioned by Williams et al. (1998), of the 147 species examined, 57 are newly recognized, while 19 of 34 genera are new or involve resurrected names from the nineteenth century. Referring to *Leptocarpus*, a genus that in the older treatment encompassed ten species from Australia, and one each from New Zealand, Southeastern Asia, and Chile, now is considered to consist of three species. Former members of the genus are now dispersed in *Apodasmia*, *Dapsilanthus*, and *Stenotalis*. Flavonoid profiles now follow more closely the taxonomy of the genera rather than their geographic location.

An interesting result of this taxonomic house cleaning is the disposition of the Chilean taxon, which is now considered as *Apodasmia chilensis* (Gay) B. G. Briggs & L. A. S. Johnson. *Apodasmia* is considered to consist of three species, the Chilean taxon *A. chilensis* just noted, *A. brownii* (Hook. f.) B. G. Briggs & L. A. S. Johnson from South Australia, Tasmania, and Victoria, and *A. ceramophila* from Western Australia. A remarkable finding is the presence of a very unusual compound, gossypetin 7-methyl ether 3-*O*-glucoside-8-*O*-galactoside [321] (the sugars may be reversed) in both the Chilean and southeastern Australian species. It would seem remarkable were this very unusual compound to occur by chance in two taxa separated by such distance; a close relationship between these two species seems a reasonable possibility, one that might well be addressed by macromolecular means.

The overall generic flavonoid distribution from the two geographic areas can best be appreciated by referring to the tabulated data from the recent Williams et al. (1998) paper and the earlier study by Harborne (1979). This compilation, with the compounds rearranged in increasing structural complexity, appears as Table 4.3. Several significant differences are evident among which is the very strong predominance of flavones in Australian species, 0.57–0.17 in the case of luteolin and a

Table 4.3 Distribution of flavonoids in Restionaceae in Australia and Africa (from Williams et al., 1998)

Compound	Frequency of compound in species from	
	Australia	Africa
Apigenin	0.01	0.02
Luteolin	0.57	0.17
Chrysoeriol	0.08	0.05
Tricin	0.17	nd[a]
Hypolaetin	0.48	nd
C-Glycosylflavones	0.16	0.26
Kaempferol	0.09	0.10
Herbacetin 4′-methyl	nd	0.07
Quercetin	0.30	0.31
Quercetin 3-methyl	0.01	nd
Isorhamnetin	0.09	nd
Myricetin	0.09	0.22
Larycitrin	nd	0.24
Syringetin	nd	0.21
Gossypetin	0.09	0.05
Gossypetin 7-methyl	0.03	0.10
Proanthocyanidins	0.04	0.88
Sulfated derivatives	0.27	0.02

[a] nd = not detected.

complete shutout in the case of hypolaetin, 0.48–0.0, and tricin, 0.17–0.0. Although taxa in both areas have the capacity to make flavonols that lack O-methylation, that is, kaempferol, quercetin, myricetin, and gossypetin about equally, taxa from South Africa have a far greater tendency to make O-methylated derivatives of those same flavonols. This is particularly pronounced in the formation of larycitrin and syringetin, which are not known from genera in Australia. The formation of isorhamnetin (quercetin 3′-methyl ether) [322] and tricin (5,7,4′-trihydroxy-3′,5′-dimethoxyflavone) [323] happens only in Australian taxa, however, in contrast to the compounds just mentioned. A study of the *O*-methyltransferases in these taxa might uncover some interesting differences in substrate specificity as in the case of *Chrysosplenium* (De Luca and Ibrahim, 1985a, b). Proanthocyanidins are clearly a major component of South African taxa (0.88) as opposed to Australian (0.04). In contrast, sulfated derivatives are much more common in Australian taxa (0.27) than in African ones (0.02). The chemical information clearly supports the newly restructured taxonomy of the family.

Williams et al. (1998) presented a short discussion of the distribution of the O-methylated flavonoids in Restionaceae, with reference to the occurrence of isorhamnetin in members of the related families, Anarthriaceae, Hopkinsiaceae, Lyginiaceae, and Ecdeiocoleaceae. They suggested that the absence of isorhamnetin from African members of Restionaceae, therefore, represents an advanced feature, and that the presence of hypolaetin in a large fraction of Australian taxa is an advanced character. The difficulty in determining polarity in flavonoid characters is well known. Suggestions as to the degree to which a given flavonoid, or flavonoid profile, reflects a primitive (pleisiomorphic) state or an advanced (apomorphic) state remain only speculative until some reliable phylogeny of the taxa involved has been achieved. In the present case, a phylogeny of the families, as well as the genera within Restionaceae, is absolutely essential. When an understanding of the phylogenetic relationships among the taxa has been achieved, it would then become possible to infer the order of evolutionary events underlying the flavonoid profiles. Until then, the patterns of flavonoid occurrence in the Restionaceae can serve only phenetically. Efforts to use DNA phylogenies to assess the relationships among groups of secondary metabolites, primarily iridoids, were reviewed recently by Grayer et al. (1999). Discussions involving flavonoid profiles and sequence-based phylogenies can be found in Soltis et al. (1993) concerning Saxifragaceae s. str., where it was evident that several structural features have been gained and/or lost a number of times, and a discussion from our group concerning flavonoid and sequence data as they relate to the evolutionary position of *Itea* and *Pterostemmon*, two genera whose status within Saxifragaceae is of interest (J. Y. Yang et al., 1998).

The flavonoid situation in Restionaceae provides an opportunity to comment on the importance of establishing phylogenetic relationships within a given taxon. In addition to affording a better understanding of evolutionary relationships within the group itself, a reliable phylogeny allows one to investigate the changes that occur with regard to some particular feature or suite of features, the general topic of character evolution. This applies no less to microchemicals than it does to morphological or anatomical features. In a broad sense, it would be possible to infer how often a particular compound or compound type has appeared or disappeared within the evolutionary history of the taxon. Our current level of understanding of the enzymes and genetic control of the flavonoid pathway, for example, would provide valuable insights into which gene or genes might be involved in the processes, specifically, which might be likely points of control of the pathway. In the case of Restionaceae, a generic phylogeny would provide an opportunity to judge the relative level of advancement of O-methylation, and thereby provide a possibly definitive answer to the question of whether the appearance of isorhamnetin in Australian genera represents a gain or a loss. Observations of that sort would not only advance our understanding of the evolution of a particular character within the family, but would also almost certainly provide a better view of a major flavonoid biosynthetic process. The same argument could be made about the enzyme(s) responsible for the formation of the 8-hydroxylated flavone hypolaetin, which was observed exclusively in Australian members of the family.

A well-resolved species phylogeny could also provide useful information on evolution of flavonoid structural features unique to one or a group of species. Linder and Mann (1998) have described such a study for 37 species in the restionaceous genus *Thamnochortus.* What is needed is a detailed chemical analysis of this genus to see if any chemical patterns exist that parallel patterns based on the morphological data. Their observation that a strong relationship exists between species occurrence and rainfall would provide an opportunity to examine the influence of drought on the formation of cuticular components such as long-chain hydrocarbon derivatives or nonpolar flavonoids.

4.2.3 **Villarsia** *(Menyanthaceae)*

Another example of an Australian–South African disjunction involves members of *Villarsia*, one of the five genera that comprise Menyanthaceae. The family, although comparatively small, offers a number of interesting distributional problems. Three of the genera are monotypic, *Menyanthes* (*M. trifoliata* L.) is circumboreal; *Faurea cristi-galli* (Menzies ex Hook.) Mak. occurs in eastern Asia and western North America; while *Liparophyllum gunnii* J. D. Hooker is known from Tasmania and New Zealand. The largest genus, *Nymphoides*, consists of about 35 species and is nearly cosmopolitan. The genus of interest here, *Villarsia*, consists of one species in the Cape Province of South Africa, *V. capensis* (Houtt.) Merrill, one species in Southeastern Asia, and perhaps a dozen in Australia.

An examination of the flavonoids of *Villarsia* (Bohm et al., 1986) revealed a heterogeneous array of flavonol glycosides, based upon kaempferol, quercetin, and isorhamnetin, with each species exhibiting a unique assortment. Surprisingly, there were greater similarities between the pigment profile of *V. capensis* and species from eastern Australia, *V. exaltata* (Soland. ex Sims) G. Don from Tasmania and *V. reniformis* R. Br. from New South Wales, than with species native to

Western Australia, as one might have expected. This observation was in agreement with conclusions drawn on the basis of morphological differences (Ornduff, 1974).

4.2.4 Icacinaceae

One of the most serious contributors to chemical ecology-chemical geography has been the Brazilian natural product chemist Otto Gottlieb (see Gottlieb and Kubitzki, 1983; Gottlieb, 1986, 1990). It has been his goal to examine chemical profiles of target taxa, not just from the point of view of presence or absence of a suite of compounds, but rather to investigate the levels of complexity involved in the processes by which the compounds are elaborated. Among other things, his work has involved searches for regular changes in oxidation level of a set of compounds, or for situations where there has been a switch between different biochemical pathways (e.g., shikimate to acetate/malonate), or differences in the way plants protect component compounds from degradation. Following these analyses is not always easy, with point scores being tallied for certain structural aspects within each class of compounds and the calculation of various indices of evolutionary advancement. Detailed consideration is given to the steps involved in forming end products, with recognition given not only to compounds that are specialized or "advanced" because they have had substituents added to the base structure, but also to those that have been modified in ways that make them appear simpler then the base structure, that is, the loss of a functional group originally present. An example of this is seen below, where loss of a carbon atom through oxidation leads to a derived (advanced) molecule.

The example to be described, admittedly one whose chemistry is difficult, is, nonetheless, typical of the approach. In the case of Icacinaceae, Kaplan et al. (1991) studied the increase in complexity of terpenoid compounds of selected members of the family as a function of where, in the geographic range of the family, the various genera occur. Although the work was set in a taxonomic context—using chemical features to assess the proper placement of the family—our interest lies in the chemical changes that appear to be associated with geography.

Icacinaceae are a moderate-sized, primarily tropical family consisting of 52 genera, many of which are monotypic, and 300 species. Additional background on taxonomic problems surrounding the family can be found in the 1991 paper by Kaplan et al. Chemical data have been used by Dahlgren (1980) to assess these relationships with related families, but the application involved simple presence or absence data (iridoids) and did not touch upon the dynamic nature of the pathways involved.

The genera and relevant chemistry of Icacinaceae involved in this example are given in Fig. 4.8, using the format of Kaplan et al. (1991) wherein the four sets of genera, grouped according to chemical similarity, are listed according to geographic areas. The geographic areas represented range from eastern Australia (Queensland) westward through Melanesia, southeastern Asia, India, and Africa to South America.

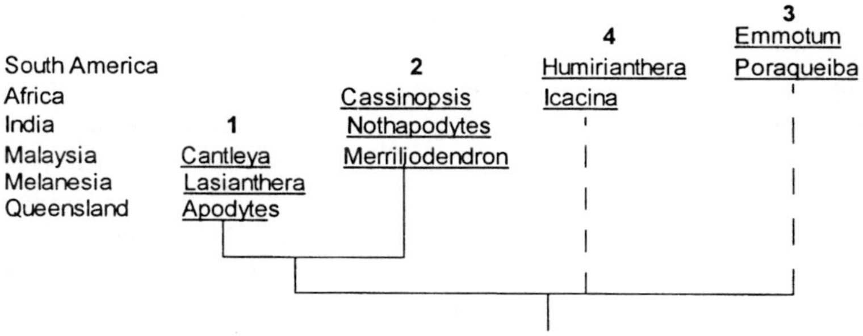

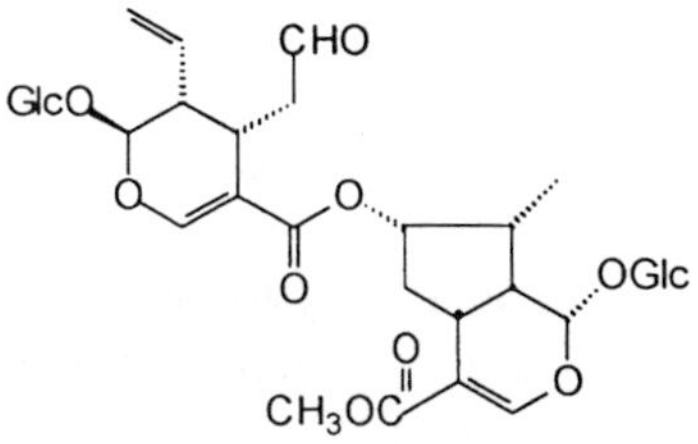

[324]

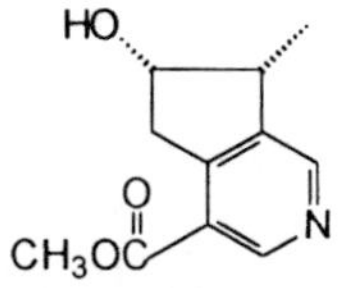

[325]

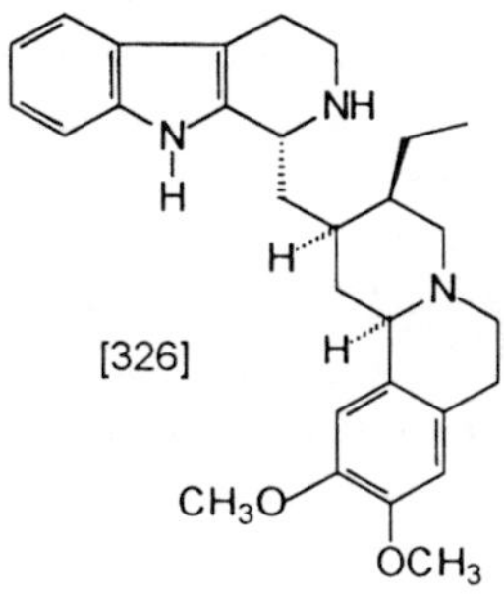

[326]

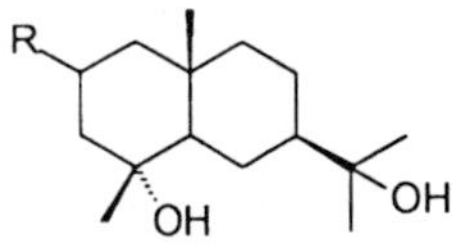

[327] R = alpha-OH
[328] R = beta-OH

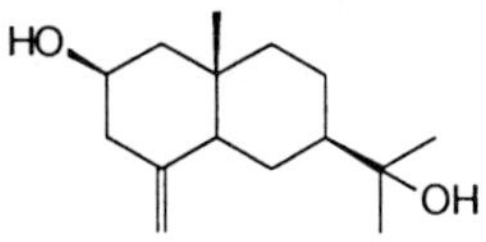

[329]

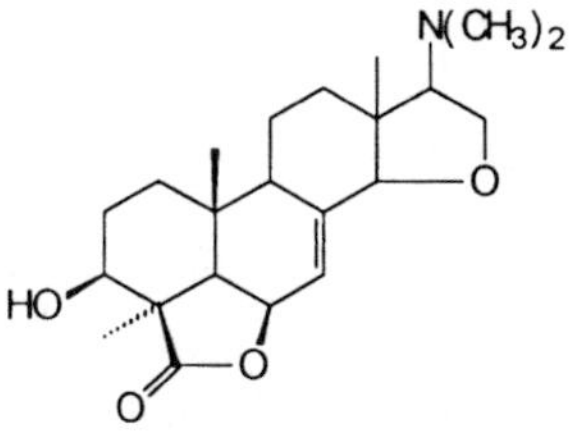

[330]

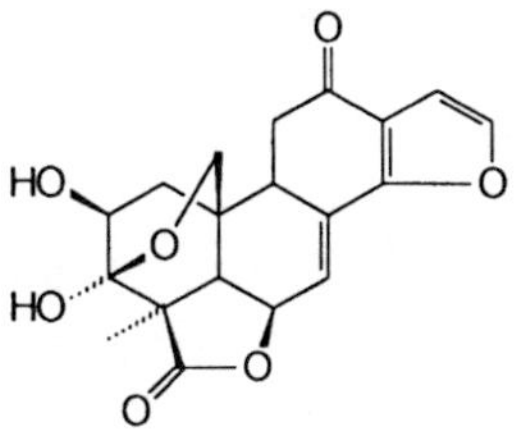

[331]

Fig. 4.8 Compounds 324–331, terpenoids and alkaloids from some Icacinaceae

Group **1** consists of *Apodytes* from Queensland, the monotypic *Cantleya*, native to Borneo, Malaysia, and Sumatra, and *Lasianthera* from New Caledonia. The terpenoid chemistry of this group is comparatively simple with cantleyoside [324], an ester of loganin and secologanic acid, and cantleyine [325], an artifact obtained from the former, having been isolated from all three genera. Group **2** genera, for which appropriate data are available, are *Merriliodendron*, a monotypic genus, from Malaysia, Melanesia, and Micronesia, *Nothapodytes* and the "distantly related" *Cassinopsis* from India. The compounds from Group **2** are more complex, having resulted from condensation of *seco*-loganin with tryptamine and/or DOPA fragments as illustrated in deoxytubulosin [326].

Group **3** consists of two South American genera: *Poraqueiba* and *Emmotum*. The presence of *seco*-loganoside and its methyl ester have been reported from members of the former, but it is with these genera that the chemical evolutionary process appears to have turned its attention to the production of sesquiterpenes based on the eudesmane system [327], [328], and [329]. Further elaboration of terpenoid biosynthesis is seen in the compounds of Group **4**, which include a complex array of diterpenoid derivatives, including the alkaloids icaceine [330] and de-*N*-methylicaceine [330 less one of the *N*-methyl groups]. One of the underlying ideas in Gottlieb's analyses is the increasing level of oxygenation as taxa evolve (become more "advanced"). Perhaps the reader can appreciate this idea better by examining icacinone, compound [331], with reference to the number of oxygen atoms involved. A more subtle point concerning that molecule is that it consists of only 19 carbon atoms rather than the 20 that form the diterpene system. Loss of the carbon is attributed to yet another oxidation resulting in the elimination of the angular methyl group at C-13 (Kaplan et al., 1991).

Despite the chemical sophistication of the Icacinaceae study, and others like it, it remains very difficult to assess the significance of the conclusions without some clear idea of phylogeny. Molecular genetic data that indicate evolutionary relationships among the genera involved would allow interpretation of the secondary chemical data from a much safer vantage point. As pointed out a few paragraphs earlier, this criticism, more a suggestion really, could be made about any set of relationships based upon secondary chemical data.

4.3 North Pacific

Few disjunctions have attracted more attention than those involving taxa whose ranges are disjunct between eastern Asia and eastern North America or between eastern Asia, western North America, and eastern North America. A substantial literature has accumulated on this subject, including the historically important work of Asa Gray (1846, 1859, 1878), a large number of descriptive works (e.g., Hu, 1935; Li, 1952, 1972; Wolfe and Leopold, 1967; Graham, 1972a, b; Hara, 1972; Wolfe, 1975; Hsü, 1983; Hong, 1993; N. S. Lee et al., 1996), a thoughtful history of the subject beginning at the time of Linnaeus by Boufford and Spongberg (1983), and a recent review of relevant molecular phylogenies by Wen (1999). Additional insights and critical comments on several issues relating to the similarities between the eastern Asian and eastern North American floras can be found in Tiffney (1985). As the reader can easily see from the *partial* list of publications above, the subject is a massive one, indeed. It is not possible to discuss all applications of secondary metabolite data to these disjunctions in the space available here. Therefore, the following examples have been selected to represent a reasonable cross-section of the information available.

4.3.1 Glehnia *(Apiaceae)*

Glehnia, a taxon restricted to eastern Asia and western North America, consists of either two species, the Asian *G. littoralis* Fr. Schmidt ex Miq. and the North American *G. leiocarpa* Mathias, or two subspecies of *G. littoralis*, subsp. *littoralis* and subsp. *leiocarpa*. The Asian taxon occurs on sandy beaches of China and Japan, with the North American taxon occupying the same type of habitat from northern California to the northeastern coast of the Queen Charlotte Islands, British Columbia, and west to Kodiak Island (Fig. 4.9). Both taxa are diploids with $n=11$.

Itoh et al. (1997) showed that leaves, fruits, and rhizomes of *G. littoralis* accumulate a rich array of furanocoumarins [332–339] as well as the polyacetylenic compound panaxynol [340] (see Fig. 4.10 for structures). A second acetylenic compound, falcarindiol [341], has also been reported from this species (Satoh et al., 1996). Examination of plants collected throughout the range of the species in Japan revealed two distinct chemical races, a northern form with imperatorin [335], isoimperatorin [337], and 8-geranyloxypsoralen [338] as the major coumarins, and a low ratio of panaxynol to total coumarin content (ratio less than 2), and a southern form characterized by the presence of imperatorin as the major furanocoumarin and a much higher ratio of panaxynol to total coumarin (ratio 1:10). Mizutani et al. (1993) concluded that the level of genetic variation between the northern and southern forms was quite low based upon restriction fragment-length polymorphism analyses.

The availability of North American specimens of *Glehnia* made it possible to compare coumarin and acetylene profiles of this widely disjunct system (Hiraoka et al., 2002). Analysis of plant material collected from four widely separated sites, that is, northern California, central Oregon, northern Washington, and northeastern Queen Charlotte Islands, British Columbia, revealed a profile characterized by low levels of furanocoumarins and a comparatively high level of panaxynol. The results are similar to those that characterize the southern Japanese form. The North American

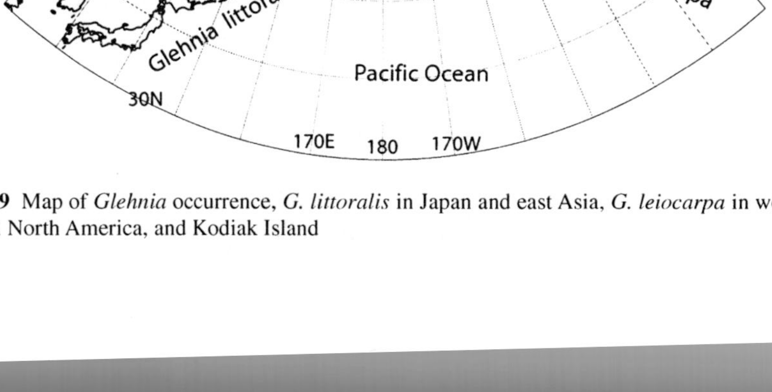

Fig. 4.9 Map of *Glehnia* occurrence, *G. littoralis* in Japan and east Asia, *G. leiocarpa* in western coastal North America, and Kodiak Island

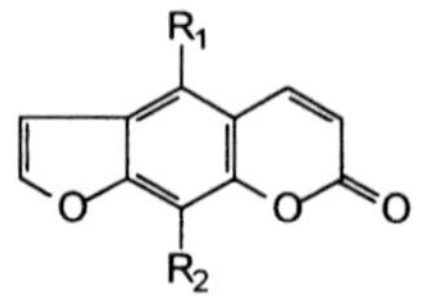

[332] $R_1 = R_2 = H$ Psoralen
[333] $R_1 = H, R_2 = OCH_3$ Xanthotoxin
[334] $R_1 = OCH_3, R_2 = H$ Bergapten
[335] $R_1 = H, R_2 = DMA$ Imperatorin
[336] $R_1 = OCH_3, R_2 = DMA$ Phellopterin
[337] $R_1 = DMA, R_2 = H$ Isoimperatorin
[338] $R_1 = GER, R_2 = H$ 8-Geranyloxypsoralin
[339] $R_1 = H, R_2 = GER$ Bergamottin
DMA = 3,3-Dimethylallyloxy (Prenyloxy)
GER = Geranyloxy

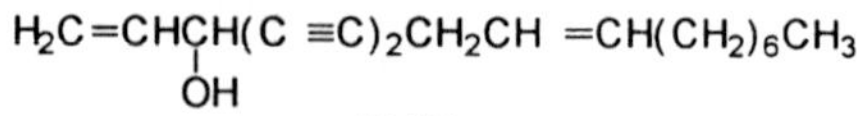

[340]

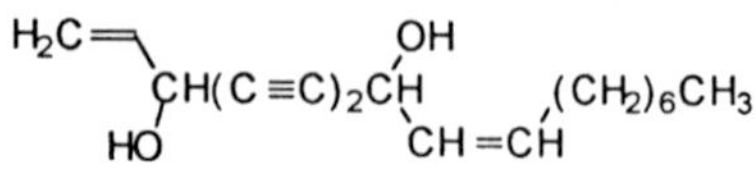

[341]

Fig. 4.10 Compounds 332–341, coumarins and acetylene derivatives from *Glehnia*

material is quite different, however, in the absence of imperatorin, phellopterin, and the 5- and 8-geranyloxypsoralens. Three of these, imperatorin [335], phellopterin [336], and 8-geranyloxypsoralen [338], are common compounds in fruits of *Glehnia* from Japan. Of critical importance in understanding the evolutionary relationships between the Asian and North American taxa would be the inclusion of material from both sides of the Pacific Ocean in gene sequence-based study.

4.3.2 Agastache *(Lamiaceae)*

Agastache sect. *Agastache* consists of seven species distributed across North America, with one species in China, Japan, Korea, and southeastern Russia representing the eastern Asian region. *Agastache nepetoides* (L.) Kuntze and *A. scrophulariifolia* (Willd.) Kuntze occur in eastern North America, *A. foeniculum* (Pursh) Kuntze occurs in central United States and Canada, with *A. urticifolia* (Benth.) Kuntze, *A. occidentalis* (Piper) Heller, *A. parvifolia* Eastwood, and *A. cusickii* (Greenman) Heller from the west. Vogelmann and Gastony (1987) noted that the western species are morphologically very similar and "... are probably conspecific at least in part." *Agastache rugosa* (Fisch. & Meyer) Kuntze, is the Asian species.

Isolation and identification of the flavonoids of all members of the section (Vogelmann, 1984) revealed a comparatively simple profile of apigenin and luteolin 7-*O*-glucosides, acacetin 7-*O*-glucoside, diosmetin 7-*O*-glucoside, and an unidentified aglycone. The Asian species exhibited a profile identical to those observed in the central and western North American species. *Agastache nepetoides* and *A. scrophulariifolia* exhibited the simplest profiles, but they were differentiated on the basis of the apparent lack of O-methylation capacity of the latter. The similarity of the eastern Asian species and the bulk of the North American species is interesting insofar as the two sets of taxa are thought to have been isolated since the Bering land bridge became submerged, possibly as long ago as 12 mya (Sanders, 1979).

A more recent electrophoretic analysis of 40 populations representing all eight species provided meaningful information on 11 enzyme systems (Vogelmann and Gastony, 1987). The Asian species differs from the North American species at only two of the 15 loci examined, which also represents a comparatively small level of differentiation for taxa separated by this length of time. This apparent close relationship is not supported by crossing studies (Vogelmann, 1985), which demonstrated that artificial interspecific crosses yielded only sterile offspring.

4.3.3 Cladothamneae (Ericaceae)

Tribe Cladothamneae of Ericaceae (subfam. Rhododendroideae) consists of four species that have at one time or another been placed variously in four genera: *Botryostege* Stapf, *Cladothamnus* Bong., *Elliottia* Muhlenb. Ex Elliott, and *Tripetaleia* Siebold & Zucc. For the purposes of the flavonoid study (Bohm et al., 1978), the nomenclature of Copeland (1943) was used, wherein *Cladothamnus* was taken to consist of the single species *C. pyroliflorus*, the western North American member of the group, and three species of *Tripetaleia*, *T. bracteata,* and *T. paniculata* from Japan, and *T.* (*Elliottia*) *racemosa* from Georgia and adjacent South Carolina. [Note that the final suggestion by S. Brim and P. F. Stevens (co-authors of the Bohm et al. 1978 work) was that similarities suggested that the four species are best recognized as belonging to a single genus, *Elliottia.*] The taxonomy of the subfamilies and tribes of Ericaceae has been reviewed by P. F. Stevens (1971).

The flavonoid profiles of all four species are based upon the common flavonols, such as kaempferol, quercetin, isorhamnetin, and myricetin. Kaempferol and quercetin derivatives are the most common constituents, myricetin derivatives much less so, while isorhametin was seen only in *T. paniculata.* Although the aglycone chemistry of the tribe is comparatively simple, the glycoside profiles of the four species exhibit several interesting differences. The most common glycosides are, as one might expect, 3-*O*-mono- and 3-*O*-diglycosides, with the usual sugars being present, such as arabinose, glucose, galactose, glucuronic acid, and rhamnose, although not all taxa contain the same sugars. Nor do all taxa exhibit the same degree of glycosylation, wherein lie the main distinguishing features within the tribe. Kaempferol and quercetin 3-*O*-rutinosides are among the major components of *Elliottia* species, whereas they were not seen in *Cladothamnus* (Bohm and Saleh, 1972; Bohm et al., 1978). *Tripetaleia* (E.) *paniculata* is unique within the group in accumulating quercetin 3-*O*-arabinosylrhamnoside. *Cladothamnus*, in contrast, accumulates kaempferol and quercetin 3-*O*-galactoside-7-*O*-rhamnosides and 3-*O*-arabinoside-7-*O*-rhamnosides. Quercetin 7-*O*-rhamnoside was also detected in extracts of *C. pyroliflorus.* This compound might have been an intermediate in the formation of the 3,7-diglycosides, although it would seem more likely that the

3-*O*-glycoside would be formed first. Alternatively, it might simply be the product of the 7-*O*-glucosyltransferase operating on quercetin rather than on one of the quercetin 3-*O*-monoglycosides. The key feature in these observations is the presence of the 3,7-diglycosylation products instead of 3-*O*-diglycosides in *Cladothamnus*. These results were discussed in terms of "diglycoside replacement," where it was concluded that the set of compounds involved represented the operation of enzymes that characterize this genus (or species, depending upon one's taxonomic view). It seems reasonable to suggest that this level of glycosylation represents a derived situation relative to the more common sequence of two glycosylations at position-3 of the flavonols (one on the flavonoid hydroxyl group and a second on one of the sugar hydroxyl groups), although again the matter of the need for a reliable phylogeny arises. At any rate, the occurrence of these compounds clearly sets this taxon off from the others. The other three taxa also are distinguishable from each other on the basis of their respective glycosylation patterns.

4.3.4 Tiarella *(Saxifragaceae)*

Tiarella consists of three species, *T. cordifolia* L., which grows widely in eastern North America; *T. polyphylla* D. Don, which is native to eastern Asia; and the western North American *T. trifoliata* L. *Tiarella trifoliata* is considered by some to comprise three varieties based upon leaf morphology: var. *trifoliata*, var. *unifoliata*, and var. *laciniata*. The flavonoid profiles of these three species of *Tiarella* were found to be based upon the common flavonols, that is, kaempferol, quercetin, and myricetin, and the flavone luteolin (Soltis and Bohm, 1984). The glycosidic derivatives identified included 3-*O*- and 7-*O*-mono-, 3-*O*- and 3,7-di-*O*-, and 3-*O*-triglycosides of the flavonols, and a 7-*O*-glucoside of the only flavone observed, luteolin. Picman and Bohm (1982) had described the flavonoid profiles of the *T. trifoliata* complex (all varieties). The distribution of compounds in the three species is presented in Table 4.4. The only compounds held in common by the three species are kaempferol, quercetin, and myricetin 3-*O*-glucosides and quercetin 3-*O*-rutinoside. All other compounds exhibited variable occurrences and apparent concentrations (estimated by chromatographic spot size). *Tiarella polyphylla* displayed the simplest profile, which consisted of only six compounds. *Tiarella trifoliata* exhibited the most complex profile, lacking only the 3-*O*-rhamnosides and triglycosides of quercetin and myricetin.

The glycoside profile of *T. trifoliata* was the most complex of the three taxa in its possession of flavonol 7-*O*-glucosides, 3,7-di-*O*-glucosides, and the flavonol 3-*O*-glucosylxylosides. It is interesting to note that the array of flavonoids reported from Asian members of tribe Cladothamneae (Ericaceae) is also the simplest of the four taxa involved in that study. In contrast, Vogelmann (1983) observed simpler flavonoid profiles in North American members of the genus *Agastache* (Lamiaceae) compared to the Asian counterparts.

Table 4.4 Flavonoid profiles of the disjunct species of *Tiarella* (from Soltis and Bohm, 1984)

Compound	*Tiarella* species[a]		
	CORD	TRIF	POLY
K, Q, M			
3-Glc	+	+	+
3-Gal	t	+	–
3-Rhm	t	–	–
Kaempferol			
7-Glc	–	+	–
3-Glc-Rhm	+	+	–
3-Xyl-Glc	–	+	–
3,7-DiGlc	–	+	–
3-TriGlys	+	+	–
Quercetin			
7-Glc	–	+	–
3-Glc-Rhm	+	+	+
3-Xyl-Glc	–	+	–
3-TriGlys	+	–	–
Myricetin			
3-Glc-Rhm	+	?	–
3-TriGlys	+	–	–
Others			
Luteolin	–	+	–
Lute. 7-Glc	–	–	+
Unknown-1	+	–	–
Unknown-2	–	–	+

[a] CORD = *T. cordifolia* (eastern North America; TRIF = *T. trifoliata* (western North America); POLY = *T. polyphylla* (eastern Asia).

4.3.5 Chrysosplenium *(Saxifragaceae)*

Chrysosplenium was discussed above where differences in flavonoid profiles of *C. glechomaefolium* among populations along a transect ranging from north central California to northern Washington were discussed (Bohm and Collins, 1979). *Chrysosplenium* also appears later in the section on polar disjunctions.

Chrysosplenium also plays a significant role in the present discussion owing to its widespread distribution in the Northern Hemisphere, with representatives in eastern North America, western North America, and eastern Asia. The genus is currently thought to consist of about 60 species (Mabberley, 1997, p. 157). Franchet (1890, 1891) considered that the species fall naturally into two groups based upon leaf presentation, that is, alternate (sect. *Alternifolium*) or opposite (sect. *Oppositifolium*). In his monograph of the genus, however, Hara (1957) did not recognize this formal breakdown. Recent molecular evidence supports this sectional treatment, at least in the Japanese species (Nakazawa et al., 1997).

Several investigations of flavonoids of *Chrysosplenium* have revealed a rich array of flavonols having both extra-oxygenation and a moderately high level of

O-methylation. A particularly interesting observation is that members of the alternate-leafed group (Franchet's sect. *Alternifolia*) accumulate a variety of flavonols, among which are several characterized by the presence of 6-oxygenation, for example, *C. tetrandrum* (Bohm et al., 1977), while the opposite-leafed group (Franchet's sect. *Oppositifolia*) accumulate a similar array of compounds along with flavonols that feature oxygenation at C-2′. As per the author, 2′-oxygenated flavonoids have so far been reported only from opposite-leafed species, although about half of the species in *Chrysosplenium* have not been thoroughly examined or, in many cases, not examined at all for flavonoids. As the examination of *C. glechomaefolium* revealed, not all members of an opposite-leafed species necessarily have these compounds, although most appear to be capable of making the 6-oxygenated flavonols (but see *C. valdivicum* below for an important exception). However, *C. americanum* (Collins et al., 1981) from the eastern North America, *C. glechomaefolium* (Bohm and Collins, 1979) from the western United States, and several opposite-leafed species from Japan (Jay and Voirin, 1976; Arisawa et al., 1991, 1992, 1993a, b, 1997, and earlier work by Morita and co-workers cited therein) accumulate 2′-oxygenated flavonoids.

There have been several suggestions above that the significance of relationships based on secondary metabolites can only be assessed when phylogenetic relationships within the taxon in question have become established. A paper by Soltis et al. (2001) has now provided the opportunity to do just that. Sequence information from the *mat*K gene was obtained for 28 species (33 collections) representing 13 of the 17 series of the genus defined by Hara (1957). The data were subjected to phylogenetic analysis, from which two main clades emerged that corresponded to the two groups of species recognized by Hara as sections *Oppositifolia* and *Alternifolia.* Further, the analyses revealed that eastern Asia was the most likely area of origin of the genus and that differentiation into the two groups was a comparatively early event in its evolutionary history and that it is likely to have preceded major dispersal events. Subsequent divergence of the alternate-leafed group yielded two subgroups, one of which comprises taxa native to the Himalayan region, and one that comprises Japanese species plus the two circumboreal species *C. alternifolium* L. and *C. tetrandrum* Makino. At least one migration out of eastern Asia was necessary to account for the establishment of the circumboreal taxa. The clade comprising sect. *Oppositifolia* consisted of three subclades: (1) Asian species; (2) two Asian species (*C. ramosum* Maxim. and *C. delavayi* Franch.) in addition to one of the South American species (*C. valdivicum* Hook.); and (3) a group of species that occurs in Asia (*C. grayanum* Maxim. and *C. pseudofauriei* H. Lev.), Europe (*C. oppositifolium* L.), and North America (*C. americanum* Schwein. ex Hook. and *C. glechomaefolium* Nutt. ex Torr. & A. Gray). The North American species are sisters, and together are the sister to the European species. In turn, the North American and European species constitute the sister group to the Asian species. This set of relationships requires a second migration out of eastern Asia.

Hara (1957) considered the South American species, *C. valdivicum* and *C. macranthum* Hook., to be ancestral in the genus. This was not borne out by the *mat*K sequence data. Rather, *C. valdivicum* (*C. macranthum* was not studied), emerged as part of a clade consisting otherwise of Asian species. A disjunction of this magnitude suggests that the South American species, or an ancestor, got to their present locations by means of long-distance dispersal. Whether this disjunction occurred as a result of direct long-distance dispersal from eastern Asia, or whether it represents the results of migration from Asia and then southward migration along the western Cordillera of North and South America, with subsequent disappearance of intermediate populations, cannot be resolved from the available data. Neither is it possible to say when dispersal might have occurred. Whether it was an ancient or comparatively recent event cannot be estimated with any degree of certainty at the moment, owing to the lack of a reliable "clock" for *mat*K divergence.

With the suggestion that the South American species are not ancestors, it is possible to comment on the evolutionary history of the flavonoid profiles of the genus. Instead of the depauperate flavonoid profile of *C. valdivicum* representing the ancestral condition, it would now appear that the simple array of pigments resulted from loss of some biosynthetic capacities. Several changes appear to have occurred: (1) reduction in overall level of substitution, that is, number of hydroxyl groups; (2) loss of the capacity to effect 2′-hydroxylation, which appears to be a feature of sect. *Oppositifolia* to which the species belongs; and (3) possible change in the level of O-methylation. The apparent loss of 2′-hydroxylation requires some comment. Although 2′-hydroxylation appears to be limited to members of sect. *Oppositifolia*, not all members of the section have been examined for flavonoids. Thus, it is entirely possible that some member of that section, which could not form 2′-hydroxyflavonoids, gave rise to the ancestral South American taxon. The third point, having to do with lower level of O-methylation, may simply be a ramification of the first listed change, that is, fewer substitutions on the flavonoid nucleus provide fewer sites for O-methylation.

The relationships between the North American species of sect. *Oppositifolia* and their closest relative in Japan led Soltis et al. (2001) to an interesting conclusion. The observation that the eastern and western North American species emerged as a monophyletic group parallels the finding with other taxa that share the eastern North America–western North America–eastern Asian disjunction. The same pattern of relationships was observed in a study of *Aesculus* (Hippocastanaceae, horse chestnuts) (Xiang et al., 1998a) and members of several other genera, including *Aralia* sect. *Aralia*, *Boykinia*, *Calycanthus*, *Cornus*, *Tiarella*, and *Trautvetteria* (Xiang et al., 1998b). In all of these, the North American members are monophyletic, suggesting that the eastern North American member of each set is more closely related to the western North American member than it is to the eastern Asian member. This contrasts with the classical view (Gray, 1846) that the eastern Asian and eastern North American members of a species pair are most closely related.

4.3.6 **Iris setosa** *(Iridaceae)*

Iris setosa Pallas occurs widely in subarctic parts of the Northern Hemisphere. Several varieties of *I. setosa* have been recognized over the years: var. *canadensis* M. Foster (syn. *I.* hookeri Penny) in the Canadian north, and vars. *setosa*, *nasuensis*

Hara, and *hondoensis* Honda in Japan. An examination of the polyphenolic constituents of these four taxa revealed clear-cut differences between North American and Japanese populations (Iwashima and Ootani, 1995). The overall profile of the species was shown to consist of *C*-glycosylflavones based upon apigenin and luteolin, secondary *O*-glycosides of some of these, O-methylated flavones, and the *C*-glucosylxanthones mangiferin and isomangiferin (see Fig. 2.38, page 63). The Japanese varieties *hondoensis* and *nasuensis* exhibited identical profiles, whereas the profile of var. *setosa* lacked some of the secondary *O*-glycosides and an O-glycosyl derivative of the xanthones. The profile of the North American material was very different, in that it lacked apigenin mono-*C*-glycosyl derivatives, the secondarily O-methylated derivatives, and the O-methylated flavones. An O-glycosylated xanthone derivative was also present which was not observed in the Japanese plants. The authors suggested that the polyphenolic data could be taken to support recognition of *Iris hookeri* rather than to continue recognition of this taxon at the level of variety. No comments were made vis-à-vis evolutionary relationships among the four taxa.

4.3.7 Humulus lupulus *(Cannabaceae)*

The resins derived from female flowers of *Humulus lupulus* L., known commonly as hops, provide the bitter and aromatic components so essential for the brewing of beer. According to Mabberley (1997), the genus consists of two species, *H. lupulus* of northern temperate regions and *H. japonicus* Sieb. & Zucc. of temperate eastern Asia. Owing to intense selection for flavor and yield characteristics, many cultivated varieties of *H. lupulus* have been developed. Although there is opinion to the contrary, J. F. Stevens et al. (2000), whose chemical study is discussed below, prefer the infraspecific taxa formally described by Small (1978). Thus, wild hops of Europe are considered to be *H. lupulus* var. *lupulus*, while North American hops are segregated into three varieties primarily on the basis of geography: *H. lupulus* var. *lupuloides* E. Small from central and eastern North America, *H. lupulus* var. *pubescens* E. Small from midwestern United States, and *H. lupulus* var. *neomexicanus* Nelson and Cockerell of western North America. These are often referred to generally as "wild American hops." According to Stevens et al. (2000), the wild hops of "Japan and perhaps eastern mainland Asia" can be referred to as *H. lupulus* var. *cordifolius* (Miquel) Maximowicz.

In addition to the bitter acids and essential oils, the flowers of hops offer a rich array of polyphenolic compounds, primarily chalcones and their accompanying flavanones, many of which are prenylated derivatives (Stevens et al., 1997, 1999a, b). The most prominent flavonoid in all plants studied was xanthohumol [342] (3′-prenyl-6′-*O*-methylchalconaringenin; chalconaringenin is 2′,4′,6′,4-tetrahydroxychalcone) (see Fig. 4.11 for structures 342–346). Several additional chalcones—variously adorned with *O*-methyl and/or *C*-prenyl functions—were also encountered, along with their respective flavanones. Three new compounds were described in the Stevens et al.

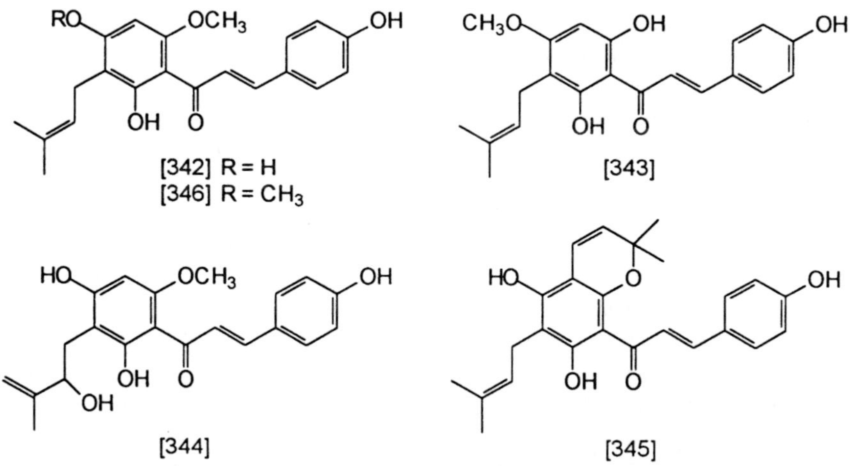

Fig. 4.11 Compounds 342–346, polyphenolic compounds from *Humulus lupulus*, hops

(2000) paper: xanthogalenol [343] (the common name derives from the cultivar "Galena" used in the study), xanthohumol D [344]; and xanthohumol E [345]. Of chemotaxonomic significance was the finding that the three 4′-*O*-methylchalcones observed in the survey (>120 plants), xanthogalenol, 4′-*O*-methylxanthohumol [346], and 4′,6′-di-*O*-methylchalconaringenin, occurred only in wild *H. lupulus* var. *cordifolius* plants collected in the Missouri-Mississippi River Basin and in their descendants (cultivars derived from var. *cordifolius*) and in Japanese wild plants. The absence of these compounds from European and southeastern North American members of the species suggests, as pointed out by Stevens et al. (2000), that at least two separate lineages of H. *lupulus* exist. The North American members are considered to resemble the ancestral form, which means that 4′-O-methylation is an ancestral feature that was subsequently lost by European hops (or by the ancestor from which the European line arose). Although there were points of difference, the results obtained in the flavonoid survey are in overall agreement with those obtained in a study of restriction fragment-length polymorphisms of ribosomal DNA (Pillay and Kenny, 1996).

4.4 South Pacific

4.4.1 Eucryphia *(Eucryphiaceae)*

Eucryphia, the sole genus in Eucryphiaceae, consists of six (Mabberley, 1997, p. 270) or seven species (Wollenweber et al., 2000) and can be found occurring on the Australian mainland, in Tasmania, and in southern South America. Specifically, *E. cordifolia* Cav. and *E. glutinosa* Cav. occur in Chile, *E. lucida* (Labill.)

Baillon. and *E. milliganii* Hook. f. occur in Tasmania, and *E. moorei* F. Muell., *E. jinksii* P. I. Forst., and *E. wilkiei* B. Hyland occur in Australia. The first two named Australian species are temperate rainforest species; *E. wilkiei* is an outlier known from higher elevations in tropical northeastern Queensland.

The genus has attracted a good deal more chemical attention than might have been expected for a group its size, including a study of the principal aroma constituent of leatherwood (*E. lucida*) honey, which was shown to be 3,7-dimethyl-1,5,7-octatrien-3-ol [347] (Rowland et al., 1995) (see Fig. 4.12 for structures 347–356). Our concern with the genus, however, lies with the patterns of distribution of flavonoids, including several structurally uncommon ones. The earliest interest in flavonoids of *Eucryphia* appears to have been that of E. C. Bate-Smith (1962), who observed an unusual yellow-fluorescent spot in an extract of *E. glutinosa* examined as part of his chemotaxonomic survey of the dicots. Subsequent work established the structure of the compound responsible as quercetin 5-methyl ether (azaleatin) [348] (Bate-Smith et al., 1966). In the following year, these workers (Bate-Smith et al., 1967) described a survey of the two Chilean species and three Australian species. (Note: the earlier workers considered the genus to consist of five species.) Reference to Table 4.5 illustrates the striking differences among the species. The capacity to synthesize both azaleatin and caryotin (quercetin 3,5-dimethyl ether) [349], the rare 5-methyl ethers of quercetin, as well as very similar arrays of quercetin glycosides, plus the two unidentified flavonoids, by both Chilean species speaks clearly for their close

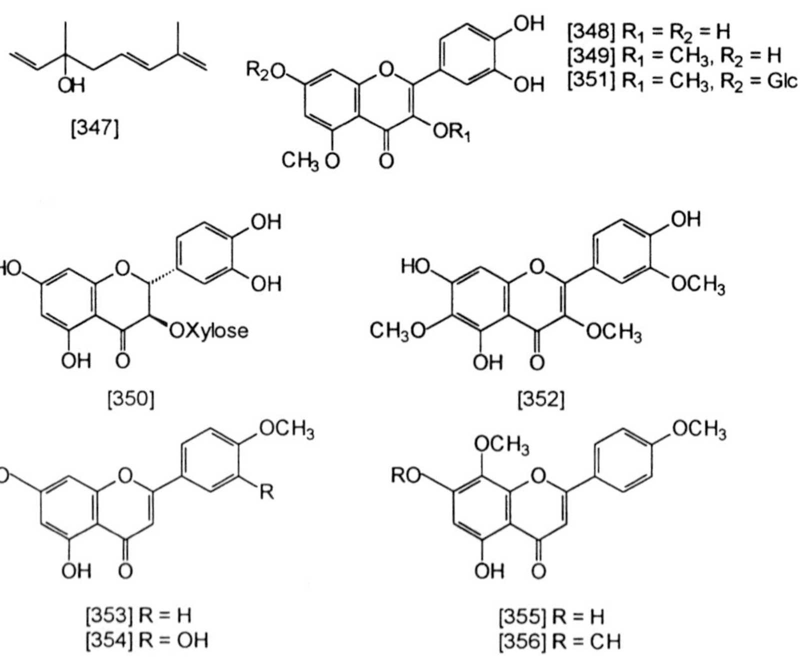

Fig. 4.12 Compounds 347–356, a terpene alcohol and flavonoids from *Eucryphia*

Table 4.5 Occurrence of flavonoids in *Eucryphia* (from Bate-Smith, 1967)

Flavonoid[b]	Chilean species[a]		Australian species[a]		
	COR	GLU	MOO	MIL	LUC
Caryatin	+	+	–	–	–
Azaleatin 3-Gal	–	+	–	–	–
Azaleatin 3-DiGlc	+	+	–	–	–
Azaleatin 3-GalAra	+	–	–	–	–
Unidentified flavanone	+	+	–	–	–
Quercetin 3-Gal	+	+	–	+	–
Quercetin 3-Rhm	+	+	–	–	–
Quercetin 3-DiGly	+	+	–	–	–
Quercetin 3-TriGly	–	–	+	–	–
Dihydroquercetin 3-Gly[c]	–	–	–	+	–
Kaempferol 3,7-DiMe	–	–	–	–	+
Unidentified flavan	+	+	–	–	–

[a] COR=*E. cordifolia*; GLU=*E. glutinosa*; MOO=*E. moorei*; MIL=*E. milliganii*; LUC=*E. lucida*.
[b] Gal=galactose; Glc=glucose; Ara=arabinose; Rhm=rhamnose; DiMe=dimethyl (ether).
[c] Dihydrokaempferol and dihydroquercetin were subsequently reported from *E. cordifolia* (Tschesche et al., 1979).

relationship. By contrast, the two Tasmanian and single Australian species studied exhibited extremely simple flavonoid profiles showing none of the "specialized" compounds present in the Chilean species. This conclusion has had to be tempered somewhat by more recent research involving the Chilean species. Tschesche et al. (1979) identified the 3-*O*-rhamnosides of dihydroquercetin and dihydrokaempferol from *E. cordifolia.* In a more recent paper, dihydroquercetin 3-*O*-xyloside [350], caryatin 7-*O*-glucoside [351], and jaceidin 5-*O*-glucoside (jaceidin is 5,7,4′-trihydroxy-3,6,3′-trimethoxyflavone) [352] were obtained from twigs of *E. glutinosa* (Sepulveda-Boza et al., 1993).

The most recent contribution to the chemistry of *Eucryphia* came from the work of Wollenweber et al. (2000) who, not surprisingly, examined the flavonoid components of glandular exudates of all members of genus, which in the present view comprises seven species. The tables appear to be completely turned with this subset of *Eucryphia* flavonoids. In contrast to the richness of profiles exhibited by the Chilean taxa with regard to their polar components, their exudate chemistries are by far the simplest seen in the genus; *E. cordifolia* afforded a single compound, apigenin-7,4′-dimethyl ether [353], and only in trace amounts. The other South American species, *E. glutinosa*, yielded only two compounds, apigenin-7,4′-dimethyl ether and luteolin-7,4′-dimethyl ether [354]. Apigenin-7,4′-dimethyl ether was also detected in the two Tasmanian species and in the two mainland species, such as *E. jinksii* and *E. moorei.* The richest arrays of aglycones came from the two Tasmanian species with a profile based on O-methylated derivatives of apigenin, luteolin, kaempferol, and quercetin, 17 in *E. lucida*, and 13 in *E. milliganii. Eucryphia jinksii* exhibited a somewhat simpler profile, based as well on both flavones and flavonols, but clearly distinguished from all other species in the genus by the capacity to make the 8-oxygenated flavones isoscutellarein-8,4′-dimethyl ether [355] and 7,8,4′-trimethyl ether [356]. *Eucryphia moorei* exhibited a profile somewhat simpler than that of *E. jinksii* but, as mentioned, lacked flavones with the isoscutellarein oxygenation pattern. *Eucryphia wilkiei* exhibited a profile consisting of a single, unidentified aglycone, seen as well in *E. lucida* (Tasmania) and *E. jinksii* (mainland Australia). It was also unique in the genus in having flavonol glycosides as exudate components.

Numerical treatments of the flavonoid data were performed using nonmetric multidimensional scaling (NMDS), and unweighted pair group method with arithmetic mean (UPGMA) or average linkage clustering of Bray-Curtis dissimilarities. The NMDS treatment resulted in clear separation of *E. wilkiei*, but the other six taxa appeared scattered, although geographically close pairs, for the most part, appeared closer to each other than to other species. The UPGMA treatment of Bray-Curtis values resulted in clear separation of *E. wilkiei* from all other species and the following pairings, *E. cordifolia-E. glutinosa* (Chilean species); *E. jinksii-E. moorei* (mainland Australia); and *E. lucida-E. millinganii* (Tasmania). Cladistic analysis of the flavonoid data failed to resolve any relationships. A recent cladistic analysis using morphological data (Taylor and Hill, 1996) suggested that *E. lucida* and *E. milliganii* are sister taxa, which is supported by the phytochemical data. Neither the relationship of *E. wilkiei* to *E. lucida* and *E. milliganii*, as suggested by the cladistic analysis of Taylor and Hill (1996), nor to *E. jinksii*, as viewed by Forster and Hyland (1997), is supported by the flavonoid data. Because of incongruities between relationships suggested by the phytochemical data and those that arose from cladistic analysis of morphological data, one is once again inclined to suggest that gene-sequence studies might be of value in this system. In addition to the obvious phylogenetic insights that might be gained from such a study, some ideas of what gains and losses have occurred in the flavonoid biosynthetic pathway that resulted in the significantly different profiles of compounds observed within the genus.

4.4.2 Blennospermatinae (Asteraceae)

The pattern of distribution of the four genera that comprise this subtribe is one of the more complex ones that we encounter in this review, in that it involves western North America, South America, including the Falklands and the Juan Fernandez Islands; Australia, including Tasmania; New Zealand, including Stewart Island, the sub-Antarctic Campbell and Auckland Islands, and New Guinea. The four genera involved are the monotypic western North American *Crocidium*, *Blennosperma* with two Californian species, and one species disjunct in Chile, *Ischnea* with four species endemic to New Guinea (Swenson, 1994), and *Abrotanella*, which consists of 19 species distributed in the other areas listed. The subtribe has been the subject of recent biosystematic and geographic study (Swenson, 1995a, b; Swenson and Bremer, 1997). From these data, it is clear that the subtribe represents a monophyletic group.

In a paper describing pollen-wall ultrastructure, Skvarla and Turner (1966) stated that the flavonoids of *Blennosperma* and *Crocidium* exhibited different chromatographic behavior and were likely to be isoflavones. An examination of the flavonoids of *Crocidium* and all three species (four taxa) comprising *Blennosperma* showed a high level of similarity (Ornduff et al., 1973b), but isoflavones were not detected. The pigment profiles were shown to consist of a quercetin 3-*O*-glucoside (but not the normal β-glucopyranoside) and a quercetin 3-*O*-galactoside as major components of all taxa. A "quercetin 3-*O*-polyglucoside" was observed in all species of *Blennosperma* (present in *B. nanum* S. F. Blake var. *nanum,* but not seen in var. *robustum* J. T. Howell) but only scattered in populations of *Crocidium.* Luteolin 7-*O*-glucoside was observed in some specimens (three out of five) of *Crocidium* from California, but not in plants from Oregon or British Columbia (one collection each), or in *B. nanum* var. *robustum.* It is interesting that the flavonoid profile of *B.* chilense Lessing is identical to that observed for the Californian taxa, *B. bakeri* C. B. Heiser and *B. nanum* var. *nanum.*

An important indication of relationships among these taxa came from a study of chromosome numbers. Ornduff (1963, 1964) suggested that *B. chilense*, with $2n=32$, is an amphidiploid that originated in California from a cross between *B. nanum* ($2n=14$) and *B. bakeri* ($2n=18$), was dispersed to Chile, and subsequently became extinct in California. The combination of habitat, moist places, and vernal pools in California, and the sticky nature of the cypselas would provide the opportunity for dispersal by migrating waterfowl. A similar situation exists with *Lasthenia kunthii* (Lessing) Hook. & Arn. and *L. glaberrima* DC. (Asteraceae). The former, the only non-North American member of the genus, occurs in Chile, while the latter occurs in the Central Valley of California in and around vernal pools (Ornduff, 1966). The flavonoid profiles of the two species were shown to be very similar (Bohm et al., 1974).

A close relationship among *Blennosperma*, *Crocidium*, and *Ischnea* was indicated by a detailed study of morphological features of the entire subtribe. Members of those three genera comprised one of major clades. Species of *Abrotanella* were partitioned into two clades, seven species in one, the remaining 12 in the other (Swenson, 1995a). Swenson et al. (1999) noted that the flavonoid profiles of *Abrotanella* and *Ischnea* have the most in common, with only a single compound, the quercetin glucoside isomer, being shared by all tested species. It may be of significance that *Abrotanella* (six of seven species) and *Ischnea* have delphinidin 3-*O*-glucoside in common with *Crocidium.* The disparate flavonoid profiles of these genera would seem to suggest that considerable differentiation has occurred since divergence from a common ancestor.

4.4.3 Alexa *and* Castanospermum *(Fabaceae)*

This next example involves a set of distantly disjunct genera of legumes that possess unusual polyhydroxylated alkaloids in common. Representative of these compounds are the indolizine castanospermine [357] (see Fig. 4.13 for structures 357–359) and the pyrrolizidine derivative, australine [358]. The genera involved are

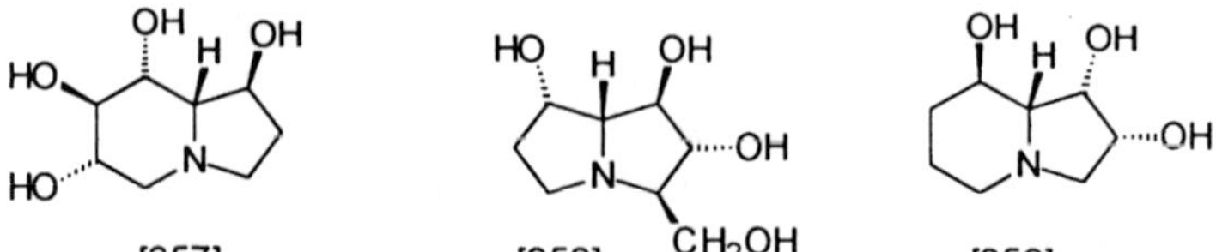

[357] [358] [359]

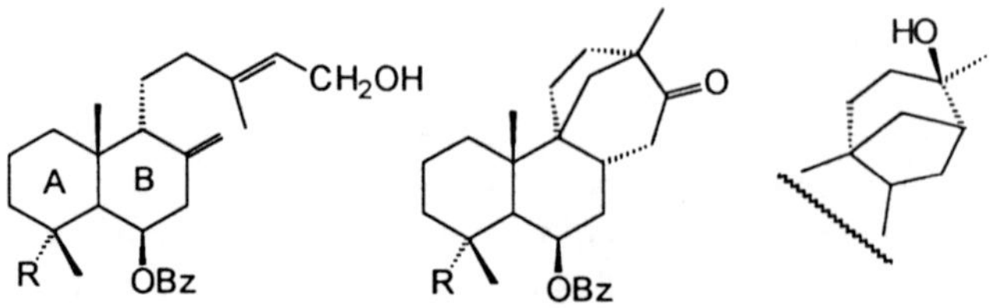

[360] R = COOH
[363] R = CH_2OH
Bz = benzoyl

[361] R = COOH
[364] R = CH_2OH
Bz = benzoyl

[362] A and B rings as in 360 and 361

Fig. 4.13 Compounds 357 and 358 from *Castanospermum* and *Alexa.* Compounds 359–364, diterpenes from *Scoparia dulcis*

Castanospermum, an Australian genus that consists of the single species *C. australe* A. Cunn. & C. Fraser ex Hook., and *Alexa*, a genus of eight or so species from South America (Kaplan, 1995); both belong to the tribe Sophoreae of Papilionoideae. Other, related genera will be commented upon below.

Castanospermine was obtained from *Castanospermum australe* (Hohenschutz et al., 1981), while australine was isolated from the same species by Molyneux et al. (1988). Nash et al. (1988) identified these compounds as components of *Alexa.* Related compounds that differ in the number of hydroxyl groups, and respective stereochemistries, have been isolated from these taxa as well. It is of interest that swainsonine [359], closely related to castanospermine, has been identified (Colegate et al., 1979) as a component of *Swainsona canescens* (Benth.) F. Muell. *Swainsona* is a genus of some 50 species (Mabberley, 1997, p. 692), found only in Australia and on the south island of New Zealand (one species). Swainsonine has also been isolated from North American species of *Astragalus* and *Oxytropis* (Colgate et al., 1979; Molyneux and James, 1982). Neither of these genera is represented in the Australian flora. Kaplan (1995) suggested that these chemical data point to a "… common ancestor thriving on the old South Pangaean continent…" from which *Castanospermum* and *Alexa* arose. Kaplan goes on to suggest that ancestors of *Astragalus* and *Oxytropis* may have occupied the "continent" as well. The authors were not precise in their estimates of the time scale involved in these events. Given that the age of the legumes has been reasonably estimated at about 50 million years, based on fossil evidence, it ought to have been possible to narrow the possibilities for the time of separation of the ancestors of the taxa involved. In any event, it seems unlikely that compounds of the sort seen in these taxa developed independently in several unrelated ancestral (or modern) taxa, although one can never rule out such a possibility completely. Biosynthetic studies, including comparison of the enzymes responsible for one or more of the steps, would be helpful. Again, DNA sequence divergence data would be of critical importance.

4.4.4 Scoparia *(Scrophulariaceae)*

Mabberley (1997, p. 651) lists *Scoparia* as a genus consisting of 20 species in tropical America, and singles out *S. dulcis* L. as a pantropical weed. T. Hayashi et al. (1991, 1993) reported that plants collected in Taiwan, China, and Thailand were characterized by a set of compounds that had not been reported from plants collected in Paraguay, and therefore represented a new chemotype in the species. Examination of a collection of *S. dulcis* from Indonesia revealed the Paraguayan array of compounds. All of the compounds identified in these studies were shown to be diterpene derivatives, acids in the case of the material from Paraguay as shown in structures, such as scoparic acid [360], scopadulcic acid [361], and scopadulcin [362]. The newly identified compounds, scopadiol [363] and scopadulciol [364] (see Fig. 4.13 for structures 360–364), share the same basic carbon skeleton with the compounds isolated from the Paraguayan material, but differ in the level of oxidation of C-18, exhibiting either the primary alcohol function ($-CH_2OH$) or the carboxyl group (–COOH).

4.4.5 Fuchsia *(Onagraceae)*

The genus *Fuchsia* consists of perhaps 100 species, enjoys a wide distribution, and offers a variety of systematic and evolutionary challenges. In addition to attracting purely botanical interest, *Fuchsia* is also of considerable horticultural value. Our interest in the genus concerns sect. *Skinnera* which consists of three New Zealand species, *F. excorticata* L. f., *F. perscandens* Cockayne & Allan, and *F. procumbens* R. Cunn., and the Tahitian endemic *F. cyrtandroides* Moore. A fourth taxon, *F.* x *colensoi* Hook. f. occurs wherever the ranges of *F. excorticata* and *F. perscandens* overlap.

Two groups of workers have investigated the flavonoids of this group of species, Averett et al. (1986) and Williams and Garnock-Jones (1986). Both groups reported kaempferol and quercetin 3-*O*-mono- and 3-*O*-diglycosides and a set of flavone derivatives. Although there is overall agreement in the classes of compounds reported in these two papers, there are some differences: the number of flavonol glycosides present in each species, the number of flavones represented, and the nature of the flavone sulfate derivatives. It is the pattern of occurrence of the flavonoid classes, flavonol versus flavone, and the presence or absence of sulfated flavone derivatives, rather than the individual compounds, that provide important clues to relationships among the species.

One of the noteworthy observations arising from these studies is the scarcity of flavones in the genus, and the possibility that their presence may represent the primitive condition (Averett et al., 1986). Thus, it is of interest that sect. *Skinnera* is characterized by the uniform presence of the flavones apigenin and luteolin, at least according to the Averett et al. study. The three New Zealand species are further distinguished by the presence of sulfate derivatives of the flavones, such as 7-*O*-sulfates according to Averett et al. (1986) and 7-*O*-glucuronidesulfates according to Williams and Garnock-Jones (1986). The fourth member of sect. *Skinnera, F. cyrtandroides*, has only one flavone, apigenin 7-*O*-glucoside according to Averett et al. (1986), as well as an array of flavonol glycosides in line with the other members of the section. Averett et al. (1986) also pointed out that the simplified flavonoid profile of *F. cyrtandroides* is in accord with the idea that it is the most recently derived species in the section (Raven, 1979). This would also fit with the comparatively young age of Tahiti, estimated to be about 2 million years by means of potassium-argon dating (Dymond, 1975). Also, the small fleshy fruits of *F. cyrtandroides*, the westerly wind currents, and the similar distribution patterns of other species would argue in favor of long-distance dispersal from New Zealand.

A cladistic analysis of morphological and chemical features revealed the relationships pictured in Fig. 4.14-A (Williams and Garnock-Jones, 1986), wherein the Tahitian species, *F. cyrtandroides*, is sister to the clade comprising the three New Zealand species. This defines the New Zealand clade as monophyletic based upon gynodioecy and the presence of sulfates. This arrangement suggests that the Tahitian and New Zealand species shared a common ancestor, which, presumably, no longer exists. Crisci and Berry (1990) reexamined this system using additional morphological characters as well as flavonoid data. However, they

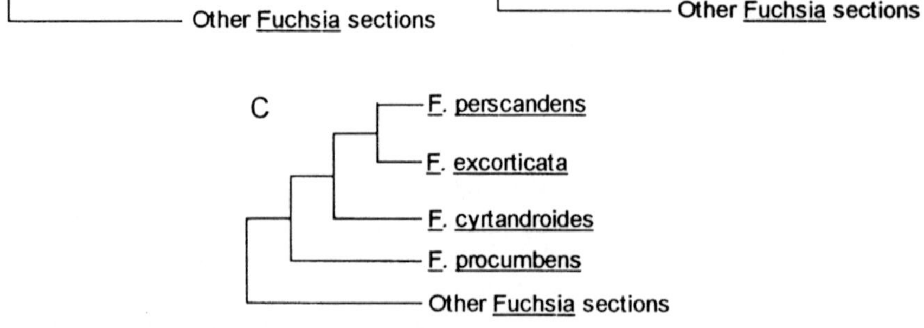

Fig. 4.14 Cladograms from Pacific Island *Fuchsia* study

chose to treat the flavonoid data in terms of classes of compounds, that is, sulfated versus nonsulfated compounds, rather than as individual compounds (thus lessening the impact of differences in flavonol glycoside composition). This treatment resulted in two most parsimonious trees, cladograms B and C in Fig. 4.14. In cladogram B, *F. cyrtandroides* retains its position as sister to the New Zealand clade, but the positions of *F. perscandens* and *F. excorticata* are exchanged. In cladogram C, *F. cyrtandroides* is part of an "inner" clade rather than sister to the New Zealand taxa, suggesting that the Tahitian species was derived from within the New Zealand species. Additional insights came from restriction fragment analysis of chloroplast DNA (Sytsma et al., 1991). The single, well-supported, tree that emerged from that analysis is identical to cladogram C, which is one of the two most parsimonious trees that emerged from the study of Crisci and Berry (1990), thus providing additional support for the emergence of *F. cyrtandroides* from within the New Zealand group. Sytsma et al. (1991) presented an evolutionary scenario accounting for the origin of *Fuchsia* sect. *Skinnera.* An abbreviated version follows.

Fuchsia is thought to have originated in warm, temperate forests of South America during the Eocene or Oligocene (Raven and Axelrod, 1974; Berry, 1982). Originally, Raven (1972) suggested that *Fuchsia* colonized New Zealand by long-distance dispersal. More recent microfossil finds suggest a different possibility. The discovery of fossil pollen in New Zealand dated as late Oligocene and early Miocene (25–30 mya) (Pocknall and Mildenhall, 1984) coupled with the identification of *Fuchsia* pollen in Oligocene and Miocene deposits in the Murray Basin of southeastern Australia (Berry et al., 1990), point to the possibility of a much older connection for *Fuchsia* between South America, Antarctica, and Australasia. With the establishment of *Fuchsia* in Australia, the stage was set for its migration across the Tasman Sea to New Zealand, followed by its eventual extinction in Australia. *Fuchsia cyrtandroides*, or an immediate ancestor, is then thought to have differentiated from the *F. excorticata* and *F. perscandens* clade (or ancestor). A large number of cpDNA synapomorphs (19) that define the *F. excorticata* – *F. perscandens* lineage and the number observed in the *F. cyrtandroides* DNA (20) suggest a divergence well before the appearance of Tahiti (2 mya) and Society Islands (4 mya). Sytsma et al. (1991) estimated, on the basis of base changes in the DNA, that *F. cyrtandroides* separated from its sister clade about 10 mya, well before its future home became available. Those workers also mentioned the possibility that *F. cyrtandroides* might have existed on other islands in the southwestern Pacific. In this scenario, the absence of sulfated flavonoids in *F. cyrtandroides* was interpreted as a loss, the basal position in the section being their presence. Similarly, the loss of gynodioecy in *F. cyrtandroides* was interpreted as a derived feature. Careful reading of the Sytsma et al. (1991) paper would repay individuals interested in breeding-system evolution, a topic beyond the scope of this treatment. As a concluding remark, it can be pointed out that, despite inconsistencies in the flavonoid data, this study represents a model of synthesis in the manner in which information from a variety of disciplines has been incorporated.

4.4.6 Nothofagus *(Nothofagaceae)*

If one were to select a single example from the plant geography literature that best encapsulates the history of the subject, one would be hard pressed to find a better one than the story of the southern beeches. In fact, van Steenis (1971, 1972) begins the title of his paper with the words, "*Nothofagus*, key genus in plant geography ...." *Nothofagus* consists of 35 species found in temperate South America (Argentina and Chile), New Zealand, Australia (including Tasmania), New Guinea, and New Caledonia. Writing in the introduction to his recent work on the molecular systematics of the genus, Paul Manos (1997) noted that explanations for this disjunct distribution have varied over the years and have depended to a large extent upon the current theory of the Earth, that is, whether continents were fixed (the stabilist perspective), or whether they moved (the mobilist perspective).

This is not the place for a detailed discussion of the evolutionary history of *Nothofagus.* Entry into the older literature can be found in papers describing studies of morphological information (Hill and Jordan, 1993), isozyme data (Haase, 1992), and DNA sequences (Martin and Dowd, 1993; Linder and Crisp, 1995; Manos, 1997; Setoguchi et al., 1997). Important insights into problems of *Nothofagus* biogeography, based upon a variety of data sources, have been discussed by Swenson and Hill (2001) and Swenson et al. (2001a, b).

Pollen has played a major role in understanding the history of and relationships within *Nothofagus.* An extensive study of pollen revealed eight types, four of which characterize extant species (Dettman et al., 1990). Analysis of the various data sets from the abode studies resulted in the emergence of four groups, each of which was characterized by one of the pollen types. This is in agreement with the most recent taxonomic treatment of the genus in which each of the fours groups is considered as a subgenus (Hill and Read, 1991). The occurrence of these types is as follows: subgen. *Nothofagus*, characterized by pollen type "fusca A" occurs in South America; subgen. *Fuscospora*, with pollen type "fusca B," occurs in South America, New Zealand, and Tasmania; subgen. *Lophozonia*, with pollen type "menziesii", occurs in South America, New Zealand, and Australia (including Tasmania); and subgen. *Brassospora*, with pollen type "brassii", consists of species restricted to New Guinea and New Caledonia.

The only secondary chemical data available for *Nothofagus* that relates to the geographic distribution of any of the species is a study of exudate flavonoids of four species of subgen. *Fuscospora* (Wollenweber et al., 2003). Structures were determined for compounds isolated from: *N. alessandri* Espin., *N. fusca* (Hook.) Oerst., *N. gunnii* (Hook.) Oerst., and *N. solandri* (Hook.) Oerst. *Nothofagus alessandri* is a Chilean species, *N. gunnii* occurs on Tasmania, and *N. fusca* and *N. solandri* are New Zealand species. The compounds identified are the flavonol galangin (3,5,7-trihydroxyflavone) [365] (see Fig. 4.15 for structures 365–371) and its 3- and 7-monomethyl ethers, 8-hydroxygalangin [366] and its 3- and 8-monomethyl, 3,7- and 7,8-dimethyl, and 3,7,8-trimethylethers,

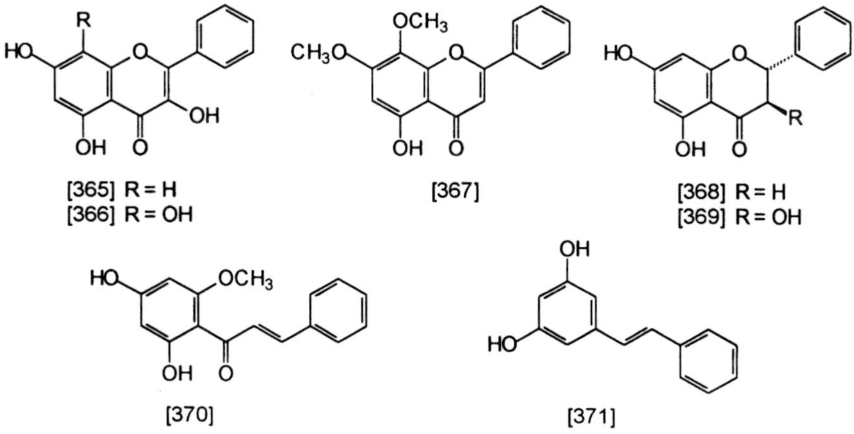

Fig. 4.15 Compounds 365–371, flavonoids from *Nothofagus*

5-hydroxy-7,8-dimethoxyflavone [367], the flavanone pinocembrin [368], the dihydroflavonol pinobanksin [369], 2′,4′-dihydroxy-6′-methoxychalcone [370], and the stilbene pinosylvin [371].

There is a remarkable homogeneity of pigment occurrence among the four taxa, the major difference being the restriction of 5-hydroxy-7,8-dimethoxyflavone to the New Zealand species, *N. solandri.* This species is also distinguishable from the others by a more complex mixture of 8-hydroxygalangin methyl ethers. The key observation here is the virtual identify of profiles of the Chilean species with the profile exhibited by the New Zealand species suggesting that essentially no divergence in flavonoid biosynthetic capacities has occurred during the time these taxa have been separated from their common (?) ancestor. As just noted, the flavonoid profile of *N. solandri* is somewhat more complex than those of the other species, but only by degree and not by major shifts in biosynthetic activities. On the other hand, *N. gunnii*, the Tasmanian representative, was found to exhibit the simplest pigment profile, with about half the number of galangin derivatives seen in the other taxa. The data are presented in Table 4.6. It is difficult to assess the significance of data such as these, especially in the face of the powerful genetic tools used by Manos and others in this, and other, work. As commented in other places in this review, an examination of the comparative enzymology, and/or sequence divergence of the genes encoding the enzymes responsible for producing these compounds could be very useful. For example, if a time of divergence for the Chilean and New Zealand members of subgen. *Fuscospora* could be estimated from the DNA data, it might indicate the time period during which the flavonoid pathway enzymes had *not* diverged appreciably, as appears to be the case with *N. alessandri* and *N. fusco*, but have diverged, as in the case of *N. gunnii* and *N. solandri.* The O-methyltransferase system seems a likely candidate for such comparative studies.

Table 4.6 Flavonoid profiles of *Nothofagus* subsp. *Fuscospora*

Compound	*Nothofagus* species[a]				
	ALE	FU1	FU2	GUN	SOL
Galangin	+	+	+	+	+
3-Methyl ether	+	v[b]	v	−	+
7-Methyl ether	v	v	+	+	+
8-Hydroxygalangin:					
3-Methyl ether	+	+	+	−	−
8-Methyl ether	+	+	+	+	+
3,7-Dimethyl ether	−	−	−	−	+
3,8-Dimethyl ether	+	+	+	−	+
7,8-Dimethyl ether	−	−	−	−	+
3,7,8-Trimethyl ether	−	−	−	−	+
5-Hydroxy-7,8-dimethoxy flavone	−	−	−	−	+
Pinocembrin	v	+	+	+	+
7-Methyl ether	−	−	−	−	+
3-Acetate	−	−	−	−	+
2′,4′-Dihydroxy-6′-methoxychalcone	v	+	+	v	−
Pinosylvin	+	+	+	+	v
vydroxy-7, 8-d					

[a] ALE=*N. alessandri*; FU1=*N. fusca*; FU2=*N. fusca*; GUN=*N. gunnii*; SOL=*N. solandri*.
[b] Variable among populations.

4.5 Western Pacific: Eastern Asia, Japan, and the Philippines

4.5.1 Euchresta *(Fabaceae)*

Euchresta is a genus of legumes consisting of five species native to the Southeastern Asian mainland, Japan, the Philippine Islands, and parts of Indonesia. It has been the subject of morphological and pollen studies as well as an examination of its flavonoid chemistry, which is what concerns us here (Matsuura et al., 1994). These workers undertook the chemical study in an effort to resolve the issue of relationships in the face of differing views based on the published studies of Ohashi and Sohma (1970) and Chen et al., (1992). An extensive series of papers dealing with the flavonoid chemistry of the entire genus has appeared (see Matsuura et al., 1994 for complete citations). The genus proved to be a source of both apigenin *O*-glycosides and *C*-glycosides, in addition to an abundant source of *C*-prenylated flavanones, for example, [375], *C*-prenylated isoflavones, [376], and *C*-prenylated coumaronochromones (picture a prenyl group at C-3′ in 374) (see Fig. 4.16 for structures). In addition to flavonoids based upon the normal 2′,4′,6′,4-tetrahydroxychalcone, certain members of the genus can also make the 6′-deoxychalcone from which the 5-deoxyflavanone was formed. Following the formation of the two types of flavanones [372 and 5-deoxy derivatives), biosynthetic activities continue in all species to produce isoflavones [373 and 5-deoxy derivatives], and in four species to produce the corresponding coumaranochromones, for example, [374]. A particularly rich source

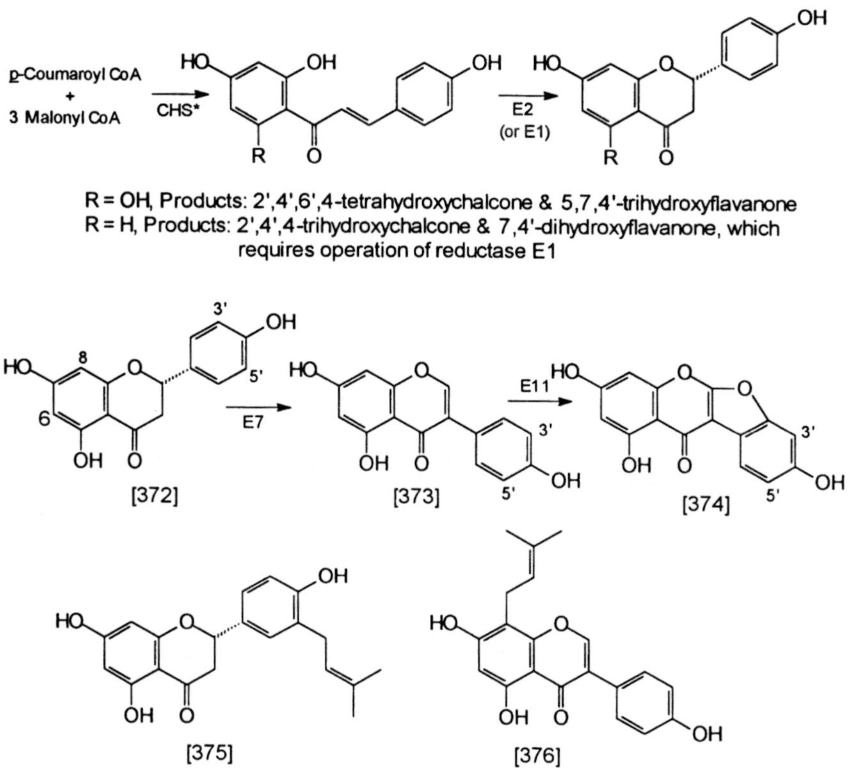

E1 Reductase
E2 Chalcone isomerase
E3 Flavanone 6-prenyltransferase
E4 Flavanone 8-prenyltransferase
E5 Flavanone 3'-prenyltransferase
E6 Flavanone 5'-prenyltransferase
E7 Isoflavone synthase
E8 Isoflavone 6-prenyltransferase
E9 Isoflavone 8-prenyltransferase
E10 Isoflavone 3'-prenyltransferase
E11 Coumaronochromone synthase
E12 Coumaronochromone 6-prenyltransferase
E13 Coumaronochromone 8-prenyltransferase
E14 Coumaronochromone 3'-prenyltransferase
*CHS Chalcone synthase (functional in all species)

Fig. 4.16 Compounds 372–376, flavonoids from *Euchresta*

of chemical features in the genus involved is the C-prenylation of flavanones (both types), isoflavones, and coumaronochromones. In some instances, flavonoids with as many as three *C*-prenyl groups were identified. The enzymes involved in these reactions are listed in Fig. 4.16, using the notation from the original paper (Matsuura et al., 1994). Based upon the compounds isolated from each species, it was possible for those workers to set up a data array showing the presumed presence of each of the enzyme systems. That information, along with the area of origin of the species, appears in Table 4.7. Using the morphologically related *Sophora tonkinensis* Gagnepain as outgroup, the authors subjected the data set to parsimony analysis using the Wagner method. Results of this analysis are illustrated in Fig. 4.17. The authors also determined trends in the ratio of flavone and *O*- and *C*-glycosides based

Table 4.7 Occurrence pattern of presumed enzymes in the biosynthesis of *Euchresta flavonoids* (from Matsuura et al., 1994)

Taxon	Origin	Enzymes (correspond to names in Fig. 4.16)													
		1	2	3	4	5	6	7	8	9	10	11	12	13	14
E. japonica	Japan	+	+	+	+	+		+	+	+	+	+	+	+	+
E. formosana	Taiwan, The Philippines	+	+	+	+	+	+	+				+	+	+	+
E. tubulosa	China		+	+	+	+		+	+	+	+	+	+	+	+
E. longiracemosa	China		+	+	+	+	+		+	+	+	+	+	+	+
E. horsfieldii	China	+	+		+	+		+							
E. horsfieldii	Thailand	+	+	+	+			+	+	+					
Sophora tonkinensis	Outgroup	+	+		+	+	+	+							

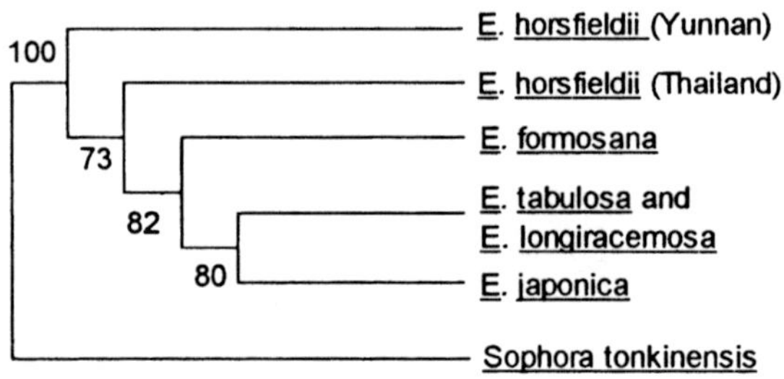

Fig. 4.17 Relationships among species of *Euchresta*

upon the idea, as argued by Stafford (1990), that *C*-glycosides represent a more primitive character than *O*-glycosides. Thus, a high ratio of *C*- to *O*-glycosides would indicate a taxon closer to the ancestral type. This analysis yielded results that were in accord with the relationships suggested by the prenylflavonoid phylogeny.

Matsuura et al. (1994) concluded from their analyses that *E. horsfieldii* (not in I.P.N.I. list) is the ancestral taxon in this genus (or is most closely related to the ancestral taxon) and that the center of origin for the group is southeastern China. From that area, the species migrated, with little in the way of morphological change, into Thailand and then Indonesia. Radiation with accompanying morphological (and chemical) diversification occurred toward the east, which gave rise to the Philippine taxon, and (generally) northward to yield the northern Chinese, Taiwanese, and Japanese taxa.

4.5.2 Heterotropa (=Asarum)*(Aristolochiaceae)*

N. Hayashi et al. (1984) described a study of phenylpropene derivatives in 44 species of *Heterotropa* Morren & Dcne. growing in Japan. According to Mabberley (1997, p. 339), *Heterotropa* is a synonym of *Asarum* L., which consists of some 70 North Temperate species, 30 of which occur in Japan. Regardless of the generic

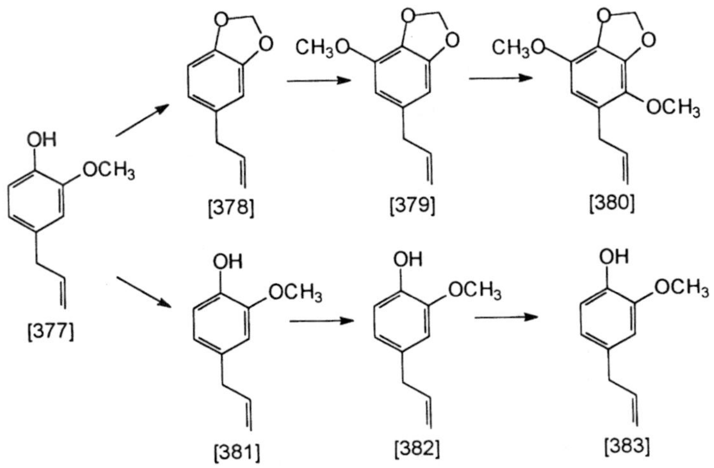

Fig. 4.18 Compounds 377–383, phenylpropanoids derivatives from *Heterotropa*

name, the distribution of the compounds of note provided an interesting view of colonization of the Japanese islands. The compounds are arranged in Fig. 4.18 to indicate their biosynthetic relationships and follow the presentation of Hayashi et al. (1984). The upper sequence of reactions, eugenol [377] to safrole [378] to myristicin [379] to apiole [380], involves first the formation of the methylenedioxy function, which requires oxidation of the *O*-methyl group in eugenol followed by cyclization. Formation of the methylenedioxy system does not occur in the lower-reaction sequence, where the pathway proceeds from eugenol to methyleugenol [381] to elemicin [382] to 1-allyl-2,3,4,5-tetramethoxybenzene [383]. Subsequent events in each pathway involve ring oxidation and O-methylation, in each case leading to the same level of oxidation, four substituted phenolic groups in each terminal product. Mapping the distribution of these compounds reveals an interesting difference in the profiles of the terminal products of the two pathways. Apiole, which contains the methylenedioxy ring system, was observed only in species collected from Okinawa, the Yaeyama Islands, and Taiwan. The terminal compound in the nonmethylenedioxy pathway, 1-allyl-2,3,4,5-tetramethoxybenzene, was observed primarily in species form Central Honshu but with a high concentration in "*H. costata*" from western Honshu and the island of Shioku (see Fig. 4.19). These observations were rationalized in terms of two routes of migration from the Chinese mainland: species with the methylenedioxy-based compounds migrated via a southern route; species with the nonmethylenedioxy-based compounds migrated via a northern route. Those authors stated that, "It is well known that *Heterotropa* species are originated from Yunnan or Szechwan Province in China." It would have added immeasurably to the work had the authors included information on the phenylpropene chemistry of species from those areas. They noted, however, that species of North American *Hexastylis* (also included in *Asarum* according to Mabberley) contains methyl eugenol, safrole, some terpenes, and elemicin (Hayashi et al., 1983).

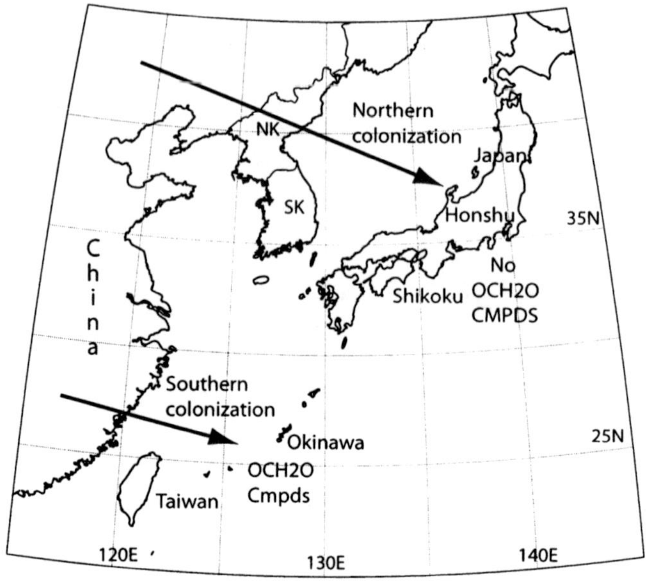

Fig. 4.19 Map showing occurrence of *Heterotropa* in Japan. Compounds bearing the methylenedioxy group occur on Taiwan, Okinawa, and associated islands. Compounds that lack the methylenedioxy group occur in Japan. Arrows show possible routes of colonization of the two types

4.5.3 Rhododendron *(Ericaceae)*

Rhododendron is a large genus of some 850 species with a nearly worldwide distribution, with areas of particular richness in the Himalayas, southeastern Asia, and New Guinea. A single species is known from Australia (Mabberley, 1997, p. 614). In addition to the large number of naturally occurring species, many horticulturally important cultivars have been developed owing to the attractive evergreen foliage and large range of flower size and color.

A study of pigment profiles within *Rhododendron* sect. *Virerya*, a reputedly monophyletic group of 300 species from the Malesian-Australian region, showed the presence of a comparatively simple array of aglycones (Harborne, 1986) with quercetin derivatives present in most species, kaempferol derivatives in several, and myricetin in a scattering of species. Present in lower frequencies than in other comparable groups of *Rhododendron* species were azaleatin [quercetin 5-methyl ether, structure 348] and caryatin [quercetin 3,5-dimethyl ether, structure 349]; gossypetin [8-hydroxyquercetin, structure 319] was not observed in species from Malesian-Australian region and North America. Harborne (1986) summarized frequencies of flavonoid aglycone occurrences in the genus as a whole; his summary is presented in Table 4.8.

Harborne (1986) noted that the absence of gossypetin from leaves of sect. *Vireya* reflects its absence in the flowers and that such absence may be correlated with

Table 4.8 Percentage flavonoid aglycone occurrence in species of *Rhododendron* sect. *Vireya* (from Harborne, 1986)

Flavonoid	Southeastern Asia	Malesian-Australian region	North America
Quercetin	100	94	100
Myricetin	51	35	35
Kaempferol	23	44	82
Azealatin	34	15	12
Caryotin	10	4	12
Gossypetin	76	0	0
Dihydroflavonols	68	13	18
Species studied	206	52	18

differences in pollination vectors. This would be in line with the morphological differences that characterize members of this section as well. Harborne went on to point out that the most significant feature of the survey is the correlation of pigment profiles with geography. The differences in profiles between species from southeastern Asia and those from the Malesian-Australian region are similar to the differences between the profiles in southeastern Asian species and North American species. These differences support the view that the genus originated in southeastern Asia, the Himalayas, or central China, and radiated outward from there. The species from the Malesian-Australian region would represent one set of derived species, and another from North America. Much additional work is needed to put these speculations on firmer ground. Needless to say, this would seem an excellent system in which to examine phylogenetic relationships using gene-sequence data.

4.5.4 Gesneriaceae

Gesneriaceae is a large tropical and subtropical family, many members of which are prized for their spectacular and commercially important floral displays, for example, *Gesneria*, *Gloxinia*, *Saintpaulia*, *Sinningia*, and *Streptocarpus*. The family is of interest to us here owing to chemical differences between the two major subfamilies, Gesnerioideae, primarily a New World taxon with representatives in southeastern Australasia (hence, its rather arbitrary inclusion in this section), and Cyrtandroideae, which is an Old World taxon. An unusual feature of the colored pigments of a *Gesneria* species was first reported by Robinson et al. in 1934, who identified an anthocyanidin derivative that lacked the hydroxyl group normally present at C-3. The compound, apigeninidin [384], which belongs to a class of compounds referred to 3-deoxyanthocyanidins, can be compared to the normal anthocyanidin pelargonidin [385] (see Fig. 4.20 for structures). The B-ring deoxy analog, luteolinidin [386], was discovered in a member of Gesneriaceae some years later (Harborne, 1960). A subsequent survey (Harborne, 1966), reported detection of 3-deoxyanthocyanins in 18 species representing 11 genera of Gesnerioideae. From that same work we also learned that these pigments were absent from 25 species

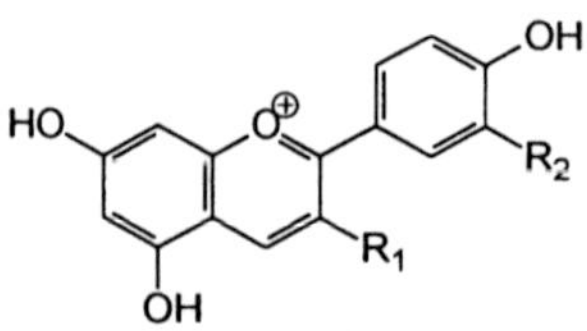

[384] $R_1 = R_2 = H$
[385] $R_1 = OH, R_2 = H$
[386] $R_1 = H, R_2 = OH$

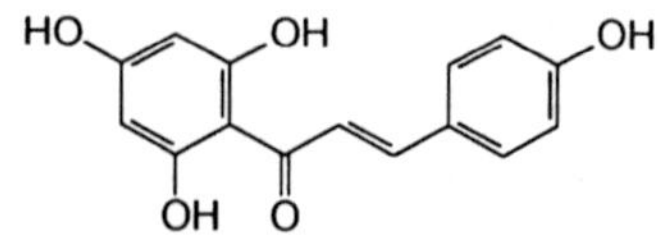

[387]

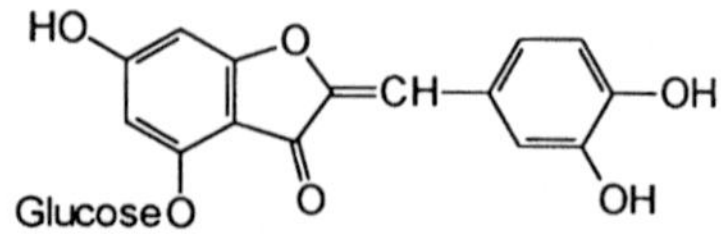

[388]

Fig. 4.20 Compounds 384–388, anthocyanidin derivatives and anthochlors from Gesneriaceae

representing eight genera of Cyrtandroideae. Studies of Cyrtandroideae, however, revealed the presence of anthochlors (chalcones and aurones), a class of pigments not detected in Gesnerioideae. Continuing his survey, Harborne (1967a) reported 3-deoxyanthocyanins in a further 11 species from seven additional genera, and anthochlors in six more species of Cyrtandroideae. Among the anthochlors identified were chalcononaringenin [387] representing the chalcones, and cernuoside [388] representing the aurones.

Chapter 5
Wide Disjunctions

This chapter presents examples of taxa that enjoy very wide distributions that may involve several continents or, in the case of marine algae, several oceans. In addition to the algae, examples include higher plants, lichens, and bryophytes. Owing to the extensive literature involving chemistry of marine algae, the examples included represent only a sample. We begin with some widespread examples from the vascular plant chemical literature.

5.1 Vascular Plants

5.1.1 Coriariaceae

Coriaria is the sole genus in Coriariaceae, a family that enjoys a wide and interesting distribution, including China, Japan, Taiwan, Tibet, New Zealand, and Mexico. Several questions surround an understanding of this little family, including its size, its phylogenetic position in the angiosperm scheme of things, and how it has come to occupy the places where it can now be found. Flavonoids of 12 species of *Coriaria* were studied as a possible aid to the problem of ordinal affinities of the family (Bohm and Ornduff, 1981). Plants examined in that study represented much of the range of the genus as noted above. The pigment profile, which consisted of common kaempferol and quercetin mono- and diglycosides and a common flavanonenaringenin 7-*O*-glucoside, was largely homogeneous among the species. Flavonoid data offered no insights into relationships, including the three groups defined by Good (1930) based upon floral presentation. Recent information on ordinal relationships of Coriariaceae based on DNA sequence information can be found in a paper by Wagstaff and Dawson (2000).

5.1.2 Gentianaceae

The following example comes from the work of Gottlieb and Kubitzki (1983) on the use of xanthones as indicators of relationships within Gentianaceae, a family consisting of 78 genera and 1225 species, according to Mabberley (1997, p. 297), and one that he defines as cosmopolitan in distribution. Xanthones are a well-known class of secondary metabolite, many of which have been isolated from members of Gentianaceae. They are biosynthesized through the condensation of a benzoyl CoA derivative with three "acetate" units (actually, three malonyl CoA units), followed by ring closure via phenol coupling, as illustrated in Fig. 5.1. Two routes of ring closure are possible so that the fundamental, or base structures, that result are 1,3,5,6-tetrahydroxyxanthone [389] and 1,3,6,7-tetrahydroxyxanthone [390]. Adaptations to this system can involve elaboration of either ring, or in several cases, both. Xanthones are known in members of Gentianaceae that have all positions on the A-ring oxygenated, others are known that have all positions on the B-ring oxygenated. Still others are known that have lost one or more, or in some cases even most, of the hydroxyl groups from the base structures.

With this preamble, we now look at the application of these structural alterations in a search for trends within Gentianaceae, as outlined by Gottlieb and Kubitzki (1983) in their paper on "Ecogeographical phytochemistry." First, these workers determined relative specialization values depending on the oxygenation patterns of the two rings, that is, the level of oxygenation of each ring. (Note that O-methylation does not affect the oxidation level of the ring and is thus not a consideration.) Points are accumulated for each ring by counting the number of steps necessary to achieve the oxidation level in any given compound. The total number of points accumulated for each compound is then divided by the number of compounds that possess that pattern to provide a measure of "evolutionary advancement" for each species, EA_A for A-ring patterns, EA_B for the B ring. A plot of these parameters against each other resulted

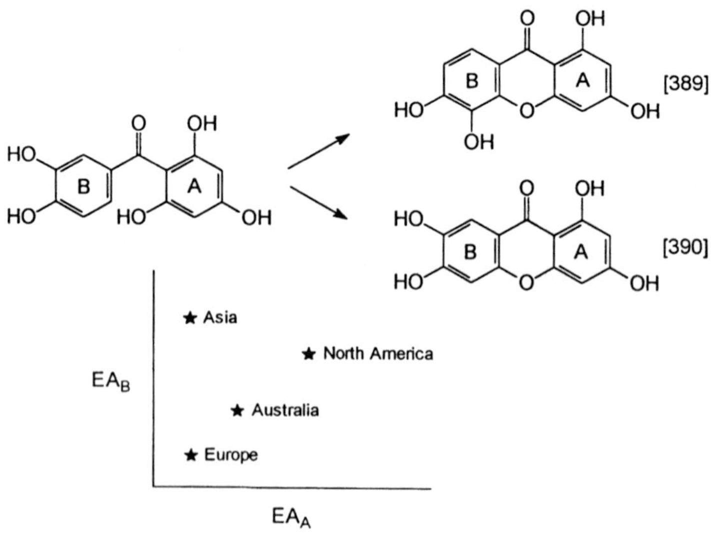

Fig. 5.1 Compounds 389 and 390; xanthones from Gentianaceae showing their formation; and a graph of oxidation indices

in the graphing of values seen in Fig. 5.1. The correlation of evolutionary advancement values, without reference to individual taxa, indicates that the European taxa have substitution patterns most closely resembling the base structures, the Australian specimens are a little more complex (advanced, specialized?), and the Asian and North American taxa display highly derived B- and A-rings, respectively. This in turn suggests that the xanthone containing Gentianaceae originated in temperate Europe, and their spread into new areas was accompanied by structural changes associated with one ring or the other.

Other examples in the Gottlieb and Kubitzki (1983) paper, following similar statistical methodology and logic, concern chemical changes in the evolution of *Aniba* (Lauraceae) in the Amazon, and associated river basins; and relationships among genera of Papilionoideae as a function of their accumulated alkaloids. These are complex examples that would repay careful study by interested readers.

5.1.3 **Hordeum** *(Poaceae)*

Hordeum vulgare L. and several closely related species have been important elements in agriculture for a very long time. Mabberley (1997, p. 346) informs us that barley was first harvested some 11,000 years ago. In the 1960s, Fröst and his colleagues undertook studies of the flavonoids of barley and its relatives as a possible source of information in tracking routes of movement of this valuable food plant (Fröst and Asker, 1973, 1977; Fröst and Holm, 1971, 1972, 1977; Fröst et al., 1975, 1977). The early survey studies revealed the existence of three clear-cut chemical races—called **A**, **B**, and **C**—based upon their flavonoid spot profiles. Analysis of 1424 local varieties was made, along with samples of *H. spontaneum* and *H. agriocrithon* (Fröst and Holm, 1975). Chemotypes A and B (there was further differentiation of type **B** into two subforms but this fact does not influence the overall picture) were observed in varieties from throughout much of the world, including collections made in Europe, Asia, the Middle East, Africa, and North America. Race **C**, on the other hand, was observed primarily in varieties collected in Ethiopia (238 of 279 varieties tested). These workers surmised that race **C** had evolved from race **B**, followed by selection. Structural studies of the flavonoids revealed the profiles to consist of O- and C-glycosyl derivatives of the common flavones, apigenin, luteolin, and chrysoeriol. The bulk of the variation observed results from the nature of the sugar attached to the 6-*C*-glucosyl flavones (Fröst et al., 1977).

5.1.4 **Pteridium aquilinum** *sens. lat. (Pteridaceae)*

The toxic properties of *Pteridium aquilinum* L., the common bracken fern, have been known to humankind for a very long time. Two principal causes of trouble are its carcinogenic properties (M. Saito et al., 1975; I. A. Evans, 1976; Hirono, 1986) and, unrelated, the presence of a potent thiaminase I, which is the causative agent in neuropathies observed in stock animals that have eaten bracken (W. C. Evans, 1986; Fenwick, 1988). A large literature has accumulated on the subject of the carcinogenic compounds of bracken, among which are several that treat the chemical structure of the compounds (e.g., Niwa et al., 1983; Ojika et al., 1987; K. Saito et al., 1990), their action on DNA (Ojika et al., 1989), and analytical applications (Alonso-Amelot et al., 1992). The main carcinogen from bracken, ptaquiloside [391], and its conversion product, pterosin-B [392], are illustrated in Fig. 5.2.

Our interest focuses on more recent work by Alonso-Amelot et al. (1995) that deals with the comparative dynamics of ptaquiloside and pterosin-B in two varieties of *P. aquilinum*, var. *caudatum* and var. *arachnoideum*, both collected in Venezuela. Whereas both varieties exhibit both compounds, the quantities present were shown to be markedly different. Thus, in newly emerged crosiers, which have the highest concentrations of these compounds among growth stages, the amount of ptaquiloside in var. *caudatum* varied in the range 1.98–3.9 mg/g biomass, whereas var. *arachnoideum* exhibited concentrations in the range 0.032–0.66. The concentrations of these compounds drops as the plant ages, but the relative differences persist. According to N. Tanaka et al. (1993), neither ptaquiloside nor pterosin-B was present in var. *esculentum* collected in New Zealand (cited as *Pteridium esculentum* in that paper). There is a reference in that paper, however, to work published in the *New Zealand Veterinary Journal* [Smith et al., Vol. 36: 56 (1988)] not only concerning the identification of ptaquiloside as the component in *P. esculentum* responsible for cattle poisoning in New Zealand but also noting its regional variation.

Other, simpler compounds appear to vary among varieties of bracken as well. Both *trans-o*-coumaric acid [393] and the related coumarin [394] have been reported from var. *caudatum* (Bohm and Tryon, 1967; Alonso-Amelot et al., 1995), but not from other varieties that have been tested: var. *aquilinum* from Europe and England, var. *latiusculum* from Europe, North America, and Japan, var. *pubescens* from western North America and Mexico, and var. *esculentum* from Australia and New Zealand (Alonso-Amelot et al., 1995).

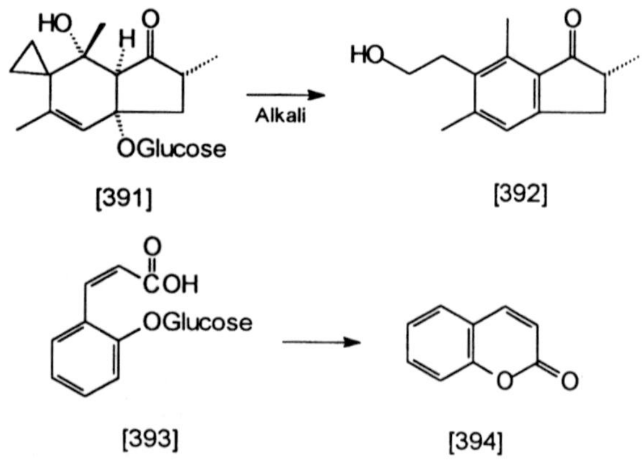

Fig. 5.2 Compounds 391–394, constituents of *Pteridium aquilinum*

Although this next example does not fit into the category of wide disjunctions, it does fall within the realm of bracken biology, if one takes a broad view. A recent study by Alonso-Amelot et al. (2004) revealed interesting responses of bracken species to different environments. Those workers measured concentration differences in two taxa, *P. caudatum* (L.) Maxon and *P. arachnoideum* (Kaulf), that occur naturally over different elevational gradients, sea level to 3200 m for the former, 1800–3200 m for the latter. High-molecular-weight phenols (HMP) and low-molecular weight phenols (LMP) were measured in samples collected along gradients for each taxon. Synthesis and accumulation of LMP was found to be largely independent of elevation, whereas HMP varied with elevation in both taxa. The authors suggested two explanations for the observed results, either higher production from UV-B from HMP than LMP, or differences in activity of light-activated steps in biosynthetic steps leading to the two types of phenols.

5.2 Nonvascular Plants

5.2.1 Pseudevernia furfuracea *(Parmeliaceae)*

The next several entries feature examples from the lichen literature. Study of chemical constituents of lichens has proved an invaluable source of information on relationships within this widespread group of organisms. The first example, *Pseudevernia furfuracea* (= *Parmelia furfuracea*) involves a disjunction across the Atlantic Ocean, but other examples in this set enjoy much wider—and often unusual—disjunctions. This example is included here in order to keep the lichen material in one place.

Hale (1956, 1968) reported different lichen acid profiles from specimens of *P. furfuracea* that were morphologically identical. European material afforded two compounds, olivetic acid [395] and its phenolic coupling derivative physodic acid [396] (see Fig. 5.3 for structures 395–405). Plant material in North America, however, yielded only lecanoric acid [397], which is not closely related biosynthetically to the compounds from the European plants, although, of course, they all belong to the same general class of metabolites. Based upon the chemical differences between materials from the two sources, a suggestion was made that the two entities should be recognized as different species (Hawksworth and Chapman, 1971; Hawksworth, 1976). Other information suggested that the taxonomic problem might not be so easily resolved. C. F. Culberson (1965) had reported that 0.5% of individuals in a Spanish population had acids of both types.

5.2.2 Rhizoplaca melanophthalma *(Lecanoraceae)*

Rhizoplaca melanophthalma (Ram.) Leuck. & Poelt seems to be a somewhat more straightforward case of the existence of chemical races. Earlier work had shown

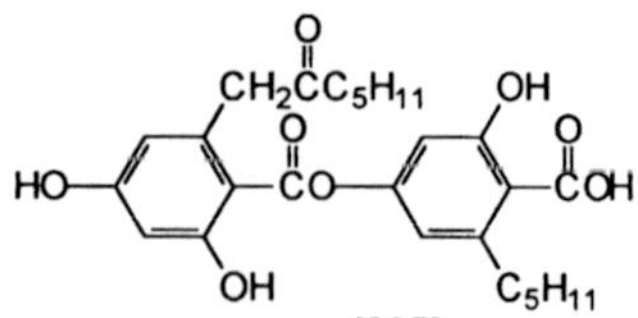

[395]

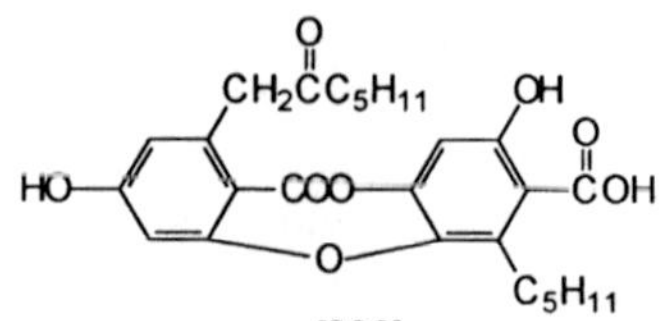

[396]

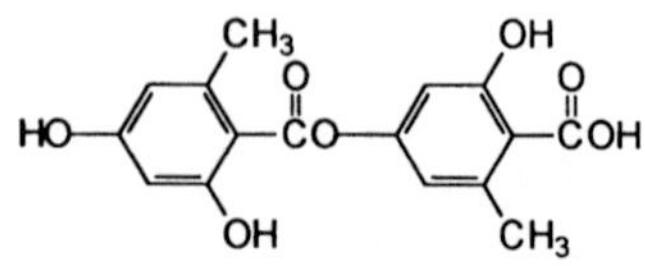

[397]

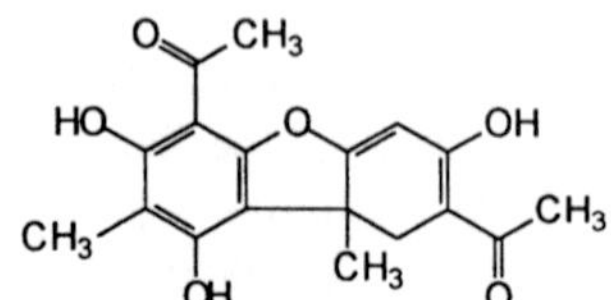

[398]

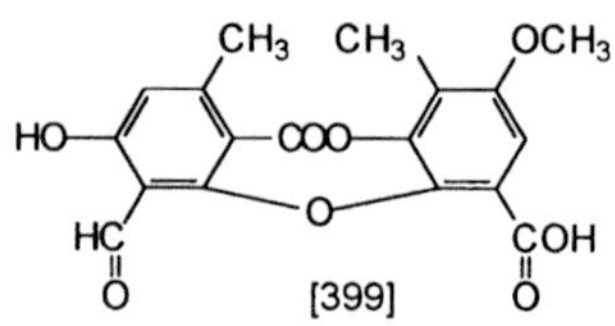

[399]

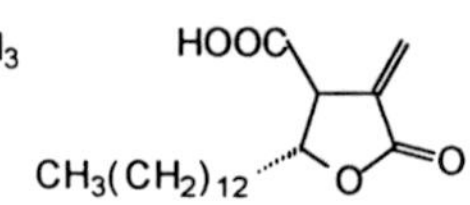

[400]

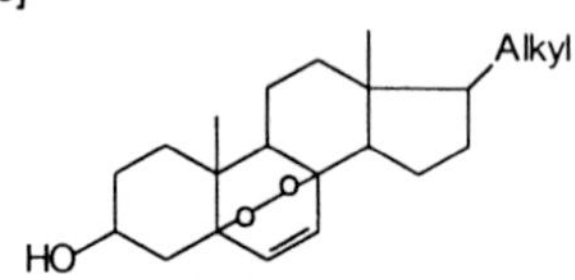

[401]

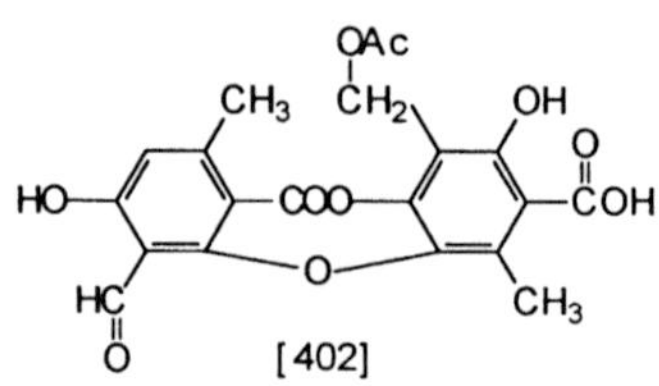

[402]

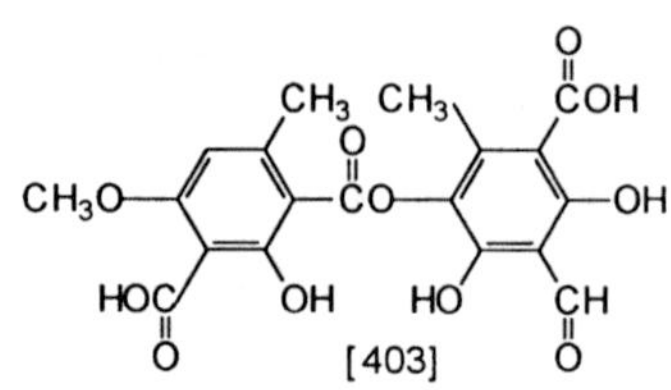

[403]

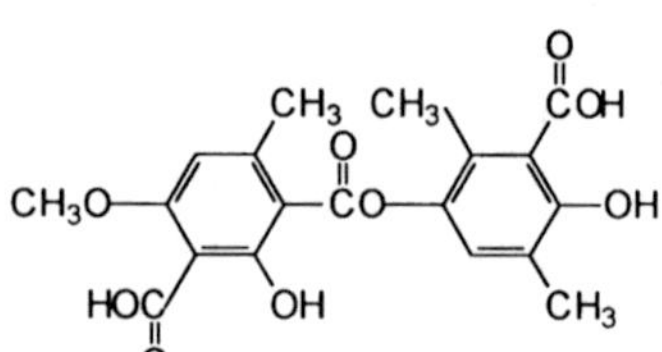

[404]

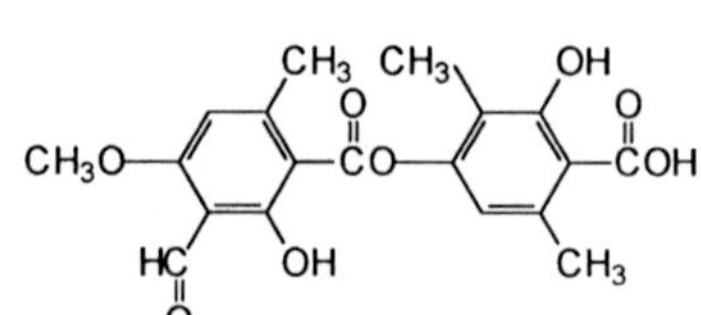

[405]

Fig. 5.3 Compounds 395–405, lichen acids from species of *Pseudevernia*, *Rhizoplaca*, and *Thamnolia*

that plant material collected in Chile afforded (−)-usnic acid [398] as, presumably, the sole component (Huneck and Follmann, 1964). Soon thereafter, a specimen from the Alps was shown by Eigler and Poelt (1965) to accumulate both (−)-usnic acid and psoromic acid [399], compounds that have also been reported from Canadian accessions (Huneck et al., 1968; Thompson et al., 1969). A more recent contribution to the list of chemical races also involves material collected in Chile. Thus, Piovano et al. (1997) not only reported both usnic and psoramic acids, in agreement with the earlier results but also identified protolichesterinic acid [400], which they suggested might be the unidentified compound mentioned by Follmann and Huneck (1972), plus two compounds hitherto not reported from this species, namely, ergosterol peroxide [401] and physodalic acid [402]. Those workers suggested that these differences might reflect the organism's response to different microhabitats, referring to Kershaw's (1985) discussion of the effects of microhabitat differences on lichen physiology.

5.2.3 Thamnolia vermicularis *(Icmadophilaceae)*

Thamnolia vermicularis is an arctic-alpine, fruticose, soil-dwelling lichen that enjoys a very wide distribution. The species consists of two chemical strains, one predominantly found in the Northern Hemisphere and the other in the Southern Hemisphere, with most populations from intermediate regions exhibiting intermediate chemical types. Sato (1965) showed that the southern strain accumulates thamnolic acid [403], whereas the northern strain is characterized by two compounds, squamatic acid [404] and baeomycic acid [405]. Populations from Greenland, Novaya Zemlya, and Svalbard exhibited only the northern profile. Two populations from South America exhibited only the southern type, while collections from New Zealand showed "nearly pure" southern type.

5.2.4 Lobaria pulmonaria *(Stictaceae)*

Lobaria pulmonaria (L.) Hoffm. var. *meridonalis* (Vain) Zahlbr. is a widespread lichen characterized by a lobed structure similar in appearance to alveolar tissue. As González et al. (1994) pointed out in their introduction to a discussion of the chemistry of this lichen, its specific epithet reflects its use as a treatment for pulmonary tuberculosis following the Doctrine of Signatures. The varietal name was appended much later to specimens collected in the Philippines.

The present chemical study was done on plants collected from *Pinus canariensis* growing at several sites in the Canary Islands. The major constituents were identified as the depsidones stictic acid [406], norstictic acid [407], and constictic acid [408] (see Fig. 5.4 for structures 406–411). All of these compounds have been identified from Asian plants. Lesser components from the Canary Island plants were identified as methylstictic acid [409], hypostictic acid [410], and cryptostictic acid [411], all of which were reported for the first time in that paper (González et al., 1994). Those workers suggested the possibility that these latter three compounds may be "found randomly in this species …" and thus have no taxonomic significance. They left the door ajar, however, recognizing that this array of pigments may be "the consequence of the geographical or ecological distribution of the area where the taxon under study was collected."

5.2.5 Bazzania trilobata *(Lepidoziaceae)*

The bryophyte literature provides many examples of disjunct distributions often involving sites that defy easy explanations for their existence. Nonetheless, they

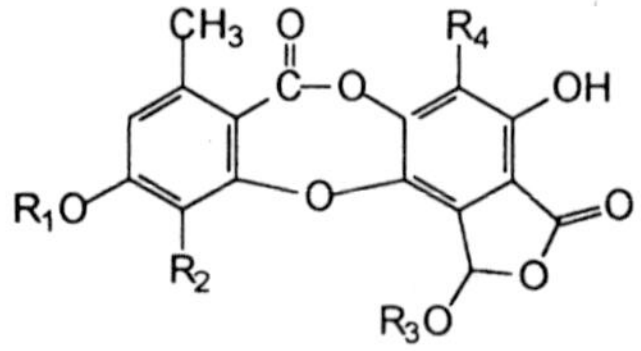

Acid	R_1	R_2	R_3	R_4
[406] Stictic	CH_3	CHO	H	CH_3
[407] Norstictic	H	CHO	H	CH_3
[408] Constictic	CH_3	CHO	H	CH_2OH
[409] Methylstictic	CH_3	CHO	CH_3	CH_3
[410] Hypostictic	CH_3	CH_3	H	CH_3
[411] Cryptostictic	CH_3	CHO	H	CH_3

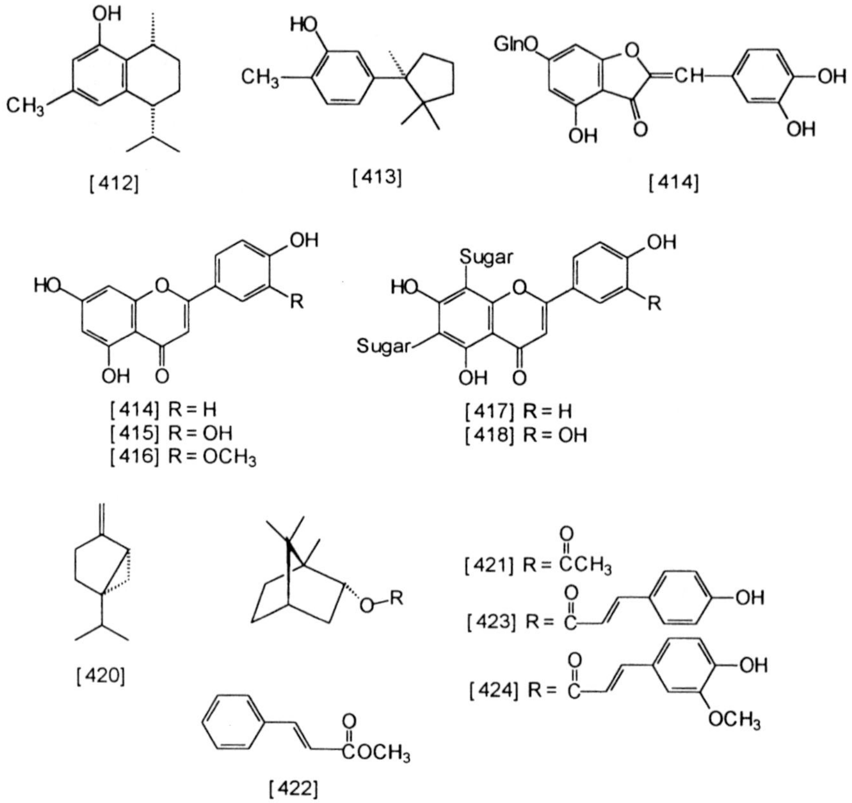

Fig. 5.4 Compounds 406–424, lichen acids and other compounds from species of *Lobaria*, *Bazzania*, and *Conocephalum*

represent interesting and challenging problems for the bryogeographer. One of the approaches to understanding possible evolutionary relationships among the various populations of these systems has been the study of their micromolecular components. The examples given below demonstrate admirably the rich array of chemical characters that exist and their potential usefulness.

Species of *Bazzania* have been shown to be a rich source of terpene derivatives. *Bazzania trilobata* is no exception as evidenced by the identification of no less than 44 compounds from European collections, and 29 from plants collected in North America (Warmers and König, 1999). The major differences between plants from the two sides of the Atlantic were the presence of (+)-*cis*-2-hydroxycalamenene [412] from the former, and (−)-δ-cuparenol [413] from the latter (see Fig. 5.4 for structures). This difference is interesting, in that it is akin to the observations of Konecny et al. (1985) that this liverwort collected in Czechoslovakia afforded compound [412], whereas material obtained from Japan yielded compound [413].

5.2.6 Conocephalum conicum *(Conocephalaceae)*

The first indication that *Conocephalum conicum* (L.) Dum., a thallous liverwort widely distributed in the Northern Hemisphere, consisted of geographically separated chemical races came from study of two widely separated collections by Markham et al. (1976). Plant material from Germany (Saarland) and the United States (Washington State) exhibited not only several flavonoid derivatives in common but also compounds unique to each site. The arrays of compounds held in common were identified as *O*-glycosides of apigenin [414], luteolin [415], and luteolin 3′-methyl ether (chrysoeriol) [416], and the flavone di-*C*-glycosides apigenin 6,8-di-*C*-glucooside (vicenin-2) [417] and luteolin 6,8-di-*C*-glucoside (lucenin-2) [418] (see Fig. 5.4 for structures). The O-linked sugar was glucuronic acid in all cases, with additional rhamnose residues in some. The distribution of these compounds in shown in Table 5.1.

Subsequently, Porter (1981) expanded the survey by looking at specimens collected over a much larger part of the range of the species: 23 sites in Europe, five in North America, and six in Asia. The majority of compounds identified were again revealed as derivatives of apigenin, luteolin, and chrysoeriol, with the major

Table 5.1 Comparison of flavones of *Conocephalum conicum* specimens from North America and Germany

Flavone[a]	North America	Germany
Vicenin-2	+	+
Lucenin-2	+	+
Apigenin 7-Gln	+	+
Luteolin 7-Gln	+	+
Chrysoeriol 7-Gln	+	+
Apigenin 7-Gln-4′-Rhm	+	+
Luteolin 7-Gln-4′-Rhm	+	+
Chrysoeriol 7-Gln-4′-Rhm	+	+
Apigenin 7,4′-diGln	−	+
Luteolin 7,4′-diGln	−	+
Apigenin 7-diGln-4′-Rhm	−	+
Luteolin 7,3′-diGln	+	−
Luteolin 7-Gln-3′,4′-diRhm	+	−
Two Luteolin 7-Gln derivatives	+	−

[a] Vicinin-2 and lucenin-2 = apigenin and luteolin 6,8-di *C*-glucosides; Gln = glucuronic acid; Rhm = rhamnose.

differences being additional combinations of glucuronic acid and rhamnose in the O-linked glycosides, and the addition of *C*-glycosylflavones involving arabinose and rhamnose. All *C*-glycosylflavones identified in the expanded study had sugar substitutions at both positions C-6 and C-8. Newly described in that study was the aurone aureusidin 6-*O*-glucuronide [419], which was seen a single specimen from France. Although the aurone does not play any role in the immediate story, it is interesting to note that it has also been identified in a specimen of *C. supradecompositum* from Japan.

Vicenin-2, apigenin, and chrysoeriol 7-*O*-glucorinides, and apigenin 7,4-di-*O*-glucuronide were essentially ubiquitous, and thus of no use in defining races. Presence and absence data involving other compounds, however, suggested distinct regional specializations outlined as follows: (1) Other than vicenin-2, *C*-glycosylflavone profiles are clearly specialized; lucenin-2 occurs only in North American plants, the violanthin-isoviolanthin pair (apigenin-6-*C*-glucoside-8-*C*-rhamnoside and the isomer with the sugars reversed) occurs only in Japanese and North American plants, shaftoside (apigenin-6-*C*-glucoside-8-*C*-arabinoside) occurs only in Asian plants, and chrysoeriol 6,8-di-*C*-glucoside occurs only in European and North American plants. (2) Luteolin is absent from all but a very few European samples, whereas it is well represented in the other areas. (3) Apigenin 7-*O*-diglucuronide and chrysoeriol 7,4′-di-*O*-glucuronide were not observed in Japanese plants. (4) Two samples from Bulgaria were characterized by a compound thought to be a derivative of acacetin (apigenin 4′-methyl ether). Porter viewed the comparative richness of the North American profiles as a possible indicator of origin, or "original watershed of the species."

In addition to the greater regional differences, some additional fine structure emerged from Porter's wider survey. There was indication that plants with simpler flavonoid profiles were often less robust than those with a fuller complement of pigments. Populations from both Great Britain and Germany showed this sort of correlation. Biochemical differences (electrophoretic data) correlated with morphological differences in populations of this taxon in Poland (Szweykowski and Bobowicz, 1979). Two flavonoid races appear to exist in Japan, with material from the island of Shikoku exhibiting a comparatively simple profile, whereas material from Honshu and Kyushu, which have very similar profiles, is characterized by a richer array of compounds that includes a unique acylated derivative of apigenin 7,4′-di-*O*-glucuronide.

One of the distinguishing features of liverworts is their capacity to produce a variety of volatile, often very fragrant, oils that occur in oil bodies distributed over much of the surface of the organism. Several of these volatile compounds figure prominently in defining geographically different chemotypes in several countries. Toyota et al. (1997) examined 280 specimens of *C. conicum* collected at Kamikatsu-cho and Katsuura-cho, both Katsuura-gun, Tokushima. Three chemotypes were identified based upon their major components. Chemotype-**I**-accumulated (−)-sabinene [420], chemotype-**II**-accumulated (+)-bornyl acetate [421], and chemotype **III** was characterized by methyl cinnamate [422] (see Fig. 5.4 for structures). In addition to a number of compounds known from previous studies of *C. conicum*, three compounds new to the species were also reported: *E*- and *Z*-isomers (N.B., *trans* and *cis*, respectively) of (+)-bornyl *p*-coumarate [423 is the *E*-form], and the *E*-isomer of (+)-bornyl ferulate [424]. These compounds appear to be restricted to chemotypes **I** and **II**. Chemotype I is the most widely distributed, whereas chemotype II tends to be coastal. Chemotype **III** is restricted to forests in mountainous regions. It is interesting to note that a collection of *C. conicum* corresponding to chemotype **III** has also been reported from North America (Illinois) (Wood et al., 1996). It was also reported in that paper that specimens from northwestern California (near Humboldt) did not exhibit any trace of methyl *p*-cinnamate.

Additional information on terpene-based chemotypes of *C. conicum* comes from a recent paper by Melching and König (1999), who studied plant material collected in southern Germany (Adelberg, near Göppingen). Major components of the essential oil fraction, illustrated in Fig. 5.5, were shown to be the known (−)-selin-11-en-4-ol [425] and conocephalenol [426]. The latter compound was first isolated from *C. conicum* collected in Scotland, and later found in a sample of the plant from Germany (Asakawa, 1995). In the newer work, three novel brasilane-type sesquiterpenes were identified: brasila-5.10-diene [427], brasila-5(10),6-diene [428], and brasila-1(6),5(10)-diene [429]. The compounds present were presilphiperfolan-1-ol [430], the aromatic compound 3,4-dimethoxyphenylstyrene [431], and a number of other nonsesquiterpene derivatives. The presence of the styrene derivative is interesting, insofar as it has

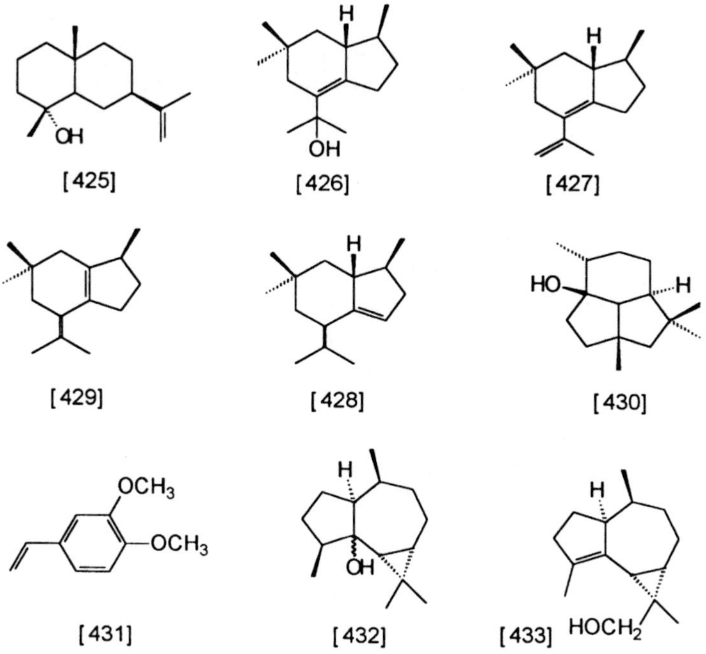

Fig. 5.5 Compounds 425–433, essential-oil components from *Conocephalum conicum*

also been identified as a component of a collection of a different chemotype from northern Germany. Further variations on the sesquiterpene theme were reported by Melching et al. (1999) from a specimen collected near Vorarlberg, Austria; two compounds of the aromadendrene type were identified: (−)-aromadendran-5-ol [432] and (+)-aromadendr-4-en-12-ol [433].

5.2.7 Adelanthus decipiens *(Adelanthaceae)*

Adelanthus decipiens (Hook.) Mitt. is another liverwort that exists in widely separated locales, in this instance, northern Europe and the Southern Hemisphere (Asakawa and Inoue, 1987). It reaches its northern limit in western Great Britain. Rycroft et al. (1998) studied the chloroform-soluble components of three specimens collected from different sites in Scotland and herbarium specimens representing sites in Wales, Ireland, Colombia, and Ecuador. The procedure employed by these workers involves extraction of the dry plant material with deuterochloroform ($CDCl_3$) followed directly by analysis by proton NMR. This procedure yields a "fingerprint" of the individual specimens as well as provides important clues to the structure of the major component(s) (Rycroft, 1996). An added benefit is the capacity to work with very small amounts of plant material, often less than 100 mg. Coupling this extraction procedure with gas chromatography coupled with mass spectroscopy (GC-MS) analysis provides a powerful method as shown by the acquisition of interpretable spectra from a sample of *A. decipiens* weighing 7 mg. The use of small samples taken from herbarium specimens, an approach discussed by Phillipson (1982), enables the study of material from sites otherwise difficult of access.

Specimens of *Adelanthus* yielded a moderately complex array of phenolic compounds based either upon acetophenone or naphthalene, although not all compounds were seen in each specimen. Structures of the compounds identified are shown in Fig. 5.6. The geographic origin of the plant specimens from which individual compounds came is presented in Table 5.2. Plants from all areas tested have the capacity to make naphthalene derivatives, with these the only type seen is the specimen from Wales. Although all other specimens exhibited acetophenone derivatives, there is a good deal of variation with regard to which compounds are accumulated and what their relative concentrations are. Even in specimens from the same general area, the first two listed specimens from Scotland, for example, differences in relative concentrations of certain were noted. It is always tempting to excuse differences (inter- as well as intrasite) on the basis of small sample size, but that possibility is not likely to carry much weight in the present situation in view of the array of compounds from the 7-mg sample from Wales. In short, the data likely represent a fair appraisal of the accumulation tendencies of plants from these areas. It would be interesting to learn how much variation there is among individuals within populations and what variation there may be between seasons. With the efficient methods employed by Rycroft and his coworkers, this information should not be difficult to obtain.

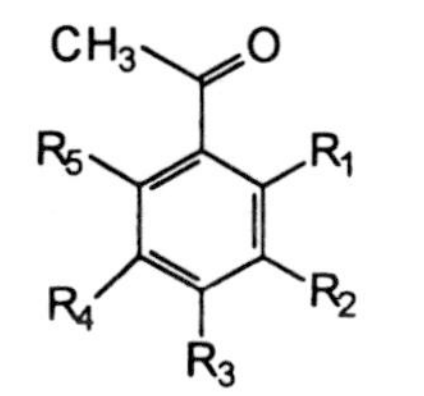

	R_1	R_2	R_3	R_4	R_5
[434]	OH	H	OCH_3	H	OCH_3
[435]	OH	OCH_3	OCH_3	H	OCH_3
[436]	OH	OCH_3	OCH_3	OCH_3	OCH_3
[437]	OCH_3	OCH_3	OCH_3	H	OCH_3
[438]	OCH_3	OCH_3	OCH_3	OCH_3	OCH_3

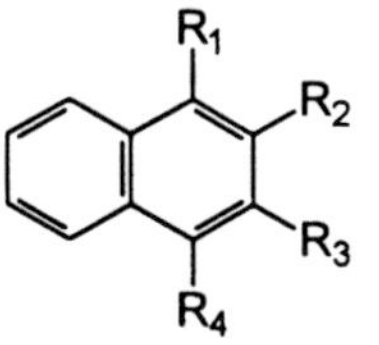

[439]	OCH_3	OCH_3	OCH_3	H	
[440]	OCH_3	OCH_3	H	OCH_3	
[441]	OCH_3	OCH_3	OCH_3	OCH_3	
[442]	OCH_3	OCH_3	$-OCH_2O-$		"wettstein-A"
[443]	OCH_3	$-OCH_2O-$		OCH_3	"wettstein-B"

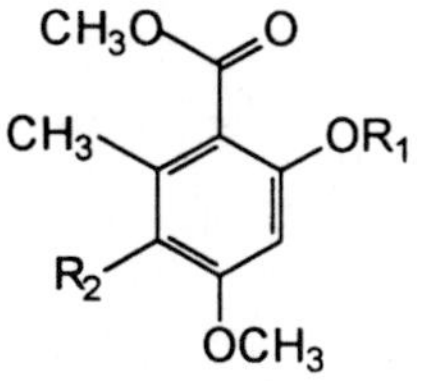

[444] $R_1 = R_2 = H$
[445] $R_1 = CH_3$, $R_2 = H$
[446] $R_1 = CH_3$, $R_2 = OCH_3$

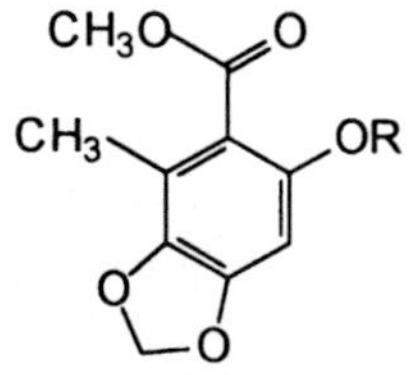

[447] $R = CH_3$
[448] $R = H$

[449]

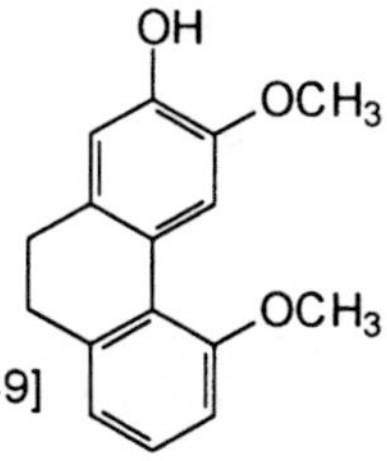

[450]

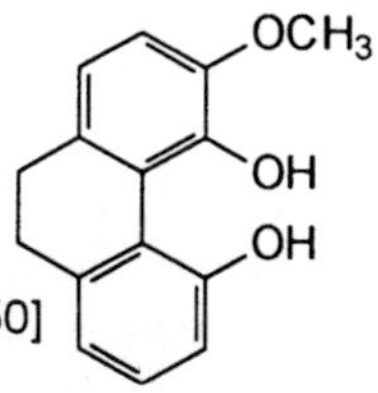

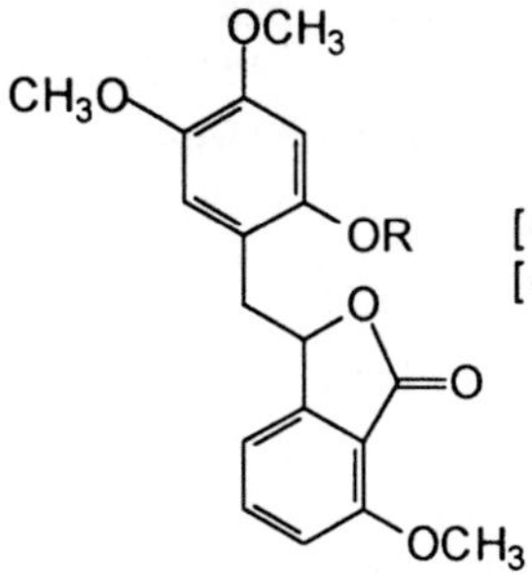

[451] R = H
[452] R = Ac

[453]

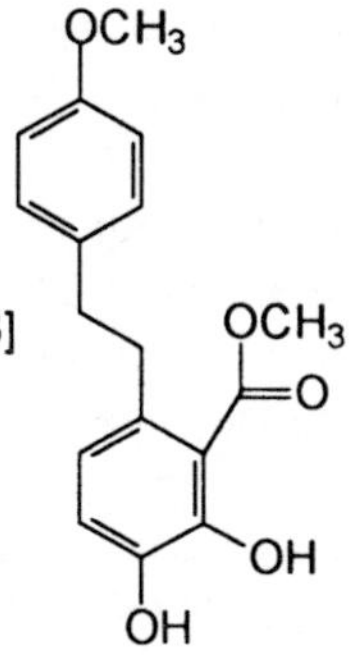

Fig. 5.6 Compounds 434–453, acetophenone and naphthalene derivatives from *Adelanthus*

5.2.8 Plagiochila *(Plagiochilaceae)*

The first example, involving a disjunction of at least 2000 km, features the liverwort *Plagiochila killarniensis* Pears. This species, which reaches its northernmost extent of range in Britain, was collected for the present study from several sites in Scotland

Table 5.2 Acetophenone and naphthalene derivatives from *Adelanthus decipiens* (after Rycroft et al., 1998b)

Compound	No.[b]	Source of plant material[a]							
		Sc1	Sc2	Sc3	Wal	Ire	Co1	Co2	Ecu
Acetophenone									
2-Hydroxy-4, 6-dimethoxy	[434]	–	–	–	–	–	–	0.11c	–
2-Hydroxy-3,4, 6-trimethoxy	[435]	0.2	0.1	0.03	–	–	3.5	0.03	0.4
2-Hydroxy-3,4,5, 6-tetramethoxy	[436]	2.9	0.6	0.4	–	0.09	–	–	0.7
2-Hydroxy-2,3,4, 6-tetramethoxy	[437]	2.0	0.01	–	–	0.02	–	–	–
Pentamethoxy	[438]	0.1	0.03	0.01	–	–	–	–	2.0
Naphthalene									
1,2,3-Trimethoxy	[439]	0.1	0.01	0.01	–	–	0.15	0.10	0.4
1,2,4-Trimethoxy	[440]	0.1	0.01	0.005	–	–	7.6	–	–
1,2,3,4-Tetramethoxy	[441]	0.1	0.01	0.02	–	–	0.07	0.07	0.1
1,2-Dimethoxy-3, 4-methylenedioxy	[442]	12.0	1.7	1.1	0.10	0.25	1.4	0.09	2.2
1,4-Dimethoxy-2, 3-methylenedioxy	[443]	3.3	0.5	0.3	0.07	0.15	0.3	0.04	0.3

[a] Si1, Sc2, and Sc3 = Scotland; Wal = Wales; Ire = Ireland; Co1 and Co2 = Colombia; Ecu = Ecuador.
[b] Structures appear in Fig. 5.6.
[c] Concentrations expressed in mM, determined from NMR spectra.

(e.g., Isle of Mull) and on the island of Terceira in the Azores. The Scottish plants lacked sporophytes, whereas the plants in the Azores, likely growing under more favorable conditions, bore sporophytes and were abundant.

Rycroft et al. (1999) identified the major components of plants from six locations in western Scotland and four from the Azores using nuclear magnetic resonance (NMR) fingerprinting and GC-MS. The terpene β-phellandrene [129], which may be responsible for the aroma of material crushed in the field, was detected in all specimens. The major components, which appear in Fig. 5.6, were shown to be methyl everninate [444], the four methyl orcellinate derivatives [445–448], the two 9,10-dihydrophenanthrene derivatives [449] and [450], the newly described phthalide "killarniensolide" [451], and the bibenzyl [453]. Methyl everninate was the major compound in all 10 specimens; other compounds were more varied in their occurrence. Killarniensolide was not isolated as such but was detected when extracts were acetylated yielding, among other compounds, [452]. The presence of the bibenzyl compound [453] in more than trace amounts in *P. killarniensis* raises the possibility that it represents contamination from *P. spinulosa* with which it was growing at the one site.

Two compounds, although present in material from both Scotland and the Azores, were present in significantly different amounts in the two. Compound [448], methyl 2-methyl-3,4-methylenedioxy-6-hydroxybenzoate, was present in the six Scottish specimens to the level of 22, 13, 14, 10, 12, and 10%, as compared to 2, 2, 3, and 2% for the specimens from the Azores. Similarly with regard to compound [447], the concentrations from Scottish sites varied from 15 to 29% compared with levels that did not exceed 1% in samples from the Azores.

A more recent paper by Rycroft and Cole (2001) described a study of *Plagiochila rutilans* Lindenb. From Bolivia, Brazil, and Costa Rica (all freshly collected) and Cuba and Ecuador (dried specimens). Material from Cuba had been described by Huneck et al. (1984) to contain, among other compounds, 1-(3,4-dihydroxy-5-methoxyphenyl)-3-methylbut-2-ene. In the recent paper, Rycroft and Cole present evidence that the correct structure of the compound is 1-(2,5-dihydroxy-3-methoxyphenyl)-3-methylbut-2-ene, which can also be called 2-methoxy-6-prenylhydroquinone. The corresponding quinone was also observed as a minor constituent.

Recent work identifying *Plagiochila retrorsa* Gottsche from collections made in the Azores and Madeira establishes a significant range extension for a taxon, known under several other names, which occurs in the southern Appalachian Mountains and in Costa Rica. Rycroft et al. (2001) described morphological and phytochemical characteristics of representative specimens of this liverwort. Phytochemically, *P. retrorsa* belongs to the 9,10-dihydrophenanthrene chemotype species (major structural type is that of compounds 449 and 450).

5.2.9 Marchantia *(Marchantiaceae)*

Bibenzyl derivatives were met above with *Plagiochila killarniensis.* Some liverworts, as we will now see, take this type of structure a step or so further to yield dimeric compounds known as cyclic bis(bibenzyls). These compounds have played a role in understanding relationships within the genus *Marchantia* from which work has emerged interesting geographic patterns. Asakawa et al. (1987) reported geographical differences in profiles of *M. polymorpha* plants collected in Japan compared to others collected in France and India. The major compound in Japanese plants, called marchantin-A, is shown as structure [454], whereas the major compound from plants collected in the other two areas, known as marchantin-E, is shown as structure [455] (see Fig. 5.7). The structural difference between these two compounds is small, but two biosynthetic steps are required, one to establish oxygen at the position marked "R" in the figure, and a second to place the methyl group on that oxygen. This, of course, implies two enzyme steps.

Marchantia brachiata (Sw.) Schiffn. is a liverwort that also enjoys a wide distribution. It is included here to offset the idea that liverworts of wide occurrence need to exhibit different chemistries in order to attract attention. Nagashima et al. (1999) identified four compounds from plants collected in Ecuador: 3,4-dimethoxystyrylbenzene [456], 2,4,5-trimethoxystyrylbenzene [457], β-caryophyllene [458], and bicyclogermacrene [459]. In the words of those workers, there was "… ostensibly no metabolic differences between *Marchantia brachiata* collected in the Netherlands Antilles, Saba, and in Ecuador."

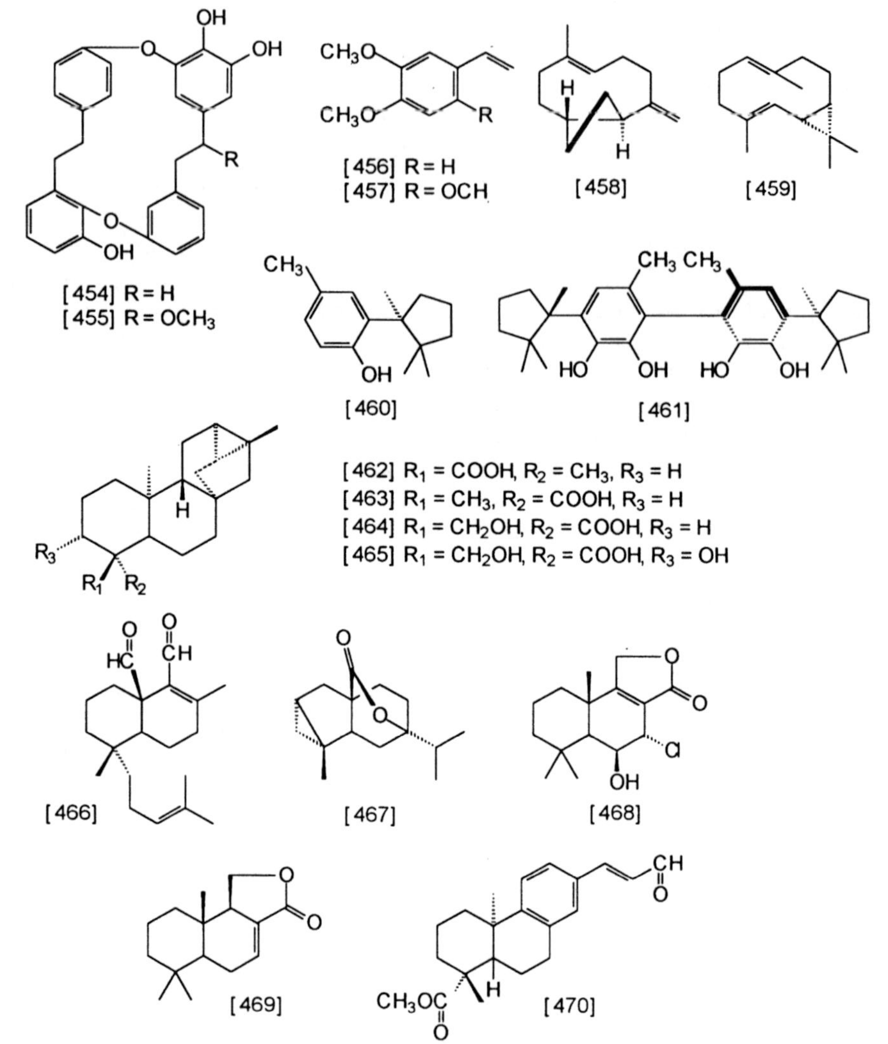

Fig. 5.7 Compounds 454–470, various compounds from species of *Marchantia*, *Mastigophora* and *Makinoa*

5.2.10 Mastigophora diclados *(Mastigopharaceae)*

Mastigophora diclados (Brid.) Nees, a liverwort that occurs in the mountains of tropical Asia, has been studied for its chemical constituents by Asakawa and his colleagues. Accessions from two disjunct sites in Malaysia were examined, one from Sabah State (1900–2500 m on Mt. Kinabalu) and one in West Malaysia from Pahang State (1500 m on Mt. Reskit). The East Malaysian material afforded compounds of the sort here exemplified by α-herbertenol [460] and mastifophorene-A [461] (see Fig. 5.7 for structures 459–465). Variations on the hybertane-type involved the number of hydroxyl groups on the aromatic ring, which ranged from none to three. The second set of compounds, the mastigophorones, varied in the nature of the interaromatic ring linkage (Asakawa et al., 1991). Subsequently, Leong and Harrison (1997) investigated a sample of the liverwort from West Malaysia. Neither of the compound types described from eastern plants was observed. Three other compounds were observed, however, and shown to be the known *ent*-trachyloban-18-oic acid [462] and ent-trachyloban-19-oic acid [463], and the new *ent*-18-hydroxytrachyloban-19-oic acid [464]. As in many, if not most, of the situations described in this review, no explanation for the chemical differences between the two populations is evident (or was offered). Trachylobane derivatives are rare natural products, but one, ent-3β,18-dihydroxytrachyloban-19-oic acid [465], has been reported from the liverwort *Jungermannia exsertifolia* Steph. subsp. *cordifolia* (Dum.) Vána by Harrison and Asakawa (1989). It is of interest to note that herbertane-type sesquiterpenes, for example, structures based on [460], have also been reported from *M. diclados* collected in Taiwan (Chau and Wu, 1987).

5.2.11 Makinoa crispata *(Makinoaceae)*

Makinoa crispata (Steph.) Miyake from Japan was shown by Hashimoto et al. (1989) to contain diterpene derivatives of the sort illustrated as [466–469] (see Fig. 5.7 for structures). More recently, Liu and Wu (1997) reported the presence of the rearranged abietane-type diterpenoid derivative "makanin" [470] from plant material of *M. crispata* collected on Taiwan. Of note was the apparent absence of any of these compounds in the Japanese plants.

5.2.12 Brown Algae

Members of Phaeophyta, the brown algae, have been shown to accumulate, in considerable amounts in some cases, polyphenolic compounds called phlorotannins. These oligomeric compounds are constructed from phloroglucinol (1,3,5-trihydroxybenzene) units linked by ether and/or carbon-carbon bonds. The reader is referred to a review of these compounds written by Ragan and Glombitza (1986). Extensive studies of brown algae, mostly involving northern temperate taxa, have established the important part that phlorotannins play in defending plants against herbivory (Estes and Steinberg, 1988; Steinberg, 1984, 1985, 1986, 1988). One of the conclusions reached was that the comparatively low level of phlorotannin production in many northern kelps and fucoids was due to the contribution made by sea otters in controlling the number of herbivores. In that scenario, the algae are considered to be under lower selection pressure to produce high levels of feeding deterrent chemicals. Estes and Steinberg (1988) suggested that the levels of phlorotannins in

Southern Hemisphere members of these algal groups ought to be higher because, in the absence of otters, the plants had to rely solely upon chemical defenses against herbivores. This suggestion was tested by examining 25 species of algae collected in New Zealand and Australian waters. The median value for the Australasian species was 6.20% total phenolics (dry wt.) in contrast to 1.33% for North American algae (25 species) (Steinberg, 1989).

5.2.13 Cystoseira amentacea *(Cystoseiraceae)*

Cystoseira amentacea Bory var. *stricta* Montagne is a brown alga widely distributed along the coast of the Mediterranean Sea. The complex terpene chemistry of this species has been reviewed by Valls et al. (1996) who, in the same paper, described a study of the alga collected at eight sites along the western French Riviera from Sausset les Pins at the western end of the transect (43°20′N, 5°5′E) to Le Trayas at the eastern end [Le Trayas lies ca. 10 km to the northeast of Boulouris (43°25′N, 6°49′E), which was also one of the collection sites]. Plant material from all eight sites gave the same array of compounds: [471], known as cystoketal, and [472], both of which were known, and the two compounds new to this taxon, [473 and 474]. Plants collected farther to the east near Nice (43°42′N, 7°16′E), however, were shown to contain the two rearranged meroditerpenes mediterraneol-A [475] and cystoseirol-D [476] (see Fig. 5.8 for structures and Fig. 5.9 for a map of the area).

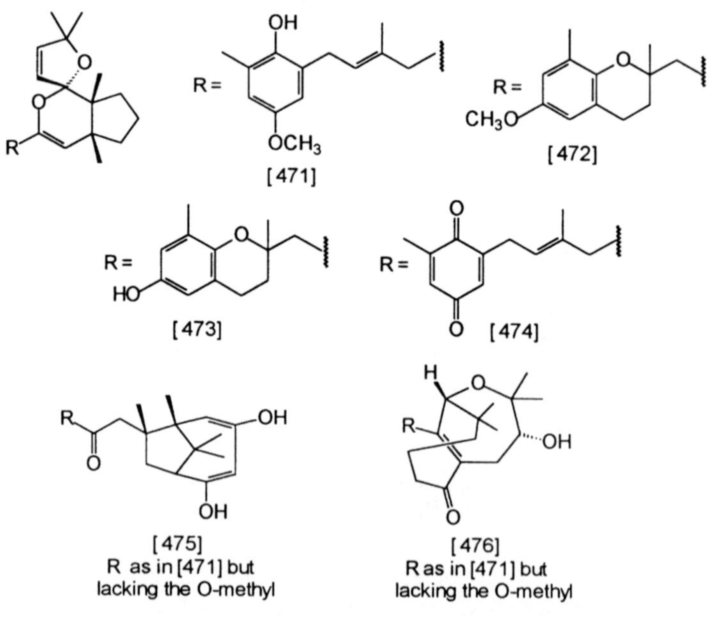

Fig. 5.8 Compounds 471–476, terpene derivatives from *Cystoseira amentacea*

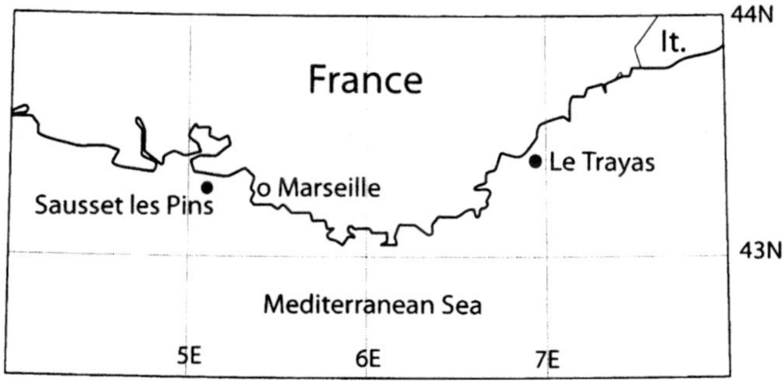

Fig. 5.9 Map showing eastern (Le Trayas) and western (Sausset les Pins) ends of a transect in the study of *Cystoseira amentacea*

5.2.14 Plocamium *(Plocamiaceae)*

Several genera of red algae (Rhodophyta) synthesize a variety of halogenated terpenes that, in addition to exhibiting interesting biological activities, have been found to be useful in chemotaxonomic studies. Several examples involving species of *Plocamium*, *Portieria*, and *Laurencia* are discussed below. The very extensive chemistry of the Rhodophyta has been reviewed by several authors, including Fenical (1975) and Faulkner (1987, 1988, 1990, 1991), in a series of more or less regular summaries. Owing to the ready availability of these reports, the examples below represent only a sampling.

Before looking at the chemistry of these taxa, it is prudent to comment briefly on taxonomic problems associated with these organisms. Several factors contribute to the difficulties in species definition in the red algae, not the least of which is the limited number of morphological features available to the systematist. Structural distinctions can be blurred further because of the amount of convergence that has occurred, much in excess of that commonly seen in vascular plants. Also lacking in many instances has been the sufficient sampling to enable limits of morphological variation to be assessed. In fact, many taxa have not been revisited taxonomically since the initial collections were made, with the attendant problem that species and generic concepts may have changed dramatically in the meantime. Another difficulty in assessing the taxonomic significance of different chemical profiles in some organisms was the tendency of some natural product chemists to accept the names of organisms on faith and not concern themselves with taxonomic nuances such as preparing voucher specimens. This is much less a problem nowadays than it was in the past owing to increased awareness of these problems by practicing natural product chemists, as well as increasing cooperation between chemists and marine biologists in addressing the possible significance of the chemical data. With these considerations in mind, we can turn our attention to the first example.

One of the more thoroughly studied rhodophyte genera is *Plocamium* (Plocamiaceae), from which highly halogenated cyclic and acyclic compounds have been obtained. Extensive studies of the genus have shown the existence of both more or less local as well as very widespread patterns of variation. For example, König et al. (1999) recorded variation in the existence of both cyclic and acyclic compounds in *P. harmatum* J. Agardh collected from three areas on the Great Barrier Reef of Australia. Specimens were collected from six sites representing northern, central, and southern parts of the reef: Milne Reef and Telford Reef from the Cairns Section (Cairns lies at 16°51′S, 145°43′E); Orpheus Island (18°40′S, 146°30′E), Rib Reef, and Potter Reef from the Central Section; and Heron Island (23°25′S, 151°55′E) from the Mackay/Capricorn Section (Fig. 5.10). This species proved to be a rich source of compounds, whose structures appear in Fig. 5.11, marked [477–487].

The terpene profile of plants from the three areas exhibited a good deal of variation, ranging from the richest arrays seen in plants from the northern sites, through highly heterogeneous arrays in plants from the central part of the range, to a comparatively simple array in plants from the Heron Island site. Only in the case of northern region were identical chemical profiles observed in plants from different sites (Milne and Telford Reefs). The data appear in Table 5.3.

Tests of biological activity revealed that compound [480], whose complete structure was determined to be (1*S*,2*S*,4*R*,5*R*,1′*E*)-2-bromo-1-bromomethyl-1,4-dichloro-5-(2′-chloroethenyl)-5-methylcyclohexane by X-ray crystallographic analysis, is very toxic to *Chlorella fusca* under laboratory conditions. The other compounds

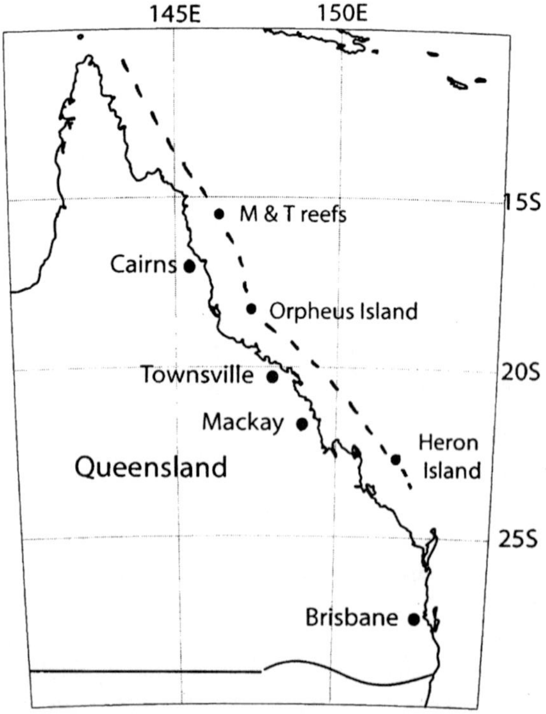

Fig. 5.10 Map of the Great Barrier Reef, Australia

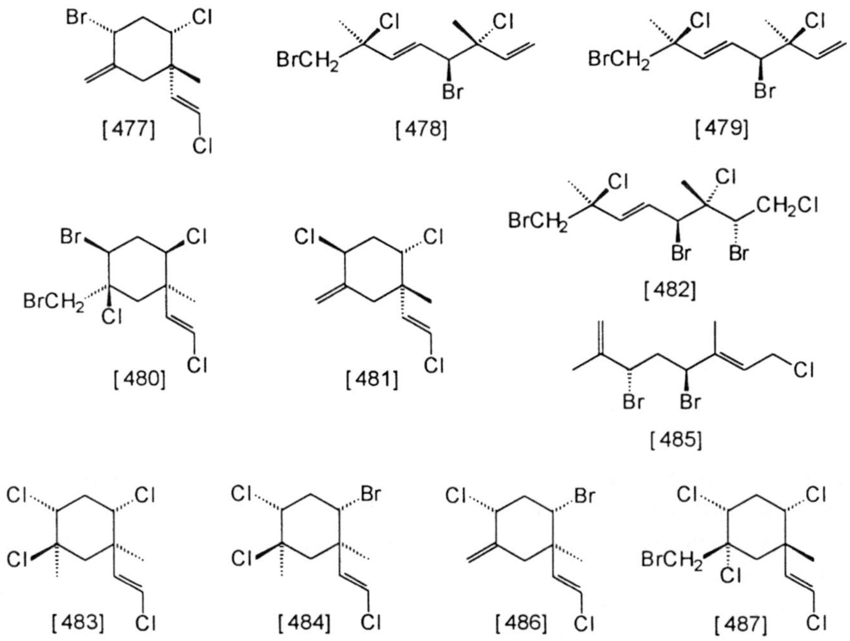

Fig. 5.11 Compounds 477–487, halogenated terpenes from *Plocamium*

Table 5.3 Distribution of halogenated monoterpenes in *Plocamium hamatum* from the Great Barrier Reef (from König et al., 1999)

Source of plants	Compounds identified[a]
Cairns Section	
Milne Reef	[477] – [482]
Telford Reef	[477] – [482]
Central Section	
Orpheus Island	[483]
Rib Reef	[478] [479] [483] – [485]
Potter Reef	[478] [479] [482] [486]
Mackay/Capricorn Section	
Heron Island	[478] [479]

[a] See Fig. 5.11 for structures.

were less active. Biological activity in nature of one or more compounds in this taxon is suggested by the observation of Prof. G. Forbes of James Cooke University (communication to König et al., 1999) that *P. hamatum* is the only alga not eaten by the local green sea turtle (*Chelonia midas*).

The second example features *P. cartilagineum*, a cold-water species that enjoys a wide distribution (Gabrielson and Scagel, 1989). Patterns of variation for halogenated monoterpenes have been reported for this species collected from several locations that include Atlantic, Pacific, and Antarctic waters. Both cyclic and acyclic

Table 5.4 Occurrence of acyclic and cyclic polyhalogenated monoterpenes in *Plocamium cartilagineum*

Collection site	Terpene class		Reference
	Acyclic	Cyclic	
Cadiz, Spain	0	2[a]	Gonzalez et al. (1978)
Isle of Wight, UK	0	5	Higgs et al. (1977)
Overton, South Wales, UK	3	3	Higgs et al. (1977)
La Jolla, California	12	0	Mynderse and Faulkner (1975)
Monterey Bay, California	1	1	Crews (1977)
Whidby Island, Washington	2	0	Crews (1977)
Elephant Island, Antarctica	0	2	San-Martin and Rovirosa (1985)
James Island, Antarctica	2	4	Stierle and Sims (1979)
Covadonga Roadstead, Chile	1	0	San-Martin and Rovirosa (1986)

[a] Number of compounds identified belonging to each class.

terpenes have been identified, with individual profiles consisting of either or both types in the same specimen as illustrated in Table 5.4.

Variation within a given area has also been documented as exemplified by the findings of San-Martin and Rovirosa (1986), who studied the chemistry of *P. cartilagineum* collected from six sites along the coast of Chile spanning a distance of ca. 1600km. Thirteen compounds were identified that differed in degree of halogenation and stereochemistry. Clear-cut geographical differences were noted based primarily on the presence or absence of mertensene [484], violacene [487], and compound [481] (see Fig. 5.11 for structures 477–487). A sample of the qualitative data involving these three compounds can be found in Table 5.5. Earlier studies of *P. violaceum* by Crews et al. (1977) had shown that chemical differences between life stages of an alga also exist. Following that lead, Rovirosa et al. (1988) examined the composition of halogenated monoterpene fractions from carposporophytes, tetrasporophytes, and gametophytes of *P. cartilagineum* collected from two sites: La Boca and Quintay (both Chile). Analysis of the Boca material confirmed the presence of only chlorinated monoterpenes, and that there were only relatively minor quantitative differences in chemistry among algal life stages. The Quintay population, however, tested positive for both bromine-containing

Table 5.5 Variation in selected halogenated monoterpenes in *Plocamium cartilagineum* collected from sites in Chile (from San-Martin and Rovirosa, 1986)

Sites (Région)	Terpenes[a]		
	[484]	[487]	[481]
La Bova (VI)	nd	nd	5.5[b]
Punta de Perros (VI)	nd	nd	5.7
El Tabo (V)	12.8	40.4	nd
Montemar (V)	11.6	24.0	nd
La Herradura (IV)	18.7	19.3	nd
Chiloé (X)	nd	nd	nd

[a] [484] = Mertensene; [487] = Violacene; see Fig. 5.11.
[b] Percentage of total halogenated terpenes; nd = not detected.

compounds, and showed significant quantitative differences between carposporophytes and tetrasporophytes. In the case of mertensene, there was a decrease from 26.2 to 6.4%; with violacene there was an increase from 26 to 47.7%.

Other studies of *P. cartilagineum* have been reported, almost all of which revealed some level of variation among populations. An example involves plants collected near the Isle of Wight (50°41′N, 1°5′W), which yielded five cyclic monoterpenes having both chlorine and bromine as substituents (Higgs et al., 1977), with structures of the general nature seen above, for example, 477. Plants collected in the vicinity of James Island on the Antarctic Peninsula (>60°S) (Stierle and Sims, 1979), and from two sites on the eastern coast of Tasmania (Mayfield Bay and Schouten Beach) (Jongaramruong and Blackman, 2000) accumulated significant amounts of acyclic monoterpenes of the sort illustrated in Fig. 5.11.

It is obvious that production of polyhalogenated monoterpenes is a characteristic feature of the red algae. Within this group of natural products, one sees an almost bewildering degree of structural variation involving the nature of the halogen atom present, the number of halogen atoms present, the stereochemistry of the substituents, and whether the parent compound is cyclic or acyclic. It is difficult, possibly even impossible, to assign chemotaxonomic significance to the presence or absence of any given compound, a situation that can only be remedied when detailed information on their biosynthesis becomes available. Is there a specific enzyme responsible for the placement of each halogen atom at each specific location? Likely not. The fundamental terpene molecule is quite reactive by itself, but the stereochemistry of the products does suggest highly specific surfaces (enzyme faces) upon which the various reactions might occur. Despite the absence of definitive biosynthetic information, we can still appreciate the remarkable chemical synthetic capacities of these organisms. If the differences in chemical profiles do nothing other than suggest further study of possible evolutionary relationships, then the chemicals have performed a useful function. Examples from other genera follow.

5.2.15 **Portieria hornemanii** *(Rhodophyllidaceae)*

Widely distributed in the Pacific, *Portieria hornemanii* (Lyngbye) Silva (synonyms: *Chondrococcus hornemannii*, *Desmia hornemanii*) has provided natural-product chemists with another interesting assortment of secondary metabolites, including more additions to the list of halogenated terpenes. Studies by Gunatilaka et al. (1999) revealed structures of two compounds from plants collected from a variety of reef sites on Guam and the Mariana Islands. Characteristic of the chemistry of this species are the cyclic compounds, apakaochtodene-A [488], apakaochtodene-B [489] and ochtodene [490] (see Fig. 5.12 for structures 488–492). This species also produces large numbers of acyclic halogenated terpenes represented here by (2Z)-6-bromo-3-chloromethyl-1,7-dichloro-7-methylocta-2-ene [491], identified from collections from Nelly Bay, Great Barrier Reef (A. D. Wright et al., 1990, 1991), and (2Z)-1,6-dichloro-3-chloromethyl-7-methylocta-2,6-diene [492], also from the Great Barrier

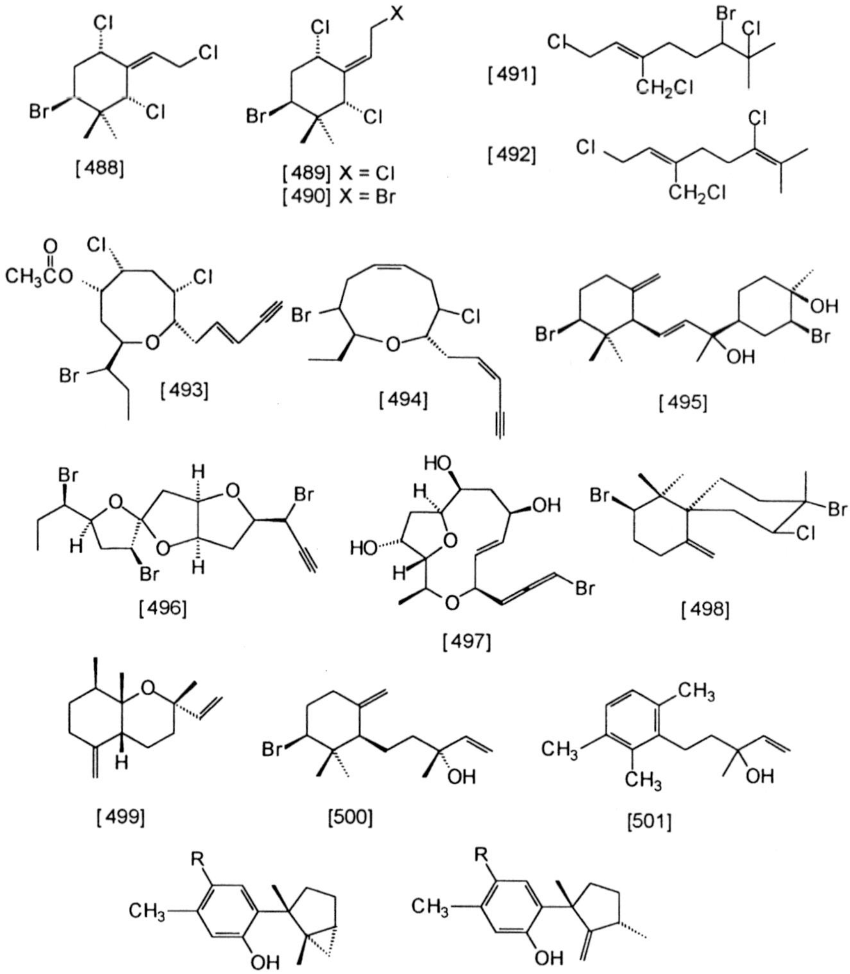

Fig. 5.12 Compounds 488–505, halogenated terpenes from *Portieria* and *Laurencia* species

Reef (Coll and Wright, 1987). Several related compounds have also been identified from plants collected on Amami Island (north of Okinawa) by Ichikawa et al. (1974).

5.2.16 Laurencia *(Rhodomelaceae)*

Laurencia, which has also attracted a lot of attention because of its chemical constituents, is a comparatively large genus. In addition to examples of geographically distinct chemotypes, the examples below illustrate the considerable structural variety that occurs in members of the genus. The first example features *L. obtusa* (Huds.) Lamour., a taxon known from the western Pacific, Indian, and Atlantic Oceans, and the Mediterranean Sea. Caccamese et al. (1981) studied lipid-soluble components of *L. obtusa* collected from four sites on the eastern coast of Sicily spanning a distance of about 75 km from Castelluccio and Brucoli (the latter at 37°17′N, 15°11′E), south to Capo Murro di Porco (just south of Siracusa) and further to Portopalo on the southeastern tip of the island (Fig. 5.13). The profiles of lipid-soluble compounds from the four collections, as determined by GC-MS, were very different, although some of the lesser compounds appeared to be shared between some sites. The sample from Castelluccio, however, exhibited a compound, identified as laurencienyne [493], not seen in plants from the other three sites. It was accompanied by a compound judged, on the basis of its identical mass spectrum, to be the *Z*-isomer. A second compound from the Castelluccio site was identified as obtusenyne [494]. Plants from Brucoli afforded compound [495], known as obtusadiol, as the major component. Plants from the most southerly site, Portopalo, also afforded obtusenyne, and probably its *E*-isomer, but there was no evidence for laurencienyne [493]. The major compound from plants from Capo Murro di Porco was identified as obtusin [496] (See Fig. 5.12 for structures 493–496). Among the other compounds identified from all sites were *n*-heptadecane and cholesterol. Additional information on sesquiterpenoids from *L. obtusa* can be found in Amico et al. (1991).

Those authors also demonstrated that neither sexual stage nor age of plant had any effect on the qualitative composition of the lipid fractions. They did comment, however, on the possibility that the secondary chemical output of plants at these different sites might reflect local environmental conditions, especially considering that the distance between the sites at Castelluccio and Brucoli is only a few kilometers. They pointed to the work of Howard et al. (1980), who had demonstrated such effects on the chemical content of two species of *Laurencia* in California over

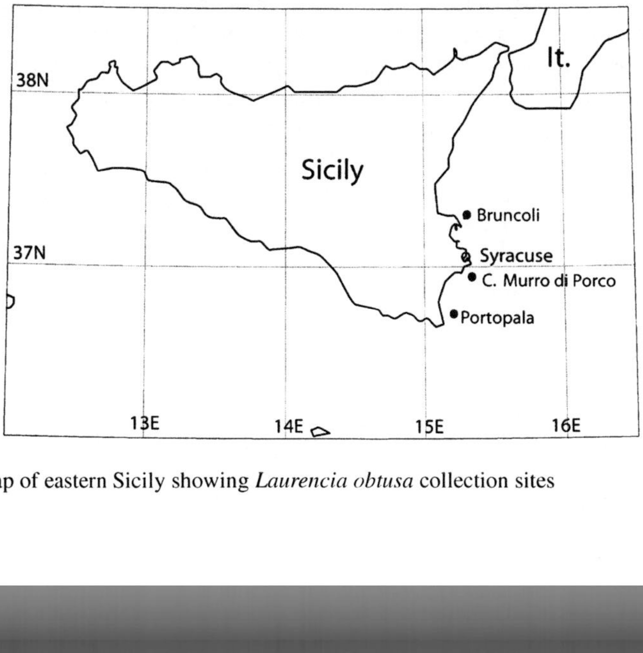

Fig. 5.13 Map of eastern Sicily showing *Laurencia obtusa* collection sites

comparable distances. That situation will be discussed with below, but first we will look at some other work involving *L. obtusa.*

Collections of *L. obtusa* from sites in the eastern Mediterranean Sea, in the Gulf of Suez, and in the Canary Islands have provided additional insights into chemical complexity of this taxon. One of the most structurally complex groups of algal products are the allenes, compounds characterized by the presence of a carbon atom connected to each of its neighbors by a double bond. Representative of this class of compounds is structure [497] (see Fig. 5.12 for structures 497–499) isolated from plants collected at Kas (36°12′N, 29°38′E) in southwestern Turkey (Oztunc et al., 1991). Plants collected near Hurghada, Egypt, which lies about 40 km south of the Gulf of Suez, afforded an array of compounds, including several akin to obtusane [498] (Ayyad et al., 1990), and a new sesquiterpene ether [499] (Ayyad et al., 1994). Finally, we can cite the work of J. D. Martin et al. (1989) who reported halogenated compounds built on the obtusane skeleton from collections made near the islands of Grand Canary and Lazarote in the Canary Islands. Only minor differences in the profiles of these compounds were observed between those two sites.

Laurencia synderae Dawson was collected from two sites, one near La Jolla (the type locality) and one near Santa Catalina Island, and examined for their terpenoid constituents (Howard et al., 1980). The La Jolla material afforded β-synderol [500] as the major component, whereas the Catalina Island alga gave the bromine-free catabolic product of β-synerol [501] as the major compound. The formation of the latter compound can be rationalized as an elimination of halogen followed by rearrangement of the ring system and subsequent aromatization. Culturing *L. synderae* (grown from spores) under various sets of conditions, including temperature, photoperiod, agitation, or source of seawater, did not alter the plant's capacity to form β-synderol as the major product. No trace of the rearranged compound was detected.

In a related study from the same laboratory (Howard et al., 1980), terpenoids were examined from *L. pacifica* Kylin sampled from four sites, two in the vicinity of La Jolla (North Bird Rock and Dike Rock) and one each near Carmel (Stillwater Cove), and Dana Point in Orange County. Plants collected from Dike Rock afforded a complex mixture from which the four compounds identified as [502–505] were obtained (see Fig. 5.12 for structures). Plants from Stillwalter Cove had the same four compounds but they were present in different proportions, whereas material from Dana Point had only the two bromine-containing compounds. In the case of all of these collections, material grown in culture gave the same profiles, although proportions were sometimes different. Those workers concluded that the halogenated terpene profiles in this species are little affected by external physical factors. In other words, these common garden-type growth experiments suggest that the terpene chemistry is under genetic control.

Additional support for this suggestion came from a study of *L. distichophylla* J. Agardh collected off the northeastern coast of New Zealand by Blunt et al. (1984). These workers examined the sesquiterpene chemistry of plants collected at three depths: (1) low intertidal to upper subtidal; (2) mid-intertidal; and (3) upper intertidal. Chromatographic fingerprints of the latter two collections were identical, but the profile of the deep-gathered plants differed in both the number of compounds present, as well as in their structure, although all compounds from this taxon were based upon the same fundamental carbon skeleton. Compounds obtained from the upper tidal sites were shown to be debromoisolarinterol [506], debromoaplysin [507], α-bromocuparene [508], the ether [509], and the newly described cuparene derivative, isolaurenisol [510]. Plants from the lowest site afforded only two compounds, isolaurinterol [511] and allolaurinterol [512] (see Fig. 5.14 for structures 506–518). Cholesterol was identified from plants collected at all sites. It is tempting to speculate that the presence of some herbivore living at lower depths has driven selection of a different set of compounds that serve as feeding deterrents in this part of the alga's range. An alternative explanation, which ought to be borne in mind with all of these chemically differentiated populations, is that one is dealing with cryptic species. It would be of interest to combine the study of chemicals, chromosome structure and number, life cycle features, and, ideally, some gene-sequence studies, on some system of this sort in order to determine how useful, or reliable, the secondary chemistry might be.

On the basis of morphological and secondary metabolic differences, Masuda et al. (1997c) concluded that an alga previously known as *Laurencia obtusa* (Hudson) Lamouroux var. *snackeyi* (Weber-van Bosse) Yamada (earlier name was *L. paniculata*

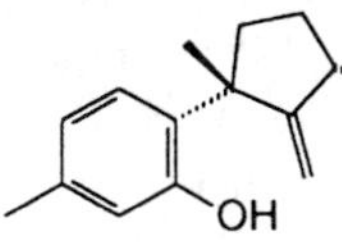

[506]

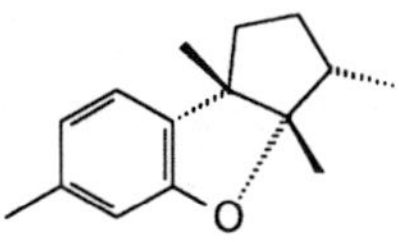

[507]

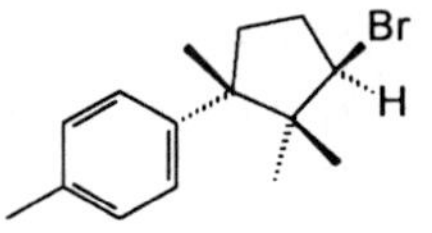

[508]

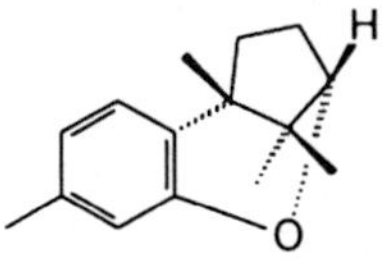

[509]

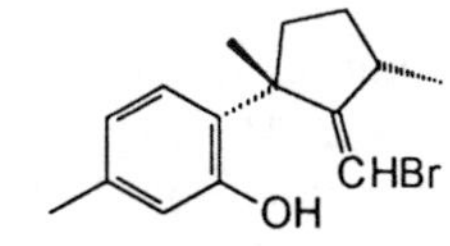

[510]

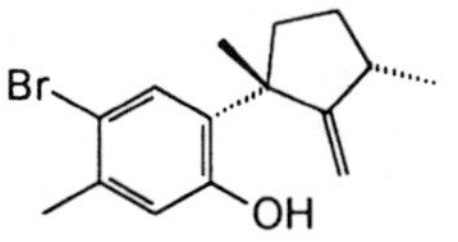

[511]

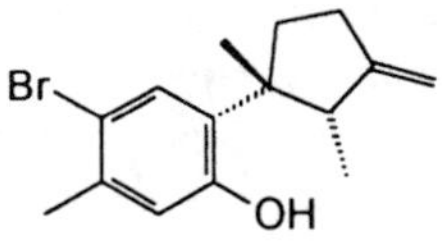

[512]

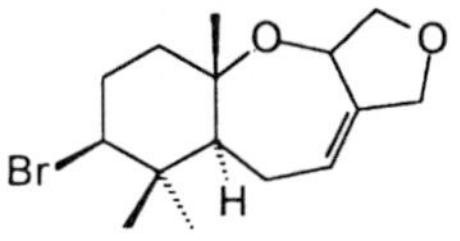

[513]

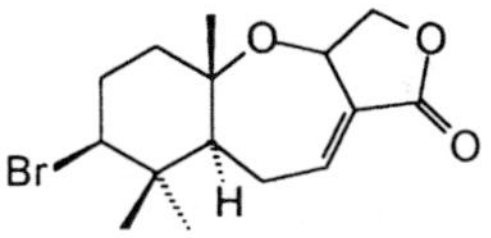

[514]

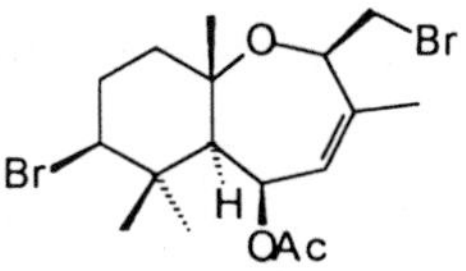

[515]

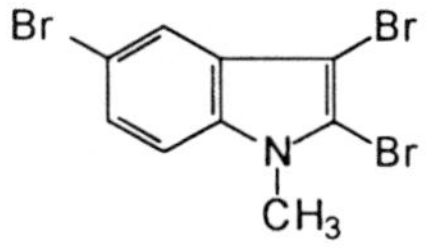

[516]

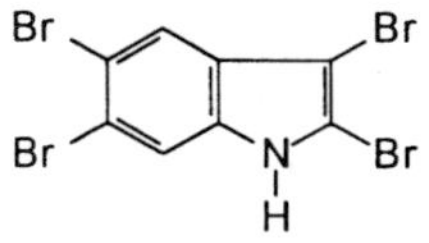

[517]

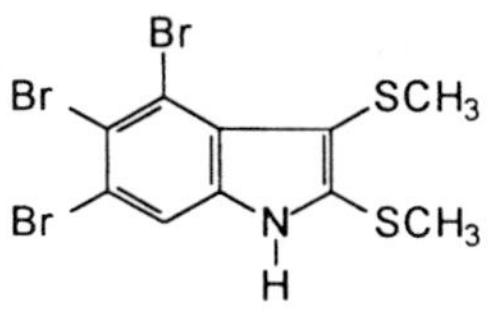

[518]

Fig. 5.14 Compounds 506–518, more terpenes from *Laurencia* species

J. Agardh f. *snackeyi* Weber-van Bosse) should be accorded recognition as *L. snackeyi* (Weber-van Bosse) *stat. nov.* Plants collected from sites in Viet Nam yielded the two halogenated sequiterpenoids palisadin-A [513] and aplysistatin [514]. Plants collected near Pulau Sipanggau, in Malaysia, however, added a third compound, 5-acetoxypalisadin-B [515], to the array. At the time of writing, at least, those compounds characterized this species.

Laurencia brongniartii J. Agardh, an alga of tropical and subtropical waters, has been subjected to chemical analysis by several groups of workers, who studied plants from the Caribbean Sea (Carter et al., 1978), Japanese waters (J. Tanaka et al., 1988, 1989), and Taiwanese waters (Erickson, 1983). This species elaborates a set of brominated indole derivatives, some of which also carry sulfur-containing groups. Representative structures of tribrominated, tetrabrominated, and sulfur-containing compounds are shown as structures [516, 517, and 518], respectively, in Fig. 5.14. Populations from the three areas exhibited different profiles based upon these indole derivatives.

Another example of biochemical diversity within this genus came from study of *L. majuscula* (Harvey) Lucas from widely separated populations. The major compounds identified from all populations were derivatives of the spirosesquiterpene chamigrane system, that is, compounds built on the carbon skeleton seen in compound [498]. Populations sampled were from Okina-shima, Kochi Prefecture, Japan (Suzuki and Kurosawa, 1978; Suzuki et al., 1979, 1987); Castelluccio, eastern Sicily (Caccamese et al., 1986, 1987), Western Australia (Capon et al., 1988), northern Queensland (Coll and Wright, 1989; A. D. Wright et al., 1990), and the Ryuku Islands (Masuda et al., 1997b). The major compounds differ among themselves by variation involving substitution, both position and degree, and oxidation level, that is, hydroxyl, epoxy, and unsaturations; the fundamental ring structure, and presumably its biosynthetic origin, is the same throughout.

A final example of extensive chemical variation, this time involving a much more limited range of sampling, features *L. nipponica* Yamada. At least nine different sesquiterpene-based races of this species have been identified from Japanese waters (Masuda et al., 1997a; Abe et al., 1999).

5.2.17 Codium *(Codiaceae)*

As part of a study of fatty acids in the Mediterranean taxa, *Codium dwarkense* Borgesen and *C. taylorii* P. Silva), Dembitsky et al. (2003) reviewed the extensive literature that exists. One of the main observations relative to the present review was the close similarity of profiles from specimens of *C. fragile* obtained from widely separated sites in the Black Sea and in the Pacific Ocean near Japan, or from sites on the coast of France and those from the eastern coast of Russia. Some exceptions were noted, however, including specimens of *C. fragile* from Australia. These observations served to strengthen taxonomic ties considered to exist between entities collected at these sites and referred to *C. fragile*. Another observation to emerge was the grouping of geographically separated species that have been considered related on the basis of other data.

Chapter 6
Oceanic Islands

Over the past decade or so, the topic of origin and evolution of oceanic island biota has almost become a growth industry. So much information is now available on island geology, continental origins of island endemic taxa, and, in the case of multi-island archipelagos, relationships of closely related taxa between and among individual islands. Thus, introductory information for the following island groups will be kept to a minimum.

6.1 The Galapagos Islands

From an evolutionary perspective, the Galapágos Islands are likely the most historically famous archipelago on the face of the Earth. As every student of biology knows, it was on the Galapagos that Darwin is thought to have perceived the full force of evolution at work. Some think that this may be an oversimplification of the history of the subject, as has been masterfully discussed by Quammen (1996) in his *The Song of the Dodo.* Nonetheless, examples of divergence on these islands have continued to fascinate and attract biologists over the years, the results of which can be found in several comprehensive volumes: *Darwin's Island. A Natural History of the Galapágos* (Thornton, 1971); *The Galapágos: Proceedings of the Symposia of the Galapágos International Scientific Project* (Bowman, 1966); and *Patterns of Evolution in Galapagos Organisms* (Bowman et al., 1983).

The Galapagos Islands (renamed the Archipiélago de Colón in 1892 by the Republic of Ecuador) lie at approximately 90°W longitude and are centered on the Equator with Hood (Española) lying at 1° 20′S, and Darwin at 1° 40′N. The shortest distance from the South American coast is about 935 km. Thornton (1971) describes the archipelago as consisting of four fairly large islands, 11 smaller ones, and numerous smaller islets and rocks. The largest island, Albermarle (Isabela), is approximately 75 miles long with an area that exceeds the sum of areas of all the rest (Fig. 6.1).

The Galapagos archipelago consists of a series of volcanoes rising above the Galapagos platform, which lies on the Nazca Plate, which is moving eastward such that the entire assemblage will eventually disappear into the Chile-Peru Trench

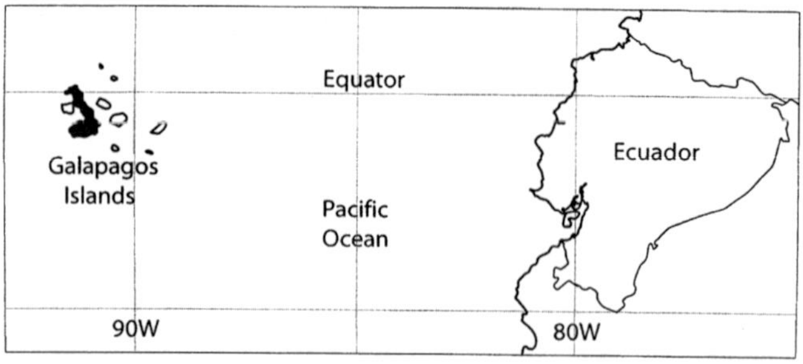

Fig. 6.1 Location of the Galapagos Islands in relation to the South American mainland. The darkened island is Albermerle, Isla Isabela, in the Ecuadorian system

(Cox, 1983). Cox (1983) suggested that the earliest of the present islands emerged about 3–5 mya. Simkin (1984) placed the age of the easternmost islands at about 3.3 million. A more recent study suggests a much longer timeline, 5–9 million years according to dredge findings (Christie et al., 1992). Those authors discuss evidence suggesting that there has been volcanic activity in the region for the past 15–20 million years and that it is likely that islands could have existed throughout the entire time of hotspot activity, 80–90 million years. At the other end of the scale, the age of the westernmost island is given as 0.50 ± 0.08 million years (Cox, 1983). Volcanic activity continues at the present time.

Porter (1983) discussed the vascular flora of the archipelago, both species numbers and possible means of dispersal. A conservative estimate of species endemism was given as 37%. Despite their significance in the biological world and their unusual flora, surprisingly little in the way of comparative (bio) chemical work seems to have been done on endemic taxa. There are important exceptions to this generalization, perhaps the most noteworthy of which in the botanical realm is the work on cotton by Jonathan Wendel and his colleagues (Wendel and Percy, 1991). Only a few studies have dealt with secondary metabolites. We begin with an examination of cyanogenic glycosides.

6.1.1 Cyanogenic Plants

A question that has concerned evolutionary biologists for many years involves the suggestion that defense mechanisms displayed by a continental species (plant or animal) would eventually be lost should that species become established on an island. The reason for the loss would be the presumed absence on the island of herbivores, or pathogens, against which the organism's defense mechanism had been selected in its original continental environment [see Carlquist (1974, 1980) for discussions of this phenomenon]. Among the defensive strategies commonly found in plants is the production of a variety of chemical substances that are either offensive to the senses of herbivores or toxic. In response to the idea that these chemicals would be lost in a shift to an island lifestyle, the occurrence of secondary plant chemicals in island plants was reviewed leading to the conclusion that losses were minimal or not apparent in the vast majority of cases (Bohm, 1998b).

One of the few examples where the loss-in-new-local hypothesis has been tested involves a detailed examination of cyanogenic plants on the Galapagos Islands. Adersen et al. (1988) examined 475 specimens using either fresh material tested in the field, or herbarium specimens available in Copenhagen using the picric acid color test (picric acid-soaked filter paper turns brown in the presence of HCN). Of 97 freshly collected species, 24 were considered to be strongly cyanogenic (10 mg HCN kg^{-1} plant material), while 27 additional species were scored as weakly to moderately cyanogenic (2.5–10 mg HCN kg^{-1} plant material). Taking the analysis a step further, these workers observed that only 45% of species endemic to the islands were HCN-positive, as compared to 62% of the species that occur on the archipelago and on the mainland. When herbarium specimens were tested, only 17.5% of endemic species gave a positive color test compared to 18.6% for the non-endemics. This dramatic drop in reactivity with herbarium specimens is likely the result of the breakdown of cyanogenic glycosides during long standing. The reduction of cyanogenesis in the Galapágos flora, relative to the mainland, can be taken as an indication that selection pressures that keep the level of toxic compounds on the mainland high may not be functioning to the same level of intensity on the islands.

A more recent study provides additional evidence for reduction of defenses in an island setting. In a study of herbivory on Santa Cruz Island, one of the California Channel Islands, Bowen and Van Vuren (1997) found that endemic plants had lower levels of defensive chemicals, as well as reduced mechanical defenses, that is, spines and thickened cuticles, compared to related taxa on the mainland (Santa Ynez Mountains). Feeding trials demonstrated that sheep, which are not natural inhabitants on the Channel Islands, grazed the endemics to a greater extent than they did mainland plants. The putatively defensive chemicals studied were expressed as total phenols and total tannins. Although cyanogenic compounds were not included in the study, three of the genera involved, *Cercocarpus*, *Heteromeles*, and *Prunus*, are members of Rosaceae, a family with many HCN-yielding species.

Returning to cyanogenic compounds for the moment, it is interesting to look at the glycoside profile in *Passiflora foetida* L., a species with a comparatively wide distribution, including the Galapagos Islands and Réunion Island in the Indian Ocean. The Galapagos material has been accorded varietal status, *P. foetida* var. *galapagensis* Killip, on the basis of differences in morphology of stipules, glands, and hairs, and appears as such in the *Flora of the Galapagos Islands* (Wiggins, 1971). This taxonomic recognition has not been universally accepted (Lawesson, 1988), however. Andersen et al. (1998) studied the cyanogenic glycosides of *P. foetida* from both locations. Plants from the Galapagos Islands, grown from seed, yielded tetraphyllin-A [519] (see Fig. 6.2 for structures 519–524), tetraphyllin-B [520], tetraphyllin-B sulfate [521], and volkenin [522], and related cyclopentene-based compounds. Material from Réunion Island afforded tetraphyllin-B, its sulfate, and volkenin, but was distinguished from the Galapagos material by the accumulation of linamarin [523],

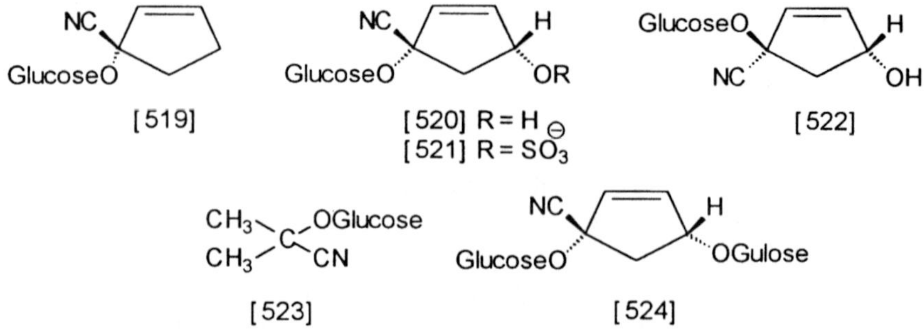

Fig. 6.2 Compounds 519–524, cyanogenic glycosides from *Passiflora* species

a valine-derived cyanogenic glucoside. The finding of cyclopentene-derived as well as common α-amino acid-derived cyanogenic glycosides in the same plant species opens up an interesting area for further investigation. It would be of some significance to determine if both types of cyanogens are produced by the same set of enzymes. If that should be the case, it would lead to the question of why valine is used by the Réunion plants and not by the Galapagos plants. If the two compound types are not produced by the same enzymes, it would be necessary to determine where in the pathway the differences lie.

The presence of an unusual cyclopentene-based cyanogenic glycoside in *Passiflora colinvauxii* Wiggins, an endemic species of the Galapagos Islands, further restricted to Indefatigable Island (Santa Cruz), has been used to argue for a close relationship between this species and the two species from mainland Ecuador, *P. biflora* Lam. and *P. punctata* L. (Adersen et al., 1993). The structure of "passibiflorin" is shown as [524]. Although the presence of this unusual diglycoside (gulose derivatives are rare in nature) in the island species certainly is a good indicator of relationship with the mainland species, it is not possible to state with certainly that the derivation of the endemic species occurred directly from a mainland species or whether there is (or was) an intermediate.

It might be mentioned in passing that variation in cyanogenesis has also been reported for members of *Turnera* sect. *Canaligerae* of Turneraceae (Shore and Obrist, 1992). The cyanogenic glycosides identified from Turneraceae are based upon the cyclopentene ring system, as are those we have just seen in the related Passifloraceae. Geographic patterning was not discussed in that paper, however.

6.1.2 Lycopersicon cheesmanii *(Solanaceae)*

The existence of a species of tomato on the Galapagos Islands was first noted by Charles Darwin. Despite a moderate level of variation, *Lycopersicon cheesmanii* Riley is considered the only tomato species on the islands. All populations examined exhibited morphological and physiological characteristics that clearly distinguish it from species of *Lycopersicon* native to mainland South America. There has been sufficient morphological differentiation, however, to justify recognition of *L. cheesmanii* f. *minus* (Hook. f.) C. H. Müll. An electrophoretic study of four enzyme systems revealed low genetic variation within populations, but quite high levels between populations. The electrophoretic profile of *L. cheesmanii* showed greatest similarity to *L. pimpinellifolium* (Jusl.) Mill., a species native to northwestern Peru (Rick and Fobes, 1975). Further support for the suggestion that *L. cheesmanii* and *L. pimpinellifolium* are closely related came from chloroplast DNA restriction fragment analysis (Palmer and Zamir, 1982), which showed that these two taxa comprise a clade that was a sister group to *L. esculentum* L. Because we are primarily interested in small molecules in this review, it is interesting that the three species involved in this study are characterized by red-orange fruit suggesting similar carotene biosynthetic pathways.

6.1.3 **Scalesia** *(Asteraceae)*

Scalesia Arn. is a genus of 15 species endemic to the Galapágos Islands (Schilling et al., 1994). The presence of trifid pales, gummy resin, and a tetraploid chromosome number of $2n=68$ for all members of the genus have led to the suggestion that *Scalesia* is monophyletic (Eliasson, 1974). Schilling et al. (1994) undertook a restriction site analysis of chloroplast DNA isolated from *S. pedunculata* Hook. f. and from species of *Pappobolus* and *Viguiera*, genera thought to be closely related to *Scalesia.* Three species each of *Encelia* and *Flourensia* were used as outgroups. In addition to confirming the position of *Scalesia* within Helianthinae, the DNA also identified *Pappobolus* as its sister group, although there were some cautionary comments included in their discussion. Nonetheless, the authors used the sequence differences, 0.19%, to estimate time of divergence of *Scalesia* from *Pappobolus.* Taking the estimated rates of evolutionary change for cpDNA to be in the range of 0.03–0.19%, the time of divergence between *Pappobolus* and *Scalesia* lies between 1.9 and 6.3 million years. This period of time is well within the estimated age of the Galapagos Islands. Another interesting point was made by Schilling et al. (1994), namely, that the time accords well with the estimated time of appearance of the land bridge between North America and South America at about 3 mya. Establishment of a bridge between these two landmasses would have provided a means of overland migration of the ancestor common to *Pappobolus* and *Scalesia* from Mexico, which has been suggested as the region of origin of Helianthinae.

6.2 The Hawaiian Islands

The Hawaiian Islands have, arguably, attracted more scientific attention than any other archipelago on the planet. Their unique flora has offered, and continues to offer, a wide spectrum of challenging problems associated with one of the highest levels of species endemism (for such a limited land area) anywhere in the world. According to Wagner et al. (1990, 1999), writing in the *Manual of the Flowering Plants of Hawaii*, there are 32 endemic genera (31 dicot, 1 monocot), accounting for 15% of island genera, 850 endemic species (89%), and 1094 endemic taxa overall (91%). From a slightly different perspective, there is 94% endemism in dicot taxa and 73% in monocot taxa. The flora consists of 146 families (none endemic), 649 genera, and 1817 species. Fosberg (1948) estimated that the affinities of the Hawaiian flora were 40.1% Indo-Pacific, 16.5% Austral, 18.3% American, 12.5% Pantropical, 2.6% Boreal, and 10.3% obscure.

Three major contributions dealing with the Hawaiian flora have appeared in recent years to which the interested reader is referred for detailed descriptions of the flora and discussions of evolutionary relationships. The two volume flora by Wagner et al. (1990, 1999) was referred to above. Volume 1 includes a long introduction that deals with many aspects of the islands, including geology, climate, and vegetation. *Hawaiian Biogeography: Evolution on a Hot Spot Archipelago*, edited by Wagner and Funk (1995), provides detailed summaries of relationships and evolutionary history of several of the major groups of vascular plants on the islands. This volume also features an up-to-date treatment of the islands' geology. The third is a book dealing specifically with the silversword alliance (*Argyroxiphium*, *Dubautia*, and *Wilkesia*) edited by Carlquist et al. (2003).

The Hawaiian archipelago can be thought of as consisting of two groups of islands, the high islands that comprise the "tourist group," and a series of lesser islands extending roughly northwestward from Kauai. The islands have resulted from the movement of the Pacific Plate over a "hot spot" located at approximately 19°N, 155° 30′W (Clague and Dalrymple, 1987). The main group of islands (maximum estimated ages in parentheses in millions of years) consists of Hawaii Island (0.43); Maui (1.32); Kahoolawe (1.03); Lanai (1.28); Molokai (1.9); Oahu (3.7); Niihau (4.89); and Kauai (5.1) (Carson and Clague, 1995). The complex consisting of Maui, Kahoolawe, Lanai, and Molokai is frequently referred to as "Maui Nui," or big Maui, since these four formed a single landmass during Pleistocene glaciation. Hawaii Island (known locally as the Big Island) is actively volcanic; volcanoes on the other islands range from dormant to extinct awaii. Beyond Kauai, the oldest of the extant tall islands, lies a chain of degraded volcanic islands (e.g., Nihoa, Necker, Gardner Pinnacles, Laysan, and Lisianski) and reefs stretching over 3000 km to Midway Island (28°12′N, 177°24′W). Beyond Midway lies the Emperor Chain, a series of seamounts that extends some 2400 km in a more northerly direction from the Daikakuji Seamount to Meiji Seamount (Carson and Clague, 1995).

A closer look at the age of the Hawaiian Islands is useful. In attempting to estimate the time of arrival of propagules on the islands, it is important to realize that there have been receptive high islands for a much longer time than is indicated by the age of Kauai (ca. 5.1 million). As Clague and Dalrymple (1987) point out, the Leeward Islands range from about 7 to 28 million years old. Attempts to address the age of arrival of a propagule and rate of subsequent genetic divergence of new species have, until comparatively recently, been mostly speculative. The measurement of genetic divergence in terms of changes in gene sequences, coupled with the knowledge of how fast a genome is thought to undergo mutation, provides a much more reliable means of estimating ages of island species. Fleischer et al. (1998) discuss molecular-rate calibration using three examples from the animal literature.

The results of studies of secondary metabolites of Hawaiian endemics are primarily useful in assessing levels of variation within taxa, but some generalizations relating to relationships with likely ancestors can be made. We start our survey with a genus well known to North Americans, *Bidens*, commonly called beggars' ticks.

6.2.1 **Bidens** *(Asteraceae)*

As we have seen often in this review, the study of secondary metabolites does not usually allow definitive statements to be made about the origin of an island endemic; such information can nonetheless be useful in suggesting possible relationships, as well as helping to establish the uniformity, or otherwise, of the taxa in question. Much the same situation exists for the examples in this section. Two studies of secondary metabolites in the Hawaiian species of *Bidens* have been described, one involving polyacetylenes, and the other involving flavonoids. The investigation of the polyacetylene (syn.: polyyne) components involved an examination of 19 species and six subspecies (the entire set *sensu* Ganders and Nagata, 1990) resulting in the identification of 18 compounds (Marchant et al., 1984). All taxa accumulated acetylenes in their roots, but only 13 contained these compounds in their leaves as well. In general, each taxon exhibited a unique array of compounds and, with a single exception, showed no interpopulational variation. The exception was *B. torta* Sherff, where each population had a unique array. (*Bidens torta* exhibited a unique array of flavonoids as well; see below.) The overall profile of acetylenes exhibited by the Hawaiian species was quite uniform, which the authors took as additional support for all species having been derived from a single ancestral immigrant (Ganders et al., 2000). Most compounds are known elsewhere in the genus, but one, 2-(2-phenylethyne-1-yl)-5-acetoxymethyl thiophene [525] (see Fig. 6.3 for structures 525–535), observed in all tested specimens, is otherwise unknown in the genus.

An examination of the flavonoids of 33 populations of *Bidens*, representing 26 taxa (all species and all but one subspecies), was described by Ganders et al. (1990). The overall pigment chemistry is in general agreement with reports of flavonoids from other species of *Bidens.* Aurones based on sulfuretin [526] (see Fig. 6.3 for structures) and maritimetin [527]; chalcones based on butein [528] and okanin [529]; the flavanone eriodictyol [530]; the flavones, apigenin [531] and luteolin [532]; and the flavonols kaempferol [533] and quercetin [534] were identified and shown to occur as a variety of *O*-glycosides in most species. O-Methylated derivatives of some of the aglycones were also seen, but their distribution was more sporadic. It was noted above that *B. torta* was the only taxon tested for acetylenes

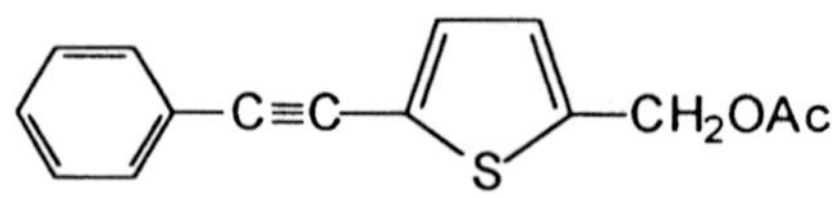

[525]

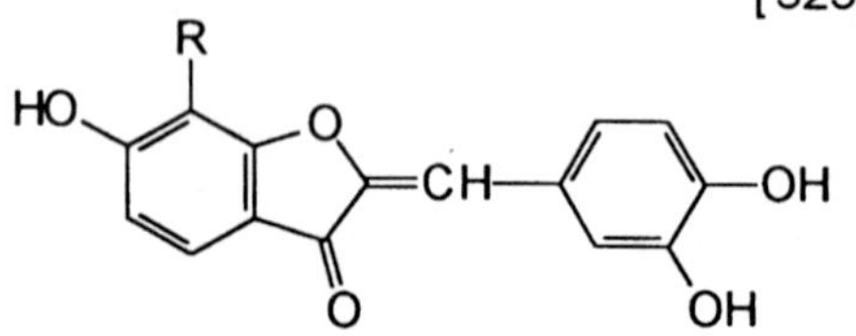

[526] R = H
[527] R = OH

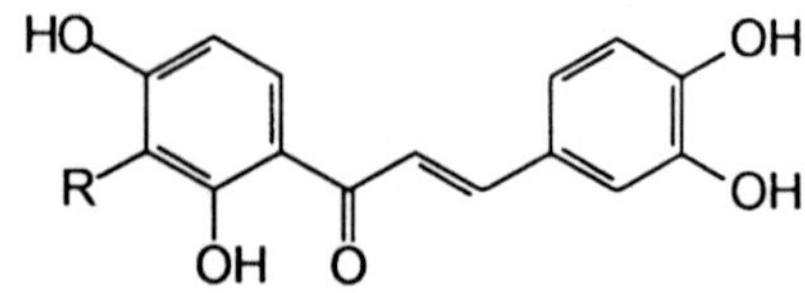

[528] R = H
[529] R = OH

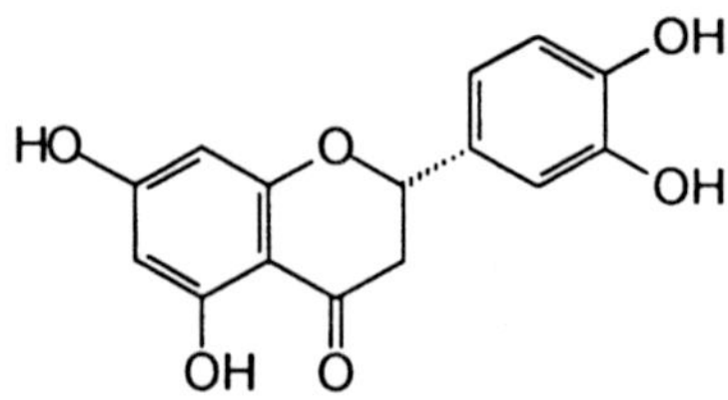

[530]

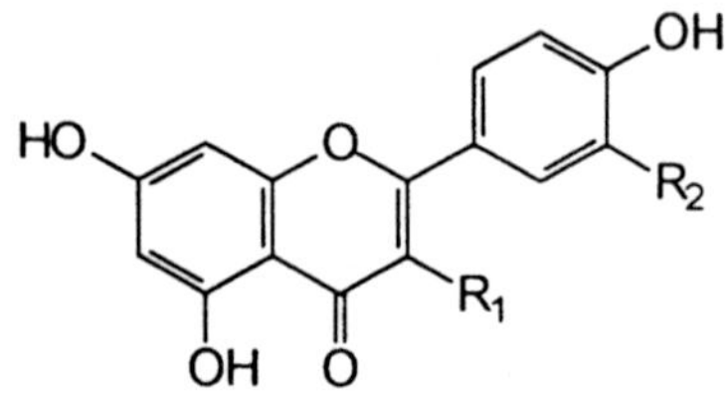

[531] $R_1 = R_2 = H$
[532] $R_1 = H, R_2 = OH$
[533] $R_1 = OH, R_2 = H$
[534] $R_1 = R_2 = OH$

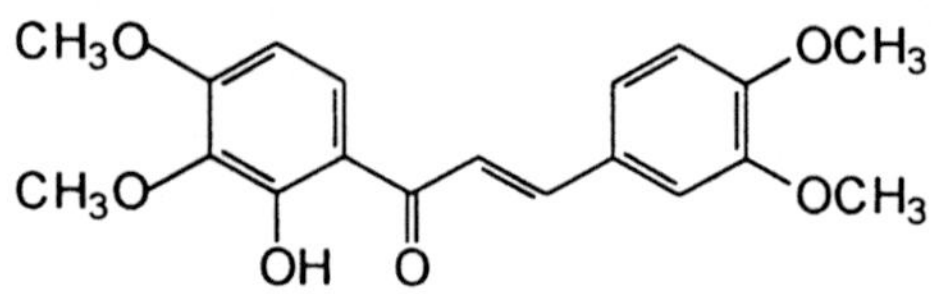

[535]

Fig. 6.3 Compounds 525–535, a thiophene derivative and flavonoids from Hawaiian *Bidens*

that exhibited profile differences among populations. The flavonoid chemistry of *B. torta* was also shown to be somewhat more complex than the other taxa, in that it exhibited a more highly developed capacity for flavonoid O-methylation resulting in the accumulation of a suite of okanin mono-, di-, tri-, and tetramethyl ether derivatives, for example, [535], (McCormick et al., 1984).

Perhaps the most noteworthy aspect of the flavonoid study, however, was the high level of variation exhibited across the range of collections. Simply put, there was no relationship between taxonomic position and flavonoid profile. Differences between individuals within the same population (even adjacent plants!) often exceeded differences between other pairs of taxa. The level of variation present within this group of taxa represents one of the most extreme cases of flavonoid variation in the present author's experience [see Bohm (1987) for a review of flavonoid variation].

6.2.2 Cyperus rotundus *(Cyperaceae)*

Cyperus rotundus L. is a weedy species, native to India, but widely distributed in countries on the Pacific Rim and islands in the Pacific Basin. Commonly referred to as "purple nut sedge," it has been known in the Hawaiian Islands since the middle of the nineteenth century. In addition to its weedy nature, the taxon has attracted attention because of the antifebrile activity of its rhizomes. Chemical studies have disclosed the presence of several sesquiterpene derivatives, some of which have been implicated in the plant's medicinal use (cyperene and cyperinerol) (Wagner et al., 1990, p. 1399). Our interest in this species is the existence of several chemotypes with interesting patterns of occurrence involving Pacific Rim countries and several oceanic islands, including the Hawaiian Islands, islands in the southern Pacific, and the Philippines.

Sesquiterpenes identified from the sedge include cyperene [536], cyperotundone [537], patchoulenone [538], patchoulenyl acetate [539], sugeonyl acetate [540], and ß-selinene [541] (see Fig. 6.4 for structures 536–541). Several chemotypes involving these compounds have been identified, although there is overlap in the countries involved in some instances (Komai et al., 1991). The "H-type" occurs in Japan and in the Kalapana area of southeastern Hawaii (Island of Hawaii). Plants exhibiting the "M-type" chemical phenotype occur in Japan, Taiwan, southern China, Hong Kong, and Vietnam. "O-Type" plants occur in Japan; Taiwan; the Philippines; Thailand; Indonesia; the Hawaiian Islands (Hawaii, Kauai, Maui, and Oahu); southern Pacific Islands, including Guam, Samoa, Saipan, Palau, and Tonga; Australia; and southern United States. The "K-type" plants are known from four of the Hawaiian Islands (Hawaii, Kauai, Maui, and Oahu), southern California, and Mexico. Despite the presence of both H- and O-chemotypes, the predominant chemotype in the Hawaiian Islands is the K-type. Selected occurrence data are summarized in Table 6.1.

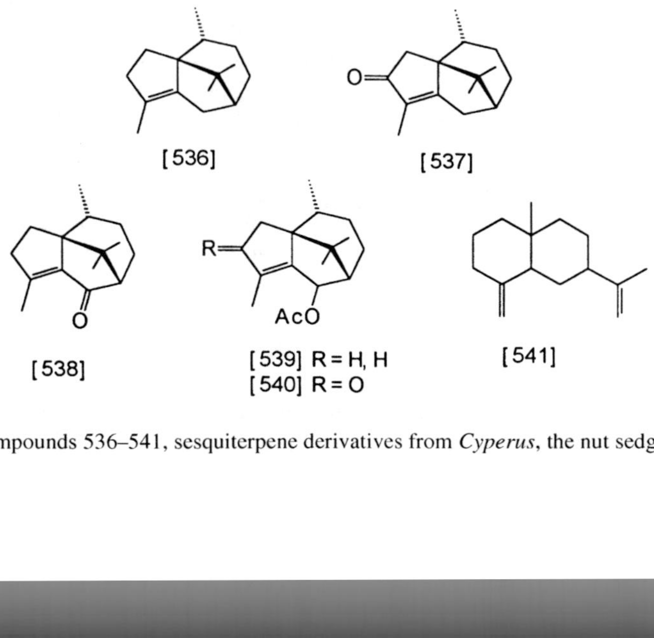

Fig. 6.4 Compounds 536–541, sesquiterpene derivatives from *Cyperus*, the nut sedge

Table 6.1 Distribution of sesquiterpene derivatives in populations of *Cyperus rotundas* (from Komai and Tang, 1989)

Compound	Chemotype[a]				
	H	O	M	K	OH
Cyperene	nd	30.8[b]	7.2	28.7	20.7
β-Selinene	18.5	trace	17.8	nd	nd
Cyperotundone	nd	13.1	19.4	8.8	25.0
Patchoulenyl acetate	nd	trace	trace	8.0	trace
Sugenyl acetate	nd	trace	trace	6.9	trace
Patchoulenone	trace	nd	trace	nd	nd

[a] See text for description. "OH" is the O-type found in the Hawaiian Islands.
[b] Values are percentage total GLC peak area; nd = not detected.

An interesting situation surrounds the H-type on the Island of Hawaii. Komai and Tang (1989) suggested that the similarity of the H-type from the Island of Hawaii and from Japan point to a relatively recent arrival on the Hawaii Island; although they do not offer an estimate of the time involved. Owing to the high tourist density in the area of active volcanism on the Island of Hawaii, it would not be surprising that propagules could have been transported there by that route. It would be of interest to reexamine this situation on the island at the present time, especially in light of the very extensive volcanic disturbance in the Kalapana area over the past several years.

6.2.3 Hesperomannia *(Asteraceae)*

As recorded in the *Manual of the Flowering Plants of Hawaii* (Wagner et al., 1990, pp. 323–326), *Hesperomannia* (Asteraceae: Vernonieae) comprises three species, two of which *H. arborescens* A. Gray and *H. arbuscula* Hillebr. are listed as endangered, while the third, *H. lydgatei* C. Forbes, is described as being vulnerable. The only information on secondary metabolites of *Hesperomannia* of which the author is aware of is the presence of an array of common flavonoids in *H. arborescens*: quercetin 3-*O*-glucoside, 3-*O*-glucuronide, and 3-*O*-rutinoside (Bohm and Stuessy, 1995), and the flavanone eriodictyol 7-*O*-glucoside (Bohm, unpublished data). These compounds are so widespread in the plant kingdom, and indeed in Asteraceae (Bohm and Stuessy, 2001), that their presence in this species offers no clue to the possible evolutionary relationships.

Some insights into relationships of *Hesperomannia* did come from macromolecular studies by H.-G. Kim et al. (1998) whose work, based on the chloroplast *ndh*F gene, suggested that the closest affinity of the genus was among African members of *Vernonia*, and that the time of divergence lies in the range 14–27 million years. This length of time would be sufficient for ancestors to have moved from Africa, likely through the availability of stepping stones, to islands in the Pacific. No likely intermediate taxa are known, however. The flavonoids of *H. arborescens* are of such general nature that they do not help in sorting out relationships.

6.2.4 Metrosideros *(Myrtaceae)*

Metrosideros consists of 50 species, about half of which comprise each of the two subgenera, subgen. *Mearnsia*, which occurs in New Zealand, New Caledonia, and New Guinea, but does not figure in the present example, and subgen. *Metrosideros*, whose members occur from New Zealand and its sub-Antarctic islands north to the Bonin Islands, east to Pitcairn, and north to the Hawaiian Islands. New Zealand appears to be the center of diversity, and is the source of the only known fossil pollen of the genus, aged at late Paleocene/early Eocene (Mildenhall, 1980). As S. D. Wright et al. (2000) point out in the introduction to their work, members of subgen. *Metrosideros* are well equipped for long-distance travel; seeds are small and light, viable for at least 6 hours at –30°C, and can tolerate saltwater for at least a month.

Five species of *Metrosideros* are endemic on the Hawaiian Islands, one of which, *M. polymorpha* Gaud., is the very common ‘ohi‘a lehua. The variation implied by the specific epithet has been accommodated by the recognition of eight varieties. A survey of collections of *M. polymorpha* (four varieties by our reckoning) from four islands revealed a remarkable level of homogeneity with few exceptions. The only significant differences noted lay with the collections from Molokai where a higher level of B-ring hydroxylation was observed (Bohm and Yang, unpublished observations). The usual problems plague this sort of study, a limited sample size (both in numbers of sites and numbers of individuals), lack of information on how local environmental conditions may influence pigment profile, and the lack of detailed information on other species, notably *M. colina* (Forster and Forster) A. Gray from the Marquesas, which has been shown by DNA studies to be closely related to the Hawaiian taxon (S. D. Wright et al., 2000, 2001).

6.2.5 Silversword alliance (Asteraceae)

The “silversword alliance” (Carr, 1985) consists of the three endemic Hawaiian genera, *Argyroxiphium* DC, *Dubautia* Gaudich., and *Wilkesia* A. Gray. The silversword alliance plus 13 genera from western North America comprise subtribe Madiinae (Asteraceae; Helenieae). The North American members of the subtribe are commonly referred to as tarweeds owing to the resinous exudate that occurs on aerial parts of most species of the subtribe.

The term “silversword” refers the sword-like, white, tomentose leaves that characterize both subspecies of *Argyroxiphium sandwicensis* DC. *Argyroxiphium* consists of five species, one of which is thought to be extinct (*A. virescens* Hillebr.), limited to higher elevation sites on the islands of Hawaii and Maui. *Wilkesia* consists of two species, one fairly common and one very rare, restricted to drier areas of western Kauai. *Dubautia* is the largest of the three with 21 species (33 taxa in all) (Carr, 1985) that grow in a wide range of habitats throughout the islands. In addition to detailed morphological analysis that resulted in the floristic treatment, there have been cytogenetic (Carr and Kyhos, 1981, 1986), enzyme electrophoretic studies (Witter and Carr, 1988), crossing studies (Carr, 1995), and DNA sequence-based studies (Carr et al., 1996) of the alliance.

The first suggestion that the silverswords had a connection with North American Heliantheae tribe Madiinae, came from Gray (1852), who was familiar with *Argyroxiphium.* This idea was rejected by Keck (1936), a noted expert on North American Madiinae, who failed to see the morphological similarities and, not surprisingly, was dissuaded by the distances involved that a colonizer would have to travel from the North American mainland. The issue attracted the attention of Carlquist (1959a, b, 1966, 1967, 1983), who presented detailed anatomical evidence that clearly pointed to a close relationship of the Hawaiian taxa to the tarweeds.

A North American origin for the silverswords clearly emerged from the application of DNA-sequence techniques to members of the alliance and representative Californian species (Baldwin et al., 1990, 1991). The conclusions from these studies, based on chloroplast restriction site analysis were subsequently corroborated by ITS nuclear ribosomal DNA studies. A comprehensive review of these studies can be found in Carlquist et al. (2003).

Extensive studies of flavonoids of mainland and Hawaiian Madiinae have revealed a high degree of similarity in terms of both the components of the sticky glandular material (the "tar" of the tarweeds) and vacuolar constituents (Bohm and Fong, 1990; Crins and Bohm, 1990; Bohm et al., 1992; Bohm, 1999). The exudate profiles consist of, in various combinations, flavanones, flavones, and flavonols with varying degrees of extra oxygenation (substitution at C-6 or C-6 and C-8) plus a comparatively high level of O-methylation. Profiles of vacuolar flavonoids—those characterized by glycosidic linkages—are also similar in both sets of species.

Qualitative and quantitative studies of vacuolar and glandular flavonoids of *Wilkesia gymnoxiphium* A. Gray from several populations showed only minor levels of variation (Bohm and Fong, 1990; Yang and Bohm, unpublished data). Five specimens of each *Dubautia scabra* (DC.) Keck subsp. *scabra* and *D. ciliolata* (DC.) Keck subsp. *glutinosa* G. Carr also showed species homogeneity (Crins et al., 1988a).

Perhaps the most useful contribution made by flavonoids in this group of plants, however, was the assistance they provided in studies of natural hybridization between *Dubautia scabra* and *D. ciliolata* (Crins et al., 1988a). *Dubautia scabra* is a pioneer plant that colonizes new lava moderately; *D. ciliolata* occurs on somewhat older lava. Where newer flows overlap older ones, one can often find plants with features intermediate between the two species. This phenomenon is readily observed in the vicinity of Kilauea Volcano on the island of Hawaii, where one finds *D. scabra* subsp. *scabra* on newer substrate and *D. ciliolata* subsp. *glutinosa* on the older substrate. Since the two taxa exhibit different flavonoid profiles, and we know that flavonoid profiles are inherited in an additive fashion, hybrids would be expected to exhibit flavonoid markers inherited from each parent. A major advantage in this study was the availability of known F_1 hybrids that had been produced from controlled crosses at the University of Hawaii (by G. D. Carr). Analysis of the flavonoids of individual plants revealed that hybrid plants of several generations existed in the area, some clearly first generation, others that appear to have arisen through backcrossing with one of the parental species. Unfortunately, there were too few individuals available in the field to allow us to pursue the study. Ideally, one would have liked to study the flavonoid profiles of individuals obtained from controlled backcrosses.

6.2.6 **Vaccinium** ***(Ericaceae)***

Vaccinium is represented on the islands by three endemic species, *V. calycinum*, *V. dentatum*, and *V. reticulatum.* Following the discovery of a high level of within-population variation in the flavonoids of *Bidens*, as noted, above, an examination of other Hawaiian taxa was undertaken to see if similar patterns (or lack thereof) of variation were characteristic of island endemics. Analysis of extensive collections of *V. calycinum* and *V. reticulatum* from several islands for flavonoids and condensed tannins (Bohm and Koupai-Abyazani, 1994) revealed qualitatively identical profiles. Some quantitative variation was evident, but no efforts were made to determine if this were caused by local environmental factors or reflected genetic differences. The profiles observed were of the same general sort seen in North American species of blueberry, in agreement with observations, based on DNA sequences, that the island species are most likely related to mainland blueberries (Powell and Kron, 2002). The DNA study revealed that the island species very likely resulted from a single colonization event. Diversification into the three current species has not affected the species' flavonoid biosynthetic apparatus.

6.2.7 **Zanthoxylum**

In a study of insecticidal properties of *Zanthoxylum* (Rutaceae) endemic to the Hawaiian Islands, Marr and Tang (1992) noted a good deal of variation among individuals of the three species available to them, *Z. dipetalum* H. Mann, *Z. hawaiiense* Hillebr. and *Z. kauaense* A. Gray, as well as among the species themselves. Significant differences in toxicity (based upon egg hatch assay using *Dacus dorsalis* Hendel, the oriental fruit fly) were observed among the species, 12 of 47 individuals of *Z. kauaense* yielded toxic extracts, whereas only one of 12 individuals of *Z. dipetalum* was toxic. Twenty-one individuals representing *Z. hawaiiense* were tested but none of them was toxic.

Variation within a species was examined by looking at the chemistry of *Z. dipetalum* specimens collected from sites on three islands, Hawaii, Kauai, and Oahu. Of the seven compounds detected in this comparison, only two, caryophyllene [542] and humulene [543] (see Fig. 6.5 for structures 542–546), were present in all individuals. Trees from a site on Oahu were unique in their possession of

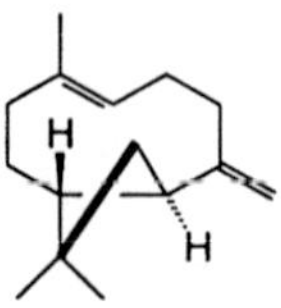

[542]

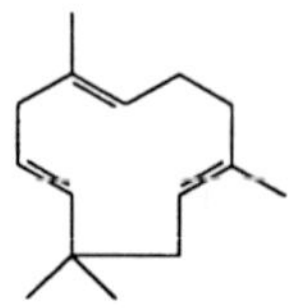

[543]

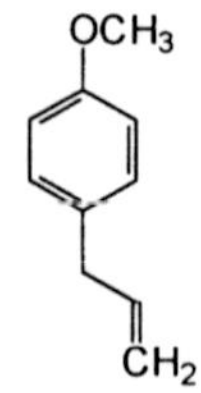

[544]

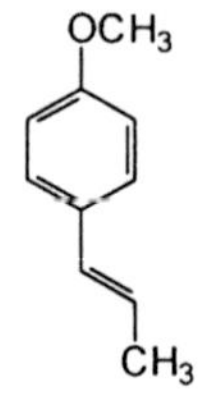

[545]

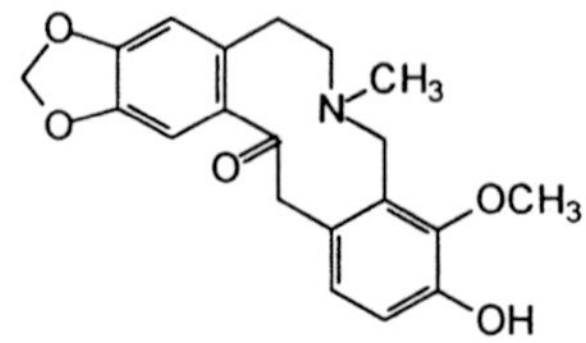

[546]

Fig. 6.5 Compounds 542–546, terpene derivatives from *Zanthoxylum*

estragole [544] and anethole [545]. These data are presented in Table 6.2. Included in this table are data for both leaves and pericarps of the specimen collected on Oahu, which data are included to indicate a further level of complexity in this system. The leaves of this specimen gave the only positive test for toxicity recorded for this species, yet the pericarp extract from this sample, with very nearly the same composition, was nontoxic. An immediate explanation for this sort of biological activity is evasive, although some possibilities come to mind: (1) some synergism exists within the peculiar assemblage of compounds in the leaves; (2) the unknown compound in the pericarp tissue somehow inhibits the activity; and (3) some other compound(s) may be responsible for the observed differences. Whatever the explanation, the authors point out the importance of using as large a sample as possible to assess biological activity. It is clear that selection of a single individual of either *Z. dipetalum* or *Z. kauaense* could have resulted in misleading results.

Table 6.2 Interisland variation in volatile components of *Zanthoxylem dipetalum* (after Marr and Tang, 1992)

Compound	Island			
	Hawaii	Kauai	Oahu (L)[a]	Oahu (P)[a]
Limonene	2[b]	22	–	–
Estragole	–	–	4	14
Anethole	–	–	57	68
2-Undecanone	11	12	–	–
Caryophyllene	45	40	33	13
Humulene	9	7	5	2
2-Tridecanone	16	6	–	–
Unknown	–	–	–	2

[a] L=Leaves; P=Pericarp.

[b] Values are percentage of compound in hexane extract (GLC).

Differences in alkaloid composition between samples of *Z. dipetalum* from different islands have also been reported (Arslanian et al., 1990). These workers found that a tree from Oahu (anethole/estragole chemotype) lacked thalicitrine [546] (see Fig. 6.5), whereas a tree from Kauai and one from Hawaii (both 2-undecanone/2-tridecanone chemotype) possessed the compound.

6.3 Southern Pacific

6.3.1 Piper methysticum *(Piperaceae)*

This next example involves the well-known plant "kawa." A psychoactive beverage made from the roots of this plant is used widely in the islands of the southwestern Pacific Ocean either for ritualistic or routine consumption. Kava is the common name for *Piper methysticum* Forst. f. from which several compounds responsible for the pharmacological activity have been isolated and identified. Representative structures of the family of styrylpyrones, commonly called "kavalactones," are given in Fig. 6.6. The compounds are based upon a carbon skeleton consisting of a styryl function (C_6C_2) attached to a six-membered lactone ring. The fundamental compound, kawain, is shown as structure [547]. Structural variants include

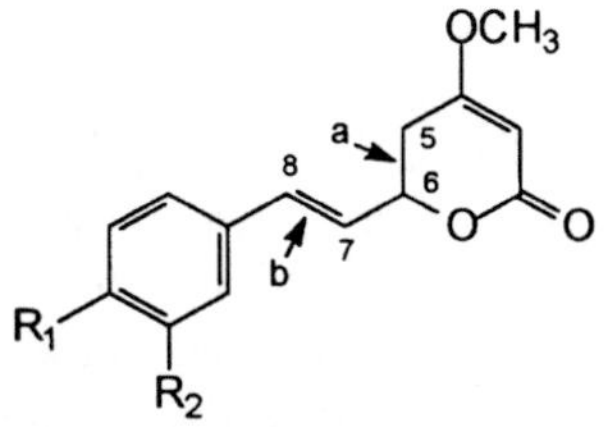

[547]	$R_1 = R_2 = H$	Kavain
[548]	$R_1 = R_2 = H$, "b" sat'd.	7,8-Dihydrokavain
[549]	$R_1 = R_2 = H$, "a" unsat'd.	5,6-Dihydrokavain
[550]	$R_1 = OCH_3$, $R_2 = H$, "a" unsat'd.	Yangonin
[551]	$R_1 = R_2 = OCH_2O$	Methysticin
[552]	$R_1 = R_2 = OCH_2O$, "a" sat'd.	Dihydromethysticin

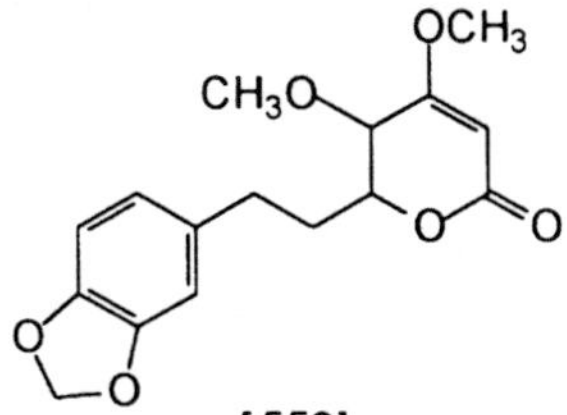

[553]

Fig. 6.6 Compounds 547–553, kavalactones from *Piper methysticum*

compounds with the styryl double bond reduced (making it the 2-phenylethyl function) as seen in 7,8-dihydrokawain [548], compounds with a second double bond in the lactone moiety, as seen in 5,6-dehydrokawain [549], are also known, and compounds in which the aromatic ring is substituted, such as in yangonin [550], which bears a methoxy function at the *p*-position. Several kavalactones are characterized by the methylenedioxy function, as seen in methysticin [551] and in dihydromethysticin [552]. Other members of the family combine these functional groups in various ways. The six compounds illustrated, however, are the most commonly encountered ones, often constituting 95% of the total lactone content of a particular cultivar.

Kava has been the subject of detailed study by several workers, whose interests have included usage of the preparations in Hawaii (Titcomb, 1948); pyrone analysis (Young et al., 1966); physiology of action of the constituents (Hänsel, 1968); monographic study (Chew, 1972); cultivation (Lebot and Cabalion, 1986); origin of the Oceanian plant (Lebot and Lévesque, 1989); genetic control of kavalactone chemotypes (Lebot and Lévesque, 1996a); and relationships with the related *P. wichmannii* C. DC. (Lebot and Lévesque, 1996b). Comprehensive citation lists can be found in the papers by Lebot and Cabalion (1986) and Lebot and Lévesque (1989).

Lebot and Lévesque's 1989 work was based upon an exhaustive collection of plant material. *Piper wichmannii* was obtained from Papua New Guinea, the Solomon Islands, and Vanuatu, which comprises its natural range. *Piper methysticum* was collected from cultivated plots on three islands representing Micronesia, eight representing Melanesia, and 24 from Polynesia. In all, more than 240 individual plant acquisitions were subjected to chemical and morphological analysis.

The first problem addressed by Lebot and Lévesque (1989), in their comprehensive ethnobotanical study, dealt with the taxonomic status of *P. methysticum* and *P. wichmannii* C. DC. Problems that had to be dealt with included the nearly total absence of female plants, even in cultivation, and the complete lack of sexual reproduction of *P. methysticum* in cultivation, and consequently, the absence of seeds. *Piper wichmannii*, on the other hand, sets normal seed that germinate and develop normally. Furthermore, morphological differences are slight, certainly not sufficient for specific recognition. Both species appear to be decaploids with $2n=10x=130$ chromosomes (Jose and Sharma, 1985; Okada, 1986). The high ploidy level in *P. methysticum* cannot be taken as explanation for the sterility of the cultivars, since *P. wichmannii*, also decaploid, reproduces normally. Electrophoretic studies of both species (Lebot et al., 1991; Lebot and Lévesque, 1996b) revealed a reduced level of genetic variability within *P. methysticum*; only four of eight enzyme systems examined were polymorphic, whereas all systems tested in *P. wichmannii* were polymorphic. These studies provided important information relevant to the area of origin of *P. methysticum.* Multivariate analysis of presence/absence of "electromorphs" indicated that *P. wichmannii* from the Western Province of Papua New Guinea are very different from those of *P. methysticum,* suggesting that the cultivated form did not originate from that area. The closest "match" came from cultivated plants collected in Vanuatu and southern Papua New Guinea, an observation that the authors took to indicate that cultivated kava might have had its origin in that general area.

Chemical data also support a close relationship between the two taxa. *Piper wichmannii* is the only other member of the genus that produces styrylpyrones in abundance. There has been a report that one of the minor lactones, 5-methoxy-5,6-dihydromethysticin [553], occurs in *P. sanctum* (Sengupta and Ray, 1987). Lebot and Lévesque (1996b) concluded that *P. wichmannii* comprises the "seed-bearing wild forms" and that *P. methystichum* represents an assemblage of cultivated forms that have been selected for their psychoactive properties over a considerable period of time. It is useful to mention that Lebot and Lévesque (1996b) suggested a classification that would recognize the status of both the wild and cultivated forms of kava. Since *P. methysticum* has nomenclatural priority, no problems should arise from recognizing the sterile, cultivated form as *P. methysticum* Forst. f. var. *methysticum*, and the fertile, wild plants as *P. methysticum* Forst. f. var. *wichmannii* (DC) Lebot *stat. nov.*

In an effort to map the migration of various peoples throughout the southwestern Pacific, Lebot and Lévesque (1989) amassed a large database concerning chemical and morphological differences among collections of *P. methysticum* from throughout the area. [Our attention must obviously focus on the chemical variants, but the significance of the data for understanding human migrations in the Pacific Basin is substantial.] Quantitative analysis of a large number of individuals resulted in the recognition of several "chemotypes." Unfortunately, the term "chemotype" is used in two different ways in the paper, leading to some confusion. The first usage refers to the raw data from high-performance liquid chromatography (HPLC) analysis of kavalactone composition of individual collections, which should probably be better referred to as "profiles," and again as a means of identifying groupings of cultivars based on multivariate analysis of their chemical and morphological features, the true "chemotypes."

Six kavalactone profiles (the author's usage) were observed based upon quantitative differences in the constitution of individual components. These profiles were expressed as a six-digit number representing the decreasing order of the concentration of the respective compounds, for example, "123456," where 1=demethoxyyangonin, 2=dihydrokavain, 3=yangonin, 4=kavain, 5=dihydromethysticin, and 6=methysticin. For example, taking a specimen at random, cultivar "Matakaro" from Fiji was described as having profile "246351," representing a mixture of those components in roughly the following proportions (rounded): 30:24:18:10:10:9. In all, nine chemotypes (the author's usage) were documented. Four chemotypes were found to characterize kava from Melanesia, all of which represent *P. wichmannii*, Chemotypes **A**, **B**, **C**, and **D**. For example, Chemotypes **A** and **B** have very high concentrations of dihydromethysticin (compound No. 5) and dihydrokavain (No. 2), intermediate concentrations of demethoxyyangonin (No. 1), methysticin (No. 6), and yangonin (No. 3), and a low concentration of kavain (No. 4) (hence its profile designation of "521634"). Without going any further into the technical details, suffice it to say that a significant geographical distribution emerged from these analyses: chemotype **E** (526431 or 526341) occurs in Vanuatu, Tonga, Wallis, Fatu Hiva, Oahu, and Pohnpei; chemotype **F** (256431 or 254613) occurs only in Vanuatu and Papua New Guinea; chemotype **G** (246531 or 264531), used for the preparation of a daily beverage, occurs on Vanuatu and Wallis; chemotype **H**, which, incidentally, is described as the most palatable, occurs only in Vanuatu and Western Samoa; and chemotype **I** is not only grown mainly in Fiji, but also occurs on Tonga, Samoa, the Cook Islands, Tahiti, the Hawaiian Islands, and Pohnpei. Chemotypes **E–I** all represent *P. methysticum.* Despite the complexity of this system, not in the least surprising in view of the migration history of humankind in the area, significant patterns of distribution of secondary metabolites emerged from a combination of quantitative methods (HPLC) and multivariate statistical analysis. There is no question that this is labor-intensive work, but the results, as was also seen in studies of conifer terpenes described earlier in the review, are clearly worth the effort.

Two recent papers might be of interest to workers dealing with kavalactones. Dharmaratne et al. (2002) describe chromatographic and spectroscopic data for 13 of the more common kavalactone derivatives. Siméoni and Lebot (2002) attempted to identify factors responsible for kavalactone content and chemotype using plants under cultivation on several islands of Vanuatu. The content of kavalactone appears to be determined primarily by the local environment or by the agricultural methods used by local farmers. Growth studies over a period of several years showed that 10-month-old juvenile plants accumulate about 3% (dry weight) of kavalactone with the concentration rising to 8% after 18 months. At this age the concentration seems, for the most part, to stabilize.

6.3.2 Degeneria *(Degeneriaceae)*

The Fiji Archipelago lies astride longitude 180°E/W (between ca. 177°E and ca. 179°W) and between ca. 16° and 19° S latitude in the southern Pacific Ocean. The archipelago consists of two main islands, Viti Levu and, to the northeast, Vanua Levu, along with many smaller islands, islets, and reefs (Fig. 6.7). The plants of interest belong to the genus *Degeneria*, which, until a few years ago, was thought to consist of the single species *D. vitiensis* I.W. Bailey & A.C. Smith (I. W. Bailey and Smith, 1942). Based upon extensive reinvestigation of *Degeneria* on the islands, Miller (1988) defined a second species *D. roseiflora* J. M. Miller, based upon several morphological features and a distinct floral pigmentation, as indicated by the new taxon's name. Although no chemical analyses were performed on floral tissue of the new species, it seems reasonable to suggest that the presence of anthocyanin pigments can be credited as a chemical distinction between the two species.

The distribution of the two species is of interest insofar as they appear to be more or less restricted in occurrence: *D. vitiensis* occurs on Viti Levu, whereas *D. roseiflora* occurs on Vanua Levu and adjacent Taveuni. Both species occur on older substrates on the larger islands, whereas *D. roseiflora* is found on younger substrates on Taveuni. A. C. Smith (1981, 1985) thought that its occurrence on the smaller island might have been the result of short-distance dispersal via birds or bats.

According to Takhtajan (1969) and Cronquist (1981) Degeneriaceae represents an ancient lineage within the Magnoliales. The "old enigma" referred to by

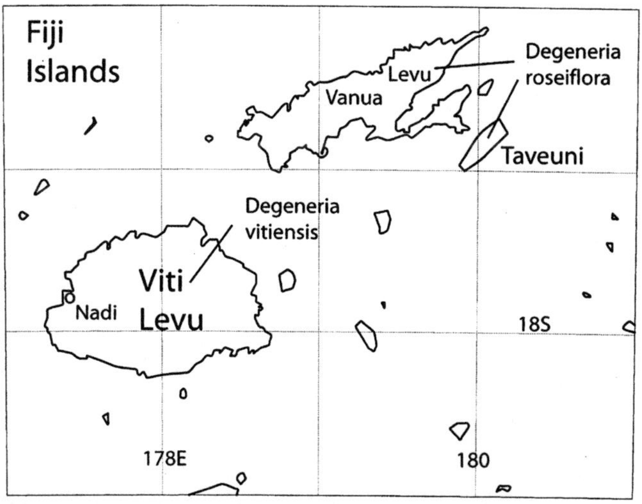

Fig. 6.7 Map of the Fiji Islands showing the occurrence of the two species of *Degeneria*

Miller (1989) is the question of how such an ancient organism has come to exist on a comparatively new oceanic island. Long-distance dispersal is an unlikely explanation for the present situation. Dispersal over shorter distances from island to island seems a reasonable possibility, with subsequent disappearance of many, or apparently all in the present case, of the intermediate stepping stones. Invocation of island hopping, after all, has many precedents. As Miller (1989) pointed out, "obviously some vascular plant lineages have survived," some of which may have histories that reach back to the Antarctic flora. Needless to say, deep phylogenetic analyses would be helpful in addressing this question.

6.4 Juan Fernandez Islands

The Juan Fernandez Islands, Chile, lie east of the South American mainland at approximately 33°30′S, 80°W (Fig. 6.8). The archipelago consists of two main islands, Masatierra, which lies about 660 km from the mainland, and Masafuera, which lies 150 km further west. A third, small island, Santa Clara, lies off the western end of Masatierra and probably was attached to Masatierra at one time. The ages of the islands have been determined using potassium-argon dating methods to be: 5.8 ± 2.1 million years for Santa Clara; 4.23 ± 0.16 million years for Masatierra; and 2.44 ± 0.14 million years for Masafuera (Stuessy et al., 1984).

The vascular flora of the Juan Fernandez Islands consists of 361 species and includes 53 ferns, 65 monocots, and 243 dicots. There are 126 endemic species, 12 endemic genera, and one endemic family, Lactoridaceae (Stuessy et al., 1992). A considerable amount of work has gone into attempts to establish evolutionary

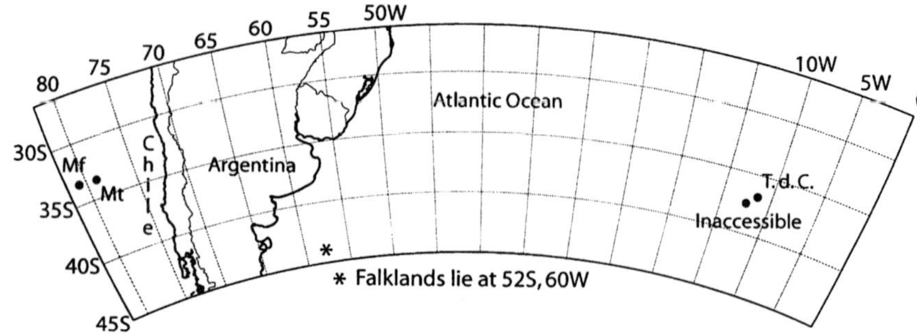

Fig. 6.8 Map of the Juan Fernandez Islands (Mt = Masatierra, Mf = Masafuera), Tristan da Cunha (T.d.C.), and Inaccessible Island in relation to South America. The location of the Falkland Islands (Islas Malvinas) is included

relationships among endemic taxa on the Juan Fernandez Islands and likely ancestral species on the mainland. Much of these efforts have been joint ventures of botanists from North American universities, the Universidad de Concepción, Chile, and, more recently, the University of Vienna, with the overall aim of preserving this highly unusual area from further devastation at the hand of mankind (or the feet of goats). In several of the systems described below, we have the luxury of both micro- and macromolecular information which, in addition to providing as many sources of data as possible, offer an opportunity to assess the relative contribution of each type to the problems at hand.

6.4.1 Dendroseris *(Asteraceae)*

Dendroseris D. Don. (Asteraceae: Lactuceae), with 11 species, is the largest endemic genus on the Juan Fernandez Islands. It also counts among its members some of the rarest species in the archipelago; Stuessy et al. (1998) reported that only three plants of *D. neriifolia* (Dcne.) Hook. & Arn. existed at the time of their discussion of plant conservation problems in the islands. Sanders et al. (1987) placed the species in three subgenera, *Dendroseris* with four species, *Phoenicoseris* with three, and *Rea* with four. Eight of the species are restricted to the older island, while three species, one from each subgenus, occur only on the younger island: *D. macrophylla* D. Don (subgen. *Dendroseris*), *D. regia* Skottsb. (subgen. *Phoenicoseris*), and *D. gigantea* Johow (subgen. *Rea*).

The flavonoid profiles of all species were studied by Pacheco et al. (1991a), who identified a series of apigenin and luteolin 7-*O*-mono- and 7-*O*-diglucosides and a quercetin 3-*O*-mono- and 3-*O*-diglucoside. The somewhat unusual chemistry of these compounds, three of the apigenin derivatives were 7-*O*-monoglucosides and four of the luteolin derivatives were 7-*O*-monoglucosides, suggesting an array of isomers for each, has been commented on (Bohm, 1998b). Differences among the subgenera with regard to flavone profiles were not very significant, whereas quercetin (flavonol) derivatives were not seen in any of the species of subgen. *Phoenicoseris.* Other differences are relatively minor, accountable by operation (or lack thereof) of a single enzyme. For example, all four species in subgen. *Dendroseris* appear to be unable to make apigenin diglucosides, but at least some of the members of the other two subgenera can. However, without knowing precisely what the structures of these compounds are, it is impossible to draw any conclusions. It is useful to note, however, that each of the Masafueran species exhibits a flavonoid profile closely resembling at least one of the species from the respective subgenera from Masatierra. At most, there would appear to be single enzyme differences involved.

In an effort to locate a likely, or at least possible, mainland ancestral type, these workers determined the flavonoid profiles of representatives of a number of candidate genera. Their results pointed to *Hieracium* and *Hypochaeris* as possibilities, with their choice going to a species of the former from Chile. More recent work using macromolecular techniques suggested closer ties to *Sonchus* s.l., or possibly *Embergeria* from New Zealand (Bohm, 1998b, p. 252; parenthetical remark added by T. F. Stuessy, volume editor). It is prudent to point out that, even with detailed flavonoid chemistry of possible ancestral taxa, comparatively few additional insights would be gained owing to the simple and homogeneous flavonoid chemistry of these genera (Bohm and Stuessy, 2001).

Insights into evolutionary relationships of *Dendroseris* have been gained through application of macromolecular tools. The morphological studies of Sanders et al. (1987) provided the background for all subsequent investigations, first by uniting in one genus the 11 taxa that had, prior to then, been dispersed in four genera, and then by providing the first explicit phylogeny of the genus. Electrophoretic patterns within the genus were studied by Crawford et al. (1987), who found comparatively low level of genetic variation compared to other insular endemics (De Joode and Wendel, 1992). Esselman et al. (2000) described a study of random amplified polymorphic DNA (RAPD) markers, while a study by Crawford et al. (1992a) involved a restriction-site analysis of chloroplast and nuclear ribosomal DNA. Although four species were missing from that work, owing to limited quantities of plant material, results were encouraging in that the emerging phylogeny was very similar to that obtained from the other data sets. The sampling problem was partially overcome when Sang et al. (1994) employed the polymerase chain reaction (PCR) technique to amplify DNA (using internal transcribed spacer (ITS) region of nuclear ribosomal DNA) making it possible to include two additional species. Again, subgenera *Dendroseris* and *Phoenicoseris* emerged as strongly supported clades. The situation with regard to members of subgen. *Rea* is less straightforward.

6.4.2 Robinsonia *(Asteraceae)*

Robinsonia (Asteraceae: Senecioneae), the second largest endemic genus on the islands, consists of seven species assorted into two subgenera, subgen. *Rhetinodendron,* consisting solely of *R. berteroi* (Hemsl.) Sanders, Stuessy & Marticorena (formerly the monotypic genus *Rhetinodendron*), and subgen. *Robinsonia*, which consists of three sections, sect. *Robinsonia* with two species, sect. *Eleutherolepis* with three species, and sect. *Symphyochaeta* with one species (Sanders et al., 1987). *Robinsonia masafuerae* Skottsb. is the only member of this genus present on the younger island.

Examination of allozyme diversity of four species of *Robinsonia* (Crawford et al., 1992b) allowed an estimation of time of divergence of the four species ranging from 1.7 to 5.5 million years, which is in line with the estimated age of the older island, Masatierra, on which the four species examined occur. Restriction-site mutations in chloroplast DNA and the intergenic spacer region (IGS) of nuclear ribosomal DNA (Crawford et al., 1993) revealed mutations characterizing individual species, but provided no information on phylogeny. Study of ITS sequences, however, resulted in a well-resolved cladogram very similar to the tree generated from morphological information (Sang et al., 1995). The information also indicated that the genus is monophyletic.

Pacheco et al. (1985) examined 54 populations representing all species of *Robinsonia* for flavonoids; comparatively little variation was observed. The profiles of *R. berteroi* and *R. macrocephala* were identical and the simplest seen in the genus consisting of quercetin 3-*O*-mono- and digalactosides. *Robinsonia gayana* Dcne. and *R. thurifera* Dcne., which comprise sect. *Robinsonia*, exhibited those two compounds plus quercetin 7-*O*-glucoside. The greatest differences were observed in the three species that form sect. *Eleutherolepis*, each of which exhibited a different array based on flavones, flavonols, flavanones, and dihydroflavonols, although not all of these aglycones were present in all of the three species. Pacheco et al. (1985) concluded that evolutionary advancement within the genus had been accompanied by the accumulation of more complex flavonoid arrays, suggesting that in addition to making aglycones not seen in the other species, these three also accumulated—sequestered to use their term—pathway intermediates not seen in the other taxa, that is, the flavanone and dihydroflavonol derivatives. Owing to the comparatively simple pigment profiles exhibited by these taxa, it is not possible to do more than say that the compounds observed are in general similar to the flavonoid chemistry of *Senecio*, which is in agreement with the suggestion made on the basis of macromolecular information that *Robinsonia* has arisen from a self-incompatible species of *Senecio* from the South American mainland.

6.4.3 Erigeron *(Asteraceae)*

Six endemic species of *Erigeron* occur on the Juan Fernandez Islands, five of which occur only on the younger island (Masafuera): *E. ingae* Skottsb., *E. luteoviridis* Skottsb., *E. rupicola* Philippi, *E. stuessyi* Valdebenito, and *E. turricola* Skottsb. The sixth species, known from both islands, is *E. fernandezianus* (Colla) Harling. Flavonoid profiles of the entire group, as well as several likely relatives from the mainland, were described by Valdebenito et al. (1992b). The profiles were based upon glycosides of quercetin, quercetagetin (syn. 6-hydroxyquercetin), apigenin, luteolin, and 6-*C*-glucosylacacetin in a variety of combinations. There was insufficient variation within species to render any definitive statements unwise, although the overall patterns were in general agreement with relationships suggested by the morphological data. An unusual flavonoid occurrence will be discussed in detail below.

The combination of morphological data with the fact that five of the six species occur on the younger island (Masafuera) was taken by those workers to suggest that colonization of the younger island occurred first, followed by differentiation and ultimate migration of *E. fernandezianus* (or an ancestor) to Masatierra. The flavonoid data do not support this scenario, suggesting instead that evolution of *E. fernandezianus* involved colonization of Masatierra from the mainland, with subsequent colonization of Masafuera. The flavonoid profile of *E. fernandezianus* from Masatierra has, in addition to compounds present in other individuals of the species, such as luteolin 7-*O*-diglucoside [554] and 6-*C*-glucosylacacetin 7-*O*-diglucoside [555] (see Fig.6.9 for structures). These two compounds are not present in *E. fernandezianus* from Masafuera, but are present in four of the five species of *Erigeron* from the mainland. The most parsimonious explanation would involve colonization of Masatierra from the mainland with subsequent expansion of its range to Masafuera accompanied by loss of the capacity to make the two compounds in question. Morphological data suggest, however, that *E. fernandezianus* originated on Masafuera, along with the other five species, and subsequently migrated to Masatierra. This would require that the individuals that colonized Masatierra regain the capacity to make the two compounds. In order for these plants to regain the capacity to make both these compounds, three reactions are necessary: (1) placement of the "outer" glucose moiety to form luteolin 7-*O*-diglucoside; (2) establishment of the 6-*C*-glucosyl linkage; and (3) activation of the enzyme necessary to O-methylate apigenin to form acacetin. The total loss of the genes responsible for controlling these reactions, in the colonization of Masafuera from the mainland, and their seemingly miraculous reappearance in the Masatierra plants seems a rather unlikely process. It is much more likely that, given the route of evolution supported by morphology, that the genes involved were not lost, but simply silenced. Upon establishment of the new populations on Masatierra, the suite of genes was released from whatever repression was responsible, with the result that the full suite of compounds was again produced.

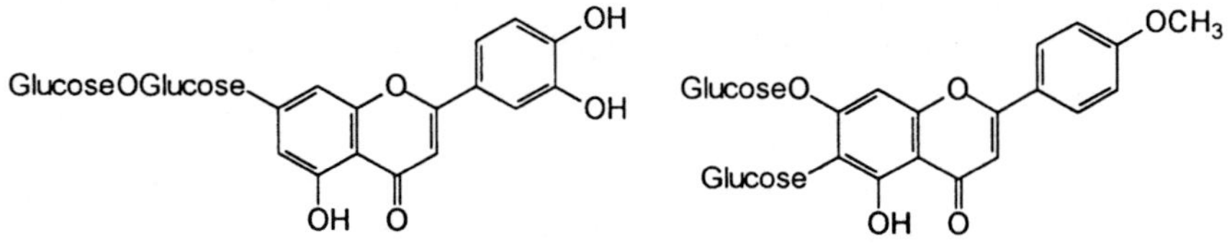

[554] [555]

Fig. 6.9 Compounds 554 and 555, the odd flavonoids from *Erigeron*

It is not possible to do much more than speculate on this problem, but some additional information might help to shed some additional light on the subject. In particular, a rigorously supported DNA phylogeny should tell if *E. fernandezianus* made the trip from the mainland to the islands independently, that is, *Erigeron* on the Juan Fernandez Islands is *not* monophyletic. Unfortunately, no macromolecular studies appear to have been done on this set of species (T. F. Stuessy, personal communication). If the genus on the islands should prove to be monophyletic, and the route of evolution suggested by morphology is correct, then a study of the genes controlling flavonoid biosynthesis could provide important clues as to how these processes are moderated. It would, of course, be interesting to learn if some ecological factor is involved in the turning-off and turning-on processes.

6.4.4 Gunnera *(Gunneraceae)*

Gunnera L., the sole genus of Gunneraceae (at one time considered part of Haloragidaceae), consists of perhaps 40 species (Mabberley, 1997, p. 319). Members of *Gunnera* occur mainly in the Southern Hemisphere, with representatives known from the Hawaiian Islands and as far north as southern Mexico. Perhaps the most well-known member of the genus is *G. chilensis* Lam., which is often found as a decorative plant owing to its spectacularly large leaves. We are concerned in this chapter with three species of *Gunnera* endemic to the Juan Fernandez Islands: *G. bracteata* Steud., *G. masafurae* Skottsb., and *G. peltata* Phil., and possible ancestors on the South American mainland. A further distinction can be made with the island endemics: *G. bracteata* and *G. peltata* are endemic to Masatierra, while *G. masafuerae*, as its name implies, is known only from Masafuera, the younger island. Vegetative and morphological characteristics of mainland *Gunnera* (belonging to subgen. *Panke*) suggested *G. tinctoria* (Mol.) Mirb. as the most likely ancestral candidate for the island endemics. Identical chromosome counts ($n = 17$) for the island taxa and for *G. tinctoria* lend further support for this idea (Pacheco et al., 1993).

A detailed study of the flavonoid chemistry of the island endemics, the closely related *G. tinctoria*, and five additional species from the mainland provided additional evidence pointing toward *G. tinctoria* as the ancestral species (Pacheco et al., 1993). The flavonoid profiles of all species consisted of flavonol glycosides as major components with an unidentified flavone glycoside and several unidentified phenolic compounds (presumably not flavonoids). The pattern of distribution of the flavonol glycosides and unidentified flavones within the set of nine species proved to be extremely informative. (The phenols were ubiquitous and are not considered further.) Kaempferol glycosides were seen in neither the island species nor *G. tinctoria*, but were present, in several combinations, in the rest of the mainland taxa. The isorhamnetin glycosides showed the reverse pattern, with one exception: the island endemics and *G. tinctoria* exhibited these compounds, whereas four of the other mainland species did not. The sole exception is *G. boliviari*, which exhibited one of the isorhamnetin derivatives.

Although all tested species (and populations thereof) exhibited quercetin glycosides, the array of compounds in the individual species proved of particular interest. Six quercetin derivatives were identified: (1) 3-*O*-arabinoside, (2) 3-*O*-glucosylgalactoside, (3) 3-*O*-galactosylgalactoside, (4) 3-*O*-glucosylglucoside, (5) 3-*O*-xylosylglucoside, and (6) 3-*O*-glucoside-7-*O*-glucoside. The occurrence data are summarized in Table 6.3.

Although each taxon exhibits a unique flavonoid profile, the overall pattern similarities between the island endemics and *G. tinctoria* are striking, both in the compound types they have in common, as well as the shared absence of kaempferol glycosides. If, as the accumulated morphological and cytological evidence indicates, *G. tinctoria* is the likely ancestor of *G. peltata*, some comments are in order concerning the changes in the flavonoid profiles that appear to have occurred during the evolutionary process. Pacheco et al. (1993) discuss the changes in terms of loss or gain of single compounds, in this case, the various quercetin diglycosides. It is perhaps more instructive to examine the situation in terms of changes in biochemical steps. The first step in the biosynthesis of flavonoid diglycosides is the formation of the monoglycoside at the position in question, in this case the 3-OH group. Three quercetin 3-*O*-monoglycosides are needed to account for the array of diglycosides observed, that is, the arabinoside, the glucoside, and the galactoside. Since there are no diglycosides with arabinose, we can focus on the fates of the glucoside and the galactoside. Diglycosides are formed by transfer of a second sugar unit to the

Table 6.3 Flavonol glycoside profiles of Juan Fernandez Islands species of *Gunnera* compared to species from South America and Mexico (from Pacheco et al., 1993)

Species	Quercetin glycosides						K-Glys[a]
	1	2	3	4	5	6	
Juan Fernandez Is.							
G. bracteata							
1 site	+	+	–	–	–	–	–
6 sites	+	+	–	+	–	–	–
G. masafuerae							
2 sites	+	+	+	+		–	–
3 sites	+	+	+	+	+	+	–
G. peltata							
1 site	+	+	–	+	–	–	–
4 sites	+	+	–	+	+	–	–
Mainland							
G. tinctoria							
2 sites	+	+	+	–	–	–	–
2 sites	+	+	–	+	–	–	–
5 sites	+	+	+	+	–	–	–
G. boliviari	+	+	–	+	–	–	+
G. boliviana	+	+	–	–	–	–	+
G. margaretae	+	+	–	–	–	–	+
G. peruviana	+	+	–	+	–	–	+
G. mexicana	+	+	–	+	–	–	+

[a] Presence or absence of any kaempferol glycoside.

appropriate position on the O-bound sugar residue (the inner sugar). All four taxa in question have the capacity to make quercetin 3-*O*-diglucoside, which allows us to remove it from further consideration. The putative ancestral taxon, *G. tinctoria*, has the capacity to convert its 3-*O*-galactoside into 3-*O*-digalactoside, but *G. peltata* apparently does not. *Gunnera peltata*, interestingly, has the capacity to transfer a xylose unit to its quercetin 3-*O*-glucoside, resulting in the formation of quercetin 3-O-xylosylglucoside. This represents gain of a feature by which these two taxa can be distinguished. Thus, *G. peltata* has lost the capacity to make the digalactoside, but has gained the capacity to make the xylosylglucoside, a second feature by which they can be distinguished. This argument is based on the assumption that there are two enzymes involved in attaching the outer sugar, one specific for glucose, which both taxa have, and one in *G. peltata* specific for xylose. If this were the case, then alterations in the glycosyltransferase systems have occurred during the migration to Masatierra and subsequent differentiation. The situation becomes even more intriguing when the diglycoside chemistry of *G. masafuerae* is considered. This Masafuera endemic, thought to have arisen from *G. peltata*, has the most complex diglycoside chemistry of the group. The first thing that can be noted is the capacity of this species to form the xylosylglucoside, which, as we have just seen, is otherwise seen only in its progenitor, *G. peltata*. This co-occurrence clearly points to a close link between the two. Somewhat perplexing, however, is the reappearance of quercetin 3-*O*-digalactoside, which, as the reader will recall, was seen in the mainland ancestor but not in *G. peltata*. It would be of interest to determine if this is simply a matter of relative amounts caused by some drop in the level of activity of the appropriate enzyme(s) in the intermediate species (*G. peltata*) with a return to the earlier level in *G. masafuerae*. Obviously, quantitative studies of enzyme activity are necessary to answer such questions. However, probably the most significant change in diglycoside biochemistry within this group of taxa is the capacity of *G. masafuerae* to make the 3,7-di-*O*-diglucoside. Whereas the above assumption concerning outer sugar specificity is speculative, the involvement of a new position of glycosylation, and thus an enzyme of different specificity, is much less so. Position specificity in flavonoid substitution reactions is a well-known phenomenon. This is also likely to be the case in a shift from transfer of a sugar to the outer position in a diglycoside to transfer to a new phenolic position, in this case, the 7-OH group. One of the difficulties in addressing problems of this sort is our comparative ignorance of the sugar transferases, particularly what happens to control these steps as evolutionary divergence occurs. It is generally thought, for example, that control of glycosylation can be upset in hybridization, even between closely related species. This may explain some of the observations in an instance of interspecific hybridization, involving the two *Gunnera* species on Masatierra.

Additional evidence indicating a close relationship between *G. bracteata* and *G. peltata* came from the observation of apparent natural hybridization between them in Quebrada Villagra on Masatierra. A comparison of morphological features of both parents and putative hybrid individuals taken along two transects showed clear-cut intermediacy in the latter (Pacheco et al., 1991b). In addition, introgressive hybridization was indicated by the presence of individuals with intermediate values when the data were used to construct a hybrid index. Flavonoid data were not useful in this system owing to the high level of variation seen in the hybrid individuals. The area surrounding Quebrada Villagra has been subject to considerable disturbance through human activities and the pressures of grazing by both feral and domestic animals. Within this area, the two *Gunnera* species have come into contact with resulting gene exchange. No other instances of hybridization were found by those authors.

6.4.5 Peperomia *(Piperaceae)*

Peperomia Ruiz & Pavon is a large genus of approximately 1000 species (Mabberley, 1997, p. 540) distributed mainly in warm and tropical areas. In this section, we consider four species that occur on the Juan Fernandez Islands, three endemics plus one that also occurs on the South American mainland. The endemic species are *P. margaritifera* Bert., from Masatierra, *P. skottsbergii* C. DC. from Masafuera and *P. berteroana* Miq., which occurs on both islands. *Peperomia fernandeziana* Miq. occurs on both islands and on the mainland. A detailed examination of the flavonoid chemistry of the island endemics, *P. fernandeziana*, and four additional mainland species was conducted by Valdebenito et al. (1992a). The complex array of compounds identified included flavones, C-glycosylflavones, and flavonols. The flavones were based on apigenin, acacetin, luteolin, luteolin 7-methyl ether, and luteolin 4′- methyl ether (diosmetin) and were obtained either as aglycones, as glycosides, or, in some cases, as sulfates. The *C*-glycosylflavones, 19 compounds in all, represented the major flavonoid components. Comparison of the flavonoid profiles of the island species revealed significant differences among them. *Peperomia margaritifera* exhibited the simplest profile characterized by a single flavonol, three *C*-glycosylflavones, and the absence of flavone *O*-glycosides. *Peperomia berteroana* also had a depauperate *C*-glycosylflavone profile, but differed from *P. margaritifera* in having a rich array of flavone *O*-glycosides. Specimens of *P. fernandeziana* from the Juan Fernandez Islands and from the mainland, though different in detail, were sufficiently similar in overall flavonoid chemistry to suggest they are indeed the same taxon. A particularly interesting observation, the significance of which we will see presently, was the finding that the pigment profiles of *P. berteroana* from the two islands were different from each other with regard to flavone *O*-glycoside, *C*-glycosylflavone, and flavone sulfate profiles.

During their study of the Juan Fernandez endemic *Peperomia*, Valdebenito et al. (1990a; 1992a) had opportunity to examine specimens of *P. tristanensis* Christoph., a taxon known only from Inaccessible Island in the South Atlantic. Inaccessible Island is one of a small group that comprises the Tristan da Cunha Islands. This group lies in the southern Atlantic Ocean at about 37°S, 12°W on the Mid-Atlantic Ridge (Fig. 6.8). The group consists of three main islands, Tristan (0.5 million years old), Inaccessible (2.9 million), and Nightingale (18 million), plus two lesser islets, Middle and Stoltenhoff. (Gough Island, sometimes included with the Tristan group, lies 350 km south–southeast of Tristan at 40°S, 10°W.)

A single population of *Peperomia* exists on Inaccessible Island, which lies approximately 40 km to the southwest of Tristan at 37°19′S,12°44′W (Preece et al., 1986). Examination of the specimens revealed that the Inaccessible material bore striking resemblance to *P. berteroana*, one of the Juan Fernandez endemics, and that the similarity between the two should be recognized by defining the Inaccessible specimens as *P. berteroana* Miq. subsp. *tristanensis* (Christoph.) Valdebenito (Valdebenito et al., 1990b).

Part of the examination of Juan Fernandez endemic species had involved an analysis of the flavonoid profiles of specimens of *Peperomia* from both islands. Recall that the profiles of *P. berteroana* from Masafuera and Masatierra were different from each other. With these data on hand, it was simple to compare those profiles with the pigment profile of *Peperomia* from Inaccessible Island. The finding that the Inaccessible profile matches exactly that of plants from Masafuera not only provides additional support for a close relationship between the subspecies, but also suggests that propagules that gave rise to the Inaccessible population originated from Masafuera. Flavonoid data for this disjunct set are presented in Table 6.4.

The connection between the Juan Fernandez Islands and the Tristan da Cunha Islands represents the first involving a Pacific Ocean-Atlantic Ocean disjunction. Owing to the comparative richness of *Peperomia* populations on the Juan Fernandez Islands, and the very limited population on Inaccessible Island, it seems entirely appropriate to suggest that the propagule(s) originated on the former. Valdebenito et al. (1990a) suggested that a possible agent for propagule delivery was the petrel, a long-distance flyer known to visit both island systems. Additional material of the Inaccessible Island population has been collected for future study of the taxon (T. F. Stuessy, personal communication, September 2001).

6.4.6 Sophora *(Fabaceae)*

Sophora, a leguminous genus of perhaps 45 species (Mabberley, 1997, p. 671), offers a number of very interesting distributional problems. Members of sect. *Edwardsia*

Table 6.4 Flavonoid profiles of *Peperomia berteroana* from the Juan Fernandez Islands and Inaccessible Island (from Valdebenito et al., 1992b)

Taxon	Flavonoids[a]				
	1–7	8	9	10	11–14
P. b. subsp. *berteroana* (MT)[b]	+	+	+	+	–
P. b. subsp. *berteroana* (MF)	+	–	–	–	+
P. b. subsp. *tristanensis*	+	–	–	–	+

[a] Compounds 1–7: acacetin, acacetin sulfate, diosmetin 7-Glc, diosmetin 7-GlcRhm, luteolin 7-DiAra, and luteolin 7-sulfate; 8 = apigenin 7-Glc-sulfate; 9 = luteolin 7-GlcRhm; 10 = diosmetin 7-Glc; 11–14: acacetin 7-Glc-sulfate, diosmetin C-GlcAra, diosmetin 7-GlcAra, and diosmetin 7-sulfate.

[b] MT = Masatierra; MF = Masafuera.

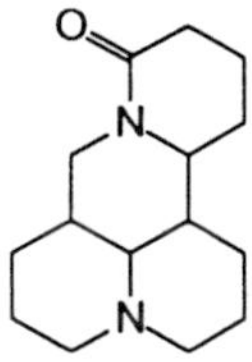

[556]

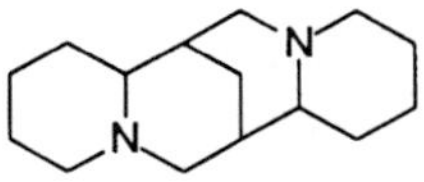

[557]

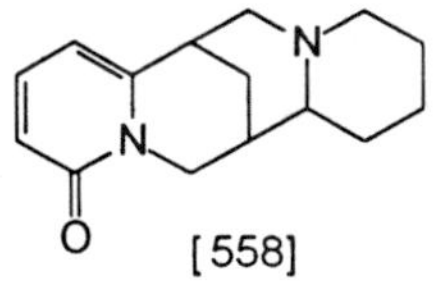

[558]

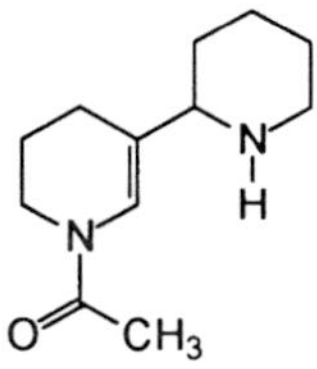

[559]

Fig. 6.10 Compounds 556–559, amine derivatives from *Sophora*

(Salisb.) Taub., numbering ten or so (Polhill, 1991), to which the Juan Fernandez Islands taxa belong, also occur on mainland South America, in Hawaii [the endemic *S. chrysophylla* (Salisb.) Seem.], Lord Howe Islands, and New Zealand (Wagner et al., 1990). There is also evidence to suggest that *S. microphylla* Ait., one of the New Zealand species, also occurs on Gough Island in the southern Atlantic Ocean (Markham and Godley, 1972). Sykes and Godley (1968) discussed transoceanic dispersal of *Sophora* and other genera.

Our interest lies in a comparative study of the alkaloids of *Sophora* species from the Juan Fernandez Islands, mainland South America, and New Zealand (Hoeneisen et al., 1993). Five species were examined, *S. masafuerana* Skottsb. and *S. fernandeziana* Skottsb. from the Juan Fernandez Islands, *S. microphylla* Ait.and *S. macrocarpa* from mainland Chile, *S. linearifolia* Griseb. from Argentina, and *S. microphylla* from New Zealand. Eleven alkaloids were isolated and identified as derivatives of matrine [556] (see Fig. 6.10 for structures 556–559), sparteine [557], and anagyrine [558]. Ammodendrine [559] was also identified. The authors concluded that the alkaloids of the Juan Fernandez Islands species were different from the other taxa and suggested that this indicated that the island species had been isolated from the continental ones "for some time, thus allowing for the evolution of different alkaloid patterns." To my reading of these data, this is an overstatement, there being only relatively minor structural differences among the taxa compared. The information could just as well be used to argue for a close relationship among all of the tested species.

6.5 Lord Howe Island

6.5.1 Carmichaelia *(Fabaceae)*

Lord Howe Island lies at 31°28′S, 159°09′E about 900 km east northeast of Sydney, New South Wales, and 1500 km northwest of Auckland, New Zealand (Fig. 6.11. The origin and evolution of this volcanic island have been described in detail by

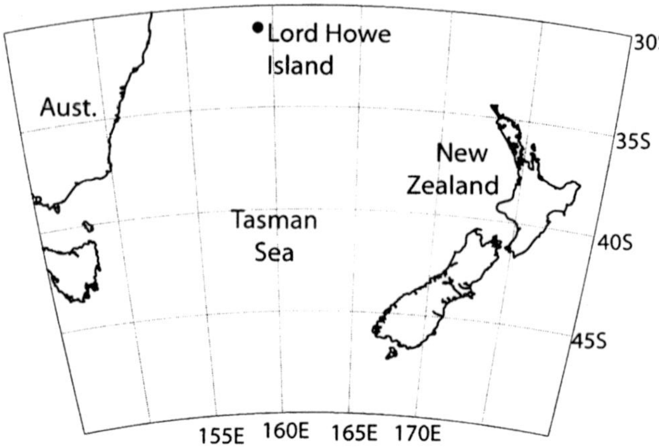

Fig. 6.11 Map showing location of Lord Howe Island

McDougall et al. (1981), who reckon it to be approximately 6 million years old. One of the taxa endemic to the island, *Carmichaelia exsul* F. Muell., has been included in at least two studies of this genus. *Carmichaelia* (Fabaceae, Papilionoideae) consists of perhaps as many as 40 species, of which all but *C. exsul* are endemic to New Zealand. In his study of the flavonoids of the genus, Purdie (1984) pointed out that species often exhibit a good deal of morphological variation, leading in many cases to unclear taxonomic limits. It was his intent to assess the usefulness of flavonoids as an additional source of characters. His study involved 31 species of *Carmichaelia* and included the Lord Howe Island endemic. He reported that the profile of the genus consists of two *C*-glycosylflavones, which were not further identified, a series of flavonol 3-*O*-, 7-*O*-, and 3,7-di-*O*-glycosides, and ubiquitous isoflavones. The flavonol aglycones were identified as kaempferol, quercetin, and isorhamnetin. Kaempferol and quercetin derivatives were observed in all taxa, whereas isorhamnetin was missing from two. Flavonol 3-*O*-glycosides were observed in all taxa; all other compounds showed distinct patterns in their occurrence. Of particular interest was the flavonoid profile of *C. exsul*, which consisted only of kaempferol and quercetin 3-*O*-glycosides and isoflavones, and was the simplest pattern observed in this study. The only other taxon that lacked isorhamnetin was *C. kirkii* Hook. f., which occurs in the east central part of the South Island (vicinity of Canterbury and Otago). *Carmichaelia kirkii*, the sole member of subgen. *Kirkiella* (as used by Purdie) is also distinguished from all other species in possessing an unidentified aglycone. The species is also unique within the genus in being a liana. The absence of isorhamnetin from both *C. exsul* and *C. kirkii* would appear to have no significance. Members of other subgenera also lack certain flavonoids in common with *C. exsul*, but the significance of the common absences and the reduced flavonoid profile in the *C. exsul* cannot be determined without a better understanding of relationships within the genus as a whole.

Pointing in that direction is the recent study of ITS sequences in *Carmichaelia*, *Clianthus*, *Montigena*, and *Swainsonia*, four genera that comprise Carmichaelinae (Wagstaff et al., 1999). [Those authors provided a node-based definition and recommended that the clade be recognized as subtribe Carmichaelinae.] Briefly, their data suggest an origin of Carmichaelinae from a North American ancestor (from Astragalinae) with subsequent diversification into several lineages, one of which was *Carmichaelia.* This divergence was thought to have accompanied the increased aridity in Australia and New Zealand during the Tertiary. *Carmichaelia* first appears in the fossil record in the late Pliocene Waipaoa Series (Oliver, 1928). Of interest to us here is the finding that ITS information places the Lord Howe Island taxon, *C. exsul*, in the midst of a group of nine species, which, interestingly, includes *C. kirkii*, whose flavonoid chemistry was commented upon above. Wagstaff et al. (1999) suggest that dispersal to Lord Howe Island has occurred "recently," but did not provide an estimate of the time involved, other than to comment on the age of the island.

6.6 Ullung Island, Korea

6.6.1 Acer *(Sapindaceae)*

Ullung Island (or Ulleung-do) is a small, volcanic island lying in the Sea of Japan, approximately 150 km east of Korea and 300 km west of Japan at 37°30′N, 130°50′E. The island has an area of only about 73 km^2 and has been available for colonization by propagules for an estimated 1.8 million years (B.-Y. Sun and Stuessy, 1998). Species belonging to 34 genera (25 families) are endemic to the island. Two endemic members of *Acer*, *A. okamotanum* Nakai and *A. takesimense* Nakai, have been examined using both micro- and macromolecular methods. Morphological data suggest that these two species have been derived from two different ancestors. ITS data (Cho et al., 1996) confirm this suggestion. *Acer okamotanum* is thought to be closely related to *A. mono* Maxim., and thus could have arisen from ancestral stock from either Korea or Japan. A close relationship between *A. takesimense* and *A. pseudosieboldianum* (Pax) Kom., based on morphological features, is strongly supported by the lack of ITS sequence divergence. It is interesting to note that Chang and Giannasi (1991), studying flavonoids of *Acer* sect. *Palmata*, suggested that *A. takesimense* and *A. pseudosieboldianum* might be better treated as conspecific. Sun and Stuessy (1998) discuss other endemic taxa of Ullung Island.

6.6.2 Cotoneaster wilsonii *(Rosaceae)*

Cotoneaster wilsonii Nakai, endemic to Ullung Island, is one of the many species of the genus that occur in eastern Asia. The flavonoid profile of this species consists primarily of *O*-glycosides of the flavonols, quercetin and isorhamnetin (quercetin 3′-methyl ether), along with *C*-glycosyl derivatives of the flavones, apigenin and luteolin (Chang and Jeon, 2003). This profile is very similar to what has been recorded for other species of Cotoneaster in Asia. This led the authors to conclude that, as in the case of other woody plants growing on the island, very little change in either morphology or chemistry has occurred during the species' existence on the island.

6.7 Tristan da Cunha and the Falklands

6.7.1 Empetrum *(Empetraceae)*

This example involves the Tristan da Cunha Archipelago, the Falkland Islands (Islas Malvinas), and mainland South America (Fig. 6.8); the plant in question is *Empetrum rubrum* Vahl ex Willd. This is one of two or three species that constitute Empetraceae (Mabberley, 1997, p. 255); the others are the circumboreal *E. nigrum* L. and *E. eamesii* Fernald & Wiegand (considered by those authors as a segregate from *E. nigrum*). The taxonomic history of the genus was reviewed in D. M. Moore et al. (1970).

As part of a chemotaxonomic study of the family, Moore et al. (1970) examined the flavonoid profiles of 18 populations of *E. rubrum*, 16 ranging from the Chilean Andes to southeastern Fuegia, and one each from the Falkland Islands and Gough Island in the Tristan da Cunha group (Fig. 6.8). (The species also occurs on the Juan Fernandez Islands but was not sampled from there.) Many compounds (spots) were observed on chromatograms, six of which were identified as quercetin, 3-*O*-arabinoside, 3-*O*-galactoside, and 3-*O*-rutinoside; gossypetin, 3-*O*-galactoside (gossypetin is 8-hydroxyquercetin); and two hydroxycinnamic acid derivatives. Additional flavonoids were among the group of unidentified spots. Chromatographic analysis of the unusual black-fruited *E. rubrum* revealed the same pigment array as normally found in the black fruits of *E. nigrum*, namely, glycosides of petunidin [560], cyanidin [561], peonidin [562], delphinidin [563], and malvidin [564] (see Fig. 6.12 for structures 560–564). Normal red-fruited *E. rubrum* lacked pigments having a trisubstituted B-ring (delphinidin and malvidin) and had only glycosides of cyanidin and peonidin.

Although not the main issue here, it is interesting to note that flavonoid profile variation in the Northern Hemisphere members of *Empetrum* shows no correlation with geography, regardless of whether *E. nigrum* is taken in the broad sense, or whether one recognizes *E. eamesii* as well. The situation with regard to *E. rubrum* is somewhat different in that some geographical patterns are evident, as is seen in the case of quercetin 3-*O*-rutinoside (commonly known as rutin) (identified as F3 in that work). Rutin is a common constituent of *E. nigrum*, including *E. eamesii*, but was observed only rarely in specimens of *E. rubrum*. However, rutin

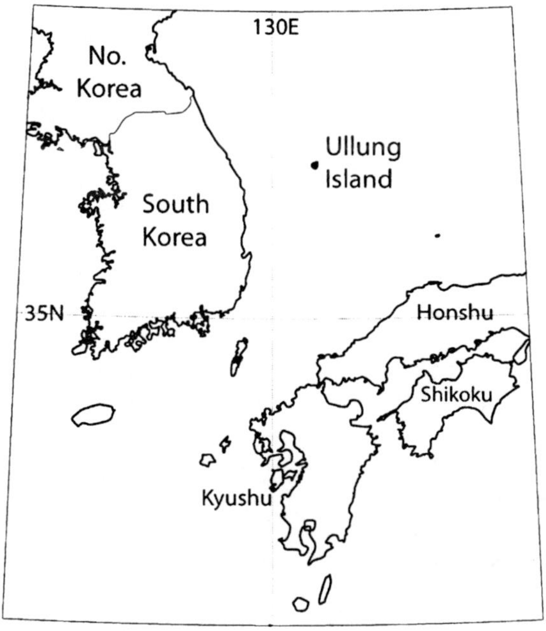

Fig. 6.12 Map showing location of Ullung Island, Korea

was observed in the two northernmost populations in the Chilean Andes (provinces of Talca and Nuble). The authors speculated that this might indicate amphitropical gene exchange with the northernmost populations. Unidentified compound F8, observed in about a third of *E. nigrum* populations, was found to be characteristic of *E. rubrum*, except that it was not seen in the two northernmost populations. Interestingly, this compound was also absent from the profile of Gough Island plants, suggesting long-distance dispersal from one of the northern populations. It is, of course, also possible that the capacity to make compound F8 was lost subsequent to colonization of Gough Island. It was also noted that the flavonoids of the Falkland Island population bore closest similarity to populations from Central and Southern Tierra del Fuego. This is in line with Moore's (1968) view that the Falkland Island flora has closest affinities with the same area.

6.8 Kerguelen Islands

6.8.1 Ranunculus *(Ranunculaceae)*

The Kerguelen Islands, which lie at about 49°30′S, 69°30′E in the southern Indian Ocean, represent perhaps the most remote island system with which we deal in this chapter. The system is made up of one main island and a number of small rocky outcrops, and it lies closer to Antarctica than to any other major landmass. Of interest in the present context are three species of *Ranunculus*: one of which, *R. moseleyi* Hook. f., is endemic. The other two species, *R. pseudotrullifolius* Skottsberg and *R. biternatus* Smith, are of circumpolar occurrence. A close relationship between these species had previously been demonstrated through a cytogenetic study (Hennion and Couderc, 1993), in which all three were shown to be hexaploids with $2n=48$.

The cytogenetic study was followed by a detailed examination of the flavonoid profiles of the three species (Gluchoff-Fiasson et al., 1994), which revealed the presence of a complex mixture of acylated quercetin di- and triglycosides. An example of the sort of compounds identified is quercetin 3-*O*-(caffeoylxylosylglucoside)-7-*O*-glucoside [565] (see Fig. 6.13), which was observed as a major component of both *R. pseudotrullifolius* and *R. moseleyi*, but present in only trace quantities in *R. biternatus*. Clear distinctions among the taxa were thus possible based on the degree of xylosylation seen in the respective taxa. The level of acylation observed in each taxon also proved to be of value (Hennion et al., 1994). Glycosides bearing xylose were predominantly seen in *R. pseudotrullifolius* and the endemic *R. moseleyi*, in both of which it occurred in compounds that were marked as "abundant" or as "major component." Three xylose-containing glycosides were observed in *R. biternatus* but were present as minor or trace components only. An analysis of the HPLC peak areas for all compounds containing xylose showed highly discriminating values, which the authors presented as "% xylosylation." Four individuals of *R. biternatus* averaged 1.1% xylosylation, with a range of 0.6–1.5%. Four individuals of *R. pseudotrullifolius* and six of *R. moseleyi* gave values of 78.5% (range 58–91%) and 89.7% (range 86–94%), respectively. A similar calculation provided a value for "% acylation." A value of 84% (78–86%) was obtained for *R. biternatus*, while values of 69.3% (range 64–81%) and 69.3% (range 46–77%) were obtained for *R. pseudotrullifolius* and *R. moseleyi*, respectively. Further differences among the three species were noted with regard to morphological characters and habitat. In general, the comparatively narrower ranges of morphological variation for *R. pseudotrullifolius* and *R. moseleyi* were very similar, whereas *R. biternatus* exhibited a much wider range of structural variation as well as exhibiting a wider range of variation in habitat ranging from aquatic sites to fell fields. *Ranunculus pseudotrullifolius* and *R. moseleyi* are much more restricted in their choice of habitats being almost entirely aquatic. The authors concluded that the combined results of cytotaxonomic, morphological, chemical, and ecological examinations support a comparatively recent colonization of the Kerguelen Islands by the two magellanic species following the most recent glacial period, some 8000 years ago, followed by differentiation of *R. moseleyi* from *R. pseudotrullifolius*.

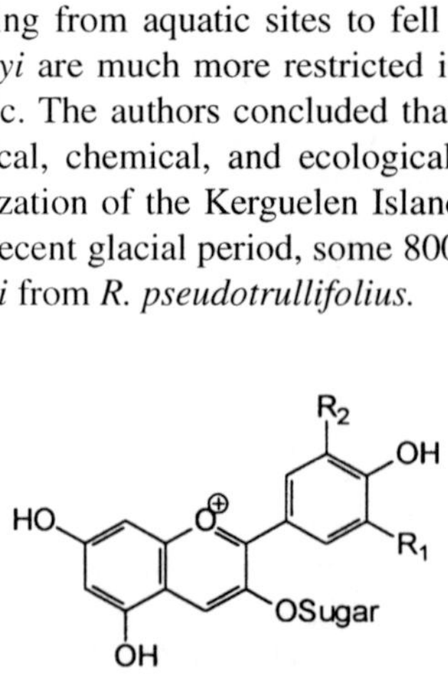

[560] $R_1 = R_2 = H$
[561] $R_1 = OH, R_2 = H$
[562] $R_1 = OCH_3, R_2 = H$
[563] $R_1 = R_2 = OH$
[564] $R_1 = R_2 = OCH_3$

Fig. 6.13 Compounds 560–564, anthocyanins from *Empetrum*

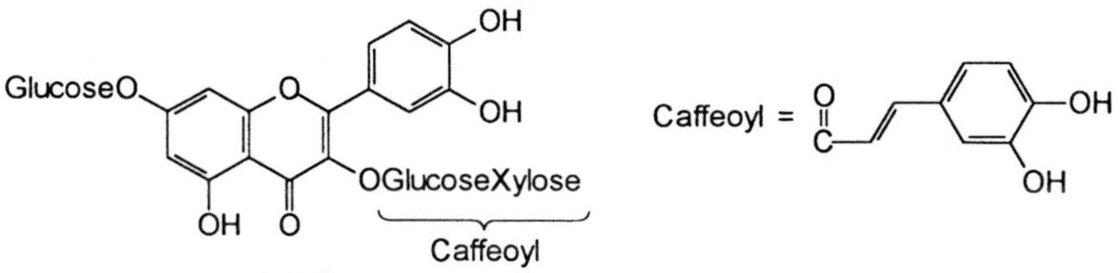

[565]

Fig. 6.14 Compound 565, the flavonol triglycosides from *Ranunculus* species on the Kerguelen Islands

6.9 Macaronesia

Macaronesia is a biogeographic region in the eastern Atlantic Ocean, consisting of the Azores, the Canary Islands, the Cape Verde Islands, Madeira, and the Selvagen Islands (Fig. 6.15). The Azores archipelago consists of nine islands centered at 38°30′N, 28°00′W, approximately 1700 km west of Lisbon. The Canary Islands archipelago lies between ca. 27°30′ and 29°30′N and ca. 18° and 13°45′W, with its easternmost island only about 100 km off the coast of Morocco. The Archipelago consists of seven main islands (from west to east): El Hierro, La Palma, La Gomera, Tenerife, Gran Canaria,

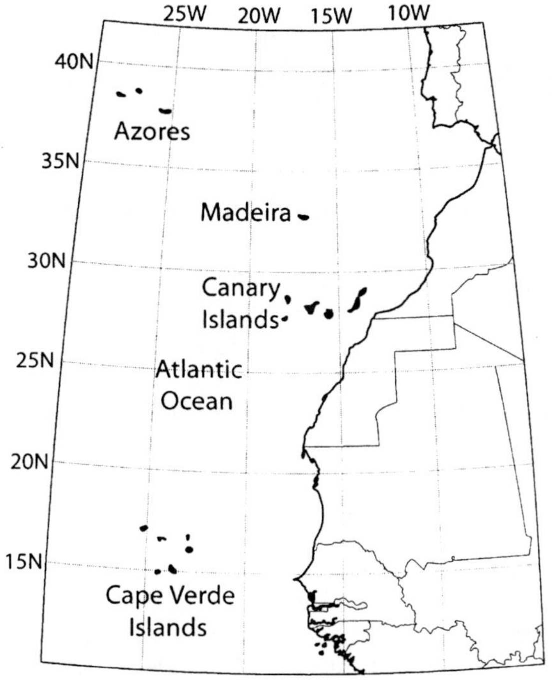

Fig. 6.15 Map of Macaronesia

Fuerteventura, and Lanzarote. The islands are volcanic and range in age from less than 1 million to about 21 million years (Carracedo, 1994). Two recent papers present detailed accounts of the geologic history of the Canary Islands along with extensive literature reviews (Anguita and Hernan, 2000; Dañobeitia and Canales, 2000). The Cape Verde archipelago consists of several small islands centered at 16°00′N, 24°00′W, to the west of Senegal. The Madeira archipelago consists of two main islands, Madeira (32°45′N, 17°W) and the much smaller Porto Santo (33°04′N, 16°20′W). The Selvagens Islands, among the smallest of the Macaronesian group, consist of two small landmasses at ca. 30°00′N, 16°00′W, which puts them roughly one-third of the way between the Canary Islands and the island of Madeira.

6.9.1 **Anomalographis madeirensis** *(Graphidaceae)*

Differences in the depsidone chemistry of the lichen *Anomalographis madeirensis* (Tav.) Kalb. have been noted by González et al. (1999), who reported identification of stictic (see Fig. 5.4 for structures), norstictic, and cryptostictic acids from a specimen collected on Tenerife. In contrast, Kalb and Hafellner (1992) reported stictic and norstictic acids from material of this species collected on Madeira, but did not find cryptostictic acid. They did find two other depsidones, however, that is, constictic and connorstictic acids. It would be interesting to know if plants from these two sites respond to differences in localized microhabitats by synthesizing different arrays of compounds similar to the case with *Ramalina siliquosa* in Wales, which was commented upon above.

6.9.2 **Dactylis glomerata** *(Poaceae)*

Dactylis is a small genus thought by some to be monotypic, consisting only of *D. glomerata* L. (cock's-foot), a Eurasian native now widely naturalized in other parts of the world. That the species is complex can be seen in the recognition of 15 diploid ($2n = 14$) members as subspecies (Stebbins and Zohary, 1959; Lumaret, 1988). In Macaronesia, several morphologically and chromosomally distinct entities are restricted to single islands, and some islands support different subspecies in different habitats. *Dactylis* has also been divided into three climatic groups, mesic-temperate, Mediterranean, and subtropical (Lumaret, 1988), each of which may have both diploid and tetraploid members.

In order to gain a better understanding of the relationships and evolutionary history of *Dactylis*, Lumaret and colleagues have undertaken extensive studies of various biochemical markers. The first of these that can be examined is the flavonoid analysis of the Atlantic part of the subtropical group (Jay and Lumaret, 1995). In all,

126 individuals were analyzed by HPLC revealing six caffeic-acid derivatives, five luteolin mono-*C*-glycosides, two luteolin di-*C*-glycosides, a tricin *O*-glycoside, and a mixture of luteolin mono-C-glycosides and apigenin di-*C*-glycosides. Principal-components analysis (PCA) revealed seven identifiable groups reckoned by their centroids: (1) coastal populations from Madeira representing subsp. *marina,* but differing in phenolic pattern from coastal Portuguese specimens of subsp. *marina* (Borrill) W. Greuter; (2) tetraploids from Madeira identified as subsp. *hylodes*; (Parker) J. Holub; (3) diploids from Tenerife identified as subsp. *smithii* (*D. smithii* Link?); (4) tetraploids from the center of La Palma; (5) tetraploids from the center of Grand Canary; and (6) and (7) populations from the coast of Portugal identified as subsp. *marina.* Clearly separated from all other groups was the population from Grand Canary. Individuals from this population exhibited a unique combination of phenolic compounds. Individuals from this population also exhibited specific alleles at several loci (Sahuquillo and Lumaret, 1995), as well as morphological features resembling both subtropical and Mediterranean members of *Dactylis.*

Populations from this area are also ecologically different, in that they grow at high elevations in dry habitats not influenced by the ocean to the degree experienced by other groups. As Jay and Lumaret (1995) pointed out, this combination of features suggests that these Grand Canary populations are likely to have been isolated for a long time. Although the other PCA-group centroids were separated from one another, the level of individual variation resulted in significant overlap. A second PCA of tetraploid individuals from ten geographic locations again resulted in clear separation of the Grand Canary plants, but with even greater overlap among the other populations. Included in this analysis were representatives from the French Alps (subsp. *reichenbachii*, nomenclature?) and from the south of France (subsp. *hispanica*; *D. hispanica* Noe ex Steud.?). These groups were clearly distinguished from both of the others, although there was a good deal of overlap among them. In their concluding remarks, the authors pointed out the contrasting influences experienced by *Dactylis* on these islands, ranging from genetic differentiation in isolation, as in the Grand Canary situation, to possible adaptive value of different combinations of phenolic compounds, as seen in the heterogeneous populations on Madeira.

A recent examination of chloroplast DNA variation has provided additional insights into the evolutionary history of cock's foot (Sahuquillo and Lumaret, 1999). Plant material was collected from six diploid and 15 tetraploid populations (121 individuals in all) representing the Canary Islands, Madeira, the coasts of Portugal and Spain, and sites in Morocco (northern and northwestern Africa). Eight restriction enzymes yielded fragments that defined two "chlorotypes," chlorotype I, considered to represent the ancestral type, and chlorotype II characterized by a 290-base-pair deletion. Chlorotype I was found to be the predominant form, occurring in the majority of continental populations and in the westernmost Macaronesian islands. It was found in both diploids from Tenerife, one from the lowland scrub and the other from the high-elevation heath. It also characterized the tetraploids endemic on Madeira and La Palma. Chlorotype II occurred in populations from the eastern end of the archipelago, that is, the islands nearest Africa, and in a few diploid and tetraploid Mediterranean populations growing at higher elevations. Chlorotype II was also observed in tetraploid populations from La Gomera (one of the western islands) and in one of the seven populations examined from Madeira. These results were taken to indicate that introgression from Mediterranean material into tropical forms has happened, possibly several times, and that repeated colonization of the islands from the mainland or between islands has likely played a major role in development of the patterns of variation observed.

An earlier paper by Jay et al. (1984) should be consulted for information on patterns of variation in *D. glomerata* representing populations from Scotland, southern France, and the Mediterranean islands of Corsica and Sardinia. The flavonoid profiles, based on mono- and di-*C*-glycosylflavones, were clearly sorted into three geographical groups, but not strictly according to subspecific limits.

6.9.3 Laurus azorica *(Lauraceae)*

Laurus azorica (Seub.) Franco (=*L. canariensis* Webb & Berth.) is an endemic species on the Azores, the Canary Islands, and Madeira, where it occurs as a component of the laurel-juniper cloud forest community. The other member of the genus is the so-called "true bay," *L. nobilis* L. A recent study by Pedro et al. (2001) described the essential-oil chemistry of *L. azorica* collected from ten populations representing five Azorean islands, Faial (3 populations), Pico (1), Sao Jorge (2), S. Miguel (2), and Terceira (2). Monoterpenes dominated the foliar-oil fraction with α-pinene (15–37%), β-pinene (9–18%), and 1,8-cineol (12–31%) as the major components. Major monoterpenes of oil from unripe berries were α-pinene (12–22%), β-pinene (7–13%), *trans*-β-ocimene (27–45%), and *cis*-β-ocimene (9–16%). Cluster analysis revealed that two clear-cut groups exist based upon the enantiomeric composition of the pinenes. The two populations from Sao Jorge were strongly differentiated from all other populations. Analysis of plant material from the Santo António site on Sao Jorge from different years revealed consistent results. Information on oil composition of plants from the Canary Islands and Madeira would have been useful in helping to assess the significance of these interpopulational (and inter-island) differences, in particular, are the populations on Sao Jorge more closely related to populations from one of the other archipelagoes, or is the difference in chemistry the result of a strictly local event or local conditions?

6.9.4 Todaroa aurea *(Apiaceae)*

Todaroa is a small genus endemic to the Canary Islands. According to González et al. (1988), who were interested in secondary metabolites of the genus, *T. aurea* Parl. was regarded "tentatively" as a single species endemic to the four westernmost islands of the archipelago. Chemical analysis of specimens collected on two islands,

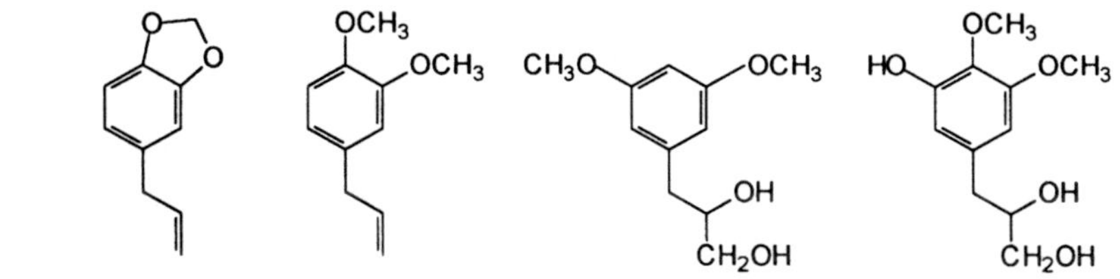

[566] [567] [568] [569]

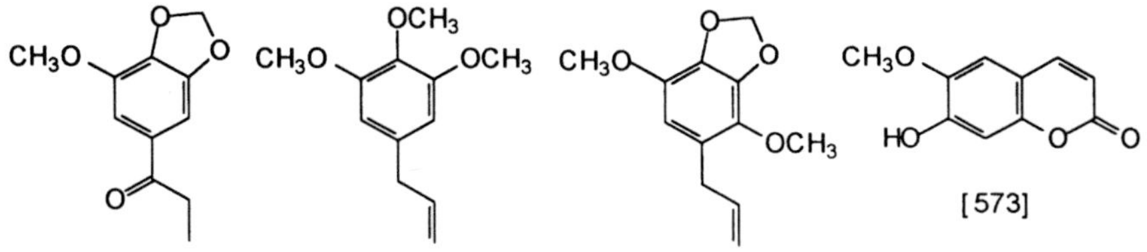

[570] [571] [572]

Fig. 6.16 Compounds 566–573, phenolic compounds from *Todaroa aurea*

Tenerife and La Palma, revealed the existence of luteolin and an array of simple phenolic derivatives as well as three known phytosterols, ß-amyrin, sitosterol, and stigmasterol. The phenols identified comprised a set of phenylpropanoids: myristicin [566] (see Fig. 6.16 for structures 566–573), methyleugenol [567], todadiol [568], todatriol [569], crocatone [570], elemicin [571], apiole [572], and the coumarin scopoletin [573]. The occurrence of these compounds is recorded in Table 6.5. The differences between the two profiles were taken by González and his co-workers

Table 6.5 Distribution of simple phenols and phytosterols in *Todaroa aurea* (from González et al., 1988)

Compound	Island	
	Tenerife	La Palma
Myristicin	+	+
Methyleugenol	+	–
Todadiol	+	+
Todatriol	+	–
β-Amyrin	+	+
Sitosterol	+	+
Stigmasterol	+	+
Luteolin	+	+
Crocatone	+	–
Scopoletin	+	+
Elemicin	–	+
Apiole	–	+

to support recognition of the plants from Tenerife as subsp. *aurea* and the plants from La Palma as subsp. *suaveolens*. Although the chemical differences were in accord with "noticeable" morphological differences between the two specimens, it is unfortunate that the sampling of these taxa was so meager. Should there be any interpopulational variation in the capacity of these taxa to express their phenolic chemistry, the recorded differences may well be misleading. Additional sampling, including several plants from several populations, would be most welcome. Specimens from other islands would be needed for a fuller view of the system.

Chapter 7
Polar Disjunctions

This chapter deals with taxa whose ranges of occurrence are disjunct between the Northern and Southern Hemispheres. Some of the examples involve taxa with related members whose distributions fit into other categories, some of which have been discussed above. *Chrysosplenium*, for example, has a distribution that involves disjunctions between Asia and North America, between eastern and western North America, and between the Northern Hemisphere and the Southern Hemisphere.

Excellent general information on amphitropical relationships involving taxa from the Pacific Coast of North America and South America appears in a symposium on the subject published in the 1963 *Quarterly Review of Biology* (Raven, 1963) with contributions from Lincoln Constance on Apiaceae, Larry Heckard on Hydrophyllaceae, Kenton Chambers and Robert Ornduff on genera in Asteraceae, and Peter Raven on an overview of floristic relationships between North America and South America.

7.1 *Ambrosia chamissonis* (Asteraceae)

Ambrosia chamissonis (Less.) Greene is a shrubby ragweed—often called beach bur—whose range is characterized by a disjunct distribution between the Pacific Coast of North America, extending from southwestern British Columbia to northern Baja California (ca. 30°–50° N), and the coast of Chile (ca. 29° and 40° S). The distribution in California is not uniform, however; there is a gap of several hundred kilometers south of Monterey where suitable habitats are absent. It was first reported in Chile in 1892 on Isla de la Mocha near Valdivia (39°46′S, 73°15′W) (Kohler, 1966; Kohler and Weisser, 1966; as cited by Payne et al., 1973). Following its introduction, it has now spread several hundred kilometers northward to occur in what constitutes the Mediterranean zone of South America. Several key reports have appeared documenting the patterns of morphological and chemical variation that characterize the taxon, making it one of the better-understood disjunctions between North America and South America.

The morphological variation within the complex is reflected in the number of taxonomic descriptions (ca. 20) that have been published defining species, subspecies, and varieties. Payne (1964), however, preferred to consider *A. chamissonis* as a large, heteromorphic species consisting of two varieties, var. *cuneifolia* (Nutt.) Payne, ranging from the mouth of the Columbia River north to British Columbia, and var. *chamissonis* from the southern part of the species' range. The latter variety consists, in turn, of two intergrading forms, forma *chamissonis* with unlobed leaves, and forma *bipinnatisecta* with pinnately decompound leaves (Payne et al., 1973).

Several reports describing different sesquiterpene derivatives from *A. chamissonis* collected from different areas suggested that secondary chemistry might be of value in helping to sort out relationships within the complex (see Geissman et al., 1973 for citations). The major compound reported in those studies was the germacranolide chamissonin [574] (see Fig. 7.1 for structures 574–580), with other collections affording costunolide [575]. More recent work, involving a wider sampling of populations throughout the range of the species, confirmed the presence of those compounds and revealed others (Geissman et al., 1973): chamisselin [576], chamissanthin (8-α-hydroxycostunolide) [577], chamissarin [578], and the two chamissonin epoxides [579] and [580]. Patterns of occurrence of these compounds have provided important information on relationships within the species.

Plants collected north of the Monterey area were shown to be variable with regard to both morphology and sesquiterpene lactone composition. Of particular significance are the sporadic occurrence of chamissonin and the appearance of costunolide as a major component in nearly half of the samples analyzed. However, there was no correlation between chemotypes and morphology. The situation south of Monterey was shown to be markedly different, where individuals proved to be morphologically uniform and to have an almost invariant chemotype characterized by a preponderance of chamissonin and a lack of costunolide. These data strongly suggested that the propagule(s) that gave rise to the southern populations came from the central part of the range of *A. chamissonis*, where one finds the greatest variation within the species. Although the sample size was small, two

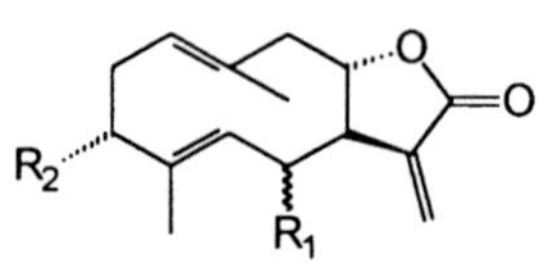

[574] $R_1 = R_2 =$ a -OH
[576] $R_1 =$ b-OH, $R_2 =$ H
[578] $R_1 =$ b-OAc, $R_2 =$ a -O H

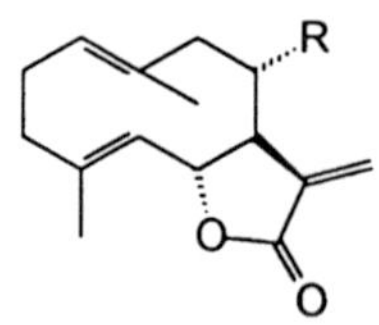

[575] R = H
[577] R = OH

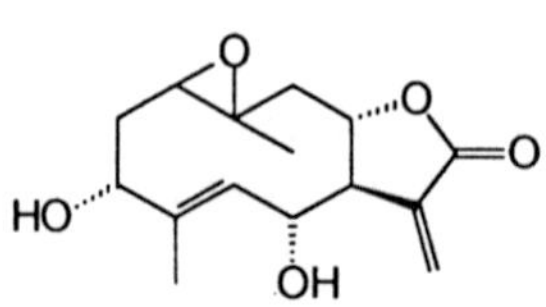

[579]

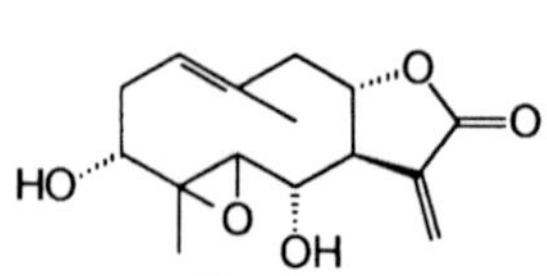

[580]

Fig. 7.1 Compounds 574–580, terpenes from *Ambrosia chamissonis*

of three specimens collected near San Francisco exhibited a chemotype closely resembling the southern populations. This appears to be a case of "founder effect" after long-distance dispersal from a northern population. Spread of the species southward from the Monterey area by local dispersal was ruled out owing to the absence of suitable habitats; south of Monterey, sandy beaches are replaced by steep cliffs for a distance of several hundred kilometers (Big Sur).

With the possible origin of the southern Californian populations accounted for, we can turn our attention to the matter of the origin of Chilean *A. chamissonis.* An examination of several populations from the Chilean coast (Nakatani et al., 1973) revealed comparatively uniform morphology and sesquiterpene lactone chemistry, a situation parallel to that just discussed for the southern California members of the species. The lactones isolated from the Chilean populations were identified as chamissonin [574], costunolide [575], and chamisselin [576]. Long-distance dispersal was again indicated, with morphology and chemistry both pointing to a source population north of the San Francisco Bay area.

Ambrosia chamissonis also exhibits a rich array of acetylenic compounds. Although not restricted to Asteraceae, polyacetylenic compounds—or polyynes—do constitute one of the characteristic secondary metabolites of the family. A comprehensive review of the structures and patterns of occurrence of these compounds, including sources other than Asteraceae, was published by Bohlmann et al. (1973). More recently, a series of review articles appeared discussing polyacetylene occurrences in Anthemideae (Christensen, 1992), Cynareae (Christensen and Lam, 1990), Heliantheae (Christensen and Lam, 1991a), and Astereae (Christensen and Lam, 1991b). Many of the acetylenic compounds identified from these taxa possess, in addition to the requisite triple bonds, other functional groups such as one or more double bonds, a hydroxyl group, an epoxide group, a halogen atom, or some combination of these, for example, chlorohydrin derivatives (Balza and Towers, 1990). Also of interest are acetylenic compounds that bear one or more sulfur atoms. One of the simplest of these is terthienyl [581] (see Fig. 7.2 for structures 581–589), a component of common garden marigolds and other species. Compounds bearing two sulfur atoms are also found in a number of species, as seen in the study of the components of *Ambrosia chamissonis.* A study of the roots of this species collected from two widely separated sites along the North American coast showed that compounds [582–584] made up 85% of the acetylenic fraction with compound [585] making up 5%. The remaining 10% of the acetylenic fraction from roots collected near San Francisco, California (Marin Co.) was made up of compounds [586 and 587]. However, material collected in extreme southwestern British Columbia (near Tsawwassen) lacked the latter two compounds but exhibited compounds [588 and 589] instead.

Although these data suggest relationships, clear-cut phylogenetic connections among the various populations must await the application of macromolecular techniques. In addition to providing insight into the evolutionary relationships, such data would also likely provide an idea of when colonization might have occurred.

[581]
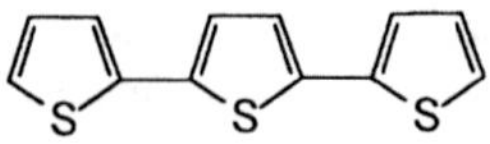

[582]
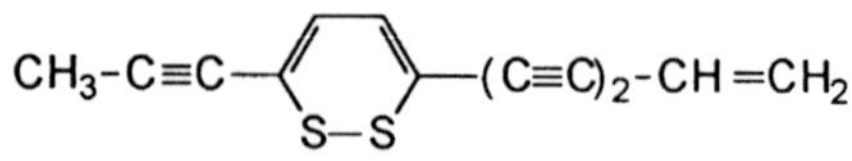

[583]
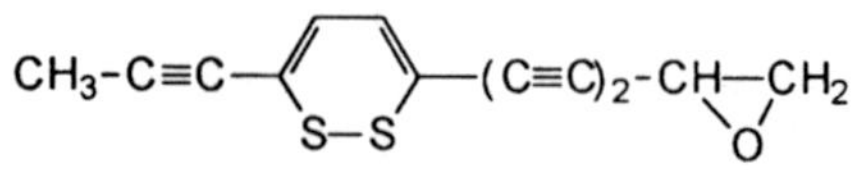

[584]
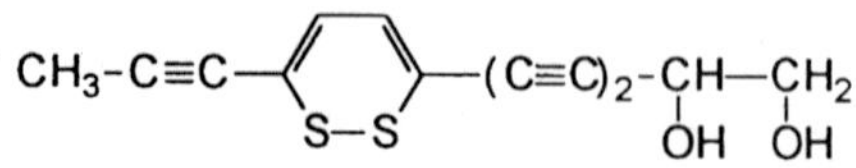

[585]
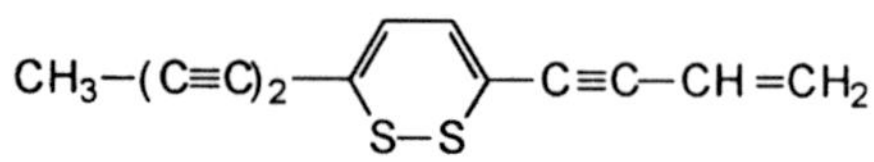

[586]
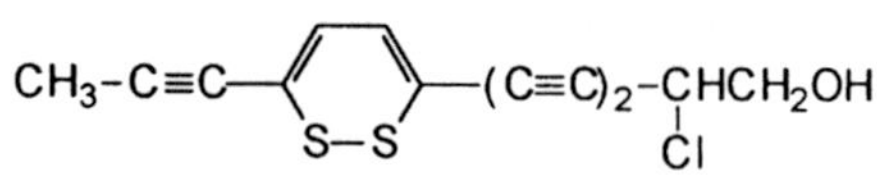

[587]
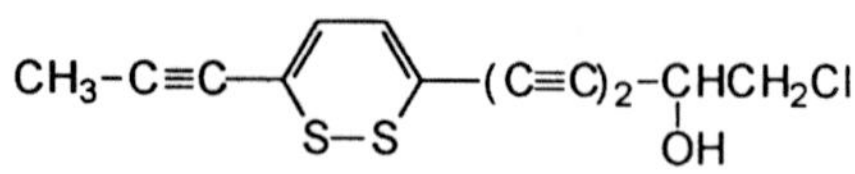

[588]
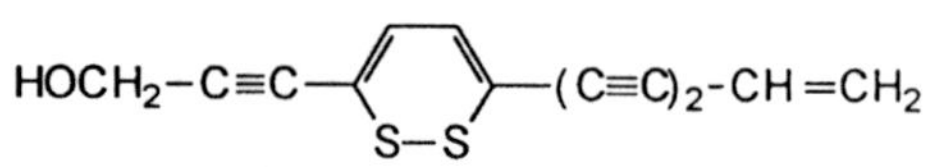

[589]
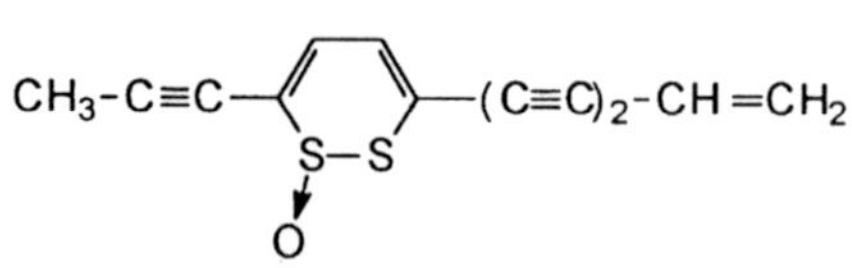

Fig. 7.2 Compounds 581–589, acetylene derivatives from *Ambrosia chamissonis*

7.2 *Blennosperma* (Asteraceae)

Blennosperma was encountered above as part of the trans-Pacific disjunction of the Blennospermatinae (pp.). Three species (four taxa) comprise this small genus, the South American *B. chilense* Less., and the western North American *B. bakeri* Heiser, *B. nanum* (Hook.) S. F. Blake var. *nanum* and *B. nanum* var. *robustum* J. Howell. The flavonoid profiles of *B. chilense*, *B. bakeri* and *B. nanum* var. *nanum* were shown to be identical (Ornduff et al., 1973b) in their possession of quercetin 3-*O*-glucoside, a quercetin 3-*O*-diglycoside (probably a rhamnosylglucoside), and luteolin 7-*O*-glucoside.

Blennosperma nanum var. *robustum* lacked the flavone, but exhibited the flavonol glycosides plus cyanidin 3-*O*-monoglycoside [structure 561] in some individuals.

7.3 *Chrysosplenium* (Saxifragaceae)

Chrysosplenium enjoys a distribution that qualifies it for inclusion in several categories in this review. Members are known from northern Europe, North America, eastern Asia (principally Japan), and extreme southern South America. The similar flavonoid chemistry of the eastern North American–western North American species pair *C. americanum* and *C. glechomaefolium* was mentioned earlier, along with comments on apparent flavonoid differences between sections *Alternifolia* and *Oppositifolia.*

Of interest here is the one South American species (there are two) for which we have some flavonoid data (Bohm and Collins, 1979). Analysis of a specimen of *C. valdivicum* W. J. Hooker afforded four compounds, three of which yielded isorhamnetin, glucose, and rhamnose on total hydrolysis and isorhamnetin 3-*O*-glucoside on partial hydrolysis, suggesting the presence of rhamnosylglucosides, for example, [590] (see Fig. 7.3 for structures 590–601). The fourth compound was identified as kaempferol 3-methyl ether [591]. There was no indication of the more highly substituted flavonols isolated from other *Chrysosplenium* species, that is, compounds possessing extra hydroxylation on both the A- and B-rings and O-methylation, for example, 5,2′-dihydroxy-3,7,4′,5′-tetramethoxyflavone [592], a compound known from *C. grayanum*, one of the Asian species. The pattern observed in *C. valdivicum* is the simplest array of flavonols reported from any member of the genus, and thus could be taken as support for Hara's (1957) view that *C. valdivicum*, with its array of "archaic" features, resembles the prototype of the genus. Following O. R. Gottlieb's (1986, 1990) ideas concerning level of oxidation and degree of "advancement" of a flavonoid profile, the picture we see in the *C. valdivicum* is of a set of compounds with an oxygenation pattern characteristic of the basic flavonoid structure, compared to the more complex flavonoids from the genus that have extra oxygenation on both the A-ring (6-oxygenation) and B-ring (2′-oxygenation), and in several cases both. The difficulty with this interpretation, as pointed out by Gornall and Bohm (1978), is that a simple oxygenation pattern may also be arrived at by loss of biosynthetic capabilities, and thus represent a derived, or specialized, situation.

A recent phylogeny of the genus (Soltis et al., 2001) revealed that the South American taxa are most closely related to species from eastern Asia. In terms of flavonoid profiles, this requires us to reappraise the significance of the simple compounds. Thus, as suggested above, the simple flavonoid glycoside profile of *C. valdivicum* appears to represent a loss of biosynthetic capacities, in particular, extra oxygenation, and not a simple profile to which structural features were added in later stages of evolution.

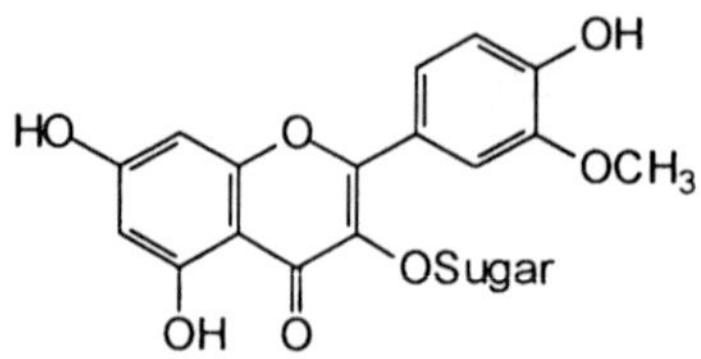

[590]

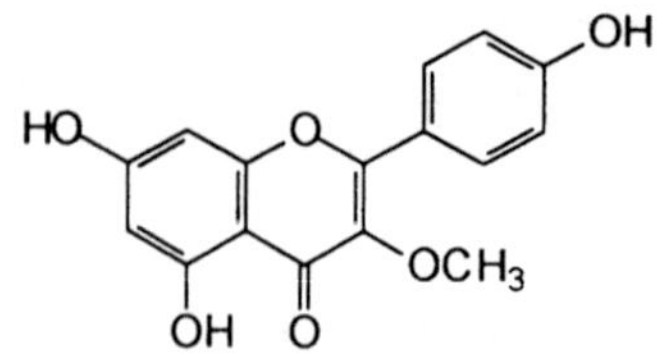

[591]

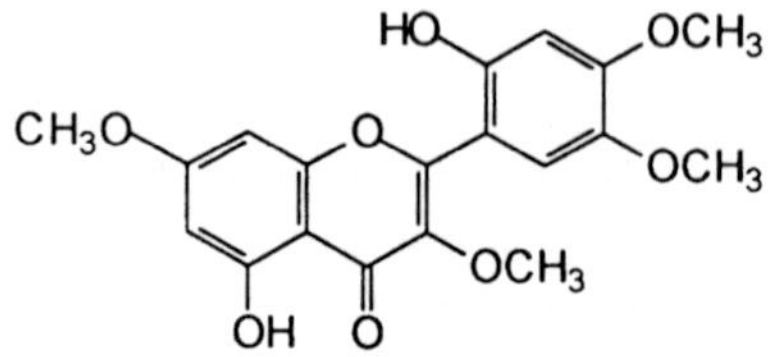

[592]

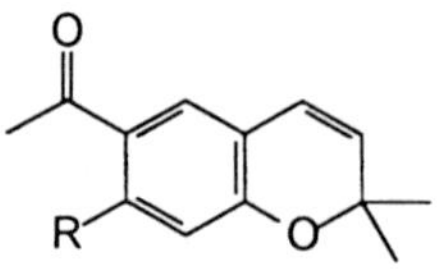

[593] R = H
[594] R = OH
[595] R = OCH_3

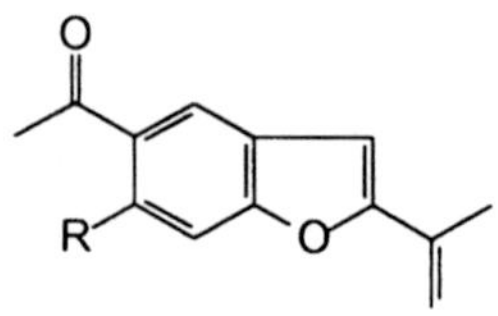

[596] R = OH
[597] R = OCH_3

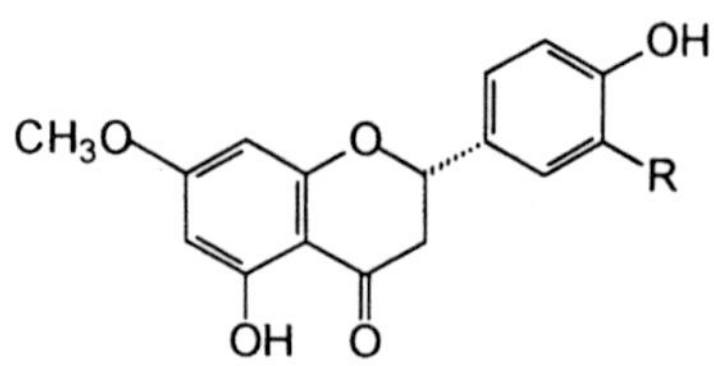

[599] R = H
[600] R = OH

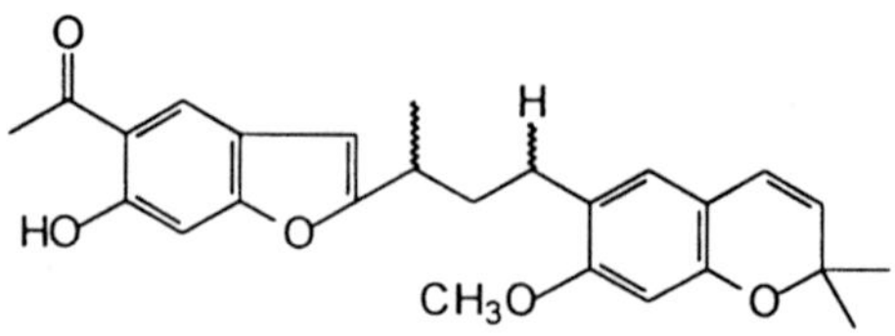

[598]

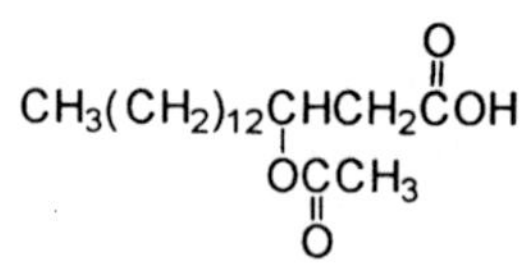

[601]

Fig. 7.3 Compounds 590–601, flavonoids of *Chrysosplenium*, and simple phenolic derivatives from *Encelia*

7.4 *Encelia canescens* (Asteraceae)

Encelia consists of some 15–19 species, depending upon one's point of view, with all but two occurring in arid and semiarid portions of southwestern North America. One of those exceptions is *Encelia canescens* Lam., which occurs in coastal Peru and Chile and in northern inland Argentina. In an examination of 19 taxa for their chromones and benzofurans, Proksch and Clark (1987) observed four groups of species. The South American taxon exhibited the same array of compounds as did nine of the North American taxa, including a dimeric compound unique to this group and to the genus. The compounds identified are the chromones [593–595] (see Fig. 7.3 for structures), the two benzofurans [596 and 597], and the unique dimeric compound [598]. In a subsequent publication (Proksch et al., 1988) dealing with leaf exudate flavonoids, these workers found marked differences between a sample of *E. canescens* and 11 other species from which flavonoids were isolated. The South American specimen yielded the two flavanones [599 and 600], but none of the flavone or flavonol derivatives was exhibited by most of the other species.

7.5 *Eschscholzia californica* (Papaveraceae)

Eschscholzia californica Cham., the California poppy (and state flower), is one of the most noticeable elements of the Californian spring flowers, often covering entire fields in near monoculture. A detailed study of the taxon's genetics, patterns of variation, and adaptive strategies was published by Cook (1962). Among his observations on genetic systems is the basis of flower color. The majority of California populations have uniformly orange petals lacking yellow margins, a condition characteristic of the *JJ* genotype. By contrast, this condition has not been observed in any Chilean populations. Yellow petals with distinct orange spots at the base result when this locus is *jj*. Populations with this apparent genotype have been found along the central Californian coast near Morro Bay and farther north, and in Oregon by Cook (1962), and by Stebbins (in Friás L. et al., 1975), as well as in coastal areas of Chile. It was surmised that a high frequency of the *J* allele is characteristic of populations living in hot, dry interior areas, and that the recessive allele is more likely to be found in cooler, more moist habitats, such as those along the coast. The genetic situation in Chile contrasts sharply with that in California, in that the *JJ* genotype is much more likely to be found in cooler areas, whereas the frequency of the *j* allele is significantly higher in populations inhabiting hot, arid habitats. Friás L. et al., (1975) suggested that this situation might have been brought about by significant recombinations resulting in the *jj* genotype becoming linked with the genes responsible for survival in the hot, dry interior of the country. Anecdotal tales of introductions of the California poppy into Chile include its use to stabilize railroad rights of way [a 1907 newspaper account cited by Friás L. et al. (1975)], although those authors also imply that the introduction might have been earlier, possibly as early as 1895. A white-flowered form is also known and is likely to have been introduced via horticultural activities.

7.6 *Krameria* (Krameriaceae)

Krameria L. consists of 15 (Mabberley, 1997, p. 383) or about 17 species (Simpson et al., 1979), with a disjunct distribution between central United States (Kansas) and northern Argentina to Chile. Two reports of lipids from flower glands of several species representing both regions have been published, the North American members by Seigler et al.

(1978), and seven species from South America by Simpson et al. (1979). In all cases, the lipid fractions consisted solely of 3-acetoxy fatty acids based on C_{16}, C_{18}, and C_{20} chains, for example, [601] (see Fig. 7.3 for structure). The authors concluded, quite reasonably, that the possession of identical profiles of unusual fatty-acid derivatives argues for a close relationship among the species of the two disjunct areas and further, that these compounds are likely to have been present in the ancestral taxon.

7.7 *Larrea* (Zygophyllaceae)

The genus *Larrea*, the creosote bush, has been the subject of extensive examination from many points of view, not the least of which is the origin of the North American–South American disjunction (see Hunziker et al., 1977 and Mabry et al., 1977 for detailed reviews). Although our primary interest involves the two disjunct taxa (or one with disjunct populations, depending upon one's taxonomic perspective), information on the entire genus is necessary to address the question of direction of evolution within the group. In the broader context, the genus consists of five species grouped into two sections on the basis of leaf and flower size. Four species occur in southern South America, and the fifth occurs in northern Mexico and the southwestern United States (jointly referred to as the "North American" taxon for convenience). Section *Larrea* consists of two species characterized by multifoliolate leaves and small flowers: *L. nitida* Cav. ($2n=26$), which occurs in western Argentina and in Chile, and *L. amieghinoi* Speg. ($2n=26$), which occurs in southern Argentina from Santa Cruz Province north to Neuquén. Section *Bifolium*, characterized by bifoliolate leaves and larger flowers, comprises the remaining three species: *L. divaricata* Cav. ($2n=26$) from the semidesert of central Argentina and disjunctly west of the Andes in isolated pockets in Chile and Peru; *L. cuneifolia* Cav. ($2n=52$) from the central desert of Argentina (the Monte); and *L. tridentata* (DC.) Colville, the North American species.

Larrea tridentata is cytologically the most complex member of the genus with diploids ($2n=26$) in the Chihuahuan Desert, tetraploids ($2n=52$) in the Sonoran Desert, and hexaploids ($2n=78$) in the Mojave Desert. Some authorities prefer to consider the North American *L. tridentata* and the South American *L. divaricata* as conspecific, a suggestion based upon close similarities in morphology, and supported by their capacity for forming fertile hybrids. Electrophoretic data, reviewed briefly by Hunziker et al. (1977), clearly distinguished among the South American species and, because of the high degree of similarity of their protein profiles, provided additional support for the hypothesis that *L. divaricata* and *L. tridentata* are closely related, if not conspecific. Moreover, the electrophoretic data suggested that *L. divaricata* from northern Patagonia is most closely related to the diploid form of *L. tridentata* from New Mexico. Comparisons of chromatographic profiles of all five taxa also revealed significant differences among the South American taxa and a high degree of similarity between *L. divaricata* and *L. tridentata*. Details of the chemical composition of *Larrea* can be found in Mabry et al. (1977).

Extraction of the surface of fresh leaves of creosote bush gave a rich array of flavonoid aglycones based on the flavonols such as kaempferol, quercetin, herbacetin, and gossypetin, and on the flavones such as apigenin and luteolin. An unusual dihydroflavonol was also obtained. Representative members of this array are illustrated in Fig. 7.4: luteolin 7,3′-dimethyl ether [602], kaempferol 3,7-dimethyl ether [603], quercetin 7,3′,4′-trimethyl ether [604], herbacetin 3,7-dimethyl ether [605], gossypetin 3,7,3′-trimethyl ether [606], and dihydromyricetin 3′,5′-dimethyl ether [607]. The array of leaf-surface flavonoids of *L. tridentata* was the most complex of any of the five species and, significantly, there were no differences among populations that exhibit different ploidy levels. In contrast, *L. divaricata* exhibited two flavonoid profiles, the more complex of which was observed in plants from Argentina; plants from Peru exhibited a decidedly simpler profile. The distribution of nonpolar compounds identified from these taxa is presented in Table 7.1.

The similarity between the profiles of the North American and Argentinean species suggests the possibility of a closer relationship between them than between any other pair of taxa in the genus, including between the Argentinean and Peruvian

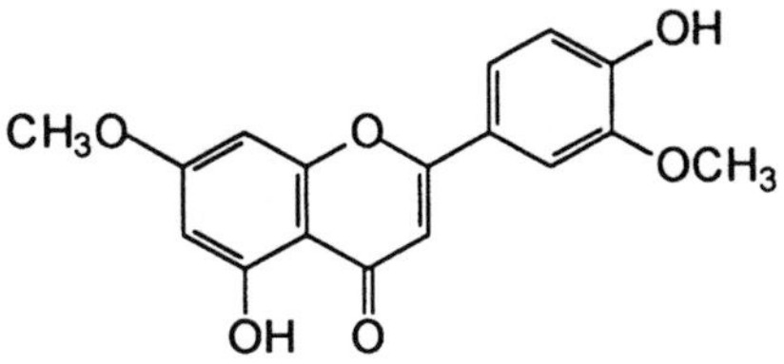

[602]

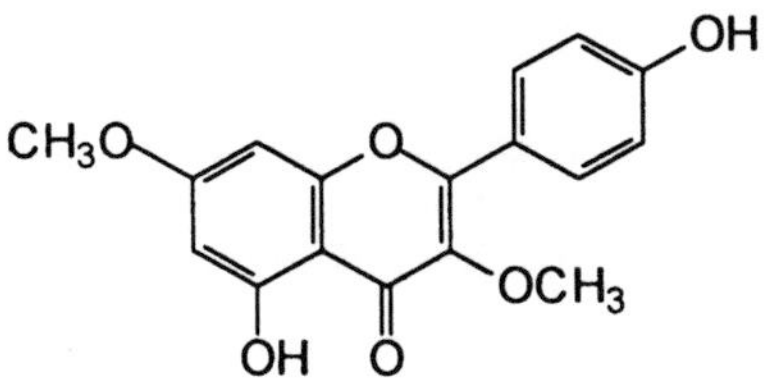

[603]

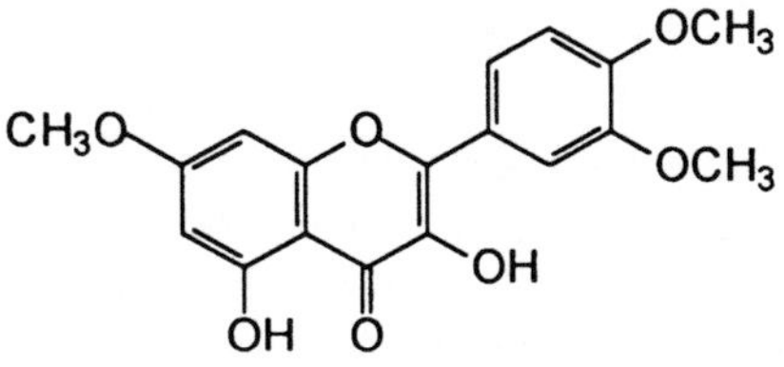

[604]

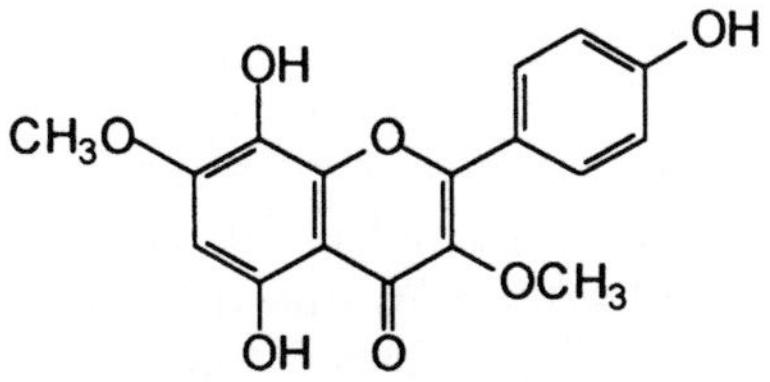

[605]

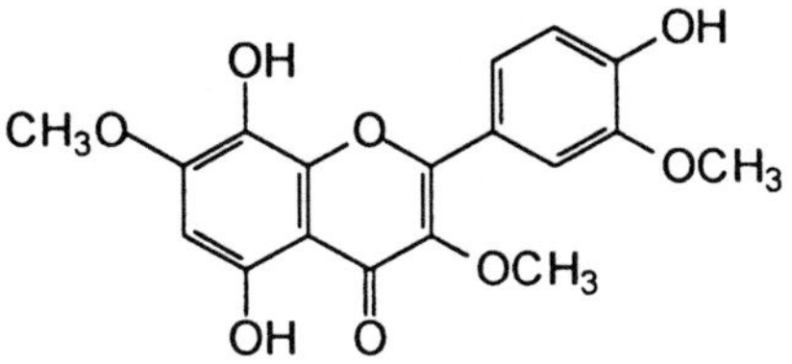

[606]

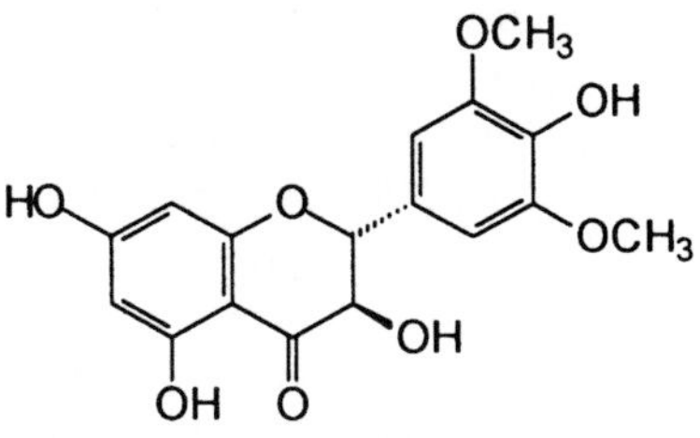

[607]

Fig. 7.4 Compounds 602–607, flavonoids from *Larrea tridentata*

Table 7.1 Flavonoid aglycones from leaf surfaces of *Larrea tridentata* and *L. divaricata* (from Mabry et al., 1977)

Flavonoid	*Larrea* taxon studied[a]		
	TRID	DIV-A	DIV-P
Flavones			
Apigenin	+	+	−
Apigenin 7-methyl ether	+	+	−
Luteolin 7-methyl ether	+	+	−
Luteolin 3′-methyl ether	+	+	−
Flavonols			
Kaempferol	+	+	+
Kaempferol 3-methyl ether	+	+	+
Kaempferol 7-methyl ether	+	+	+
Kaempferol 3,7-dimethyl ether	+	+	−
Kaempferol 3,4′-dimethyl ether	+	−	+
Quercetin 3′-methyl ether	+	+	+
Quercetin 3,7-dimethyl ether	+	+	−
Quercetin 3,3′-dimethyl ether	+	+	+
Quercetin 7,3′-dimethyl ether	+	+	−
Quer. 3,7,3′-trimethyl ether	+	+	−
Quer. 7,3′,4′-trimethyl ether	+	+	−
Quer. 3,7,3′,4′-tetramethyl ether	+	+	−
Herbacetin 3,7-dimethyl ether	+	−	+
Gossypetin 3,7-dimethyl ether	+	−	+
Goss. 3,7,3′-trimethyl ether	+	−	−
Dihydroflavonol			
Dihydromyr. 3,5′-dimethyl ether	+	+	+

[a] TRID=*L. tridentata*; DIV-A=*L. divaricata* from Argentina; DIV-P=*L. divaricata* from Peru.

specimens of *L. divaricata*. Mabry et al. (1977, and others cited therein) suggest that these results support the view that the isolation of the Peruvian population occurred much earlier than the separation of the North American species. One can surmise that the ancestral plants possessed the capacity for 8-oxygenation which gave rise to the formation of the two 8-hydroxy flavonols (herbacetin and gossypetin), which is the major structural difference by which the *L. tridentata-L. divaricata* pair can be distinguished from the other species. If, as suggested by Mabry et al. (1977), the Peruvian populations arose through migration independent of the events that gave rise to the North American taxon, the following scenario could account for the observed profile differences: (1) appearance of 8-hydroxylation in *L. divaricata*, or a *L. divaricata* prototype; (2) dispersal to Peru, 8-hydroxylation maintained, some depletion of capacity for O-methylation, and loss of capacity to make flavones; (3) independent dispersal to North America, and 8-hydroxylation maintained; (4) loss of 8-hydroxylation by Argentinean plants; (5) polyploidization in North American plants without changes to flavonoid profile.

Flavonoids glycosides of the genus were also studied in detail and again *L. tridentata*, with 17 of the 18 identified compounds, exhibited the most complex array. The close relationship between *L. tridentata* and *L. divaricata* was again borne out by the accumulation of 16 of the flavonoid glycosides by the latter. The remaining three species yielded only seven or eight glycosides each. A further distinguishing feature was the observation that sulfated flavonoids were detected only in *L. tridentata* and *L. divaricata.* While affording additional data to support the similarity of these two species, flavonoid glycosides did not provide any further insight into their evolutionary history.

Speculation on the age of *L. tridentata* in North America is based to some extent on the lack of appropriate desert habitats until the end of the Wisconsin glaciation, which has been reckoned to be about 11,000–12,000 years ago at lower elevations and about 9000 years ago at higher elevations. Exploration of rat middens revealed no material of *Larrea* older than about 11,000 years. It seems likely that propagules of *Larrea*, most likely dispersed by birds, arriving at the time of the warming and drying of the regions now occupied by deserts, found new habitats, became established, and differentiated cytologically into the races that now exist.

7.8 *Lasthenia* (Asteraceae)

Lasthenia, a genus visited before in this review, is western North American in occurrence except for *L. kunthii* (Lessing) Hook. & Arn. (the type species), which occurs in Chile. *Lasthenia kunthii* bears strong morphological similarities to *L. glaberrima* DC., one of the Californian species. The two species also have the same chromosome number ($n=5$) and form highly fertile hybrids in artificial crosses (Ornduff, 1963), all of which led Ornduff (1966) to place them together in sect. *Lasthenia.* A survey of flavonoids of the genus (Bohm et al., 1974) revealed that the pigment profiles of these two taxa were among the simplest of the genus lacking both anthochlors (aurones and chalcones) that characterize sections *Baeria* and *Burrielia*, and the patuletin (6-methoxyquercetin) derivatives seen in some members of sect. *Ptilomeris* and in sect. *Platycarpha.* The compounds observed in *L. kunthii* and *L. glaberrima* are based on quercetin, with quercetin 3-*O*-galactoside held in common. *Lasthenia glaberrima* also exhibited the 3-*O*-glucuronide and a 3-*O*-rhamnosylgalactoside, whereas *L. kunthii* afforded only one other compound, a 3-*O*-galactosylglucoside.

It is likely significant that *L. glaberrima* is self-compatible, whereas most other species of *Lasthenia* are self-incompatible. Long-distance dispersal of a single propagule of *L. glaberrima* would then have been sufficient to establish the species in Chile. It is, of course, entirely possible that propagules from one or another of the self-incompatible species could also have made the journey, but without a reproductively compatible partner their fates were sealed.

7.9 *Perityle emoryi* (Asteraceae)

Perityle is a genus consisting of ca. 63 species distributed primarily in western North America and Mexico, with one species in Chile and Peru (Karis and Ryding, 1994). The species of interest is *P. emoryi* Torr., an annual, polyploid, weedy species that occurs throughout much of the range of the genus. Specimens of this taxon from Peru, Sonora, Mexico, Baja California, Mexico, and southern California were analyzed for their flavonoid aglycones by Crins et al. (1988b), who identified a series of O-methylated flavonols and flavanones. The simplest compounds identified were kaempferol 3-mono- and 3,4′-dimethyl ethers; all others possessed 6-O-methoxylation. The sole flavanone isolated was identified as eriodictyol 7-methyl ether.

The authors of the survey attempted to infer phylogenetic relationships using several parsimony criteria. Although a reasonably parsimonious set of relationships among the members of the subtribe (the study included *Amauria*, *Pericome*, and three sections of *Perityle*) emerged from the treatment, serious problems surround this type of analysis, however, insofar as only a single set of data was involved. Adding to the problem is the possibility of highly reticulate biosynthetic pathways: the 6-oxygenation and O-methylation profiles can be arrived at by various combinations of the two processes. The authors pointed out that there is an increase in the complexity of the flavonoid profile of *P. emoryi* as one goes northward in its range. This trend has been noted in other taxa: *Arnica* (Wolf and Denford, 1983), *Chenopodium* (Crawford and Mabry, 1978), *Chrysosplenium* (Bohm and Collins, 1979), and *Phlox* (Levy, 1983). A final comment on this species' lifestyle might also be significant. *Perityle emoryi* is a self-compatible, polyploid (hexaploid and octaploid) (see Powell, 1974) that may have undergone fixation of local races, chemical in the present context, as it spread throughout the western parts of North America, Mexico, and parts of South America.

7.10 *Tradescantia* (Commelinaceae)

Tradescantia, a genus of perhaps 70 species (Mabberley, 1997, p. 720), occurs exclusively in the New World, but therein enjoys an interesting disjunct distribution. For the purpose of the present discussion we can follow Del Pero Martinez and Martinez (1993) in their description of three major areas of occurrence: (1) North America to the Mexican border; (2) Texas, New Mexico, and Mexico (excluding Baja California and much of northwestern Sonora) to the border with Guatemala; and (3) South America. Those workers examined 90 individuals representing 42 species from the entire range of the genus. A rich array of phenolic compounds was reported based upon C-glycosyl derivatives of apigenin and luteolin, flavone *O*-glycosides based upon luteolin, tricin (5,7,4′-trihydroxy-3′,5′-dimethoxyflavone) [608], and 6- and 8-hydroxyluteolin [609 and 610, respectively] (see Fig. 7.5 for structures 608–610), O-glycosides of kaempferol, quercetin, and isorhamnetin, and hydroxycinnamoyl glucose ester sulfates.

Three distinct chemotypes were observed. Taxa from each of the three areas noted above exhibited unique combinations of the compounds and compound types, although there was a degree of variation within most groups, as one might expect. The North American chemotype was characterized by very high frequency of the sulfated cinnamic-acid derivatives, a frequency of flavonol derivatives of

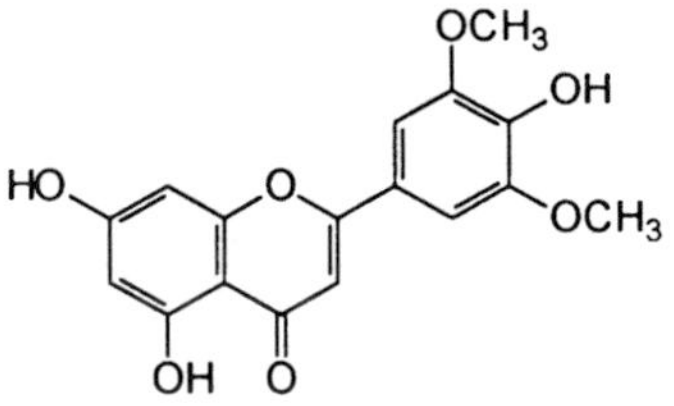

[608]

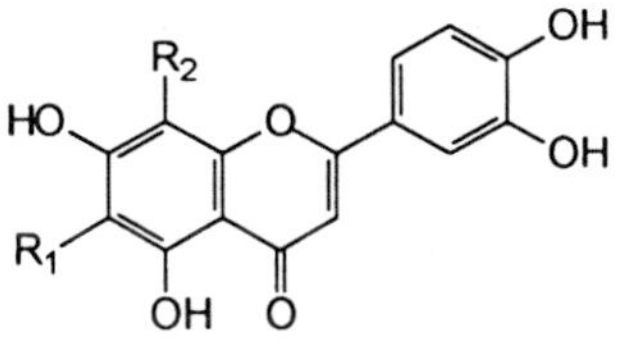

[609] R_1 = Glucose, R_2 = H
[610] R_1 = H, R_2 = Glucose

Fig. 7.5 Compounds 608–610, flavonoids of *Tradescantia*

about 0.25, a frequency of 6-hydroxylated flavones of about 0.12, a frequency of C-glycosylflavones of about 0.65, and the absence of B-ring trihydroxylated compounds. Mexican taxa exhibited 6-hydroxyflavones at a very high frequency, C-glycosylflavones at a frequency of about 0.30, and a low level of trihydroxylated compounds, but, significantly, lacked both flavonols and the sulfated cinnamic acids. Taxa from South America were rich sources of *C*-glycosylflavones (frequency of 1.0), but had only low frequencies of trihydroxylated compounds and sulfated acids, and lacked 6-hydroxylated flavones and flavonols altogether. As has been pointed out in numerous other examples in this review, these kinds of data are very difficult to interpret on their own; phylogenetic hypotheses are needed before further progress can be made.

Chapter 8
Conclusions

Some years ago I read an essay entitled, as I recall, "A sequence of saviors," that outlined the history of experimental systematics, pointing out that each new technique was at the time heralded by proponents as *the* technique that would, once and for all, set the subject on the road to objective respectability. [Unfortunately, I can recall neither the source nor the writer.] Chemotaxonomy, or chemical systematics, played its part in this sequence, as did numerical taxonomy (the two of them often together), chromosome features, breeding biology, protein electrophoresis, amino-acid sequence, and now, of course, DNA sequence analysis. Although there is a strong bias in today's systematics toward DNA, the other techniques have not lost their inherent usefulness. Indeed, the successful laboratory today will utilize whatever data that prove useful. Thus, as the examples listed above indicate, geographic patterning of secondary metabolites can often help illuminate potential relationships, or draw attention to relationships that may be incorrect. Every little bit helps.

B.A. Bohm, *The Geography of Phytochemical Races*,

Bibliography

Abbott, R. J. 1981. The keel petal colour polymorphism of *Lotus corniculatus* L. in Scotland. New Phytol. **88**: 549–553.

Abe, T., Masuda, M., Suzuki, T. and Suzuki, M. 1999. Chemical races in the red alga *Laurencia nipponica* (Rhodomelaceae, Ceramales) Phycol. Res. **47**: 87–95.

Abraham, A., Kirson, I., Glotter, E. and Lavie, D. 1968. A chemotaxonomic study of *Withania somnifera* (L.) Dun. Phytochemistry **7**: 957–962.

Abrahamson, W. G. and Solbrig, O. T. 1970. Soil preferences and variation in flavonoid pigments in species of *Aster.* Rhodora **72**: 251–263.

Adams, R. P. 1986. Geographic variation in *Juniperus silicicola* and *J. virginiana* of the southeastern United States: multivariate analysis of morphology and terpenoids. Taxon **35**: 61–75.

———**, Neisess, K. R., Parkhurst, R. M., Makhubu, L. P. and Wolde-Yohannes, L.** 1989. *Phytolacca dodecandra* (Phytolaccaceae) in Africa: geographic variation. Taxon **38**: 17–26.

———**, Parkhurst, R. M., Wolde-Yohannes, L. and Makhubu, L. P.** 1990. *Phytolacca dodecandra* (Phytolaccaceae) in Africa: geographical variation in leaf chemistry. Biochem. Syst. Ecol. **18**: 429–433.

Adersen, A., Adersen, H. and Brimer, L. 1988. Cyanogenic constituents in plants from the Galapagos Islands. Biochem. Syst. Ecol. **16**: 65–77.

———**, Brimer, L., Olsen, C. E. and Jaroszewski, J. W.** 1993. Cyanogenesis of *Passiflora colinvauxii*, a species endemic to the Galapagos Islands. Phytochemistry **33**: 365–367.

Agosta, W. 1996. Bombardier Beetles and Fever Trees. Addison-Wesley, Reading, Massachusetts, pp. 84–89.

Aitken, K., Botero, J., Zwart, R., and Teasdale, R. 1998. Detection of genetic diversity using RAPD markers in the genus *Melaleuca*. Acta Hort. (ISHS). **461**: 209–218.

Allnutt, T. R., Newton, A. C., Lara, A., Premoli, A. C., Armesto, J. J., Vergara, R. and Gardner, M. 1999. Genetic variation in *Fitzroya cupressoides* (alerce) a threatened South American conifer. Mol. Ecol. **8**: 975–987.

Alonso-Amelot, M. E., Oliveros, A. and Calcagno-Pisarelli, M. P. 2004. Phenolics and condensed tannins in relation to altitude in neotropical *Pteridium* ssp. A field study in the Venezuelan Andes. Biochem. Syst. Ecol. **32**: 969–981.

———**, Perez-Mena, M., Calcagno, M. P. and Jaimes-Espinoza, R.** 1992. Quantitation of pterosins A and B, ptaquiloside, the main carcinogen of *Pteridium aquilinum* (L.) Kuhn by high pressure liquid chromatography. Phytochem. Anal. **3**: 160–164.

———**, Rodulfo-Baechler, S. and Jaimes-Espinoza, R.** 1995. Comparative dynamics of ptaquiloside and pterosin B in the two varieties (*caudatum* and *arachnoideum*) of neotropical bracken fern [*Pteridium aquilinum* (L.) Kuhn]. Biochem. Syst. Ecol. **23**: 709–716.

Alston, R. E. 1967. Biochemical systematics. Pages 197–305 in T. Dobzhansky, M. K. Hecht and W. C. Steere (eds.) Evolutionary Biology, Vol. 1. Appleton-Century-Crofts, New York.

——— **and Turner, B. L.** 1963. Biochemical Systematics. Prentice-Hall, Engle wood Cliffs, New Jersey.

Amico, V., Caccamese, S., Neri, P., Russo, G. and Foti, M. 1991. Brasilane-type sesquiterpenoids from the Mediterraanean red alga *Laurencia obtusa.* Phytochemistry **30**: 1921–1928.

Anderberg, A. A. 1994. Tribe Inuleae. Pages 273–291 in K. Bremer (ed.) Asteraceae: Cladistics and Classification. Timber Press, Portland, OR.

Andersen, L., Adersen, A. and Jaroszewski, J. W. 1998. Cyanogenesis of *Passiflora foetida* L. Phytochemistry **47**: 1049–1050.

Angelopoulou, D., Demetzos, C. and Perdetzolou, D. 2001. An interpopulation study of the essential oils of *Cistus parviflorus* L. growing in Crete (Greece). Biochem. Syst. Ecol. **29**: 405–415.

———, **Demetzos, C. and Perdetzoglou, D.** 2002. Diurnal and seasonal variation of the essential oil labdanes and clerodanes from *Cistus monspeliensis* L. leaves. Biochem. Syst. Ecol. **30**: 189–203.

Anguita, F. and Hernan, F. 2000. The Canary Islands origin: a unifying model. J. Volcanol. Geoth. Res. **103**: 1–26.

Arisawa, M., Hayashi, T., Shimizu, M. Morita, N., Kuze, S. and Ito, Y. 1991. Isolation and cytotoxicity of two new flavonoids from *Chrysosplenium grayanum* and related flavonols. J. Nat. Prod. **54**: 898–901.

———, **Shiojima, M., Bai, H., Hayashi, T., Tezuka, Y., T. Taga, Miwa, Y., Kikuchi, T. and Morita, N.** 1992. Chrysograyanone, a novel chromone derivative from *Chrysosplenium grayanum* Maxim. Tetrahedron Lett. **33**: 5977–5980.

———, **Bai, H., Shiojima, M., Hayashi, T., Tezuka, Y., Taga, T., Miwa, Y. Ito, Y., Kikuchi, T. and Morita, N.** 1993a. Novel flavonoids from *Chrysosplenium grayanum* Maxim. Chem. Pharm. Bull. (Tokyo) **41**: 571–574.

———, **Takeshima, Y., Bai, H., Hayashi, T. and Morita, N.** 1993b. Isolation and identification of cytotoxic principles from *Chrysosplenium japonicum* Maxim. (Saxifragaceae). Shoyakugaku Zasshi **47**: 334–337.

———, **Hatashita, T., Numata, Y., Tanaka, M. and Sasaki, T.** 1997. Cytotoxic principles from *Chrysosplenium flagelliferum.* Int. J. Pharmacogn. **35**: 141–143.

Armstrong, H. E., Armstrong, F. and Horton, E. 1912. Herbage studies. 1. *Lotus corniculatus*, a cyanophoric plant. Proc. Royal Soc. Series B **84**: 471–484.

Arslanian, R. L., Mondragon, B., Stermitz, F. R. and Marr, K. L. 1990. Acyl histamines and a rare protopine type alkaloid from leaves of *Zanthoxylum dipetalum.* Biochem. Syst. Ecol. **18**: 345–347.

Asakawa, Y. 1995. Chemistry of liverworts. Pages 1–562 in W. Herz, G. W. Kirby, R. E. Moore, W. Steglich and Ch. Tamm (eds.) Progress in the Chemistry of Organic Natural Products, Vol. 65. Springer Verlag, Vienna.

——— **and Inoue, H.** 1987. Page 119 in H. Inoue (ed.) Studies on Cryptograms in Southern Peru. Tokai University Press, Tokyo. [Cited by Rycroft et al., 1998.]

———, **Lin, X., Kondo, K., and Fukuyama, Y.** 1991. Terpenoids and aromatic compounds from selected East Malaysian liverworts. Phytochemistry **30**: 4019–4024.

———, **Tori, M., Takikawa, K., Krishnamurty, H. G. and Kar, S. K.** 1987. Cyclic bis(bibenzyls) and related compounds from the liverworts *Marchantia polymorpha* and *Marchantia palmata.* Phytochemistry **26**: 1811–1816.

———, **Toyota, M., Nakaishi, E. and Taka, Y.** 1997. Distribution of terpenoids and aromatic compounds in the liverwort *Isotachis* species. J. Hattori Bot. Lab. No. **83**: 257–263.

Atwood, S. S. and Sullivan, J. T. 1943. Inheritance of a cyanogenic glycoside and its hydrolyzing enzyme in *Trifolium repens.* J. Heredity **34**: 311–320.

Averett, J. E., Hahn, W. J., Berry, P. E. and Raven, P. H. 1986. Flavonoids and flavonoid evolution in *Fuchsia* (Onagraceae) Amer. J. Bot. **73**: 1525–1534.

Axelrod, D. I. 1979. Age and origin of the Sonoran Desert vegetation. California Acad. Sci. Occas. Papers, No. 132: 1–74.

Axelton, E. A. 1967. Ponderosa pine bibliography through 1965. U. S. Forest Service Res. Paper INT-40.

Ayyad, S. E. N., Dawidar, A. A. M., Dias, H. W., Howie, R. A., Jakupovic, J., and Thomson, R. H. 1990. Three halogenated metabolites from *Laurencia obtusa*. Phytochemistry **29**: 3193–3196.

Ayyad, S. E. N., Jakupovic, J. and Abdel-Mogib, M. 1994. A sesquiterpene ether from *Laurencia obtusa*. Phytochemistry **36**: 1077–1078.

Bailey, D. K. 1970. Phytogeography and taxonomy of *Pinus* subsection *Balfourianae*. Ann. Mo. Bot. Gar. **57**: 210–249.

Bailey, I. W. and Smith, A. C. 1942. Degeneriaceae, a new family of flowering plants from Fiji. J. Arnold Arbor. **23**: 356–365.

Baker, K. and Jones, D. A. 1986. Altitude and the keel petal polymorphism of *Lotus corniculatus* L. J. Nat. Hist. **20**: 1429–1433.

Bakker, F. T., Vassiliades, D. D., Morton, C. and Savolainen, V. 1998. Phylogenetic relationships of *Biebersteinia* Stephan (Geraniaceae) inferred from *rbc*L and *atp*B sequence comparisons. Bot. J. Linn. Soc. **127**: 149–158.

Baldwin, B. G., Kyhos, D. W. and Dvolrak, J. 1990. Chloroplast DNA evolution and adaptive radiation in the Hawaiian silversword alliance (Asteraceae—Madiinae). Ann. Missouri Bot. Gard. **77**: 96–109.

———, ———, ——— **and Carr, G. D.** 1991. Chloroplast DNA evidence for a North American origin of the Hawaiian silversword alliance (Asteraceae). Proc. Nat'l. Acad. Sci. USA **88**: 1840–1843.

Balza, F. and Towers, G. H. N. 1990. Dithiacyclohexadiene chlorohydrins and related sulphur-containing polyynes from *Ambrosia chamissonis*. Phytochemistry **29**: 2901–2904.

Baradat, P., Michelozzi, R., Tognetti, M. L. and Khaldi, A. 1995. Geographical variation in the terpene composition of *Pinus halepensis* Mill. Pages 141–158 in P. Baradat, W. T. Adams, and G. Müller-Stark (eds.) Population Genetics and Genetic Conservation of Forest Trees, SPB Academic Press, Amsterdam.

Barazani, O., Cohen, Y., Fait, A., Diminshtein, S., Dudai, N., Ravid, U., Putievsky, E. and Friedman H. 2002. Chemotypic differentiation in indigenous populations of *Foeniculum vulgare* var. *vulgare* in Israel. Biochem. Syst. Ecol. **30**: 721–731.

Barbour, M. G. 1970. Seedling ecology of *Cakile maritima* along the California coast. Bull. Torrey Bot. Club **97**: 280–289.

——— **and Rodman, J. E.** 1970. Saga of the west coast sea-rockets: *Cakile edentula* ssp. *californica* and *C. maritima*. Rhodora **72**: 370–386.

Barker, W. W. 1966. Apomixis in the genus *Arnica* (Compositae). Ph.D. dissertation. University of Washington, Seattle, WA.

Barzani, O., Cohen, Y., Fait, A., Diminshtein, S., Dudai, N., Ravid, U., Putievsky, E. and Friedman, J. 2002. Chemotypic differentiation in indigenous populations of *Foeniculum vulgare* var. *vulgare* in Israel. Biochem. Syst. Ecol. **30**: 721–731.

Bate-Smith, E. C. 1962. The phenolic constituents of plants and their taxonomic significance. I. Dicotyledons. J. Linn. Soc. Bot. **58**: 95–173.

———. 1968. The phenolic constituents of plants and their taxonomic significance. II. Monocotyledons. J. Linn. Soc. Bot. **60**: 325–356.

———. 1973. Chemotaxonomy of geranium. J. Linn. Soc. Bot. **67**: 347–359.

———, **Davenport, S. M. and Harborne, J. B.** 1967. Comparative biochemistry of flavonoids. III. A correlation between chemistry and plant geography in the genus *Eucryphia*. Phytochemistry **6**: 1407–1413.

———, **Harborne, J. B. and Davenport, S. M.** 1966. Identification of quercetin 5-methyl ether (azaleatin) from *Eucryphia glutinosa*. Nature **212**: 1065.

Bayer, R. J. 1990. Investigations into the evolutionary history of the *Antennaria rosea* complex (Asteraceae: Inuleae) polyploid complex. Plant Syst. Evol. **169**: 97–110.

Beker, R., Dafni, A., Eiskowitch, D. and Ravid, U. 1989. Volatiles of two chemotypes of *Majorana syriaca* L. (Labiatae) as olfactory cues for the honeybee. Oecologia **79**: 446–451.

Behnke, H.-D. and Turner, B. L. 1971. On specific sieve-tube plastids in Caryophyllales. Taxon **20**: 731–737.

Bennett, K. D. 1995. Post-glacial dynamics of pine (*Pinus sylvestris* L.) and pinewoods in Scotland. Pages 23–39 in J. E. Aldous (ed.), Our Pinewood Heritage. Bell and Bain, Glasgow.

Bergström, G., Birgersson, G., Groth, I. and Nilsson, L. A. 1992. Floral fragrance disparity between three taxa of lady's slipper *Cypripedium calceolus* (Orchidaceae). Phytochemistry **31**: 2315–2319.

Berry, P. E. 1982. The systematics and evolution of *Fuchsia* sect. *Fuchsia* (Onagraceae). Ann. Missouri Bot. Gard. **69**: 1–198.

———**, Skvarla, J. J., Partridge, A. D. and Macphail, M. K.** 1990. *Fuchsia* pollen from the Tertiary of Australia. Australian Syst. Bot. **3**: 739–744.

Biddulph, S. F. 1944. A revision of the genus *Gaillardia*. Research Studies State College, Washington **12**: 195–256. [Cited by Heywood and Levin, 1984.]

Binns, S. E., Arnason, J. T. and Baum, B. R. 2002. Phytochemical variation within populations of *Echinacea angustifolia* (Asteraceae). Biochem. Syst. Ecol. **30**: 837–854.

Birks, H. J. B. 1989. Holocene isochrone maps of tree spreading in the British Isles. J. Biogeography **18**: 103–115.

Blanquer, A., Boira, H., Soler, V. and Perez, I. 1998. Variability of the essential oil of *Thymus piperella*. Phytochemistry **47**: 1271–1276.

Blua, M. J., Hanscom, Z. and Collier, B. D. 1988. Glucocapparin variability among four populations of *Isomeris arborea* Nutt. J. Chem. Ecol. **14**: 623–633.

Blunt, J. W., Lake, R. J. and Munro, M. H. G. 1984. Sesquiterpenes from the marine red alga *Laurencia distichophylla*. Phytochemistry **23**: 1951–1954.

Boelens, M. H. and Jimenez, R. 1991. Chemical composition of the essential oil from the gum and various parts of *Pistacia lentiscus*. L. (Mastic Gum Tree). Flavour Fragrance J. **6**: 271–275.

Bohlmann, F., Burkhardt, T. and Zdero, C. 1973. Naturally Occurring Acetylenes. London: Academic Press.

Bohm, B. A. 1979. Flavonoids of *Tolmiea menziesii*. Phytochemistry **18**: 1079–1080.

———. 1987. Intraspecific flavonoid variation. Bot. Rev. **53**: 197–279.

———. 1988. Flavonoid systematics of the Datiscaceae. Biochem. Syst. Ecol. **16**: 151–155.

———. 1998a. Introduction to Flavonoids. Harwood Academic Press, Amsterdam.

———. 1998b. Secondary compounds and evolutionary relationships of island plants. Pages 233–306 in T. F. Stuessy and M. Ono (eds.), Evolution and Speciation of Island Plants. Cambridge University Press, Cambridge, UK.

———. 1999. Major exudate flavonoids of *Dubautia arborea* (Asteraceae). Biochem. Syst. Ecol. **27**: 755–757.

——— **and Banek, H. M.** 1987. Flavonoid variation in *Lasthenia burkei*. Biochem. Syst. Ecol. **15**: 57–59.

———, ——— **and Maze, J. R.** 1984. Flavonoid variation in North American *Menziesia* (Ericaceae). Syst. Bot. **9**: 324–345.

———**, Brim, S. W., Hebda, R. J. and Stevens, P. F.** 1978. Generic limits in the tribe Cladothamneae (Ericaceae), and its position in the Rhododendroideae. J. Arnold Arbor. **59**: 311–341.

——— **and Collins, F. W.** 1979. Flavonoids of some species of *Chrysosplenium*. Biochem. Syst. Ecol. **7**: 195–201.

———, ——— **and Bose, R.** 1977. Flavonoids of *Chrysosplenium tetrandrum*. Phytochemistry **16**: 1205–1209.

——— **and Fong, C.** 1990. Nonpolar flavonoids of *Wilkesia* and *Argyroxiphium*. Phytochemistry **29**: 1175–1177.

———, ———**, Hiebert, M., Jamal, A. and Crins, W. J.** 1992. Non-polar flavonoids of *Calycadenia*, *Lagophylla* and *Madia*. Phytochemistry **31**: 1261–1263.

———**, Herring, A., Nicholls, K. W., Bohm, L. R. and Ornduff, R.** 1989. A six-year study of flavonoid distribution in a population of *Lasthenia californica* (Asteraceae). Amer. J. Bot. **76**: 157–163.

——— **and Koupai-Abyazani, M.** 1994. Flavonoids and condensed tannins from leaves of Hawaiian *Vaccinium reticulatum* and *V. calycinum* (Ericaceae). Pac. Sci. **48**: 458–463.

———, **Nicholls, K. W. and Ornduff, R.** 1986. Flavonoids of the Menyanthaceae: intra- and interfamilial relationships. Amer. J. Bot. **73**: 204–213.

——— **and Ornduff, R.** 1981. Leaf flavonoids and ordinal affinities of Coriariaceae. Syst. Bot. **6**: 15–26.

——— **and Saleh, N. A. M.** 1972. The flavonoids of *Cladothamnus pyrolaeflorus*. Can. J. Bot. **50**: 2081–2083.

———, ——— **and Ornduff, R.** 1974. The flavonoid chemistry of *Lasthenia* (Compositae). Amer. J. Bot. **61**: 551–561.

——— **and Stuessy, T. F.** 2001. Flavonoids of the Sunflower Family (Asteraceae). Springer-Verlag. Vienna.

——— **and Tryon, R. M.** 1967. Phenolic compounds in ferns. I. A survey of some ferns for cinnamic acid and benzoic acid derivatives. Can. J. Bot. **45**: 585–593.

Borokowski, B., Kowelewski, S. and Skrzypczakowa, L. 1966. Chemical composition of inflorescence of some *Arnica* species. I. Analysis of flavonoid fractions. Diss. Pharm. Pharmacol. **18**: 367–374.

Boufford, D. E. and Spongberg, S. A. 1983. Eastern Asian–eastern North American phytogeographical realtionships—A history from the time of Linnaeus to the twentieth century. Ann. Missouri Bot. Gard. **70**: 423–439.

Bowen, E. and Van Vuren, D. 1997. Insular endemic plants lack defenses against herbivores. Conserv. Biol. **11**: 1249–1254.

Bowman, E. I. (ed.) 1966. The Galapagos. Proceedings of the Symposia of the Galapagos International Scientific Project. University of California Press, Berkeley.

Bowman, R. I., Berson, M., and Leviton, A. E. (eds.) 1983. Patterns of Evolution in Galapagos Organisms. American Association for the Advancement of Science, Pacific Division. San Francisco.

Bowman, R. N. 1983. Intraspecific variability of leaf cuticle alkanes in *Sedum lanceolatum* along an elevational gradient. Biochem. Syst. Ecol. **11**: 195–198.

Bramwell, D. 1972. Endemism in the flora of the Canary Islands. Pages 141–159 in D. H. Valentine (ed.) Taxonomy, phytogeography, and evolution. Academic Press, New York.

———. 1976. The endemic flora of the Canary Islands. Pages 207–240 In Biogeography and Ecology in the Canary Islands. Dr. W. Junk, The Hague.

———. 1985. Contribución a la biogeografia de las islas Canarias. Bótanica Macaronésica **14**: 3–34.

Breckon, G. J. and Barbour, M. G. 1974. Review of North American Pacific Coast beach vegetation. Madroño **22**: 333–360.

Briggs, B. G. and Johnson, L. A. S. 1998a. New genera and species of Australian Restionaceae (Poales). Telopea **7**: 345–373.

——— **and** ——— 1998b. New combinations arising from a new classification of non-African Restionaceae. Telopea **8**: 21–33.

Briggs, D. and Walters, S. M. 1997. Plant Variation and Evolution. Ed. 3. Cambridge University Press, Cambridge.

Buil, P., Garnero, J., and Guichard, G. 1975. Contribution à la connaissance de la composition chimique de l'essence de lentisque de Provence. Riv. Ital. EPPOS Cosmet. Aerosol. **56**: 245–252 [cited by Castola et al., 2000].

Burbidge, F. W. 1903. The leaf-marking of *Arum maculatum*. Irish Nat., Dublin **XII**: 137.

Burke, B. A., Goldsby, G. and, Mudd, J. B. 1987. Polar epicuticular lipids of *Lycopersicon pennellii*. Phytochemistry **26**: 2567–2571.

Butcher, P. A., Bell, J. C. and Moran, G. F. 1992. Patterns of genetic diversity and nature of the breeding system in *Melaleuca alternifolia* (Myrtaceae). *Australian J. Bot.* **40**: 365–375.

———, **Doran, J. C., and Slee, M. U.** 1994. Intraspecific variation in leaf oils of *Melaleuca alternifolia* (Myrtaceae). Biochem. Syst. Ecol. **22**: 419–430.

Cabido, M., Ateca, N., Astegiano, M. E. and Anton, A. M. 1997. Distribution of C_3 and C_4 grasses along an altitudinal gradient in central Argentina. J. Biogeography **24**: 197–204.

Caccamese, S., R. Azzolina, R. M. Toscano and K. L. Rinehart, Jr. 1981. Variations in the halogenated metabolites of *Laurencia obtusa* from eastern Sicily. Biochem. Syst. Ecol. **9**: 241–246.

———, **Compagnini, A. and Toscana R. M.** 1986. Paciferol from the Mediterranean red alga *Laurencia majuscula*. J. Nat. Prod. **49**: 173–174.

———, ———, ———, **Nicoló, R. M. and Chapuis G.** 1987. A new labile bromoterpenoid from the red alga *Laurencia majuscula*: dehydrochloroprepacifenol. Tetrahedron **43**: 5393–5399.

Calabro, G. and P. Curro. 1974. Constituenti degli oil essenziali Nota IV. Essenza di lentisco. Essence Deriv. Agrum. **44**: 82–92.

Calamassi, R. 1986. Charactérisation de quelques provenances de *Pinus halepensis* Mill. sur la base de la structure anatomique et morphologique des aiguilles. Ann. Sci. **43**: 281–298.

Calder, J. A. and Taylor, R. L. 1956. New taxa and nomenclatural changes with respect to the flora of the Queen Charlotte Islands, British Columbia. Can. J. Bot. **43**: 1387–1400.

——— **and** ——— 1968. Flora of the Queen Charlotte Islands. Part 1. Canada Department of Agriculture, Monograph 4, Ottawa, Canada.

Camp, W. H. 1949. *Cinchona* at high altitudes in Ecuador. Brittonia **6**: 394–430.

Cañigueral, S., R. Vila, G. Vicario, X. Tomas and T. Adzet. 1994. Chemometrics and essential oil analysis: chemical polymorphism in two *Thymus* species. Biochem. Syst. Ecol. **22**: 307–315.

Capon, R. J., E. L. Ghisalberti, T. A. Mori and P. R. Jefferies. 1988. Sesquiterpenes from *Laurencia* spp. J. Nat. Prod. **51**: 1302–1304.

Carlquist, S. 1959a. Vegetative anatomy of *Dubautia, Argyroxiphium* and *Wilkesia* (Compositae). Pac. Sci. **13**: 195–210.

———. 1959b. Studies on Madiinae: anatomy, cytology and evolutionary relationships. *Aliso* **4**: 171–236.

———. 1966. The biota of long-distance dispersal. IV. Genetic systems in the flora of oceanic islands. Evolution **20**: 433–455.

———. 1967. The biota of long-distance dispersal. V. Plant dispersal to Pacific Islands. Bull. Torrey Bot. Club **94**: 129–162.

———. 1974. Island biology. Columbia University Press, New York.

———. 1980. Hawaii: A Natural History: Geology, Climate, Native Flora and Fauna above the Shoreline, Ed 2. Pacific Tropical Botanical Garden, Honolulu.

———. 1983. Intercontinental dispersal. Pages 37–47 in K. Kubitzki (ed.) Dispersal and Distribution. Paul Parey, Hamburg, Germany.

———, **Baldwin, B. G. and Carr, G. D.** (eds.) 2003. Tarweeds & Silverswords. Evolution of the Madiinae (Asteraceae). Missouri Botanical Garden Press, St. Louis, Missouri.

Carr, G. D. 1985. Monograph of the Hawaiian Madiinae (Asteraceae): *Argyroxiphium*, *Dubautia* and *Wilkesia*. Allertonia **4**: 1–123.

———. 1995. A fully fertile intergeneric hybrid derivative from *Argyroxiphium sandwicense* ssp. *macrocephalum* X *Dubautia menziesii* (Asteraceae) and its relevance to plant evolution in the Hawaiian Islands. Amer. J. Bot. **82**: 1574–1581.

——— **and Kyhos, D. W.** 1981. Adaptive radiation in the Hawaiian silversword alliance (Compositae-Madiinae). I. Cytogenetics of spontaneous hybrids. Evolution **35**: 543–556.

——— **and** ———. 1986. Adaptive radiation in the Hawaiian silversword alliance (Compositae-Madiinae). II. Cytogenetics of artificial and natural hybrids. Evolution **40**: 959–976.

———, **Baldwin, B. G and Kyhos, D. W.** 1996. Cytogenetic implications of artificial hybrids between Hawaiian silversword alliance and North American tarweeds (Asteraceae: Heliantheae—Madiinae) Amer. J. Bot. **83**: 653–660.

Carracedo, J. C. 1994. The Canary Islands: an example of structural control on the growth of large oceanic-island volcanoes. J. Volcan. Geotherm. Res. **60**: 225–241.

Carson, H. L. and Clague, D. A. 1995. Geology and biogeography of the Hawaian Islands. Pages 14–29 in W. L. Wagner and V. A. Funk (eds.), Hawaiian Biogeography: Evolution on a Hot Spot Archipelago. Smithsonian Institution Press, Washington, D.C.

Carter, G. T., K. L. Rinehart, Jr., H. Li, S. L. Kuentzel and J. L. Conner. 1978. Brominated indoles from *Laurencia brongniartii*. Tetrahedron Lett. 4479–4482.

Castola, V., A. Bighelli and Casanova, J. 2000. Intraspecific chemical variability of the essential oil of *Pistacia lentiscus* L. from Corsica. Biochem. Syst. Ecol. **28**: 79–88.

Cavagnaro, J. B. 1988. Distribution of C_3 and C_4 grasses at different altitudes in a temperate arid region of Argentina. Oecologia **76**: 273–277.

Caveney, S., Charlet, D. A., Freitag, H., Maier-Stolte, M. and Starratt, A. N. 2001. New observations on the secondary chemistry of world *Ephedra* (Ephedraceae). Amer. J. Bot. **88**: 1199–1208.

Chandra, P. and Purohit, A. N. 1980. Berberine contents and alkaloid profile of *Berberis* species from different altitudes. Biochem. Syst. Ecol. **8**: 379–380.

Chang, C.-S. and Giannasi, D. E. 1991. Foliar flavonoids of *Acer* sect. *Palmata* series *Palmata*. Syst. Bot. **16**: 225–241.

——— **and Jeon, J. I.** 2003. Leaf flavonoids in *Cotoneaster wilsonii* (Rosaceae) from the island Ulleung-do, Korea. Biochem. Syst. Ecol. **31**: 171–179.

Chau, P. and Wu, C.-L. 1987. Proc. Nat'l. Sci. Council, Republic of Korea (A) **11**: 124. [Cited by Asakawa et al., 1991].

Chazdon, R. L. 1978. Ecological aspects of the distribution of C_4 grasses in selected habitats of Costa Rica. *Biotropica* **10**: 265–269.

Chen, C., Sun, H. and Mizuno, M. 1992. On the genus *Euchresta* Benn (Leguminosae) with "Wallaces line." Acta Phytotax. Sin. **30**: 43–56.

Chew, W. L. 1972. The genus *Piper* (Piperaceae) in New Guinea, Soloman Islands and Australia. J. Arnold Arbor. **53**: 1–25.

Cho, H.-J., Kim, S., Suh, Y. and Park, C.-W. 1996. ITS sequences of some *Acer* species and phylogenetic implication. Korean J. Plant Tax. **26**: 271–291.

Christensen, L. P. 1992. Acetylenes and related compounds in Anthemideae. Phytochemistry **31**: 7–49.

———**, and Lam, J.** 1990. Acetylenes and related compounds in Cynareae. Phytochemistry **29**: 2753–2785.

——— **and** ———. 1991a. Acetylenes and related compounds in Heliantheae. Phytochemistry **30**: 11–49.

——— **and** ———. 1991b Acetylenes and related compounds in Astereae. Phytochemistry **30**: 2453–2476.

Christie, D. M., Duncan, R. A., McBirney, A. R., Richards, M. A., White, M. W., Harpp, K. S. and Fox, C. G. 1992. Drowned islands downstream from the Galapagos hotspot imply extended speciation times. Nature **355**: 246–248.

Clague, D. A. and Dalrymple, G. B. 1987. The Hawaiian-Emperor volcanic chain. Pages 1–54 in R. W. Decker, T. L. Wright and P. H. Stauffer (eds.) Volcanism in Hawaii. U. S. Geological Survey Paper 1350, U.S. Government Printing Office, Washington, D.C.

Coassini Lokar, L. and Moneghini, M. 1989. Geographical variation in the monoterpenes of *Valeriana officinalis* leaf. Biochem. Syst. Ecol. **17**: 563–567.

Colegate, S. M., Dorling, P. R. and Huxtable, C. R. 1979. A spectroscopic investigation of swainsonine: an α-mannosidase inhibitor isolated from *Swainsona canescens*, Austral. J. Chem. **32**: 2257–2264.

Coley, P. D., Bryant, J. P. and Chapin, F. S. 1985. Resource availability and plant antiherbivore defense. Science **230**: 895–899.

Colgan, N. 1903. The leaf-marking of *Arum maculatum*. Irish Nat. Dublin **XII**: 78–81.

Coll, J. C. and Wright, A. D. 1987. Tropical marine algae. I. New halogenated monoterpenes from *Chondrococcus hornemannii* (Rhodophyta, Gigartinales, Rhizophyllidaceae). Australian J. Chem. **40**:1893–1900.

——— **and** ———. 1989. Tropical marine algae. III. New sesquiterpenes from *Laurencia majuscula* (Rhodophyta, Rhodophyceae, Ceramales, Rhodomelaceae). Aust. J. Chem. **42**: 1591–1603.

Collins, F. W., De Luca, V., Ibrahim, R. K., Voirin, B. and Jay, M. 1981. Polymethylated flavonols of *Chrysosplenium americanum*. I. Identification and enzymic synthesis. Zeit. Naturforsch. **36c**: 730–736.

Compadre, C. M., Pezzuto, J. M., Kinghorn, A. D. and Kamath, S. K. 1985. Hernandulcin: an intensely sweet compound discovered by review of ancient literature. Science **227**: 417–419.

———, **Hussain, R. A., López-de Compadre, R. L., Pezzuto, J. M., and Kinghorn, A. D.** 1987. The intensely sweet sesquiterpene hernandulcin: isolation, synthesis, characterization and preliminary safety evaluation. J. Agric. Food Chem. **35**: 273–279.

Conkle, M. T., Schiller, G. and Grunwald, C. 1988. Electrophoretic analysis of diversity and phylogeny of *Pinus brutia* and closely related taxa. Syst. Bot. **13**: 411–424.

Connor, H. E. and Purdie, A. W. 1976. Triterpene methyl ether differentiation in *Chionochloa* (Gramineae). N. Z. J. Bot. **14**: 315–326.

——— **and** ———. 1981. Triterpene methyl ethers in *Chionochloa* (Gramineae): distribution in western South Island, New Zealand. N. Z. J. Bot. **19**: 161–170.

Cook, S. 1962. Genetic system, variation and adaptation in *Eschscholzia californica.* Evolution **16**: 278–299.

Cool, L. G., Power, A. B. and Zavarin, E. 1991. Variability of foliage terpenes of *Fitzroya cupressoides.* Biochem. Syst. Ecol. **19**: 421–432.

——— **and Zavarin, E.** 1992. Terpene variability of mainland *Pinus radiata.* Biochem. Syst. Ecol. **20**: 133–144.

Cope, E. A. 1983. Chemosystematic affinities of a California population of *Abies lasiocarpa.* Madroño **30**: 110–114.

Copeland, H. F. 1943. A study, anatomic and taxonomic, of the genera of the Rhododendroideae. *Amer. Mid. Nat.* **30**: 533–625.

Corkill, L. 1942. The inheritance of cyanogenesis. N. Z. J. Sci. Tech., Series B **23**: 178–193.

Cowlishaw, M. G., Bickerstaffe, R. and Connor, H. E. 1983. Intraspecific variation in the epicuticular wax composition of four species of *Chionochloa.* Biochem. Syst. Ecol. **11**: 247–259.

Cox, A. 1983. Ages of the Galapagos Islands. Pages 11–23 in R. I. Bowman, M. Berson and A. E. Leviton (eds.) Patterns of Evolution in Galápagos Organisms. AAAS, Pacific Division. San Francisco.

Crawford, D. J. 1976. Variation in seed protein profiles of *Chenopodium fremontii.* Biochem. Syst. Ecol. **4**: 169–172.

———. 1990. Plant Molecular Systematics. Macromolecular Approaches. John Wiley & Sons, New York.

——— **and Julian, E. A.** 1976. Seed protein profiles in the narrow-leaved species of *Chenopodium* of the western United States: Taxonomic value and comparison with distribution of flavonoid compounds. Amer. J. Bot. **63**: 302–308.

——— **and Mabry, T. J.** 1978. Flavonoid chemistry of *Chenopodium fremontii.* Infraspecific variation and systematic implications at the interspecific level. Biochem. Syst. Ecol. **6**: 189–192.

———, **Stuessy, T. F. and Silva O. M.** 1987. Allozyme divergence and the evolution of *Dendroseris* (Compositae: Lactuceae) on the Juan Fernandez Islands. Syst. Bot. **12**: 435–443.

———, ———, **Cosner, M. B., Haines, D. W., and Silva O. M.** 1993. Ribosomal and chloroplast DNA rerstriction site mutations and the radiation of *Robinsonia* (Asteraceae: Senecioneae) on the Juan Fernandez Islands. Plant Syst. Evol. **184**: 233–239.

———, ———, ———, ———, ——— **and Baeza, M.** 1992a, Evolution of the genus *Dendroseris* (Asteraceae: Lactuceae) on the Juan Fernandez Islands: evidence from chloroplast and ribosomal DNA. Syst. Bot. **17**: 675–681.

———, ———, **Haines, D. W., Cosner, M. B., Silva O. M. and Lopez, P.** 1992b. Allozyme diversity within and divergence among four species of *Robinsonia* (Astereaceae: Senecioneae), a genus endemic to the Juan Fernandez Islands, Chile. Amer. J. Bot. **79**: 962–969.

——— and **Wilson, H. D.** 1977. Allozyme variation in *Chenopodium fremontii.* Syst. Bot. **2**: 180–190.

Crawford, T. J. and Jones, D. A. 1986. Variation in the colour of the keel petals in *Lotus corniculatus* L. 2. Clines in Yorkshire and adjacent counties. *Watsonia* **16**: 15–19.

——— **and** ———. 1988. Variation in the colour of the keel petals in *Lotus corniculatus* L. 4. Morph distribution in the British Isles. Heredity **61**: 175–188.

Crews, P., Campbell, L. and Heron, E. 1977. Different chemical types of *Plocamium violaceum* (Rhodophyta) from the Monterey Bay Region, California. J. Phycol. **13**: 297–301.

Crins, W. J. and Bohm, B. A. 1990. Flavonoid diversity iln relation to the systematics and evolution of the tarweeds. Ann. Missouri Bot. Gard. **77**: 73–83.

———, ——— **and Carr, G. D.** 1988a. Flavonoids as indicators of hybridization in a mixed population of lava-colonizing Hawaiian tarweeds (Asteraceae: Heliantheae: Madiinae). Syst. Bot. **13**: 567–571.

———, ———, **Powell, A. M. and Guppy, C. S.** 1988b. Evolutionary considerations based on flavonoid aglycones in the Peritylinae. Biochem. Syst. Ecol. **16**: 273–278.

Crisci, J. V. and Berry, P. E. 1990. A phylogenetic re-evaluation fo the Old World species of *Fuchsia* (Onagraceae). Ann. Missouri Bot. Gard. **77**: 517–522.

Critchfield, W. B. 1957. Geographic variation in *Pinus contorta*. Maria Moors Cabot Foundation Publ. 3.

———. 1984. Crossability and relationships of Washoe pine. Madroño 31: 144–170.

——— **and Little, E. L.** 1966. The geographic distribution of the pines of the world. Misc. Publ. U. S. Dept. Agric. No. 991, 16–17 (Map 47).

Cronquist, A. 1981. An Integrated System of Classification of Flowering Plants. Columbia University Press, New York, NY.

Culberson, C. F. 1965. A note on the chemical strains of *Parmelia furfuracea*. Bryologist **68**: 435–439.

Culberson, W. L. 1967. Analysis of chemical and morphological variation in the *Ramalina siliquosa* species complex. Brittonia **19**: 333–352.

———. 1969. The behavior of the species of the *Ramalina siliquosa* group in Portugal. Öst. Bot. Zeit. **116**: 85–94.

———. 1970. Chemosystematics and ecology of lichen-forming fungi. Ann. Rev. Ecol. Syst. **1**: 153–170.

——— **and Culberson, C. F.** 1967. Habitat selection by chemically differentiated races of lichens. Science **158**: 1195–1197.

Cutler, D. F. 1969. Anatomy of the Monocotyledons. IV. Juncales. Clarendon Press, Oxford.

———. 1972. Restionaceae. Pages 73–83 in D. H. Valentine (ed.) Taxonomy, Phytogeography and Evolution. Academic Press, London.

Cutler, D. F., Brandham, P. E., Carter, S. and Harris, S. J. 1980. Morphological, anatomical, cytological and biochemical aspects of evolution in East African shrubby species of *Aloë* L. (Liliaceae). Bot. J. Linn. Soc. **80**: 293–317.

Cywnar, L. C. and MacDonald, G. M. 1987. Geographical variation of lodge pole pine in relation to population history. American Naturalist **129**: 463–469.

Daday, H. 1954a. Gene frequencies in wild populations of *Trifolium repens*.1. Distribution by lattitude. Heredity **8**: 61–78.

———. 1954b. Gene frequencies in wild populations of *Trifolium repens*. 2. Distribution by altitude. Heredity **8**: 377–384.

Dahlgren, R. 1980. A revised system of classification of the angiosperms. Bot. J. Linn. Soc. **80**: 91–124.

———, **Clifford, H. T. and Yeo, P. F.** 1985. The Families of the Monocotyledons. Springer-Verlag, Berlin.

Dallimore, W. and Jackson, A. B. 1966. A Handbook of the Coniferae and Ginkgoaceae. St. Martin's Press, New York, NY.

Dañobeitia, J. J. and Canales, J. P. 2000. Magmatic underplating in the Canary Archipelago. J. Volcan. Geotherm. Res. **103**: 27–41.

Daubenmire, R. 1978. Plant Geography with Special Reference to North America. Academic Press, New York, NY.

Dawson, C. D. R. 1941. Tetrasomic inheritance in *Lotus corniculatus* L. J. Genetics **42**: 49–72.

de Araujo, A. M. 1976. The relationship between altitude and cyanogenesis in white clover (*Trifolium repens* L.). Heredity **37**: 291–293.

Delcourt, P. A. 1980. Goshen springs: late-quaternary vegetation record for southern Alabama. Ecology **61**: 371–386.

Del Pero Martinez, M. A. and Martinez, A. J. 1993. Flavonoid distribution in *Tradescantia*. Biochem. Syst. Ecol. **21**: 255–265.

De Luca, V. and Ibrahim, R. K. 1985a. Enzymatic synthesis of polymethylated flavonols of *Chrysosplenium americanum*. I. Partial purification and some properties of S-adenosyl-L methionine: flavonol 3-, 6-, 7-, and 4′-O-methyltransferases. Arch. Biochem. Biophys. **238**: 596–605.

——— **and** ——— 1985b. Enzymatic synthesis of polymethylated flavonols of *Chrysosplenium americanum*. II. Substrate interaction and product inhibition studies of flavonol 3-, 6-, and 4′-O-methyltransferases. Arch. Biochem. Biophys. **238**: 606–618.

Dembitsky, V. M., Rezanková, H., Rezanka, T. and Hanus, L. O. 2003. Variabilitiy of the fatty acids of the marine green algae belonging to the genus *Codium*. Biochem. Syst. Ecol. **31**: 1125–1145.

Denton, M. F. 1979. Cytological and reproductive differentiation in *Sedum* section *Gormania*. Brittonia **31**:197–211.

——— **and Kerwin, J. L.** 1980. Survey of vegetative flavonoids of *Sedum* section *Gormania* (Crassulaceae). Can. J. Bot. **58**: 902–905.

DePooter, H. L., Schamp, N. M., Aboutabl, E. A., El Thoamy, S. L. and Doss, S. L. 1991. Essential oil of the leaves of three *Pistacia* species growing in Egypt. Flavour Frag. J. **6**: 229–232.

Desrochers, A. M. 1992. A biosystematic study of *Lasthenia californica* (Asteraceae). Ph.D. Dissertation, University of British Columbia.

——— **and Bohm, B. A.** 1993. Flavonoid variation in *Lasthenia californica*. Biochem. Syst. Ecol. **21**: 449–453.

——— **and** ——— 1995. Biosystematic study of *Lasthenia californica* (Asteraceae). Syst. Bot. **20**: 65–84.

Dettmann, M. E., Pocknall, D. T., and Romera, E. J. 1990. *Nothofagidites* Erdtman ex Potonie 1960; a catalogue of species with notes on the paleogeographic distribution of *Nothofagus* Bl. (southern Beech). New Zealand Geol. Surv. Palaeontol. Bull. **60**: 1–79.

Dharmaratne, H. R. W., Nanayakkara, N. P. D. and Khan, I. A. 2002. Kavalactones from *Piper methysticum*, and their ^{13}C NMR spectroscopic analyses. Phytochemistry **59**: 429–433.

Dodd, R. S., Fromard, F., Rafii, Z. A., and Blasco, F. 1995. Biodiversity among West African *Rhizophora*: foliar wax chemistry. Biochem. Syst. Ecol. **23**: 859–868.

——— **and Rafii, Z. A.** 1994. Chemical and ecological variability of *Cupressus bakeri* on Goosenest Mountain, California. Biochem. Syst. Ecol. **22**: 393–400.

——— **and** ——— 1995. Ecogeographic variation in seed fatty acids of *Austrocedrus chilensis*. Biochem. Syst. Ecol. **23**: 825–833.

Dodd, S. C. and Helenurm, K. 2000. Floral variation in *Delphinium variegatum* (Ranunculaceae). Madroño **47**: 116–126.

——— **and** ——— 2002. Genetic diversilty in *Delphinium variegatum* (Ranunculaceae): a comparison of two insular endemic subspecies and their widespread mainland relative. Amer. J. Bot. **89**: 613–622.

Dodson, C. H. 1970. The role of chemical attractants in orchid pollination. Pages 83–107 in K. L. Chambers (ed.) Biochemical Coevolution. Oregon State University Press, Corvallis.

Downie, S. R. 1988. Morphological, cytological, and flavonoid variability of the *Arnica angustifolia* aggregate (Asteraceae). Can. J. Bot. **64**: 24–39.

——— **and Denford, K. E.** 1986a. The taxonomy of *Arnica frigida* and *A. louiseana* (Asteraceae). Can. J. Bot. **64**: 1355–1372.

——— **and** ——— 1986b. The flavonoids of *Arnica frigida* and *A. louiseana* (Asteraceae). Can. J. Bot. **64**: 2748–2752.

——— **and** ——— 1988. Flavonoid variation in *Arnica* subgenus *Arctica*. Biochem. Syst. Ecol. **16**: 133–137.

Downum, K. R., Dole, J. and Rodriguez, E. 1988. Nordihyroguaiaretic acid: inter- and intrapopulational variation in the Sonoran Desert creosote bush (*Larrea tridentata*, Zygophyllaceae). Biochem. Syst. Ecol. **16**: 551–555

Dutschewska, H. and Kuzmanov, B. 1982. Chemosystematics of *Thalictrum minus*. J. Nat. Prod. **45**: 295–310.

Dymond, J. 1975. K-Ar ages of Tahiti and Moorea, Society Islands, and implications for the hot-spot model. Geology **3**: 236–240.

Echeverrigaray, S., Fracaro, F., Atti dos Santos, A. C., Paroul, N., Wasum, R. and Atti Serafini, L. 2003. Essential oil composition of south Brazilian populations of *Cunila galioides* and its relation to geographic distribution. Biochem. Syst. Ecol. **31**: 467–475.

Egerton-Warburton, L. M., Ghisalberti, E. L. and Considine, J. A. 1998. Infraspecific variability in the volatile leaf oils of *Chamelaucium uncinatum* (Myrtaceae). Biochem. Syst. Ecol. **26**: 873–888.

Eigler, G. and Poelt, J. 1965. Flechtenstoffe und Systematik der lobaten Arten der Flechtengattung *Lecanora* in der Holarktic. Österr. Bot. Zeit. **112**: 285–294.

Eliasson, U. H. 1974. Studies in Galápagos plants. XIV. The genus *Scalesia* Arn. Opera Bot. **36**: 1–117.

Erickson, K. L. 1983. Constituents of *Laurencia*. Pages 131–257 in P. J. Scheuer (ed.). Marine Natural Products. Academic Press, New York.

Esselman, E. J., Crawford, D. J., Brauner, S., Stuessy, T. F., Anderson, G. J., and Silva O. M. 2000. RAPD marker diversity within and divergence among species of *Dendroseris* (Asteraceae: Lactuceae). Amer. J. Bot. **87**: 591–596.

Estes, J. A. and Steinberg, P. D. 1988. Predation, herbivory and kelp evolution. Paleobiol. **14**: 19–36.

Evans, I. A. 1976. Relationship between bracken and cancer. Pages 105–112 in F. H. Perring and B. G. Gardiner (eds.), The biology of bracken. Linnean Society of London-Academic Press, London.

Evans, W. C. 1976. Bracken thiaminase-mediated neurotoxic syndromes. Pages 113–131 in F. H. Perring and B. G. Gardiner (eds.), The biology of bracken. Linnean Society of London-Academic Press, London.

———. 1986. The acute diseases caused by bracken in animals. Pages 121–132 in R. T. Smith and J. A. Taylor (eds.), Bracken, Ecology, Land Use and Control Technology. Parthenon Publishing, Carnforth, UK.

Faulkner, D. J. 1987, 1988, 1990, 1991. Marine natural products. Natural Product Reports **4**: 473–498; **5**: 613–663; **7**: 269–309; **8**: 97–147.

Fenical, W. 1975. Halogenation in the Rhodophyta. A review. J. Phycol. **11**: 245–259.

Fenwick, G. R. 1988. Bracken (*Pteridium aquilinum*). Toxic effects and toxic constituents. J. Sci. Food Agric. **46**: 147–173.

Fernald, M. L. 1925. Persistence of plants in unglaciated areas of boreal America. Memoirs Amer. Acad. Arts Sci. **15**: 237–342.

Fleak, L. S. 1971. Biosystematics of the *Lupinus sericeus* complex. Ph.D. dissertation, University of Missouri.

——— and Dunn, D. B. 1971. Nomenclature of the *Lupinus sericeus* complex. Trans. Missouri Acad. Sci. **5**: 85–88.

Fleischer, R. C., McIntosh, C. E. and Tarr, C. L. 1998. Evolution on a volcanic conveyor belt: using phylogeographic reconstructions and K-Ar-based ages of the Hawaiian Islands to estimate molecular evolutionary rates. Mol. Ecol. **7**: 533–545.

Flynn, T. M. and Southwell, I. A. 1987a. Cyanogenesis in the genus *Zieria*. Phytochemistry **26**: 1669–1672.

——— and ———. 1987b. Essential oil constituents of the genus *Zieria*. Phytochemistry **26**: 1673–1686.

Follmann, G. and Huneck, S. 1972. No title given. Philippia 3: 9. [Cited by Piovana et al., 1997.]

Forrest, G. I. 1980a. Geographical variation in the monoterpenes of *Pinus contorta* oleoresin. Biochem. Syst. Ecol. **8**: 343–359.

———. 1980b. Genotypic variation among native Scots pine populations in Scotland based on monoterpene analysis. Forestry **53**: 101–128.

———. 1981. Geographical variation in oleoresin monoterpene composition of *Pinus contorta* from natural stands and planted seed collections. Biochem. Syst. Ecol. **9**: 97–103.

——— 1987. A rangewide comparison of outlying and central lodgepole pine populations based on oleoresin monoterpene analysis. Biochem. Syst. Ecol. **15**: 19–30.

Forster, P. I. and Hyland, B. P. M. 1997. Two new species of *Eucryphia* Cav. (Cunoniaceae) from Queensland. *Austrobaileya* **4**: 589–596.

Fosberg, F. R. 1948. Derivation of the flora of the Hawaiian Islands. Pages 107–119 in E. C. Zimmermann (ed.), Insects of Hawaii. Vol. 1, Introduction. University of Hawaii Press, Honolulu.

Franchet, A. R. 1890. Monographie du genere *Chrysosplenium* Tourn. Nouv. Arch. Mus. Hist. Nat. (Paris) III. **2**: 87–114.

———. 1891. Monographie du genere *Chrysosplenium* Tourn. Nouv. Arch. Mus. Hist. Nat. (Paris) III. **3**: 1–32.

Friás L. D., Godoy, R., Iturra, P., Koref-Santibáñez, S., Navarro, J., Pacheco, N. and Stebbins, G. L. 1975. Polymorphism and geographic variation in flower color in Chilean populations of *Eschscholzia californica*. Plant Syst. Evol. **123**: 185–198.

Fröst, S. and Asker, S. 1973. Further studies of flavonoid patterns in barley. Hereditas **75**: 201–206.

—— and ——. 1977. Flavonoid patterns and polymorphisms in wild *Hordeum* species. Hereditas **85**: 145–150.

—— and Holm, G. 1971. Thin-layer chromatographic studies of phenolic compounds in twenty varieties of barley. Hereditas **69**: 25–34.

—— and ——. 1972. Thin-layer chromatographic studies of phenolic compounds in seventeen parental varieties of barley. Hereditas **70**: 259–264.

—— and ——. 1975. Variation of flavonoid patterns in *Hordeum spontaneum* and *Hordeum agriocrithon*. Hereditas **80**: 167–172.

—— and ——. 1977. Intraspecific variation of flavonoid patterns in primitive spring wheat. Hereditas **86**: 267–272.

——, Harborne, J. B. and King, L. 1977. Identification of the flavonoids in five chemical races of cultivated barley. Hereditas **85**: 163–168.

——, Holm, G. and Asker, S. 1975. Flavonoid patterns and the phylogeny of barley. Hereditas **79**: 133–142.

Gabrielson, P. W. and Scagel, R. F. 1989. The marine algae of British Columbia, northern Washington, and southeast Alaska: division Rhodophyta (red algae), class Rhodophyceae, order Gigartinales, families Caulacanthaceae and Plocamiaceae. Can. J. Bot. **67**: 1221–1234.

Gadek, P. A. and Quinn, C. J. 1987. Biflavones and the affinities of *Cupressus funebris*. Phytochemistry **26**: 2551–2552.

Gamisans, J. 1973. Flore de la Corse. V. Candollea **28**: 67–70.

Ganders, F. R. 1990. Altitudinal clines for cyanogenesis in introduced populations of white clover near Vancouver, Canada. Heredity **64**: 387–390.

———, **Berbee, M. and Pirseyedi, M.** 2000. ITS Base sequence phylogeny in *Bidens* (Asteraceae): evidence for the continental relatives of Hawaiian and Marquesan *Bidens*. Syst. Bot. **25**: 122–133.

———, **Bohm, B. A. and S. P. McCormick.** 1990. Flavonoid variation in Hawaiian *Bidens*. Syst. Bot. **15**: 231–239.

——— **and Nagata, K. M.** 1990. *Bidens*. Pages 267–283 in W. Wagner, D. Herbst and S. Sohmer (eds.) Manual of the Flowering Plants of Hawaii. Vol. 1. University of Hawaii Press and Bishop Museum Press, Honolulu.

Garcia-Vallejo, M. C., Garcia-Vallejo, I., and Velasco-Negueruela, A. 1990. Essential oils of the genus *Lavandula* L. in Spain. Pages 15–26 in S. C. Bhattacharya and K. L. Sethi (eds.) Proceedings of the 11th International Congress of Essential Oils, Fragrance and Flavors, New Delhi, India 1989 (Aspect Publishing, London).

Gartlan, J. S., McKey, D. B., Waterman, P. G., Mbi, C. N., and Struhsaker, T. T. 1980. A comparative study of the phytochemistry of two African Rain Forests. Biochem. Syst. Ecol. **8**: 401–422.

Geissman, T. A., Lucas, A. J., Saitoh, T., and Payne, W. W. 1973. Sesquiterpene lalctones of *Ambrosia chamissonis*. Biochem. Syst. **1**: 13–20.

Gershenzon, J. and Mabry, T. J. 1984. Sesquiterpene lactones from a Texas population of *Helianthus maximiliani.* Phytochemistry **23**: 1959–1966.

Gibbs, P. E. 1966. A revision of the genus *Genista* L. Notes Royal Bot. Gard. Edinburgh **27**: 11–99.

Gibson, P. B., Barnett, J. T. and Gillingham, J. T. 1972. Cyanoglucoside and hydrolyzing enzyme in species related to *Trifolium repens.* Crop Sci. **12**: 708–709.

Gibson, A. C. and Nobel, P. S. 1986. The Cactus Primer, Harvard University Press, Cambridge, MA.

Gil, A., Ghersa, C. M. and Perelman, S. 2002. Root thiophenes in *Tagetes minuta* L. accessions from Argentina: genetic and environmental contribution to changes in concentration and composition. Biochem. Syst. Ecol. **30**: 1–13.

Gleason, H. A. 1952. The New Britton and Brown Illustrated Flora of the Northeastern United States and Adjacent Canada, Vol. 3. New York Botanical Garden Press, New York, NY.

Glennie, C. W., Harborne, J. B., Rowley, G. D., and Marchant, C. J. 1971. Correlation between flavonoid chemistry and plant geography in the *Senecio radicans* complex. Phytochemistry **10**: 2413–2417.

Gluchoff-Fiasson, K., Fiasson, J. L., and Favre-Bonvin, J. 1994. Quercetin glycosides from Antarctic *Ranunculus* species. *Phytochemistry* **37**: 1629–1633.

Godwin, H. 1975. History of the British Flora, 2nd ed. Cambridge University Press, Cambridge.

González, A. G., Barrea, J. B., Diaz, J. G., López, L. A., and De Paz, P. P. 1988. Distribution of secondary metabolites in two subspecies of *Todaroa aurea.* Biochem. Syst. Ecol. **16**: 641–645.

———, **Bermejo Barrera, J., Ma Rodriguez Pérez, E. and Hernández Padrón, C. E.** 1994. Depsidones from *Lobaria pulmonaria* and their chemotaxonomic importance. Biochem. Syst. Ecol. **22**: 583–586.

———, **Hernández, C., Rodriguez, E. M., Alfayate, M. del C. and Bermejo, J.** 1999. Depsidones from *Anomalographis madeirensis.* Biochem. Syst. Ecol. **27**: 831–833.

Good, R. D'O. 1930. The geography of the genus *Coriaria.* New Phytologist **29**: 170–198.

Goodger, J. Q. D., Capon, R. J. and Woodrow, I. E. 2002. Cyanogenic polymorphism in *Eucalyptus polyanthemos* Schauer subsp. *vestita* L. Johnson and K. Hill (Myrtaceae). Biochem. Syst. Ecol. **30**: 617–630.

Gornall, R. J. and Bohm, B. A. 1978. Angiosperm flavonoid evolution: a reappraisal. Syst. Bot. **3**: 353–368.

Gotsiou, P., Naxakis, G. and Skoula, M. 2002. Diversity in the composition of monoterpenoids of *Origanum microphyllum* (Labiatae). Biochem. Syst. Ecol. **30**: 865–879.

Gottlieb, O. R. 1986. Micromolecular Evolution, Systematics and Ecology. Springer-Verlag, Berlin.

——— 1990. Phytochemicals: differentiation and function. Phytochemistry **29**: 1715–1724.

——— **and Kubitzki, K.** 1983. Ecogeographical phytochemistry. A novel approach to the study of plant evolution and dispersion. Naturwissen. **70**: 119–126.

Graham, A. 1972a. Outline of the origin and historical recognition of floristic affinities between Asia and eastern North America. Pages 1–16 in A. Graham (ed.) Floristics and paleofloristics of Asia and eastern North America. Elsevier Publ. Co., Amsterdam.

——— (ed.) 1972b. Floristics and paleofloristics of Asia and eastern North America. Elsevier Publ. Co., Amsterdam.

Granger, R. and Passet, J. 1973. *Thymus vulgaris* spontane de France: Races chimiques et chemotaxonomie. Phytochemistry **12**: 1683–1691.

Gray, A. 1846. Analogy between the flora of Japan and that of the United States. Amer. J. Sci. Arts, Ser. 2, **2**: 135–136.

———. 1851. Account of *Argyroxiphium*, a remarkable genus of Compositae, belonging to the mountains of the Sandwich Isles. Proc. Amer. Acad. Arts **2**: 323–325.

———. 1859. Diagnostic characters of new species of phanerogamous plants collected in Japan by Charles Wright, Botanists of the U. S. North Pacific Exploring Expedition, with observations upon the relationship of the Japanese flora to that of North America and of other parts of the northern Temperate zone. Memoirs Amer. Acad. Arts **6**: 377–453.

———. 1878. Forest geography and archaeology. A lecture delivered before the Harvard University Natural History Society. Amer. J. Sci. Arts, Ser. 3, **16**: 183–196.

Grayer, R. J., Chase, M. W. and Simmonds, M. S. J. 1999. A comparison between chemical and molecular characters for the determination of phylogentic relationships among plant families: An appreciation of Hegnauer's "Chemotaxonomie der Pflanzen." Biochem. Syst. Ecol. **27**: 369–393.

Greenham, J., Vassiliades, D. D., Harborne, J. B., Williams, C. A., Eagles, J., Grayer, R. J. and Veitch, N. C. 2001. A distinctive flavonoid chemistry for the anomalous genus *Biebersteinia.* Phytochemistry **56**: 87–91.

Greinwald, R., Veen, G., van Wyk, B.-E., Witte, L. and Czygan, F.-C. 1989. Distribution and taxonomic significance of major alkaloids in the genus *Virgilia.* Biochem. Syst. Ecol. **17**: 231–238.

Grisebach, H. and Grambow, H. J. 1968. Biosynthesis of flavonoids—XV. Occurrence and biosynthesis of flavonoids in *Datisca cannabina.* Phytochemistry **7**: 51–56.

Griffiths, A. J. F. and Ganders, F. R. 1983. Wildflower Genetics. A Field Guide for British 'Columbia and the Pacific Northwest. Flight Press, Vancouver, B. C.

Gritsanapan, W. and Griffin, W. J. 1991. Alkaloid variation within *Duboisia myoporoides.* Phytochemistry **30**: 2667–2669.

Guenet, A. and Aubanel, M. L. 1991. Contribution à la connaissance de l'absolute de Lentisque du Maroc. Riv. Ital. EPPOS (Numero Speciale) 332–347 [Cited by Castola et al., 2000].

Gunatilaka, A. A. L., V. J. Paul, P. U. Park, M. P. Puglisi, A. D. Gitler, D. S. Eggleston, R. C. Haltiwanger and D. G. I. Kingston. 1999. Apakaochtodenes A and B: two tetrahalogenated monoterpenes from the red marine alga *Portieria hornemannii.* J. Nat. Prod. **62**: 1376–1378.

Haagen-Smit, A. J., Wang, T., and Mirov, N. T. 1950. Composition of gum turpentine of *Pinus aristata, P. balfouriana, P. flexilis,* and *P. parviflora.* Amer. Pharm. Assoc. J. Sci. Ed. **39**: 254–259.

Häberlein, H. and Tschiersch, K.-P. 1998. On the occurrence of methylated and methoxylated flavonoids in *Leptospermum scoparium.* Biochem. Syst. Ecol. **26**: 97–103.

Hale, M. E. 1956. Chemical strains of the lichen *Parmelia furfuracea.* Amer. J. Bot. **43**: 456–459.

———. 1968. A synopsis of the lichen genus *Pseudevernia.* Hannan Bryologist **71**: 1–11.

Hall, G. D. and Langenheim, J. H. 1987. Geographic variation in leaf monoterpenes of *Sequoia sempervirens.* Biochem. Syst. Ecol. **15**: 31–43.

Haller, J. R. 1966. Systematic and evolutionary relationships of *Pinus torreyana.* Amer. J. Bot. **53**: 635 (abstract).

———. 1986. Taxonomy and relationships of the mainland and island populations of *Pinus torreyana.* Syst. Bot. **11**: 39–50.

Halliday, W. E. D. and Brown, A. W. A. 1943. Distribution of some important forest trees in Canada. Ecology 24: 353–373.

Hammer, S. and Ahti, T. 1990. New and interesting species of Cladonia from California. Mycotaxon 37: 335–348.

Hamrick, J. L. and Godt, M. J. W. 1989. Allozyme diversity in plant species. Pages 43–63 in A. H. D. Brown, M. T. Clegg, A. L. Kahler and B. S. Weir (eds.) Plant Population Genetics, Breeding and Germplasm Resources. Sinauer, Sunderland, MA.

——, Linhart, Y. B. and Mitton, J. B. 1979. Relationships between life history characteristics and electrophoretically detectable genetic variation in plants. Ann. Rev. Ecol. Syst. **10**: 173–200.

Hanover, J. W. 1966a. Inheritance of 3-carene concentration in *Pinus monticola.* Forest Sci. **12**: 447–450.

———. 1966b. Genetics of terpenes. I. Gene control of monoterpene levels in *Pinus monticola* Dougl. Heredity **21**: 73–84.

———. 1971. Genetics of terpenes. II. Genetic variatnces and interrelationships of monoterpene concentrations in *Pinus monticola.* Heredity **27**: 237–245.

Hänsel, R. 1968. Characterization and physiological activity of some Kava constituents. Pac. Sci. **22**: 369–373.

Hara, H. 1957. Synopsis of the genus *Chrysosplenium* L. J. Fac. Sci., Univ. Tokyo, Sect. III Botany **7**: 1–90.

——— 1972. Patterns of differentiation in flowering plants. Pages 55–60 in A. Graham (ed.) Floristics and Paleofloristics of Asia and Eastern North America. Elsevier, Amsterdam.

Harborne, J. B. 1960. Luteolinidin, a 3-deoxyanthocyanin analogue of luteolin. Chem. Ind. (London) 130.

———. 1966. Comparative biochemistry of flavonoids. II. 3-Desoxyanthocyanins and their systematic distribution in ferns and gesnerads. Phytochemistry **5**: 589–600.

———. 1967a. Comparative biochemistry of flavonoids. VI. Flavonoid patterns in the Bignoniaceae and the Gesneriaceae. Phytochemistry **6**: 1643–1651.

———. 1967b. Comparative Biochemistry of the Flavonoids. Academic Press, New York, pp. 304–314.

———. 1979. Correlations between flavonoid chemistry, anatomy and geography in the Restionaceae. Phytochemistry **18**: 1323–1327.

———. 1986. Flavonoid patterns and phytogeography: The genus *Rhododendron* section *Vireya*. Phytochemistry **25**: 1641–1643.

——— **and Clifford, H. T.** 1969. Flavonoid patterns in the Restionaceae. Gossypetin in *Restio* and a new flavone in *Hypolaena*. Phytochemistry **8**: 2071–2075.

Hare, J. D. 2002. Geographic and genetic variation in the leaf surface resin components of *Mimulus aurantiacus* from southern California. Biochem. Syst. Ecol. **30**: 281–296.

Harrison, L. J. and Asakawa, Y. 1989. 3α,18-Dihydroxytrachyloban-19-oic acid from the liverwort *Jungermannia exsertifolia* subsp. *cordifolia*. Phytochemistry **28**: 1533–1534.

Hart, R. H. and Wilsie, C. P. 1959. Inheritance of a flower character, brown keel tip, in *Lotus corniculatus* L. Agron. J. **51**: 379–380.

Hashimoto, T., Tori, M. and Asakawa, Y. 1989. Drimane-type sesquiterpenoids from the liverwort *Makinoa crispata*. Phytochemistry **28**: 3377–3381.

Hassan, M. A., Muhtadi, F. J. and Al-Badr, A. A. 1979. GLC-Mass spectroscopy of *Teucrium polium* oil. J. Pharm. Sci. **68**: 800–801.

Hattersley, P. W. 1983. The distribution of C_3 and C_4 grasses in Australia in relation to climate. Oecologia **57**: 113–128.

Hawksworth, D. L. 1976. Lichen chemotaxonomy. Pages 139–184 in D. H. Brown, D. L. Hawksworth, and R. H. Bailey (eds.), Lichenology: Progress and Problems. Academic Press, New York.

——— **and Chapman, D. S.** 1971. *Pseudevernia furfuracea* (L.) Zopf and its chemical races in the British Isles. Lichenologist **5**: 51–58.

Hayashi, N., Maeshima, K. and Komae, H. 1983. Phenol ethers of three North American *Hexastylis* species. Phytochemistry **22**: 299.

———**, Maeshima, K., Murakami, T. and Komae, H.** 1984. Chemosystematics of Japanese *Heterotropa* (Aristolochiaceae). Zeit. für Naturforsch. **39c**: 705–709.

Hayashi, T., Okamura, K., Kawasaki, M., and Morita, N. 1991. Two chemotypes of *Scoparia dulcis* in Paraguay. Phytochemistry **30**: 3617–3620.

———, ———**, Tamada, Y., Iida, A., Fujita, T. and Morita, N.** 1993. A new chemotype of *Scoparia dulcis*. Phytochemistry **32**: 349–352.

Heinsbroek, R., van Brederode, J., Besson, E., Chopin, J., van Nigtevecht, G. and Kamsteeg, J. 1980. The 2"-O-glucosylation of vitexin and isovitexin in petals of *Silene alba* is catalyzed by two different enzymes. Phytochemistry **19**: 1935–1937.

Hennion, F. and Coudec, H. 1993. Cytogenetic variability of *Ranunculus* species from Iles Kerguelen. Antarctic Sci. **5**: 37–40.

———**, Fiasson, J. L. and Gluchoff-Fiasson, K.** 1994. Morphological and phytochemical relationships between *Ranunculus* species from Iles Kerguelen. Biochem. Syst. Ecol. **22**: 533–542.

Hershkovitz, M. A. and Zimmer, E. A. 1997. On the evolutionary origin of the cacti. Taxon **46**: 217–232.

Herz, W. and Kumar, N. 1981. Heliangolides from *Helianthus maximiliani*. Phytochemistry **20**: 93–98.

——— **and Roy, S. K.** 1969. New pseudoguaianolides from *Gaillardia pulchella*. Phytochemistry **8**: 661–664.

Heusser, C. J. 1983. Vegetational history of the northwestern United States including Alaska. Pages 239–258 in S. C. Porter (ed.) Late-quaternary Environments of the United States. Vol. 1. The late Pleistocene. University of Minnesota Press, Minneapolis, Minnesota.

——— **and Flint, R. F.** 1977. Quaternary glaciations and environments of northern Isla de Chiloé, Chile. Geology **5**: 305–308.

Heywood, J. S. and Levin, D. A. 1984. Allozyme variation in *Gaillardia pulchella* and *G. amblyodon* (Compositae): Relation to morphological and chromosomal variation and to geographicalisolation. Syst. Bot. **9**: 448–457.

Hickman, J. C. and Johnson, M. P. 1969. An analysis of geographic variation in western North American *Menziesia* (Ericaceae). Madroño **20**: 1–32.

Hickman, J. C. and Johnson, M. D. 1983. An analysis of geographical variation in western North American *Menziesia* (Ericaceae). Madroño **20**: 1–32.

Hiebert, R. D. and Hamrick, J. L. 1983. Patterns and levels of genetic variation in Great Basin bristlecone pine, *Pinus longaeva*. Evolution **37**: 302–310.

Higgs, M. D., Vanderah, D. J., and Faulkner, D. J. 1977. Polyhalogenated monoterpenes from *Plocamium cartilagineum* from the British coast. Tetrahedron **33**: 2775–2780.

Higo, A., Hamman, B., Timmermann, B., Yoshioka, H., Lee, J., Mabry, T. J., and Payne, W. W. 1971. Sesquiterpene lactones from the genus *Ambrosia*. Phytochemistry **10**: 2241–2244.

Hill, R. S. and Jordan, G. J. 1993. The evolutionary history of *Nothofagus* (Nothofagaceae). Austral. Syst. Bot. **6**: 111–126.

——— **and Read, J.** 1991. A revised infragenetic classification of *Nothofagus* (Fagaceae). Bot. J. Linn. Soc. **105**: 37–72.

Hiraoka, N., Chang, J.-I., Bohm, L. R. and Bohm, B. A. 2002. Furanocoumarin and polyacetylenic compound composition of wild *Glehnia littoralis* in North America. Biochem. Syst. Ecol. **30**: 321–325.

Hirono, I. 1986. Carcinogenic principles isolated from bracken fern. CRC Crit. Rev. Toxicol. **17**: 1–22.

Hitchcock, C. L., Cronquist, A., Ownbey, M., and Thompson, J. W. 1969. Vascular Plants of the Pacific Northwest. Part I. Vascular Cryptograms, Gymnosperms, and Monocotyledons. University of Washington Press, Seattle, WA.

Hoeneisen, M., Silva, M., Wink, M., Crawford, D. J., and Stuessy, T. 1993. Alkaloids of *Sophora* of Juan Fernandez Islands and related taxa. Bol. Soc. Chil. Quim. **38**: 167–171.

Hohenschutz, L. D., Bell, E. A., Jewess, P. J., Leworthly, D. P., Pryce, R. J., Arnold, E., and Clardy, J. 1981. Castanospermine, a 1,6,7,8-tetrahydroxyoctahydroindolizine alkaloid from seeds of *Castanospermum australe*. Phytochemistry **20**: 811–814.

Homer, L. E., Leach, D. N., Lea, D., Lee, L. S., Henry, R. J., and Baverstock, P. R. 2000. Natural variation in the essential oil content of *Melaleuca alternifolia* Cheel (Myrtaceae). Biochem. Syst. Ecol. **28**: 367–382.

Hong, D.-Y. 1993. Eastern Asian–North American disjunctions and their biological significance. *Cathaya* **5**: 1–39.

Housman, D. C., Price, M. V. and Redak, R. A. 2002. Architecture of coastal and desert *Encelia farinosa* (Asteraceae): consequences of plastic and heritable variation in leaf characters. Amer. J. Bot. **89**: 1303–1310.

Howard, B. M., Nonomura, A. M. and Fenical, W. 1980. Chemotaxonomy in marine algae: secondary metabolite synthesis by *Laurencia* in unialgal culture. Biochem. Syst. Ecol. **8**: 329–336.

Hsü, J. 1983. Late cretaceous and Cenozoic vegetation in China, emphasizing their connections with North America. Ann. Missouri Bot. Gard. **70**: 490–508.

Hu, H. H. 1935. A comparison of the ligneous flora of China and eastern North America. Bull. Chinese Bot. Soc. **1**: 79–97.

Hultén, E. 1937. Outline of the history of arctic on boreal biota during the quaternary period. Bokfoerlags Aktiebolaget Thule, Stockholm, Sweden.

Huneck, S., Connolly, J. D., Harrison, L. J., Joseph, R. S. I., and Pócs, T. 1984. 1-(3, 4-Dihydroxy-5-methoxyphenyl)-3-methylbut-2-ene from the liverwort *Plagiochila rutilans*. Phytochemistry **23**: 2396–2397.

———**, Djerassi, C., Becker, D., Barber, M., von Ardenne, M., Steinfelder, K. and Tümmler, R.** 1968. Flechteninhaltsstoffe—I. Massenspectrometrie von Depsiden, Depsidonen, Depsonen, Dibenzofuranen und Diphenylbutadienen mit positiven und negativen Ionen. Tetrahedron **24**: 2707–2755.

——— **and Follmann, G.** 1964. Usnic acid in *Rhizophora*. Naturwissen. **51**: 591.

Hunt, R. S. and von Rudloff, E. 1977. Leaf-oil-terpene variation in western white pine populations of the Pacific Northwest. For. Sci. **23**: 507–516.

Hunziker, J. H., Palacios, R. A., Poggio, L., Naranjo, C. A. and Yang, T. W. 1977. Geographic distribution, morphology, hybridization, cytogenetics, and evolution. Pages 10–46 in T. J. Mabry, J. H. Hunziker and D. R. DiFeo, Jr. (eds.) Creosote Bush—Biology and Chemistry of Larrea in New World Deserts. Dowden, Hutchinson and Ross, Stroudsberg, Pennsylvania.

Ichikawa, N., Naya, Y. and Enomoto, S. 1974. New halogenated monoterpenes from *Desmia (Chondrococcus) hornemannii*. Chem. Lett. **11**: 1333–1336.

Iinuma, M., Tanaka, T., Suzuki, K. and Lang, F. A. 1994. Two biflavonoids in the farinose exudate of *Pentagramma triangularis*. Phytochemistry **35**: 1043–1047.

Iltis, H. H. 1999. Setchellanthaceae (Capparales), a new family for a relictual, glucosinolate-producing endemic of the Mexican deserts. Taxon **48**: 257–275.

Ireland, B. F., Hibbert, D. B., Goldsack, R. J., Doran, J. C. and Brophy, J. J. 2002. Chemical varition in the leaf essential oil of *Melaleuca quinquenervis* (Cav.) S. T. Blake. Biochem. Syst. Ecol. **30**: 457–470.

Itoh, A., Sasaki, K., Mizukami, H., Ohashi, H., Sakurai, T. and Hiraoka, N. 1997. Geographical variation in the furanocoumarin and polyacetylenic compound compositions of wild *Glehnia littoralis* plants. Nat. Med. **51**: 50–55.

Iwashima, T. and Ootani, S. 1995. Polyphenols in *Iris setosa* var. *canadensis* and their chemotaxonomic comparisons with three Japanese varieties. Ann. Tsukuba Bot. Gard. **14**: 35–41.

Jalas, J. and Suominen, J. 1973. Atlas Florae Europeae. 2. Gymnospermae. Helsinki, Finland.

Jay, M. and Lumaret, R. 1995. Variation in the subtropical group of *Dactylis glomerata* L.—2. Evidence from phenolic compound patterns. Biochem. Syst. Ecol. **23**: 523–531.

———**, Plenet, D., Ardouin, P., Lumaret, R. and Jacquard, P.** 1984. Flavonoid variation in seven tetraploid populations of *Dactylis glomerata*. Biochem. Syst. Ecol. **12**: 193–198.

——— **and Voirin, B.** 1976. Les flavonoides de deux especes du genre *Chrysosplenium*. Phytochemistry **15**: 517–519.

Johansson, A. K., Kuusisto, P. H., Laakso, P. H., Derome, K. K., Sepponen, P. J., Katajisto, J. K. and Kallio, H. 1997. Geographical variations in seed oils from *Rubus chamaemorus* and *Empetrum nigrum*. Phytochemistry **44**: 1421–1427.

Jones, D. A. 1972. Cyanogenic glycosides and their function. Pages 103–124 in J. B. Harborne (ed.), Phytochemical Ecology, Academic Press, New York.

———**, Compton, S. G., Crawford, T. J., Ellis, W. M. and Taylor, I. M.** 1986. Variation in the colour of the keel petals in *Lotus corniculatus* L. 3. Pollination, herbivory and seed production. Heredity **57**: 101–112.

——— **and Crawford, T. J.** 1977. Variation in the colour of the keel petals in *Lotus corniculatus* L. 1. The polymorphism in western Europe. Heredity **39**: 313–325.

Jongaramruong, J. and Blackman, A. J. 2000. Polyhalogenated monoterpenes from a Tasmanian collection of the red seaweed *Plocamium cartilagineum*. J. Nat. Prod. **63**: 272–275.

Jose, J. and Sharma, A. K. 1985. Structure and behaviour of chromosomes in *Piper* and *Peperomia* (family Piperaceae). Cytologia **50**: 301–310.

Junger, A. and Johnson, D. L. 1980. Was there a Quaternary land bridge to the Northern Channel Islands? Pages 33–39 in D. M. Power (ed.) The California Islands: Proceedings of a multidisciplinary symposium. Santa Barbara Museum of Natural History, Santa Barbara, California.

Jutila, H. M. 1996. Seed bank and emergent vascular flora of ballast areas in Reposaari, Finland. Ann. Bot. Fenn. **33**: 165–182. [Cited by Keskitalo *et al.*, 2001.]

Kakes, P. and Chardonnens, A. N. 2000. Cyanotypic frequencies in adjacent and mixed populations of *Trifolium occidentale* Coombe and *Trifolium repens* L. are regulated by different mechnisms. Biochem. Syst. Ecol. **28**: 633–649.

Kamal, A. and Sandra, P. 1994. Gas chromatography—mass spectrometry analysis of the volatile oils of two *Teucrium polium* varieties. Biochem. Syst. Ecol. **22**: 529–532.

Kamsteeg, J., van Brederode, J. and van Nigtevecht, G. 1978. Anthocyanins isolated from petals of various genotypes of the red campion. Zeit. für Naturforsch. **33c**: 475–483.

Kaplan, M. A. C. 1995. Amazonia versus Australia. Geographically distant, chemically close. In R. Seidl, O. R. Gottlieb, and M. A. C. Kaplan (eds.), Chemistry of the Amazon. Biodiversity, Natural Products, and Environmental Issues. P. American Chemical Society, Washington, D.C.

———, **Ribeiro, J., and Gottlieb, O. R.** 1991. Chemogeographical evolution of terpenoids in Icacinaceae. Phytochemistry **30**: 2671–2676.

Karnick, C. R. and Saxena, M. D. 1970. On the variability of alkaloid production in *Datura* species. Planta Med. **18**: 266–269.

Kaundun, S. S., Fady, B., and Lebreton, P. 1997. Genetic differences between *Pinus halepensis*, *Pinus brutia* and *Pinus eldarica* based on needle flavonoids. Biochem. Syst. Ecol. **25**: 553–562.

———, **Lebreton, P., and Fady, B.** 1998a. Geographic variability of *Pinus halepensis* Mill. as revealed by foliar flavonoids. Biochem. Syst. Ecol. **26**: 83–96.

———, ——— **and** ———. 1998b. Genetic variation in the needle flavonoid composition of *Pinus brutia* var. *brutia*. Biochem. Syst. Ecol. **26**: 485–494.

Keck, D. D. 1936. The Hawaiian silverswords: systematics, affinities, and phytogeographic problems of the genus *Argyroxiphium*. Occasional Papers Bernice P. Bishop Mus. **11**: 1–38.

Kershaw, K. A. 1985. Physiological Ecology of Lichens. Cambridge University Press, Cambridge.

Keskitalo, M., Linden, A. and Valkonen, J. P. T. 1998. Genetic and morphological diversity of Finnish tansy (*Tanacetum vulgare* L.). Theor. Appl. Gen. **96**: 1141–1150.

———, **Pehu, E. and Simon, J. E.** 2001. Variation in volatile compounds from tansy (*Tanacetum vulgare* L.) related to genetic and morphological differences of genotypes. Biochem. Syst. Ecol. **29**: 267–285.

Kim, H.-G., Keeley, S. C., Vroom, P. S. and Jansen, R. K. 1998. Molecular evidence for an African origin of the Hawaiian endemic *Hesperomannia* (Asteraceae). Proc. Nat'l. Acad. Sci., U.S.A. **95**: 15440–15445.

King, R. R., Calhoun, L. A., and Singh, L. A. 1988. 3,4-Di-O- and 2,3,4-tri-O-acylated glucoside esters from the glandular trichomes of non-tuberous *Solanum* species. Phytochemistry **27**: 3765–3768.

———, ———, ——— **and Boucher, A.** 1990. Sucrose esters associated with glandular trichomes of wild *Lycopersicon* species. Phytochemistry **29**: 2115–2118.

Kinloch, B. B., Westfall, R. D., and Forrest G. I. 1986. Caledonian Scots pine: origins, and genetic structure. New Phytologist **104**: 117–130.

Kirch, J., Veit, M., Wätzig, H., Greinwald, R. and Czygan, F.-C. 1995. Alkaloidal variation in *Genista lobelii* s.l. (Fabaceae). Biochem. Syst. Ecol. **23**: 635–643.

Kokkalou, E. 1988. The constituents of the essential oil from *Lavandula stoechas* growing wild in Greece. Planta Med. **47**: 58–59.

Kokkini, S., Karousou, R. and Vokou, D. 1994. Pattern of geographic variation of *Origanum vulgare* trichomes and essential oil content in Greece. Biochem. Syst. Ecol. **22**: 517–528.

———, ———, **Dardioti, A., Krigas, N., and Lanaras, T.** 1997. Autumn essential oils of Greek oregano. Phytochemistry **44**: 883–886.

——— **and Vokou, D.** 1993. The hybrid *Origanum* X *intercedens* from the island of Nisyros (SE Greece) and its parental taxa; comparative study of essential oils and distribution. Biochem. Syst. Ecol. **21**: 397–403.

Kolterman, D. A., Breckon, G. J. and Kowal, R. R. 1984. Chemotaxonomic studies in *Cnidoscolus* (Euphorbiaceae). II. Flavonoids of *C. aconitifolius*, *C. souzae*, and *C. spinosus*. Syst. Bot. **9**: 22–32.

Komai, K. and Tang, C. S. 1989. A chemotype of *Cyperus rotundus* in Hawaii. Phytochemistry **28**: 1883–1886.

———, ——— **and Nishimoto, R. K.** 1991. Chemotypes of *Cyperus rotundus* in Pacific rim and basin: distribution and inhibitory activities of their essential oils. J. Chem. Ecol. **17**: 1–8.

Konecny, K., Streibl, M., Vasickova, S., Budesinsky, M., Saman, D., Ubik, K., and Herout, V. 1985. Constituents of the liverwort *Bazzania trilobata* of Czech origin. Collect. Czech. Chem. Comm. **50**: 80–93.

König, G. M., Wright, A. D., and Linden, A. 1999. *Plocamium hamatum* and its monoterpenes: chemcal and biological investigations of the tropical marine red alga. Phytochemistry **52**: 1047–1053.

Kuzmanov, B. and Dutschewska, H. 1982. Evolutionary patterns and alkaloid biosynthesis in *Thalictrum.* J. Nat. Prod. **45**: 766–771.

Kyhos, D. W. 1967. Evidence of different adaptations of flower color variants of *Encelia farinosa* (Compositae). Madroño **21**: 49–61.

Lang, K. J. 1994. *Abies alba* Mill., differentiation of provenances and provenance groups by the monoterpene patterns in the cortex resin of twigs. Biochem. Syst. Ecol. **22**: 53–63.

Langlois, A., Mulholland, D. A., Crouch, N. R. and Grace, O. M. 2004. Aporphine alkaloid from *Papaver aculeatum* (sect. *Horrida*; Papaveraceae) of southern Africa. Biochem. Syst. Ecol. **32**: 1087–1090

Lara, A. and Villalba, R. 1993. A 3620-year temperature record from *Fitzroya cupressoides* tree rings in southern South America. Science **260**(5111): 1104–1106.

Lauranson, J. and Lebreton, P. 1991. Flavonoid variability within and between natural populations of *Pinus uncinata.* Biochem. Syst. Ecol. **19**: 659–664.

———, **Vekemans, X., Lefebvre, C., and Jay, M.** 1995. Flavonoid profiles variation in *Armeria maritima* (Mill.) Willd. Biochem. Syst. Ecol. **23**: 319–329.

Lawesson J. E. 1988. Contributions to the flora of the Galapagos Islands, Ecuador. Phytologia **65**: 228–230.

Lebot, V. and Cabalion, P. 1986. Les Kavas de Vanuatu, (cultivars de *Piper methysticum* Forster). Coll. Trav. Doc. ORSTOM, Paris. **205**: 1–260.

——— **and Lévesque, J.** 1989. The origin and distribution of kava (*Piper methysticum* Forst. F., Piperaceae): a phytochemical approach. Allertonia **5**: 223–281.

——— **and** ———. 1996a. Genetic control of kavalactone chemotypes in *Piper methysticum* cultivars. Phytochemistry **43**: 397–403.

——— **and** ———. 1996b. Evidence for conspecificity of *Piper methysticum* Forst. f. and *Piper wichmannii* C. DC. Biochem. Syst. Ecol. **24**: 775–782.

———, **Aradhya, M. K. and Manshardt, R. M.** 1991. Geographic survey of genetic variation in kava, *Piper methysticum* Forst. f. and *Piper wichmannii* DC. Pac. Sci. **45**: 169–185.

Lebreton, P., Laracine-Pittet, C., Bayet, C., and Lauranson, J. 1990. Variabilitie polyphenolique et systematique du pin sylvestre *Pinus sylvestris* L. Ann. Sci. Forest. **47**: 117–130.

Ledig, F. T. and Conkle, M. T. 1983. Gene diversity and genetic structure in a narrow endemic, Torrey pine (*Pinus torreyana* Parry ex Carr.). Evolution **37**: 79–85.

Lee, L. S., Brooks, L. O., Homer, L. E., Rossetto, M., Henry, R. J., and Baverstock, P. R. 2002. Geographic variation in the essential oils and morphology of natural populations of *Melaleuca alternifolia* (Myrtaceae). Biochem. Syst. Ecol. **30**: 343–360.

Lee, N. S., Sang, T., Crawford, D. J., Yeau, S. H., and Kim, S.-C. 1996. Molecular divergence between disjunct taxa in eastern Asia and eastern North America. Amer. J. Bot. **83**: 1373–1378.

Leong, Y.-W. and Harrison, L. J. 1997. *ent*-Trachylobane diterpenoids from the liverwort *Mastigophora diclados*. Phytochemistry **45**: 1457–1459.

Levin, D. A. 1976. Alkaloid-bearing plants: an ecogeographic perspective. Amer. Natural. **110**: 261–284.

——— **and York, Jr., B. M.** 1978. The toxicity of plant alkaloids: an ecogeographic perspective. Biochem. Syst. Ecol. **6**: 61–76.

Levy, M. 1976. Altered glycoflavone expression in induced autotetraploids of *Phlox*. Biochem. Syst. Ecol. **4**: 249–254.

———. 1977. Minimum biosynthetic-step indices as measures of comparative flavonoid affinity. Syst. Bot. **2**: 89–98.

———. 1983. Flavone variation and subspecific divergence in *Phlox pilosa* (Polemoniaceae) Syst. Bot. **8**: 118–126.

——— **and Fujii, F.** 1978. Geographic variation of flavonoids in *Phlox carolina*. Biochem. Syst. Ecol. **6**: 117–125.

——— **and Levin, D. A.** 1975. The novel flavonoid chemistry and phylogenetic origin of *Phlox floridana*. Evolution **29**: 487–499.

Li, H. L. 1952. Floristic relationships between eastern Asia and eastern North America. Trans. Amer. Phil. Soc. **42**: 371–429.

———. 1972. Eastern Asia–eastern North America species-pairs in wide-ranging genera. Pages 65–78 in A. Graham (ed.) Floristics and Paleofloristics of Asia and Eastern North America. Elsevier Publ. Co., Amsterdam.

Li, Y., Nan, P., Tsering, T., Wang, L., Liu, S., and Zhong, Y. 2004. Interpopulation variability of rhizome essential oils in *Rhodiola crenulata* from Tibet and Yunnan, China. Biochem. Syst. Ecol. **32:** 611–614.

Lincoln, D. E. 1980. Leaf resin flavonoids of *Diplacus aurantiacus*. Biochem. Syst. Ecol. **8**: 397–400.

——— **and Langenheim, J. H.** 1976. Geographic patterns of monoterpenoid composition in *Satureja douglasii*. Biochem. Syst. Ecol. **4**: 237–248.

——— **and** ———. 1979. Variation of *Satureja douglasii* monoterpenoids in relation to light intensity and herbivory. Biochem. Syst. Ecol. **7**: 289–298.

——— **and** ———. 1981. A genetic approach to monoterpenoid compositional variation in *Satureja douglasii*. Biochem. Syst. Ecol. **9**: 153–160.

Linder, H. P. and Crisp, M. D. 1995. *Nothofagus* and Pacific biogeography. Cladistics **11**: 5–32.

——— **and Mann, D. M.** 1998. The phylogeny and biogeography of *Thamnochortus* (Restionaceae). Bot. J. Linn. Soc. **128**: 319–357.

Liston, A., Rieseberg, L. H. and Elias, T. S. 1989. Morphological statsis and molecular divergence in the intercontinental disjunct genus *Datisca* (Datiscaceae). Aliso **12**: 525–388.

———, ——— **and Hanson, M. A.** 1992. Geographic patterning of chloroplast DNA variation in the genus *Datisca* (Datiscaceae). Plant Syst. Evol. **181**: 121–132.

Liu, H.-J. and Wu, C.-L. 1997. A rearranged abietane-type diterpenoid from the liverwort *Makinoa crispata*. Phytochemistry **44**: 1523–1525.

Loftus Hills, K., Bottomley, W. and Mortimer, P. I. 1953. Occurrence of nicotine together with hyoscine in *Duboisia myoporoides*. Nature **171**: 435.

———, **Bottomley, W., and Mortimer, P. I.** 1954. Variation in the main alkaloids of *Duboisia myoporoides* and *D. leichardtii*. I. *Duboisia myoporoides*. Austral. J. Appl. Sci. **5**: 258–275.

Lopes, J. N. C., Lopes, J. L. C., Vichnewski, W., Rodrigues, D. C. and Gottlieb, O. R. 1991. Chemical variability of *Vanillosmopsis erythropappa*. An. Acad. bras. Ci. **63**: 21–22.

Louda, S. M. and Rodman, J. E. 1983a. Ecological patterns in the glucosinolate content of a native mustard, *Cardamine cordifolia*, in the Rocky Mountains. J. Chem. Ecol. **9**: 397–422.

——— **and** ———. 1983b. Glucosinolae content in relation to insect herbivory and habitat for a native crucifer, *Cardamine cordifolia*. Biochem. Syst. Ecol. **13**: 199–207.

Lumaret, R. 1988. Cytology, genetics and evolution in the genus *Dactylis*. Chem. Rubber Co. Crit. Rev. Plant Sci. **7**: 55–91.

Mabberley, D. J. 1997. The Plant Book. Cambridge University Press, Cambridge, U.K.

Mabry, T. J. 1973. The chemistry of geographic races. Pure Appl. Chem. **34**: 377–400.

———, **DiFeo, Jr., D. R., Sakakibara, D. R., Bohnstedt, Jr., C. F. and Seigler, D.** 1977. The natural products chemistry of *Larrea*. Pages 115–134 in T.J. Mabry, J. H. Hunziker and D. R. DiFeo, Jr., (eds.), Creosote Bush—Biology and Chemistry of Larrea in New World Deserts. Dowden, Hutchinson and Ross, Stroudsberg, Pennsylvania.

———, **Hunziker, J. H. and DiFeo, Jr., D. R.** (eds.) 1977. Creosote Bush—Biology and Chemistry of *Larrea* in New World Deserts. Dowden, Hutchinson and Ross, Stroudsberg, Pennsylvania.

Maquire, B. 1943. A monograph of the genus *Arnica*. Brittonia **4**: 386–510.

Manheim Jr., B. S., Mulroy, T. W., Hogness, D. K., and Kerwin, J. L. 1979. Interspecific variation in leaf wax of *Dudleya*. Biochem. Syst. Ecol. **7**: 17–19.

Manninen, A.-M., Vuorinen, M. and Holopainen, J. K. 1998. Variation in growth, chemical defense, and herbivore resistance in Scots pine provenances. J. Chem. Ecol. **24**: 1315–1331.

Manos, P. S. 1997. Systematics of *Nothofagus* (Nothofagaceae) based on rDNA spacer sequences (ITS): taxonomic congruence with morphology and plastic sequences. Amer. J. Bot. **84**: 1137–1155.

Maquire, B. 1943. A monograph of the genus *Arnica*. Brittonia **4**: 386–510.

Marchant, Y. Y., Ganders, F. R., Wat, C.-K. and Towers, G. H. N. 1984. Polyacetylenes in Hawaiian *Bidens*. Biochem. Syst. Ecol. **12**: 167–178.

Markham, K. R. and Godley, E. J. 1972. Chemotaxonomic studies in *Sophora* 1. An evaluation of *Sophora microphylla* Ait. N. Z. J. Bot. **10**: 627–640.

———, **Porter, L. J., Mues, R., Zinsmeister, H. D. and Brehm, B. G.** 1976. Flavonoid variation in the liverwort *Conocephalum conicum*: evidence for geographic races. Phytochemistry **15**: 147–150.

Marr, K. L. and Tang, C. S. 1992. Volatile insecticidal compounds and chemical variability of Hawaiian *Zanthoxylum* (Rutaceae) species. Biochem. Syst. Ecol. **20**: 209–217.

Martin, J. D., Caballero, P., Fernandez, J. J., Norte, M., Perez, R. and Rodriquez, M. L. 1989. Metabolites from *Laurencia obtusa*. Phytochemistry **28**: 3365–3368.

Martin, P. G. and Dowd, J. M. 1993. Using sequences of *rbc*L to study phylogeny and biogeography of *Nothofagus* species. Austral. Syst. Bot. **6**: 441–447.

Martin, S. S., Langenheim, J. H. and Zavarin, E. 1973. Compositional variation of leaf pocket sesquiterpenes in *Trachylobium verrucosum*. Biochem. Syst. **1**: 35–37.

———, **Langenheim, J. H. and Zavarin, E.** 1974. Quantitativee variation in leaf pocket resin composition in *Hymenaea courbaril*. Biochem. Syst. Ecol. **2**: 75–87.

Mártonfi, P., Grejtovsky, A. and Repcak, M. 1994. Chemotype pattern differentiation of *Thymus pulegioides* on different substrates. Biochem. Syst. Ecol. **22**: 819–825.

Mastenbroek, O. 1983. Patterns of variation in European *Silene pratensis*. Ph.D. Dissertation, University of Utrecht.

———, **Hogeweg, P., Heringa, J., Niemann, G. J., van Nigtevecht, G. and van Brederode, J.** 1984. Isozyme variation in *Silene pratensis*: A response to different environments. Biochem. Syst. Ecol. **12**: 29–36.

———, ———, **van Brederode, J. and van Nigtevecht, G.** 1983a. A pattern analysis of the geographic distribution of flavone-glycosylating genes in *Silene pratensis*. Biochem. Syst. Ecol. **11**: 91–96.

———, **Maas, J. W., van Brederode, J., Niemann, G. J., and van Nigtevecht, G.** 1982. The geographic distribution of flavone glycosylating genes in *Silene pratensis* (Rafn.) Godron & Gren. (Caryophyllaceae). Genetica **59**: 139–144.

———, **Prentice, H. C., Kamps-Heinsbroek, R., van Brederode, J. and van Nigtevecht, G.** 1983b. Geographic trends in flavone-glycosylating genes and seed morphology in European *Silene pratensis* (Caryophyllaceae). Plant Syst. Evol. **141**: 257–271.

——— **and van Brederode, J.** 1986. The possible evolution of *Silene pratensis* as deduced from present day variation patterns. Biochem. Syst. Ecol. **14**: 165–181.

———, ———, **Niemann, G. J. and van Nigtevecht, J.** 1981. Changes in isoenzyme patterns during ontogeny of *Silene alba*. Biochem. Physiol. Pflanzen **176**: 584–589.

Mastrogiuseppe, J. D. 1976. *Pinus balfouriana*. Botanical Society of America Meeting Abstract, 57–58.

Masuda, M., Abe, T., Sato, S., Suzuki, T. and Suzuki, M. 1997a. Diversity of halogenated secondary metabolites in the red alga *Laurencia nipponica* (Rhodomelaceae, Ceramales) J. Phycol. **33**: 196–208.

———, **Itoh, T., Matsuo, Y. and Suzuki, M.** 1997b. Sesquiterpenoids of *Laurencia majuscula* (Ceramales, Rhodophyta) from the Ryuku Islands, Japan. Phycol. Res. **45**: 59–64.

———, **Takahashi, Y., Okamoto, K., Matsuo, Y. and Suzuki, M.** 1997c. Morphology and halogenated secondary metabolites of *Laurancia snackeyi* (Weber-van Bosse) stat. nov. (Ceramiales, Rhodophyta). Eur. J. Phycol. **32**: 293–301.

Mathela, C. S., Melkani, A. B., Pant, A. and Pande, C. 1988. Chemical variations in *Cymbopogon distans* and their chemosystematic implications. Biochem. Syst. Ecol. **16**: 161–165.

Matsuura, N., Iinuma, M. and Tanaka, T. 1994. Phylogenetic analysis in genus *Euchresta* based on secondary metabolites. Biochem. Syst. Ecol. **22**: 621–629.

Matsuzaki, T., Shinozaki, Y., Suhara, S., Ninomiya, M., Shigematsu, H. and Koiwai, A. 1989. Isolation of glycolipids from the surfrace lipids of *Nicotiana bigelovii* and their distribution in *Nicotiana* species. Agric. Biol. Chem. **53**: 3079–3082.

Mauseth, J. D. 1990. Continental drift, climate, and the evolution of cacti. Cact. Succ. J. (USA) **62**: 301–308.

McCormick, S. P., Bohm, B. A. and Ganders, F. R. 1984. Methylated chalcones form *Bidens torta.* Phytochemistry **23**: 2400–2401.

McDougal, K. M. and Parks, C. R. 1984. Elevational variation in foliar flavoniods of *Quercus rubra* L. Amer. J. Bot. **71**: 301–308.

——— **and** ———. 1986. Environmental and genetic components of flavonoid variation in red oak, *Quercus rubra.* Biochem. Syst. Ecol. **14**: 291–298.

McDougall, I., Embleton, J. J. and Stone, S. B. 1981. Origin and evolution of Lord Howe Island, south west Pacific Ocean. J. Geol. Soc. Austral. **28**: 155–176.

McNeill, J. and Crompton, C. W. 1978. Pollen dimorphism in *Silene alba* (Caryophyllaceae). Can. J. Bot. **56**: 1280–1286.

——— **and Prentice, H. C.** 1981. *Silene pratensis* (Rafn.) Gordon & Grens., the correct name for white campion or white cockle *Silene alba* (Miller) E. H. L. Krause. Nom. Illeg.). Taxon **30**: 27–32.

Mears, J. A. 1980a. The flavonoids of *Parthenium* L. J. Nat. Prod. **43**: 708–715.

———. 1980b. Flavonoid diversity and geographic endemism in *Parthenium.* Biochem. Syst. Ecol. **8**: 361–370.

Melching, S. and König, W. A. 1999. Sesquiterpenes from the essential oil of the liverwort *Conocephalum conicum.* Phytochemistry **51**: 517–523.

———, **Warmers, U., König, W. A. and Muhle, H.** 1999. Two aromadendrene type alcohols from the liverwort *Conocephalum conicum.* Phytochemistry **51**: 277–280.

Miller, H. E., Mabry, T. J., Turner, B. L. and Payne, W. W. 1968. Infraspecific variation of sesquiterpene lactones in *Ambrosia psilostachya* (Compositae). Amer. J. Bot. **55**: 316–324.

Miller, J. M. 1988. A new species of *Degeneria* (Degeneriaceae) from the Fiji Archipelago. J. Arnold Arbor. **69**: 231–236.

———. 1989. The archaic flowering plant family Degeneriaceae: its bearing on an old enigma. Nat. Geog. Res. **5**: 218–231.

Mirov, N. T. 1967. The genus *Pinus.* The Ronalds Press Co., New York.

Mizutani, H., Ohbayashi, K., Umetsu, K. and Hiraoka, N. 1993. Restriction fragment length polymorphisms of medicinal plants and crude drugs. II. Analysis of *Glehnia littoralis* of different geographical origin. Biol. Pharm Bull. **16**: 611–612.

Mockute, D. and Judzentiene, A. 2003. Variability of the essential oils composition of *Achillea millefolium* ssp. *millefolium* growing wild in Lithuania. Biochem. Syst. Ecol. **31**: 1033–1045.

Molyneux, R. J. and James, L. F. 1982. Loco intoxication: indolizidine alkaloids of spotted locoweed (*Astragalus lentiginosus*). Science **216**: 190–191.

———, **Benson, M., Wong, R. Y., Tropea, J. E. and Elbein, A. D.** 1988. Australine, a novel pyrrolizidine alkaloid glucosidase inhibitor from *Castanospermum australe.* J. Nat. Prod. **51**: 1198–2006.

Moore, D. M. 1968. The vascular flora of the Falkland Islands. Brit. Antarctic Sur. Bull. **60**: 1–202.

———, **Harborne, J. B. and Williams, C. A.** 1970. Chemotaxonomy, varaition and geographical distribution of the Empetraceae. Bot. J. Linn. Soc. **63**: 277–293.

Mortimer, P. I. and Wilkinson, S. 1957. Occurrence of nicotine, anabasine, and isopelletierine in *Duboisia myoporoides*. J. Chem. Soc. 3967–3970.

Muñoz-Collazos, S., Soriano-Ferrufino, J., Collins, G. J., Jean, F.-I. and Deslauriers, H. 1993. Variability in the composition of the essential oils of (sic) *Mintostachys andina* in central Bolivia. Phytochemistry **33**: 123–127.

Nagashima, F., Murakami, Y. and Asakawa, Y. 1999. Aromatic compounds from the Ecuadorian liverwort *Marchesinia brachiata*: a revision. Phytochemistry **51**: 1101–1104.

———, **Tanaka, H. and Asakawa, Y.** 1996. Sesqui- and di-terpenoids from the liverwort *Jungermannia vulcanicola*. Phytochemistry **42**: 93–96.

Nahal, I. 1962. Le pin d'Alep (*Pinus halepensis* Mill.) Etude taxinomique, phytogéographique, écologiquer et sylvicole. Ann. ENEF Nancy **19**: 473–686.

Nakatani, N., Bohnstedt, C. and Mabry, T. J. 1973. The origin of *Ambrosia chamissonis* in Chile. Biochem. Syst. **1**: 129–132.

Nakazawa, M., Wakabayashi, M., Ono, M. and Murata, J. 1997. Molecular phylogenetic analysis of *Chrysosplenium* (Saxifragaceae) in Japan. J. Plant Res. **110**: 265–274.

Nash, R. J., Fellows, L. E., Dring, J. V., Stirton, C. H., Carter, D., Hegarty, M. P. and Bell, E. A. 1988. Castanospermine in *Alexa* species. Phytochemistry **27**: 1403–1404.

Negueruela, A. 1990. Essential oils of the genus *Lavandula* L. in Spain. Pages 15–26 in S. C. Bhattacharya and K. L. Sethi (eds.), Proceedings of the 11th International Congress of Essential Oils, Fragrance and Flavors, New Delhi, India 1989. Chemistry—Analysis and Structure. Vol. 4., Aspect Publishing, London. [Cited by Skoula et al., 1996].

Nicholls, K. W. and Bohm, B. A. 1982a. Flavonoids and affinities in some North American lupines. Can. J. Bot. **61**: 708–730.

——— **and** ———. 1982b. Quantitative flavonoid variation in *Lupinus sericeus*. Biochem. Syst. Ecol. **10**: 225–231.

Niemann, G. J., Kriek, D., van Nigtevecht, G. and van Brederode, J. 1980. The significance of *Silene alba* flavonoids in the plant/environment interaction. Abstracts of IInd Congress of the Federation of European Societies of Plant Physiology, Santiago de Compostela, Spain, pp. 533–534 [Cited by Mastenbroek, 1983.]

Nilsson, M. 1959. The structure of ceroptene. Acta Chem. Scand. **13**: 750–757.

Niwa, H., Ojika, M., Wakamatsu, K., Yamada, K., Hirono, I. and Matsushita, K. 1983. Ptaquiloside, a novel norsesquiterpene glucoside from bracken, *Pteridium aquilinum* var. *latiusculum*. Tetrahedron lett. **24**: 4117–4120.

Nyffeler, R. 2002. Phylogenetic relationships in the cactus family (Cactaceae) based on evidence from *trnK/matK* and *trnL-trnF* sequences. Amer. J. Bot. **89**: 312–326.

Ohashi, H. and Sohma, K. 1970. A review of the genus *Euchresta* (Leguminosae). J. Fac. Sci. Univ. Tokyo III. **10**: 207–231.

Ojika, M., Wakamatsu, K., Niwa, H. and Yamada, K. 1987. Ptaquiloside, a potent carciogen isolated from bracken fern, *Pteridium aquilinum* var. *latiusculum*: structure elucidation based on chemical and spectral evidence, and reactions with aminoacids, nucleotides and nucleosides. Tetrahedron **43**: 5261–5274.

———, **Sugimoto, K., Okazaki, T. and Yamada, K.** 1989. Modification and cleavage of DNA by ptaquiloside, a new potent carcinogen isolated from bracken fern. J. Chem. Soc. Chem. Comm. 1775–1777.

Okada, H. 1986. Karyomorphology and relationship in some genera of Saururaceae and Piperaceae. Bot. Mag. **99**: 289–299.

Omurkamizinova, V. B., Maurel, N. D. and Bikbulatova, T. N. 1991. Flavonoids from *Biebersteinia multifida*. Khim. Prir. Soed. (5) 720–721.

Ornduff, R. 1963. Amphitropical relationships in the herbaceous flora of the Pacific Coast of North America and South America: a symposium. Experimental studies in two genera of Helenieae (Compositae): *Blennosperma* and *Lasthenia*. Quart. Rev. Biol. **38**: 141–150.

———. 1964. Biosystematics of *Blennosperma* (Compositae). Brittonia **16**: 289–295.

———. 1966. A biosystematic survey of the goldfield genus *Lasthenia* (Compositae: Helenieae). Univ. Calif. Pub. Bot. **40**: 1–92.

———. 1974. Cytotaxonomic observations on *Villarsia* (Menyanthacaee). Aust. J. Bot. **22**: 513–516.

———, **Bohm, B. A. and Saleh, N. A. M.** 1973a. Flavonoids of artificial interspecific hybrids in *Lasthenia*. Biochem. Syst. **1**: 147–151.

———, **Bohm, B. A. and Saleh, N. A. M.** 1974. Intraspecific variation of flavonoids in *Lasthenia*. Brittonia **26**: 411–420.

———, **Saleh, N. A. M. and Bohm, B. A.** 1973b. The flavonoids and affinities of *Blennosperma* and *Crocidium* (Compositae). Taxon **22**: 407–412.

Oztunc, A., Imre, S., Lotter, H. and Wagner, H. 1991. Two C-15 bromoalleles form the red alga *Laurencia obtusa*. Phytochemistry **30**: 255–258.

Pacheco, P., Crawford, D. J. and Stuessy, T. F. 1991b. Natural interspecific hybridization in *Gunnera* (Gunneraceae) of the Juan Fernandez Islands, Chile. Pac. Sci. **45**: 389–399.

———, ———, ——— **and Silva O. M.** 1985. Flavonoid evolution in *Robinsonia* (Compositae) of the Juan Fernandez Islands. Amer. J. Bot. **72**: 989–998.

———, ———, ———, and ———. 1993. Flavonoid chemistry and evolution of *Gunnera* (Gunneraceae) in the Juan Fernandez Islands, Chile. Gayana Botanica **50**: 17–30.

———, ———, ——— **and** ———. 1991a. Flavonoid evolution in *Dendroseris* (Compositae, Lactuceae) from the Juan Fernandez Islands, Chile. Amer. J. Bot. **78**: 534–543.

Palmer, J. D. and Zamir, D. 1982. Chloroplast DNA evolution and phylogenetic relationships in *Lycopersicon*. Proc. Nat'l. Acad. Sci., U.S.A. **79**: 5006–5010.

Panetsos, C. P. 1975. Natural hybridization between *Pinus halepensis* and *Pinus brutia* in Greece. Silvae Genet. **24**: 163–168.

Pangarova, T. T. and Zapeschnaya, G. G. 1974. Flavonoids in *Datisca*. Chem. Nat. Prod. **10**: 788.

Pares, J. O., Oksuz, S., Ulubelen, A. and Mabry, T. J. 1981. 6-Hydroxyflavonoids from *Pulicaria dysenterica*. Phytochemistry **20**: 2057.

Parks, C. R. and Wendel, J. F. 1990. Molecular divergence between Asian and North American species of *Liriodendron* (Magnoliaceae) with implications for interpretation of fossil floras. Amer. J. Bot. **77**: 1243–1256.

———, ———, **Sewell, M. M. and Qiu, Y.-L.** 1994. The significance of allozyme variation and introgression in the *Liriodendron tulipifera* complex (Magnoliaceae). Amer. J. Bot. **81**: 878–889.

Payne, W. W. 1964. A re-examination of the genus *Ambrosia* (Compositae). J. Arnold Arbor. **45**: 401–430.

———, **Geissman, T. A., Lucas, A. J. and Saitoh, T.** 1973. Chemosystematics and taxonomy of *Ambrosia chamissonis*. Biochem. Syst. **1**: 21–33.

Peck, M. E. 1961. A Manual of the Higher Plants of Oregon. 2nd ed. Oregon State University Press, Corvallis, OR.

Pedro, L. G., Santos, P. A. G., da Silva, J., Figueiredo, A. C., Barroso, J. G., Deans, S. G., Looman, A. and Scheffer, J. J. C. 2001. Essential oils from Azorean *Laurus azorica*. Phytochemistry **57**: 245–250.

Penfold, A. R., Morrison, F. R. and McKern, H. H. G. 1948a. studies in the Myrtaceae and their essential oils. Part I: The seasonal variation in yield and cineole content of *Melaleuca alternifolia* Cheel. Pages 5–7 in Researches on Essential Oils of the Australian Flora., Vol. I, Part I. Museum of Technology and Applied Science, Sydney.

———, ——— **and** ———. 1948b. Studies in the physiological forms of the Myrtaceae. Part II. The occurrence of physiological forms. Pages 18–19 in *Melaleuca alternifolia* Cheel. In Researches on Essential Oils of the Australian Flora., Vol. I, Part II. Museum of Technology and Applied Science, Sydney.

Pereira, S. I., Santos, P. A. G., Barroso, J. G., Figueiredo, A. C., Pedro, L. G., Salgueiro, L. R., Deans, S. G. and Scheffer, J. J. C. 2000. Chemical polymorphism of the essential oils from populations of *Thymus caespitosus* grown on the island of S. Jorge (Azores). Phytochemistry **55**: 241–246.

Perry, N. B., Brennan, N. J., Van Klink, J. W., Harris, W., Douglas, M. H., McGimpsey, J. A., Smallfield, B. M. and Anderson, R. E. 1997a. Essential oils form New Zealand manuka and kanuka: Chemotaxonomy of *Leptospermum*. Phytochemistry **44**: 1485–1494.

———, **Van Klink, J. W., Brennan, N. J., Harris, W., Anderson, R. E., Douglas, M. H. and Smallfield, B. M.** 1997b. Essential oils from New Zealand manuka and kanuka: Chemotaxonomy of *Kunzea*. Phytochemistry **45**: 1605–1612.

Phillips, W. R., Baj, N. J., Gunatilaka, A. A. L. and Kingston, D. G. I. 1996. C-Geranyl compounds from *Mimulus clevelandii*. J. Nat. Prod. **59**: 495–497.

Phillipson, J. S. 1982. Chemical investigations of herbarium material for alkaloids. Phytochemistry **21**: 2441–2456.

Picci, V., Scotti, A., Mariani, M. and Colombo, E. 1987. Composition of the volatile oil of *Pistacia lentiscus* L. of Sardinian origin. Pages 107–110 in M. Martens, A. Dalen and M. Russwurm (eds.), Flavour Science and Technology, Wiley, New York. [Cited by Castola et al., 2000.]

Picman, A. K. and Bohm, B. A. 1982. Flavonoids of the *Tiarella trifoliata* complex. Biochem. Syst. Ecol. **10**: 139–143.

Pielou, E. C. 1991. After the Ice Age: The Return of Life to Glaciated North America. University of Chicago Press, Chicago.

Pillans, N. S. 1950. Restionaceae in R. S. Adamson and T. M. Salter (eds.) Flora of the Cape Peninsula Cape Town, South Africa.

Pillay, M. and Kenny, S. T. 1996. Structure and inheritance of ribosomal DNA variants in cultivated and wild hop *Humulus lupulus* L. Theor. Appl. Gen. **93**: 333–340.

Piovano, M., Guzmán, G., Garbarino, J. A. and Chamy, M. C. 1997. *Rhizoplaca melanophthalma*, a new chemical race. Biochem. Syst. Ecol. **25**: 359–360.

Pocknall, D. T. and Mildenhall, D. C. 1984. Late Oligocene-early Miocene spores and pollen from Southland, New Zealand. New Zealand Geological Survey, Paleontology Bull. **51**: 1–66.

Polhill, R. M. 1991. Papilionoideae. Tribe 2. Sophoreae Sprengel (1818). Pages 213–230 in R. M. Polhill and P. H. Raven (eds.) Advances in legume systematics. Part 1. Proceedings of the International Legume Conference, Kew 2. Kew, U.K.

Popovic, M., Djurkovic, R., Gasic, O., Pal, B., Dutchewska, H. and Kuzmanov, B. 1992. Chemical and cytological investigation of *Thalictrum minus* from Vojvodina region. Biochem. Syst. Ecol. **20**: 255–258.

Porter, D. M. 1983. Vascular plants of the Galapagos: Origins and dispersal. Pages 33–96 in R. I. Bowman, M. Berson and A. E. Leviton (eds.) Patterns of Evolution in Galápagos Organisms. American Association for the Advancement of Science, Pacific Division. San Francisco.

Porter, L. J. 1981. Geographic races of *Conocephalum* (Marchantiales) as defined by flavonoid chemistry. Taxon **30**: 739–748.

——— **and Mabry, T. J.** 1972. Origin of the Texas Gulf Coast island populations of *Ambrosia psilostachya*: a numerical study using terpenoid data. Phytochemistry **11**: 715–723.

Powell, A. M. 1974. Taxonomy of *Perityle* section *Perityle* (Compositae—Peritylinae). Rhodora **76**: 229–306.

Powell, E. A. and Kron, K. A. 2002. Hawaiian blueberries and their relatives—a phylogenetic analysis of *Vaccinium* sections *Macropelma*, *Myrtillus*, and *Hemimyrtillus* (Ericaceae). Syst. Bot. **27**: 768–779.

Preece, R. C., Bennett, K. D. and Carter, J. R. 1986. The Quaternary palaeobotany of Innaccessible Island (Tristan da Cunha group). J. Biogeog. **13**: 1–33.

Premoli, A. C., Kitzberger, T. and Veblen, T. T. 2000. Isozyme variation and recent biogeographical history of the long-lived conifer *Fitzroya cupressoides*. J. Biogeog. **27**: 251–260.

Prentice, H. C. 1979. Numerical analysis of infraspecific variation in European *Silene alba* and *S. dioica* (Caryophyllaceae). Bot. J. Linn. Soc. **78**: 181–212.

Proksch, P. and Clark, C. 1987. Systematic implications of chromenes and benzofurans from *Encelia* (Asteraceae). Phytochemistry **26**: 171–174.

———, **Politt, U., Wollenweber, E., Wray, V. and Clark, C.** 1988. Epicuticular flavonoids from *Encelia*. Planta Med. **54**: 542–546.

Prus-Glowacki, W. and Stephan, B. R. 1994. Genetic variation of *Pinus sylvestris* from Spain in relation to other European populations. Silv. Gen. **43**: 7–14.

Purdie, A. W. 1984. Some flavonoid components of *Carmichaelia* (Papilionaceae)—a chemotaxonomic survey. N. Z. J. Bot. **22**: 7–14.

Pureswaran, D. S., R. Gries and J. H. Borden. 2004. Quantitative variation in monoterpenes in four species of conifers. Biochem. Syst. Ecol. **32**: 1109–1136.

Quammen, D. 1996. The Song of the Dodo. Scribner, New York.

Rafii, Z., Cool, L. G., Jonas, R. and Zavarin, E. 1992b. Chemical diversity in *Cupressus bakeri* 1. Megagametophyte fatty acids. Biochem. Syst. Ecol. **20**: 25–30.

———, ——— **and Zavarin, E.** 1992a. Variability of foliar mono- and sesquiterpenoids of *Cupressus bakeri*. Biochem. Syst. Ecol. **20**: 123–131.

——— **and Dodd, R. S.** 1996. Geographical variation in terpene composition of *Pinus nigra* Arn. Amer. J. Bot. **83**: 137 (Supplement).

——— **and** ———. 1998. Genetic diversity among coastal and Andean natural populations of *Araucaria araucana* (Molina) K. Koch. Biochem. Syst. Ecol. **26**: 441–451.

———, ——— **and Fromard, F.** 1996. Biogeographic variation in foliar waxes of mangrove species. Biochem. Syst. Ecol. **24**: 341–345.

Ragan, M. A. and Glombitza, K.-W. 1986. Phlorotannins, brown algal polyphenols. Prog. Phycol. Res. **4**: 129–241.

Rainey, F. 1946. Quinine hunters in Ecuador. National Geographic Magazine **89**: 341–363.

Rajakaruna, N., Siddiqui, M. Y., Whitton, J., Bohm, B. A. and Glass, A. D. M. 2003. Differential responses to Na+/K+ and Ca2+/Mg2+ in two edaphic races of *Lasthenia californica* (Asteraceae) complex: A case for parallel evolution of physiological traits. New Phytol. **157**: 93–103.

——— **and Whitton, J.** 2004. Trends in the evolution of edaphic specialists with an example of parallel evolution in the *Lasthenia california* complex. Pages 103–110 in Q. C. B. Cronk, J. Whitton, R. H. Lee and I. E. P. Taylor (eds.) Plant Adaptations: Molecular Genetics and Ecology. NRC Research Press, Ottawa, Ontario.

Raven, P. H. (ed.) 1963. Amphitropical relationships in the herbaceous flora of the Pacific Coast of North America and South America: a symposium. Amphitropical relationships in the floras of North and South America. Quart. Rev. Biol. **38**: 151–177.

———. 1972. Plant species disjunctions: A summary. Ann. Missouri Bot. Gard. **59**: 234–246.

———. 1979. A survey of reproductive biology in Onagraceae. N. Z. J. Bot. **17**: 575–593.

——— **and Axelrod, D. I.** 1974. Angiosperm biogeography and past continental movements. Ann. Missouri Bot. Gard. **61**: 539–673.

——— **and** ———. 1978. Origin and realtionships of the Californian flora. Univ. Calif. Pub. Bot. **72**: 1–134.

Reynolds, T. 1986. Contribution to the phytochemistry of the East African tetraploid shrubby aloes and their diploid allies. Bot. J. Linn. Soc. **92**: 383–392.

———. 1990. Comparative chromatographic patterns of leaf exudate components from shrubby aloes. Bot. J. Linn. Soc. **102**: 273–285.

Rezzi, S., Cavaleiro, C., Bighelli, A., Salgueiro, L., Proenca da Cunha, A. and Casanova, J. 2001. Intraspecific chemical variability of the leaf essential oil of *Juniperus phoenicea* subsp. *turbinata* from Corsica. Biochem. Syst. Ecol. **29**: 179–188.

Rick, C. M. and Fobes, J. F. 1975. Allozymes of Galapagos tomatoes: polymorphism, geographic distribution, and affinities. Evolution **29**: 443–457.

——— **and Tansley, S. D.** 1981. Genetic variation in *Solanum pennellii*: comparisons with two other sympatric tomato species. Plant Syst. Evol. **139**: 11–45.

Rizk, A. M., Hammouda, F. M., Rimpler, H. and Kamel, A. 1986. The iridoids and flavonoids of *Teucrium polium* herb. Planta Med. 87–88.

Robinson, G. M., Robinson, R. and Todd, A. R. 1934. Experiments on the synthesis of anthocyanins. Part XIX. 5-Glucosidylapigeninidin, believed to be identical with gesnerin, an anthocyanin of *Gesneria fulgens*. J. Chem. Soc. 809–813.

Robles, C. and Garzino, S. 1998. Essential oil composition of *Cistus albidus* leaves. Phytochemistry **48**: 1341–1345.

——— **and** ———. (2000) Infraspecific variability in the essential oil composition of *Cistus monspeliensis* leaves. Phytochemistry **53**: 71–75.

Rodman, J. E. 1974. Systematics and evolution of the genus *Cakile* (Cruciferae). Contrib. Gray Herb. **205**: 3–146.

———. 1976. Differentiation and migration of *Cakile* (Cruciferae): seed glucosinolate evidence. Syst. Bot. **1**: 137–148.

Rodriguez, A. E., Tingey, W. M. and Mutschler, M. A. 1993. Acylsugars of *Lycopersicon pennellii* deter settling and feeding of the green peach aphid (Homoptera: Aphidae). J. Econ. Entomol. **86**: 34–49.

Rosendahl, O. C. 1963. Trees and Shrubs of the Upper Midwest. Univ. of Minnesota Press, Minneapolis, MN.

Rossetto, M., Slade, R. W., Baverstock, P. R., Henry, R. J. and Lee, S. L. 1999. Microsatellite variation and assessment of genetic structure in tea tree (*Melaleuca alternifolia*-Myrtaceae). Mol. Ecol. **8**: 633–643.

Rovirosa, J., Moena, J. and San-Martin, A. 1988. Two chemical types of the red alga *Plocamium cartilagineum*. Biochem. Syst. Ecol. **16**: 593–595.

Rowland, C. Y., Blackman, A. J., D'Arcy, B. R. and Rintoul, G. B. 1995. Comparison of organic extractives found in leatherwood (*Eucryphia lucida*) honey and leatherwood flowers and leaves. J. Agric. Food Chem. **43**: 753–763.

Rundel, P. 1980. The ecological distribution of C_4 and C_3 grasses in the Hawaiian Islands. Oecologia **45**: 354–359.

Rycroft, D. S. 1996. Fingerprinting of plant extract using NMR spectroscopy: application to small samples of liverworts. Chem. Comm. (18) 2187–2188.

——— **and Cole, W. J.** 2001. Hydroquinone derivatives and monoterpenes from the Neotropical liverwort *Plagiochila rutilans*. Phytochemistry **57**: 479–488.

———, ———, **Aslam, N., Lamont, Y. M. and Gabriel, R.** 1999. Killarniensolide, methyl orsellinates and 9,10-dihydrophenanthrenes from the liverwort *Plagiochila killarniensis* from Scotland and the Azores. Phytochemistry **50**: 1167–1173.

———, ——— **and Rong, S.** 1998. Highly oxygenated naphthalenes and acetophenones from the liverwort *Adelanthus decipiens* from the British Isles and South America. Phytochemistry **48**: 1351–1356.

———, **Heinrichs, J., Cole, W. J. and Anton, H.** 2001. A phytochemical and morphological study of the liverwort *Plagiochila retrorsa* Gottsche, new to Europe. J. Bryology **23**: 23–34.

Rydberg, P. A. 1922. Carduales. North American Flora **33**: 1–110.

Sáez, F. 1999. Essential oil variability of *Thymus baeticus* growing wild in southeastern Spain. Biochem. Syst. Ecol. **27**: 269–276.

Sahuquillo-Balbuena, E. and Lumaret, R. 1995. Natural variation in the subtropical group of *Dactylis glomerata* L.—1. Evidence from enzyme polymorphism. Biochem. Syst. Ecol. **23**: 407–418.

——— **and** ———. 1999. Chloroplast DNA variation in *Dactylis glomerata* L. taxa endemic to the Macaronesian islands. Mol. Ecol. **8**: 1797–1803.

Saito, K., Umeda, M., Takasuki, S., Koyama, K. and Natori, S. 1990. The sesquiterpenoid carcinogen of bracken fern and some analogs from the Pteridaceae. Phytochemistry **29**: 1475–1480.

Saito, M., Umeda, M., Enomoto, M., Hatanaka, Y., Natori, S., Yoshihira, K., Fukuoka, M. and Kuroyanagi, M. 1975. Cytotoxicity and carcinogenicity of pterosins and pterosides, 1-indanone derivatives and bracken fern (*Pteridium aquilinum*). Experientia **31**: 824–831.

Saleh, N. A. M., Bohm, B. A. and Ornduff, R. 1971. Flavonoids of *Lasthenia conjugens* and *Lasthenia fremontii*. Phytochemistry **10**: 611–614.

Salgueiro, L. R., Vila, R., Tomi, F., Figueiredo, A. C., Barroso, J. G.,Cañigueral, S., Casanova, J., Proença da Cunha, A. and Adzet, T. 1997. Variability of essential oils of *Thymus caespitosus* from Portugal. Phytochemistry **45**: 307–311.

Sanders, R. W. 1979. A systematic study of *Agastache* section *Brittonastrum*. Ph.D. Dissertation, University of Texas, Austin.

———, **Stuessy, T. F., Marticorena, C. and Silva O. M.** 1987. Phytogeography and evolution of *Dendroseris* and *Robinsonia*, tree Compositae of the Juan Fernandez Islands. Opera Bot. **92**: 195–215.

Sang, T., Crawford, D. J., Kim, S.-C. and Stuessy, T. F. 1994. Radiation of the endemic genus *Dendroseris* (Asteraceae) in the Juan Fernandez Islands: evidence from sequences of the ITS regions of nuclear ribosomal DNA. Amer. J. Bot. **81**: 1494–1501.

———, ———, **Stuessy, T. F. and Silva, O. M.** 1995. ITS sequences and the phylogeny of the genus *Robinsonia* (Asteraceae). Syst. Bot. **20**: 55–64.

San-Martin, A. and Rovirosa, J. 1986. Variations in the halogenated monoterpene metabolites of *Plocamium cartilagineum* of the Chilean coast. Biochem. Syst. Ecol. **14**: 459–461.

Sargent, C. S. 1965. Manual of the Trees of North America. 2nd ed. Dover Publications Inc., New York, NY.

Sato, M. 1965. The mixture ratio of the lichen genus *Thamnolia* in New Zealand. The Bryologist **68**: 320–324.

Satoh, A., Narita, Y., Endo, N. and Nishimura, H. 1996. Constituents of *Glehnia littoralis*. Biosci. Biotech. Biochem. **60**: 152.

Saucy, F., Studer, J., Aerni, V. and Schneiter, B. 1999. Preference for acyanogenic white clover (*Trifolium repens*) in the vole *Arvicola terrestris*: I. experiments with two varieties. J. Chem. Ecol. **25**: 1441–1454.

Sauer, J. D. 1988. Plant Migration. The Dynamics of Geographic Patterning in Seed Plant Species. University of California Press, Berkeley, CA.

Schappert, P. J. and Shore, J. S. 1999. Effects of cyanogenesis polymorphism in *Turnera ulmifolia* on *Euptoieta hegesia* and potential *Anolis* predators. J. Chem. Ecol. **25**: 1455–1479.

Schiller, G., Conkle, M. T. and Grundwald, C. 1986. Local differentiation among Mediterranean populations of Aleppo pine in their isozymes. Silvae Genet. **35**: 11–19.

——— **and Grundwald, C.** 1987a. Cortex resin monoterpene composition in *Pinus brutia* provenances grown in Israel. Biochem. Syst. Ecol. **15**: 389–394.

——— **and** ———. 1987b. Resin monoterpene in range-wide provenance trials of *Pinus halepensis* in Israel. Silvae Gen. **36**: 109–114.

Schilling, E. E., Panero, J. L. and Eliasson, U. A. 1994. Evidence from chloroplast DNA restriction site analysis on the relationships of *Scalesia* (Asteraceae: Heliantheae). Amer. J. Bot. **81**: 248–254.

Schmidt, A., Bischof-Deichnik, C. and Stahl-Biskup, E. 2004. Essential oil polymorphism of *Thymus praecox* subsp. *arcticus* on the British Isles. Biochem. Syst. Ecol. **32**: 409–421.

Schwartz, A. G. and Redman, R. E. 1988. C_4 grasses from the boreal forest region of northwestern Canada. Can. J. Bot. **66**: 2424–2430.

Seaman, F. C. 1982. Sesquiterpene lactones as taxonomic characters in the Asteraceae. The Botanical Review **48**: 121–592.

——— **and Mabry, T. J.** 1979a. Sesquiterpene lactones of diploid and tetraploid *Ambrosia camphorata*. Biochem. Syst. Ecol. **7**: 3–6.

——— **and** ———. 1979b. Sesquiterpene lactones and species relationships among the shrubby *Ambrosia* taxa. Biochem. Syst. Ecol. **7**: 105–114.

Seguin, J., Muzac, I. and Ibrahim, R. K. 1998. Purification and immunological characterization of a recombinant trimethylflavonol 3′-O-methyltransferase. Phytochemistry **49**: 319–325.

Seigler, D. S., Simpson, B. B., Martin, C. and Neff, J. L. 1978. Free 3-acetoxyfatty acids in floral glands of *Krameria* species. Phytochemistry **17**: 995–996.

———, **Smith, D. M. and Mabry, T. J.** 1975. n-Alkanes from diploid, triploid and tetraploid plants of *Pityrogramma triangularis* flavonoid chemotypes. Biochem. Syst. Ecol. **3**: 5–6.

——— **and Wollenweber, E.** 1983. Chemical variation in *Notholaena standleyi*. Amer. J. Bot. **70**: 790–798.

Sengupta, S. and A. B. Ray. 1987. The chemistry of *Piper* species: a review. Fitoterapia **58**: 147–166.

Sepulveda-Boza, S., S. Delhvi and B. K. Cassels. 1993. Flavonoids from the twigs of *Eucryphia glutinosa*. Phytochemistry **32**: 1301–1303.

Setoguchi, H., Ono, M., Doi, Y., Koyama, H. and Tsuda, M. 1997. Molelcular phylogeny of *Nothofagus* (Nothofagaceae) based on the *atpB-rbcL* intergenic spacer of the chloroplast DNA. J. Plant Res. **110**: 469–484.

Sewell, M. M., Parks, C. R. and Chase, M. W. 1996. Intraspecific chloroplast DNA variation and biogeography of North American *Liriodendron* L. (Magnoliaceae). Evolution **50**: 1147–1154.

Shapiro, J. A., Steffens, J. C. and Mutschler, M. A. 1994. Acylsugars of the wild tomato *Lycopersicon pennellii* in relation to geographic distribution of the species. Biochem. Syst. Ecol. **22**: 545–561.

Shore, J. S. and Obrist, C. M. 1992. Variation in cyanogenesis within and among populations and species of *Turnera* section *Canaligerae* (Turneraceae). Biochem. Syst. Ecol. **20**: 9–15.

Siméoni, P. and Lebot, V. 2002. Identification of factors determining kavalactone content and chemotype in Kava (*Piper methysticum* Forst. f.). Biochem. Syst. Ecol. **30**: 413–424.

Simkin, T. 1984. Geology of Galapagos. Biol. J. Linn. Soc. **21**: 61–75.

Simpson, B. B., Seigler, D. S. and Neff, J. L. 1979. Lipids from the floral glands of *Krameria*. Biochem. Syst. Ecol. **7**: 193–194.

Sims, E. 1968. Sea Kale. Southern Australian Naturalist **42**: 76–77.

Sinclair, W. T., Morman, J., and Ennos, R. 1998. Multiple origins for Scots pine (*Pinus sylvestris* L.) in Scotland: evidence from mitochondrial DNA variation. Heredity **80**: 233–240.

Skoula, M., Abidi, C. and Kokkalou, E. 1996. Essential oil variation of *Lavandula stoechas* L. ssp. *stoechas* growing wild in Crete (Greece). Biochem. Syst. Ecol. **24**: 255–260.

Skvarla, J. J. and Turner, B. L. 1966. Pollen wall ultrastructure and its bearing on the systematic position of *Blennosperma* and *Crocidium* (Composistae). Amer. J. Bot. **53**: 555–563.

Small, E. 1978. A numerical and nomenclatural analysis of morpho-geographic taxa of *Humulus*. Syst. Bot. **3**: 37–76.

Smith, A. C. 1981. Pages 7–13 in Flora Vitiensis Nova: a new flora of Fiji, Vol 2. Pacific Tropic Botanical Garden, Lawai, Kauai, Hawaii.

———. 1985. Pages 37–41 in Flora Vitiensis Nova: a new flora of Fiji, Vol 3. Pacific Tropic Botanical Garden, Lawai, Kauai, Hawaii.

Smith, A. L., Campbell, C. L., Walker, D. B., Hanover, J. W. and Miller, R. O. 1988. Geographic variation in the essential oil monoterpenes of *Liriodendron tulipifera* L. Biochem. Syst. Ecol. **16**: 627–630.

Smith, D. M. 1980. Flavonoid analysis of the *Pityrogramma triangularis* complex. Bull. Torrey Bot. Club **107**: 134–145.

———**, Craig, S. P. and Santarosa, J.** 1971. Cytological and chemical variation in *Pityrogramma triangularis*. Amer. J. Bot. **58**: 292–299.

Snajberk, K. and Zavarin, E. 1986. Monoterpenoid differentiation in relation to morphology of *Pinus remota*. Biochem. Syst. Ecol. **14**: 155–163.

———, ——— **and Bailey, D.** 1979. Systematic studies of *Pinus balfouriana* based on volatile terpenoids from wood and needles and on seed morphology. Biochem. Syst. Ecol. **7**: 269–279.

———, ——— **and Debry, R.** 1982. Terpenoid and morphological variability of *Pinus quadrifolia* and the natural hybridization with *Pinus monophylla* in the San Jacinto Mountains of California. Biochem. Syst. Ecol. **10**: 121–132.

Soltis, D. E. 1984. Autotetraploidy in *Tolmeia menziesii* (Saxifragaceae). Amer. J. Bot. **71**: 1171–1174.

——— **and Bohm, B. A.** 1984. Karyology and flavonoid chemistry of the disjunct species of *Tiarella* (Saxifragaceae) Syst. Bot. **9**: 441–447.

——— **and** ———. 1986. Flavonoid chemisry of diploid and tetraploid cytotypes of *Tolmiea menziesii* (Saxifragacae). Syst. Bot. **11**: 20–25.

———**, Gitzendanner, M. A., Strenge, D. D., and Soltis, P. S.** 1997. Chloroplast DNA intraspecific phylogeography of plants from the Pacific Northwest of North America. Pl. Syst. Evol. 206: 353–373.

———, **Morgan, D. R., Grable, A. Soltis, P. S. and Kuzoff, R.** 1993. Molecular systematics of Saxifragaceae sensu stricto. Amer. J. Bot. **80**: 1056–1081.

———, **Ranker, T. A. and Ness, B.** 1989. Chloroplast DNA variation in a wild plant, *Tolmiea menziesii*. Genetics **121**: 819–826.

———, **Tago-Nakazawa, M., Xiang, Q.-Y., Kawano, S., Murata, J., Wakabayashi, M. and Hibsch-Jetter, C.** 2001. Phylogenetic relationships in *Chrysosplenium* (Saxifragaceae) based on analysis of a combined *rbcL*/*matK* sequence data set. Amer. J. Bot. **88**: 883–893.

Sorsa, M., von Schantz, M., Lokki, J. and Forsén, K. 1968. Variability of essential oil components in *Chrysanthemum vulgare* L. in Finland. Ann. Acad. Sci. Fennoscand., Series A VI, **135**: 1–13. [Cited by Keskitalo *et al.*, 2001.]

Southwell, I. A. and Armstrong, J. A. 1987. Chemical variation within the genus *Zieria*. Phytochemistry **26**: 1687–1692.

———, **Stiff, I. A. and Brophy, J. J.** 1992. Terpinolene varieties of *Melaleuca*. J. Essent. Oil Res. **4**: 363–367.

Souto-Bachiller, F. A., De Jesus-Echevarria, Cárdenas-González, O. E., Acuña-Rodriguez, M. F., Meléndez, P. A. and Romero-Ramsey, L. 1997. Terpenoid composition of *Lippia dulcis*. Phytochemistry **44**: 1077–1086.

Stafford, H. A. 1990. Flavonoid Metabolism. CRC Press, Boca Raton, Florida.

Stahl-Biskup, E. and Laakso, I. 1990. Essential oil polymorphism in Finnish *Thymus* species. Planta Med. **56**: 464–468.

Stanley, S. M. 1986. Anatomy of a regional mass extinction: Plio-Pleistocene decimation of the western Atlantic bivalve fauna. Palaios **1**: 17–36.

Star, A. E., Rösler, H., Mabry, T. J. and Smith, D. 1975a. Flavonoid and ceroptin pigments from frond exudates of *Pityrogramma triangularis*. Phytochemistry **14**: 2275–2278.

———, **Seigler, D. S., Mabry, T. J. and Smith, D. M.** 1975b. Internal flavonoid patterns of diploids and tetraploids of two exudate chemotypes of *Pityrogramma triangularis* (Kaulf.) Maxon. Biochem. Syst. Ecol. **2**: 109–112.

Stebbins, G. L. and Zohary, D. 1959. Cytogenetics and evolutionary studies in the genus *Dactylis* 1. Morphology, distribution and interrelationships of the diploid subspecies. Univ. Calif. Pub. Bot. **31**: 1–40.

Steere, W. C. 1945a. The *Cinchona* bark industry in South America. Sci. Monthly **61**: 114–126.

———. 1945b. The discovery and distribution of *Cinchona pitayensis* in Ecuador. Bull. Torr. Bot. Club **72**: 464–471.

Steinberg, P. D. 1984. Algal chemical defense against herbivores; allocation of phenolic compounds in the kelp *Alaria marginata*. Science **223**: 405–407.

———. 1985. Feeding preferences of *Tegula funebralis* and chemical defenses in marine brown algae. Ecol. Monographs **55**: 333–349.

———. 1986. Chemical defenses and the susceptibility of tropical marine algae to herbivores. Oecologia **69**: 628–630.

———. 1988. The effects of quantitative and qualitative variation in phenolic compounds on feeding in three species of marine invertebrate herbivores. J. Exper. Marine Biol. Ecol. **120**: 221–237.

———. 1989. Biogeographical variation in brown algal polyphenolics and other secondary metabolites: comparison between temperate Australasia and North America. Oecologia **78**: 373–382.

Stevens, J. F., Ivancic, M., Hsu, V. L. and Deinzer, M. L. 1997. Prenylflavonoids from *Humulus lupulus*. Phytochemistry **44**: 1575–1585.

———, **Taylor, A. W., Nickerson, G. B., Ivancic, M., Henning, J., Haunold, V. and Deinzer, M. K.** 2000. Prenylflavonoid variation in *Humulus lupulus*: distribution and taxonomic significance of xanthogalenol and 4′-O-methylxanthohumol. Phytochemistry **53**: 759–775.

———, ———, **Clawson, J. E., and Deinzer, M. L.** 1999a. Fate of xanthohumol and related prenylflavonoids from hops to beer. J. Agric. Food Chem. **47**: 2421–2428.

———, ——— **and Deinzer, M. L.** 1999b. Quantitative analysis of xanthohumol and related prenylflavonoids in hops and beer by liquid chromatography-tandem mass spectroscopy. J. Chromatog. **A832**: 97–107.

Stevens, P. F. 1971. A classification of the Ericaceae: subfamilies and tribes. Bot. J. Linn. Soc. **64**: 1–53.

Stierle, D. B. and Sims, J. J. 1979. Marine natural products—V. Polyhalogenated cyclic monoterpenes from the red alga *Plocamium cartilagineum* of Antarctica. Tetrahedron **35**: 1261–1265.

Stoutamire, W. P. 1954. *Gaillardia pulchella* and *Gaillardia aristata*: The morphological and cytological variation and the taxonomy of their wild and cultivated races. Ph.D. Dissertation, Indiana University, Bloomington.

———. 1977. Chromosomal races of *Gaillardia pulchella* (Asteraceae). Brittonia **29**: 297–309.

Stuessy, T. F., Foland, K. A., Sutter, J. F., Sanders, R. W. and Silva, O. M. 1984. Botanical and geological significance of potassium-argon dates from the Juan Fernández Islands. Science **225**: 49–51.

———, **Marticorena, C., Rodriquez, R. R., Crawford, D. J. and Silva, O. M.** 1992. Endemism in the vascular flora of the Juan Fernandez Islands. Aliso **13**: 297–307.

———, **Swenson, U., Crawford, D. J., Anderson, G. and Silva, O. M.** 1998. Plant conservation in the Juan Fernandez Archipelago, Chile. Aliso **16**: 89–101.

Sun, B.-Y. and Stuessy, T. F. 1998. Preliminary observations on the evolution of endemic angiosperms of Ullung Island, Korea. Pages 181–202 in T. F. Stuessy and M. Ono (eds.) Evolution and Speciation of Island Plants. Cambridge University Press, Cambridge, UK.

Suzuki, M. and Kurosawa, E. 1978. Two new halogenated sesquiterpenes from the red alga *Laurencia majuscula*. Tetrahedron Lett. 4805–4808.

———, **Furusaki, A., Hashiba, N. and Kurosawa, E.** 1979. The structures and absolute stereochemistry of two halogenated chamigrenes from the red alga *Laurencia majuscula* Harvey. Tetrahedron Lett. 879–882.

———, **Kurosawa, E. and Kurata, K.** 1987. Majusculone, a novel norchamigrane-type metabolite from the red alga *Laurencia majuscula* Harvey. Bull. Chem. Soc. Japan **60**: 3795–3796.

Swain, T. 1975. Evolution of flavonoid compounds. Pages 1096–1129 in J. B. Harborne, T. J. Mabry and H. Mabry (eds.) The Flavonoids, Chapman & Hall, London.

Swenson, U. 1994. The genus *Ischnea* (Asteraceae, Senecioneae) in New Guinea. Plant Syst. Evol. **191**: 247–263.

———. 1995a. Systematics of the Blennospermatinae (Asteraceae, Senecioneae). Ph.D. Dissertation, Uppsala University.

———. 1995b. Systematics of *Abrotanella*, an Amphi-pacific genus of Asteraceae (Senecioneae). Plant Syst. Evol. **197**: 149–193.

———, **Backland, A., McLoughlin, S. and Hill, R. S.** 2001a. *Nothofagus* biogeography revisited with special emphasis on the enigmatic distribution of subgenus *Brassospora* in New Caledonia. Cladistics **17**: 28–47.

——— **and Bremer, K.** 1997. Pacific biogeography of the Asteraceae genus *Abrotanella* (Senecioneae, Blennospermatinae). Syst. Bot. **22**: 493–508.

——— **and Hill, R. B.** 2001. Most parsimonious areagrams versus fossils: the case of *Nothofagus* (Nothofagaceae). Austral. J. Bot. **49**: 367–376.

———, ——— **and** ———. 2001b. Biogeography of *Nothofagus* supports the sequence of Gondwana break-up. Taxon **50**: 1025–1041.

———, **Yong, J. Y. and Bohm, B. A.** 1999. Flavonoid chemistry of Blennospermatinae, a trans-Pacific disjunct subtribe of Senecioneae (Asteraceae). Plant Syst. Evol. **216**: 231–241.

Sykes, W. R. and Godley, E. J. 1968. Transoceanic dispersal in *Sophora* and other genera. Nature **218**: 495–496.

Sytsma, K. J., Smith, J. F. and Berry, P. E. 1991. The use of chloroplast DNA to assess biogeography and evolution of morphology, breeding systems, and flavonoids in *Fuchsia* sect. *Skinnera* (Onagraceae). Syst. Bot. **16**: 257–269.

Szweykowski, J. and Bobowicz, M. A. 1979. Morphological variation of *Conocephalum conicum* (L.) Dum. (Hepaticae, Marchantiales) in Poland. Bull. Acad. Polon. Sci., Ser. Sci. Biol., CI, II, **27**: 21–35 [Cited by Porter, 1981.]

Takahashi, K., Nagahama, S., Nakashima, T. and Suenaga, H. 2003. Chemotaxonomy on the leaf constituents of *Thujopsis dolabrata* Sieb. et Zucc.—analysis of acidic extracts. Biochem. Syst. Ecol. **31**: 723–738.

Takhajan, A. 1969. Flowering Plants: Origin and Dispersal. Smithsonian Institution Press, Washington, D.C.

Tanaka, J., Higa, T., Bernardinella, G. and Jefford, C. W. 1988. Itomanindoles A and B, methylsulfinylindoles from *Laurencia brongniartii*. Tetrahedron Lett. **29**: 6091–6094.

———, ———, ——— **and** ———. 1989. Sulfur-containing polybromindoles from the red alga *Laurencia brongniartii*. Tetrahedron **45**: 7301–7310.

Tanaka, N., Yuhara, H., Wada, H., Murakami, T., Cambie, R. C. and Braggins, J. E. 1993. Phenolic compounds of *Pteridium esculentum*. Phytochemistry **32**: 1037–1039.

Taylor, F. and Hill, R. 1996. A phylogenetic analysis of Eucryphiaceae. Aust. Syst. Bot. **9**: 735–748.

Thompson, J. W., Scotter, G. W., and Ahti, T. 1969. Lichens of the Great Slave Lake Region, Northwest Territories, Canada. The Bryologist **72**: 137–177.

Thompson, P. A. 1973. Seed germination in relation to ecological and geographical distribution. Pages 93–119 in V. H. Heywood (ed.) Taxonomy and Ecology. Academic Press, London.

Thornton, I. 1971. Darwin's Islands. A Natural History of the Galapagos. The Natural History Press, Garden City, New York.

Tian, C., Nan, P., Chen, J. and Zhong. Y. 2004. Volatile composition of Chinese *Hippophae rhamnoides* and its chemotaxonomic implications. Biochem. Syst. Ecol. **32**: 431–441.

Tiffney, B. H. 1985. Perspectives on the origin of the floristic similarity between Asia and estern North America. J. Arnold Arbor. **66**: 73–94.

Timmermann, B. N. and Mabry, T. J. 1983. 6-Methoxyflavonols from disjunct populations of *Brickellia cylindracea* (Compositae). Biochem. Syst. Ecol. **11**: 37–39.

Titcomb, M. 1948. Kava in Hawaii. J. Polynesian Sociol. **57**: 105–201.

Tobe, H., Carlquist, S., and Iltis, H. H. 1999. Reproductive anatomy and relationships of *Setchellanthus caeruleus*, Setchellanthaceae. Taxon **48**: 277–283.

Tobolski, J. J. and Hanover, J. W. 1971. Genetic variation in the monoterpenes of Scotch pine. For. Sci. **17**: 293–299.

Tomás, F., Nieto, J. L., Barberán, F. A. T. and Ferreres, F. 1986. Flavonoids from *Phlomis lychnitys*. Phytochemistry **25**: 1253–1254.

Toyota, M., Koyama, H., Mizutani, M. and Asakawa, Y. 1996. (–)-*ent*-Spathulenol isolated from liverworts is an artefact. Phytochemistry **41**: 1347–1350.

———**, Saito, T., Matsunami, J. and Asakawa, Y.** 1997. A comparataive study of three chemotypes of the liverwort *Conocephalum conicum* using volatile constituents. Phytochemistry **44**: 1265–1270.

Tryon, R. M. 1962. Taxonomic fern notes. II. *Pityrogramma* (including *Trismeria*) and *Anogramma*. Contrib. Gray. Herb. **189**: 52–76.

——— **and Tryon, A. L.** 1982. Ferns and Allied Plans, with Special Reference to Tropical America, Springer-Verlag, New York.

Tschesche, R., Dehivi, S., Sepulveda, S. and Breitmaier, E. 1979. Eucryphin, a new chromones rhamnoside from the bark of *Eucryphia cordifolia*. Phytochemistry **18**: 867–869.

Turner, B. L. 1972. Chemosystematic data: their use in the study of disjunctions. Ann. Missouri Bot. Gard. **59**: 152–164.

——— **and Whalen, M.** 1975. Taxonomic study of *Gaillardia pulchella* (Asteraceae—Heliantheae). Wrightia **5**: 189–192.

Valant-Vetschera, K. M. and Wollenweber, E. 1988. Leaf flavonoids of the *Achillea millefolium* group. II. Distribution patterns of free aglycones in leaf exudates. Biochem. Syst. Ecol. **16**: 605–614.

——— **and** ——— 1996. Comparative analysis of leaf exudate flavonoids in *Achillea* sect. *Filipendulinae*. Biochem. Syst. Ecol. **24**: 435–446.

——— **and** ——— 2001. Exudate flavonoid aglycones in the alpine species of *Achillea* sect. *Ptarmica*: chemosystematics of *A. moschata* and related species (Compositae—Anthemideae). Biochem. Syst. Ecol. **29**: 149–159.

Valdebenito, H. A., Stuessy, T. F. and Crawford, D. J. 1990a. A new biogeographic connection between islands in the Atlantic and Pacific Oceans. Nature **347**: 549–550.

———, ——— **and** ———. 1990b. Synonymy in *Peperomia berteroana* (Piperaceae) results in biological disjunction between Pacific and Atlantic Oceans. Brittonia **42**: 121–124.

———, ———, ——— **and Silva, O. M.** 1992a. Evolution of *Peperomia* (Piperaceae) in the Juan Fernandez Islands, Chile. Plant Syst. Evol. **182**: 107–119.

———, ———, ——— **and** ———. 1992b. Evolution of *Erigeron* (Compositae) in the Juan Fernandez Islands, Chile. Syst. Bot. **17**: 470–480.

Valentini, G., Arnold, N. and Bellomaria, B. 1993. Etude chimique comparative des huiles essentielles de quatre populations de *Lavandula stoechas* L. Plante med. phytother. **224**: 289–299.

Valls, R., Mesguiche, V., Piovetti, L., Prost, M. and Peiffer, G. 1996. Meroditerpenes from the brown alga *Cystoseira amentacea* var. *stricta* collected off the French Mediterranean coast. Phytochemistry **41**: 1367–1371.

van Brederode, J. and Kamps-Heinsbroek, R. 1981. Structure and biosynthesis of vitexin 2″-O-xyloside in *Silene alba*. Zeit. Naturforsch. **36c**: 484–485.

——— **and van Nigtevecht, G.** 1975. Diminance relationships between allelic glycosyltransferase genes in *Melandrium*. An enzyme-kinetic approach. Theor. Appl. Genet. **46**: 353–358.

van Heerden, F. R., van Wyk, B.-E., Viljoen, A. M. and Steenkamp, P. A. 2003. Phenolic variation in wild populations of *Aspalathus linearis* (rooibos tea). Biochem. Syst. Ecol. **31**: 885–895.

van Nigtevecht, G. and van Brederode, J. 1972. Flavonoid glycosylation genes in European populations of *Melandrium album* and *Melandrium dioicum*. Genen and Phaenen **15**: 9–13.

Van Steenis, C. G. G. J. 1971. *Nothofagus*, key genus to plant geography, in time and space, living and fossil, ecology and phylogeny. Blumea **19**: 65–98.

———. 1972. *Nothofagus*, key genus to plant geography. Pages 275–288 in D. H. Valentine (ed.) Taxonomy, Phytogeography and Evolution. Academic Press, London.

Van Wyk, B.-E. and Verdoorn, G. H. 1991. Alkaloidal variation in the genus Pearsonia. Biochem. Syst. Ecol. **19**: 685–695.

Vekemans, X. 1992. Evolution of plant breeding systems: *Armeria maritima* (Mill.) Wild. as a study case. Ph.D. Thesis, Université Libre de Bruxelles.

———**, Lambert, A., and Lefebvre, C.** 1992. Isozyme variation at the population level and taxonomy of *Armeria maritima* (Mill.) Willd. Belgian J Botany **125**: 270–275.

Vellekoop, P., Buntjer, J. B., Maas, J. W. and van Brederode, J. 1996. Can the spread of agriculture in Europe be followed by tracing the spread of the weed *Silene latifolia*? A RAPD study. Theor. Appl. Gen. **92**: 1085–1090.

Vidakovic, M. 1991. Conifers—Morphology and Variation. Graficki Zavod Hrvatske, Zagreb.

Vieira, R. F., Grayer, R. J., Paton, A. and Simon, J. E. 2001. Genetic diversity of *Ocimum gratissimum* L. based on volatile oil components, flavonoids and RAPD markers. Biochem. Syst. Ecol. **29**: 287–304.

Vierhapper, F. 1919. Über echten und falschen Vikarismus. Österriche Botanische Zeitschrift **68**: 1–22.

Viljoen, A. M., van Wyk, B.-E. and Newton, L. E. 1999. Plicataloside in *Aloë*. A chemotaxonomic appraisal. Biochem. Syst. Ecol. **27**: 507–517.

———, ——— **and Van Heerden, F. R.** 2002. The chemotaxonomic value of the diglycoside anthrone homonataloside B in the genus *Aloë*. Biochem. Syst. Ecol. **33**: 35–43.

Vogelmann, J. E. 1983. A biosystematic study of *Agastache* section *Agastache* (Labiatae). Ph.D. Dissertation, Indiana State University.

———. 1984. Flavonoids of of *Agastache* section *Agastache* Biochem. Syst. Ecol. **12**: 363–366.

———. 1985. Crossing relationships among North American and eastern Asian populations of *Agastache* section *Agastache* (Labiatae). Syst. Bot. **10**: 445–452.

——— **and Gastony, G. J.** 1987. Electrophoretic enzyme analysis of North American and eastern Asian populations of *Agastache* section *Agastache* (Labiatae). Amer. J. Bot. **74**: 385–393.

Vokou, D. and Bessiere, J.-M. 1985. Volataile constituents of *Teucrium polium*. J. Nat. Prod. **48**: 498–499.

———**, Kokkini, S. and Bessiere, J.-M.** 1993. Geographic variation of Greek oregano (*Origanum vulgare* ssp. *hirtum*) essential oils. Biochem. Syst. Ecol. **21**: 287–295.

von Rudloff, E. 1962. Gas–liquid chromatography of terpenes. VI. The volatile oil of *Thuja plicata* Donn. Phytochemistry **1**: 195–202.

———. 1973. Geographical variation in the terpene composition of the leaf oil of Douglas fir. Pure. Appl. Chem. **34**: 401–410.

———. 1975. Volataile leaf oil analysis in chemosystematic studies of North American conifers. Biochem. Syst. Ecol. **2**: 131–167.

——— **and Lapp, M. S.** 1979. Populational variation in the leaf oil terpene composition of western red cedar, *Thuja plicata*. Can. J. Bot. **57**: 476–479.

——— **and** ———. 1991. Chemosystematic studies in the genus Pinus. VII. The leaf oil terpene composition of ponderosa pine, *Pinus ponderosa*. Can. J. Bot. **70**: 374–378.

———, ——— **and Yeh, F.** 1988. Chemostematic study of *Thuja plicata*: multivariate analysis of leaf oil terpene composition. Biochem. Syst. Ecol. **16**: 119–125.

Vuilleumier, F. 1971. Pleistocene changes in the fauna and flora of South America. Science **173**: 771–780.

Wagner, W. L., Herbst, D. R. and Sohmer, S. H. 1990, 1999. Manual of the Flowering Plants of Hawai'i. Vols. 1 and 2. University of Hawai'i Press and Bishop Museum Press, Honolulu.

——— **and Funk, V. A.** (eds.) 1995. Hawaiian Biogeography: Evolution on a Hot Spot Archipelago., Smithsonian Institution Press, Washington, D.C.

Wagstaff, S. J. and Dawson, M. I. 2000. Classfication, origin, and patterns of diversification of *Corynocarpus* (Corynocarpaceae) inferred from DNA sequences. Syst. Bot. **25**: 134–149.

———**, Heenan, P. B. and Sanderson, M. J.** 1999. Classification, origins, and patterns of diversification in New Zealand Carmichaelinae (Fabaceae). Amer. J. Bot. **86**: 1346–1356.

Warmers, U. and König, W. A. 1999. Sesquiterpene constituents of the liverwort *Bazzania trilobata*. Phytochemistry **52**: 99–104.

Warnock, M. J. 1990a. New taxa and combinations in North American *Delphinium* (Ranunculaceae). Phytologia **68**: 1–6.

———. 1990b. Taxonomy and ecological review of California Delphinium. Collectanea Botanica **19:** 45–74.

Wassel, G. M. and Ahmed, S. S. 1974. Chemical composition of the wild Egyptian plant *Teucrium polium*. Die Pharm. **29**: 351–352.

Watson, L. E., Elisens, W. J., and Estes, J. R. 1994. Genetic differentiation in populations of *Marshallia graminiifolia s. lat.* (Asteraceae). Biochem. Syst. Ecol. **22**: 577–584.

Watts, W. A. 1983. Vegetational history of the eastern United States 25,000 to10,000 years ago. Pages 294–310 in S. C. Porter (ed.) Late-quaternary Environments of the United States, Vol. 1. The late Pleistocene. University of Minnesota Press, Minneapolis, MN.

Wells, T. C. 1992. Population variation in North American *Menziesia* (Ericaceae). Ph.D. dissertation. University of British Columbia, Vancouver, B.C.

——— **and Bohm, B. A.** 1988. Flavonoids of *Luetkea pectinata* (Rosaceae: Spiraeoideae). Biochem. Syst. Ecol. **16**: 479–483.

——— **and** ———. 1994. Isozyme variation in North American *Menziesia* (Ericaceae). Syst. Bot. **19**: 407–423.

Wen, J. 1999. Evolution of eastern Asian and eastern North American disjunct distributions in flowering plants. Ann. Rev. Ecol. Syst. **30**: 421–455.

Wendel, J. F. and Percy, R. G. 1991. Allozyme diversity and introgression in the Galapagos Islands endemic *Gossypium darwinii* and its relationship to continental *G. barbadense*. Biochem. Syst. Ecol. **18**: 517–528.

Wentworth, T. R. 1983. Distributionsa of C_4 plants along environmental and compositional gradients in southeastern Arizona. Vegetatio **52**: 21–34.

Wessels, P. L., Holzapfel, C. W., van Wyk, B.-E. and Marais, W. 1996. Plicataloside, an O–O-diglucosylated naphthalene derivative from *Aloë plicatilis*. Phytochemistry **41**: 1547–1551.

Whittaker, R. H. 1961. Comments on the Klamath region and conifer diversity. Madroño **16**: 5.

Wiggins, I. L. 1971. Flora of the Galapagos Islands. I. L. Wiggins and D. M. Porter (eds.) Stanford University Press, Stanford, CA.

Wilkinson, R. C., Hanover, J. W., Wright, J. W. and Flake, R. H. 1971. Genetic variation in the monoterpene composition of white spruce. For. Sci. **17**: 83–90.

Williams, C. A. and Garnock-Jones, P. J. 1986. Leaf flavonoids and other phenolic glycosides and the taxonomy and phylogeny of *Fuchsia* sect. *Skinnera* (Onagraceae). Phytochemistry **25**: 2547–2549.

———, **Harborne, J. B., Greenham, J., Briggs, B. G. and Johnson, L. A. S.** 1998. Flavonoid patterns and the revised classification of Australian Restionaceae. Phytochemistry **49**: 529–552.

———, ——— **and** ———. 2000. Geographical variation in the surface flavonoids of *Pulicaria dysenterica*. Biochem. Syst. Ecol. **28**: 679–687.

Wisdom, C. S. 1985. Use of chemical variation and predation as plant defenses by *Encelia farinosa* against a specialist herbivore. J. Chem. Ecol. **11**: 1553–1565.

——— **and Rodriguez, E.** 1982. Quantitative variation of the sesquiterpene lactones and chromones of *Encelia farinosa*. Biochem. Syst. Ecol. **10**: 43–48.

Witter, M. S. and Carr, G. D. 1988. Adaptive radiation and genetic differentiation in the Hawaiian silversword alliance (Compositae: Madiinae). Evolution **42**: 1278–1287.

Wohlpart, A. and Mabry, T. J. 1968. The distribution and phylogenetic significance of the betalains with respect to the Centrospermae. Taxon **17**: 148–152.

Wolf, C. B. and Wagener, W. W. 1948. Botany of New World Cypresses. El Aliso **1**: 71–91.

Wolf, S. J. and Denford, K. E. 1983. Flavonoid variation in *Arnica cordifolia*: an apomictic polyploid complex. Biochem. Syst. Ecol. **11**: 111–114.

——— **and** ———. 1984a. Flavonoid diversity and endemism in *Arnica* subgenus *Austromontana*. Biochem. Syst. Ecol. **12**: 183–188.

——— **and** ———. 1984b. *Arnica gracilis* (Compositae), a natural hybrid between *A. latifolia* and *A. cordifolia*. Syst. Bot. **9**: 12–16.

——— **and Whitkus, R.** 1987. A numerical analysis of flavonoid variation in *Arnica* subgenus *Austromontana* (Asteraceae). Amer. J. Bot. **74**: 1577–1584.

Wolfe, J. A. 1975. Some aspects of plant geography of the Northern Hemisphere during the late Cretaceous and Tertiary. Ann. Missouri Bot. Gard. **62**: 264–279.

——— **and Leopold, E. B.** 1967. Neogene and early Quaternary vegetation of northwestern North America and northeastern Asia. Pages 193–206 in D. M. Hopkins (ed.) The Bering Land Bridge. Stanford University Press, Stanford, California.

Wollenweber, E. and Dietz, V. H. 1980. Flavonoid patterns in the farina of goldback and silverback ferns. Biochem. Syst. Ecol. **8**: 21–33.

———, **Dörr, E., Rozefelds, A. C., Minchin, P. and Forster, P. I.** 2000. Variation in flavonoid exudates in *Eucryphia* species from Australia and South America. Phytochemistry **28**: 111–118.

———, **Schober, I., Schilling, G., Arriaga-Giner, F. J. and Roitman, J. N.** 1989. A novel geranyl α-pyrone from the leaf resin of *Diplacus aurantiacus*. Phytochemistry **28**: 3493–3496.

———, **Stevens, J. F., Dörr, M. and Rozefelds, A. C.** 2003. Taxonomic significance of flavonoid variation in temperate species of *Nothofagus*. Phytochemistry **62**: 1125–1131.

Wood, W. F., Lancaster, W. C., Fisher, C. O. and Stotler, R. E. 1996. *trans*-Methyl cinnamate: the major volatile from some populations of the liverwort *Conocephalum conicum*. Phytochemistry **42**: 241–242.

Woodson, R. E., Jr. 1947. Some dynamics of leaf variation in *Asclepias tuberosa*. Ann. Missouri Bot. Gard. **34**: 353–432.

———. 1953. Biometric evidence of natural selection in *Asclepias tuberosa*. Proc. Nat'l. Acad. Sci., U.S.A. **39**: 74–79.

———. 1954 The North American species of *Asclepias* L. Ann. Missouri Bot. Gard. **41**: 1–211.

———. 1962. Butterflyweed revisited. Evolution **16**: 168–185.

———. 1964. The geography of flower color in butterflyweed. Evolution **18**: 143–163.

Wright, A. D., Coll, J. C. and Price, I. R. 1990. Tropical marine algae. VII. The chemical composition of marine algae from north Queensland waters. J. Nat. Prof. **53**: 845–861.

———, **Koenig, G. M., Sticher, O. and De Nys, R.** 1991. Five new monoterpenes from the red alga *Portieria hornemannii.* Tetrahedron **47**: 5717–5724.

Wright, S. D., Yong, C. G., Dawson, J. W., Whittaker, D. J. and Gardner, R. C. 2000. Riding the ice age El Niño? Pacific biogeography and evolution of *Metrosideros* subg. *Metrosideros* (Myrtaceae) inferred from nuclear ribosomal DNA. Proc. Nat'l. Acad. Sci., U.S.A. **97**: 4118–4123.

———, ———, **Wichman, S. R., Dawson, J. W. and Gardner, R. C.** 2001. Stepping stones to Hawaii: a trans-equatorial pathway for *Metrosideros* (Myrtaceae) inferred from nrDNA (ITS + ETS). J. Biogeog. **28**: 769–774.

Wylie, S. G., Brophy, J. J., Sarafis, V. and Hobbs, M. 1990. Volatile components of the fruits of *Pistacia lentiscus.* J. Food Sci. **55**: 1325–1326.

Xiang, Q.-Y., Crawford, D. J., Wolfe, A. D., Tang, Y.-C., and Depamphillis, C. W. 1998a. Origin and biogeography of *Aesculus* L. (Hippocastanaceae): a molecular phylogenetic perspective. Evolution **52**: 988–997.

———, **Soltis, D. E., and Soltis, P. S.** 1998b. The eastern Asian and eastern and western North American floristic disjunction: congruent phylogenic patterns in seven diverse genera. Mol. Phylogen. Evol. **10**: 178–190.

Yang, H., Duan, Y., Hu, F. and Liu, J. 2004. Lack of altitudinal trends in phytochemical constituents of *Swertia franchetiana.* Biochem. Syst. Ecol. **32**: 861–866.

Yang, J. Y., Page, J., Bohm, B. A. and Soltis, D. E. 1998. Flavonoids of *Itea* and *Pterostemmon.* Biochem. Syst. Ecol. **27**: 79–83.

Yao, Y. M. and Tigerstedt, P. M. A. 1995. Geographical variation of growth rhythm, height, and hardiness, and their relations in *Hippophae rhamnoides.* J. Am. Hortic. Soc. **120**: 691–698.

———, ——— **and Joy, P. 1992.** Variation of vitamin C concentration and character correlation between and within natural sea-buckthorn (*Hippophae rhamnoides* L.) populations. Acta Agric. Scand. **42**: 12–17.

Yatskievych, G., Windham, M. D. and Wollenweber, E. 1990. A reconsideration of the genus *Pityrogramma* (Adiantaceae) in western North America. Amer. Fern. J. **80**: 9–17.

Yeh, F. C. 1988. Isozyme variation of *Thuja plicata* (Cupressaceae) in British Columbia. Biochem. Syst. Ecol. **16**: 373–377.

Yoshioka, H., Mabry, T. J., Dennis, N. and Herz, W. 1970. Structure and stereochemistry of pulchellin B, C, E, and F. J. Organ. Chem. **35**: 627–631.

Young, R. L., Hylin, J. W., Plucknett, D. L. and Nakayama, R. T. 1966. Analysis of Kava pyrones in extracts of *Piper methysticum.* Phytochemistry **5**: 795–798.

Zaidi, F., Voirin, B., Jay, M. and Viricel, M. R. 1998. Free flavonoid aglycones from leaves of *Mentha pulegium* and *Mentha suavolens* (Labiatae) Phytochemistry **48**: 991–994.

Zavarin, E., Cool, L. G. and Snajberk, K. 1993. Geographic variability of *Pinus flexilis* xylem monoterpenes. Biochem. Syst. Ecol. **21**: 381–387.

———, **Critchfield, W. B. and Snajberk, K.** 1978. Geographic differentiation of monoterpenes from *Abies procera* and *Abies magnifica.* Biochem. Syst. Ecol. **6**: 267–278.

———, **Critchfield, W. B. and Snajberk, K.** 1993. Geographic variabilitiy of *Pinus flexilis* xylem monoterpenes. Biochem. Syst. Ecol. **21**: 381–387.

———, **Hathaway, W., Reichert, T. and Linhart, Y. B.** 1967a. Chemotaxonomic study of *Pinus torreyana* Parry turpentine. Phytochemistry **6**: 1019–1023.

———, **Rafii, Z., Cool, L. G. and Snajberk, K.** 1991. Geographic monoterpene variability of *Pinus albicaulis.* Biochem. Syst. Ecol. **19**: 147–156.

———, **Smith, L. V. and Bicho, J. G.** 1967b. Tropolones of Cupressaceae – III. Phytochemistry **6**: 1387–1394.

——— **and Snajberk, K.** 1973a. Variability of the wood monoterpenoids from *Pinus aristata.* Biochem. Syst. **1**: 39–44.

——— **and** ———. 1973b. Geographic variability of monoterpenes from cortex of *Pseudotsuga menziesii.* Pure Appl. Chem. **34**: 411–434.

——— **and** ———. 1975. *Pseudotsuga menziesii* chemical rances of California and Oregon. Biochem. Syst. Ecol. **2**: 121–129.

——— **and** ———. 1976. Geographic differentiation of cortical monoterpenoids of *Pseudotsuga macrocarpa*. Biochem. Syst. Ecol. **4**: 93–96.

——— **and** ———. 1985. Monoterpenoid and morphological differentiation within *Pinus cembroides*. Biochem. Syst. Ecol. **13**: 89–104.

———, ——— **and Bailey, D.** 1976. Variability in the essential oils of wood and foliage of *Pinus aristata* and *Pinus longaeva*. Biochem. Syst. Ecol. **4**: 81–92.

———, ———, ——— **and Rockwell, E. C.** 1982. Variability in essential oils and needle resin canals of *Pinus longaeva* from eastern California and western Nevada in relation to other members of subsection *Balfourianae*. Biochem. Syst. Ecol. **10**: 11–20.

———, ——— **and Cool, L.** 1990. Chemical differentiation in relation to the morphology of the single-needle pinions. Biochem. Syst. Ecol. **18**: 125–137.

———, ——— **and Critchfield, W. B.** 1973. Monoterpene variability of *Abies amabilis* cortical oleoresin. Biochem. Syst. **1**: 87–93.

———, ——— **and Critchfield, W. B.** 1977. Terpenoid chemosystematic studies of *Abies grandis*. Biochem. Syst. Ecol. **5**: 81–93.

———, ——— **and Fisher, J.** 1975. Geographic variability of monoterpenes from cortex of *Abies concolor*. Biochem. Syst. Ecol. **3**: 191–203.

Zhang, X. F., Hu, B. L. and Zhou, B. N. 1995. Studies on the active constituents of Tibetan herb *Biebersteinia heterostemon* Maxim. Acta Pharma. Sin. **30**: 211–214.

Zygadlo, J. A., Maestri, D. M., Lamarque, A. L., Guzman, C. A., Velasco-Negueruela, A., Pérez-Alonso, M. J., Garcia-Vallejos, M. C. and Grosso, N. R. 1996. Essential oil variability of *Minthostachys vertillata*. Biochem. Syst. Ecol. **24**: 319–323.

Plant Family Index

Algae

Codiaceae, 244
Cystoseiraceae, 234

Plocamiaceae, 236, 237

Rhodomelaceae, 240–244
Rhodophyllidaceae, 239

Angiosperms

Aizoaceae, 181–182
Anacardiaceae, 56
Anarthriaceae, 186
Apiaceae, 52, 191–192, 282–284
Araceae, 30
Araliaceae, 197
Aristolochiaceae, 212
Asclepiadaceae, 85–86
Asphodelaceae, 8, 10
Asteraceae, 5, 6, 36, 38, 68, 76, 88–92, 93–94, 95–100, 11, 121, 128–134, 140–141, 161, 202, 249, 251–252, 254, 255–256, 264–268, 285–288, 290, 295, 296
Avicenniaceae, 11

Berberidaceae, 62
Brassicaceae, 176

Cactaceae, 181–182
Calycanthaceae, 197
Cannabaceae, 198
Capparaceae, 102
Caryophyllaceae, 31–36
Chenopodiaceae, 81–84
Cistaceae, 41–43
Commelinaceae, 296–297
Coriariaceae, 217
Cornaceae, 197
Crassulaceae, 67, 100, 101
Cyperaceae, 253

Datiscaceae, 173–174
Degeneriaceae, 262–263

Ecdeiocoleaceae, 186
Elaeagnaceae, 67–68
Empetraceae, 44, 276–277
Ericaceae, 134–136, 193–194, 214–215, 257
Eucryphiaceae, 199
Euphorbiaceae, 117

Fabaceae, 7, 10–11, 25, 28, 28–30, 53–54, 76, 103–105, 182, 203–205, 210–212, 272–275
Fagaceae, 116, 156

Gentianaceae, 62, 217–219
Geraniaceae, 66
Gesneriaceae, 215–216
Gunneraceae, 268–271
Guttiferae, 4

Haloragidaceae, 268
Hippocastanaceae, 197
Hopkinsiaceae, 186

Icacinaceae, 188–190
Iridaceae, 157

Krameraceae, 291–292

Lactoridaceae, 263
Lamiaceae, 43, 47, 48–49, 56–57, 58–59, 61, 70, 77–78, 105–108, 192–193
Lauraceae, 219, 282
Lyginiaceae, 186

Magnoliaceae, 138–139
Menyanthaceae, 187
Myrtaceae, 12, 17, 19, 20, 22–23, 255

Nothofagaceae, 208

Onagraceae, 205–207
Orchidaceae, 179–180

Papaveraceae, 11, 291
Passifloraceae, 247–248
Phytolaccaceae, 7
Piperaceae, 259, 271–272
Poaceae, 23, 63, 219, 280–282
Poaceae-Argentina, 74–75
Poaceae-Hawaii, 75
Polemoniaceae, 86–88
Plumbaginaceae, 174

Ranunculaceae, 38–39, 101–102, 197, 277–278
Restionaceae, 183, 187
Rhizophoraceae, 11
Rosaceae, 44, 126, 156, 275–276
Rubiaceae, 70–71
Rutaceae, 16, 67, 257

Sapindaceae, 275
Saxifragaceae, 113–116, 126–128, 194–195, 197, 289
Scrophulariaceae, 30, 120, 205
Solanaceae, 14, 51, 61–62, 72–74, 248–249

Turneraceae, 248

Valerianaceae, 40
Verbenaceae, 126

Zygophyllaceae, 113, 292–295

Bryophytes (liverworts)

Adelanthaceae, 228

Conocephalaceae, 225–228

Makinoaceae, 233
Marchantiaceae, 231–232
Mastigophoraceae, 232–233

Plagiochilaceae, 229–231

Bryophytes (mosses)

Lepidoziaceae, 223–225

Ferns

Adiantaceae, 78, 78–80

Pteridaceae, 108, 219–221

Gymnosperms

Araucariaceae, 171

Chamaecyparis, 64
Cupressaceae, 60, 64, 118, 137–138, 159–161, 167–170

Ephedraceae, 180–181

Pinaceae, 49, 141–155, 157–159, 162–163, 164–165, 166–167

Lichens

Balantiopsidaceae, 19

Cladoniaceae, 120

Graphidiaceae, 280

Icmadophilaceae, 223

Jungermanniaceae, 65

Lecanoraceae, 221–222

Parmeliaceae, 221

Ramalinaceae, 24–25

Stictaceae, 223

Plant Genus Index

Algae

Chondrococcus hornemannii, 239
Codium species, 244
Cytoseira amentacea, 234

Desmia hornemanii, 239

Laurencia species, 240–244

Phaeophyta, 233
Plocamium cartilagineum, 237
Plocamium harmatum, 236
Portieria hornemanii, 239

Angiosperms

Acer species, 275
Achillea species, 36
Aloë species, 8, 10
Amauria, 296
Ambrosia chamissonis, 285–288
Ambrosia psilostachya, 140–141
Ambrosia species, 95–99
Apodasmia species, 185
Argyroxiphium, 255
Armeria maritima, 174
Arnica species, 128–134
Arum maculatum, 30
Asclepias tuberosa, 85–86
Aspalathus linearis, 10–11
Avicennia, 11

Berberis species, 62
Bidens species, 251–252
Biebersteinia species, 66, 67
Blennosperma species, 288
Brickellia cylindrica, 93–94

Cakile edentula, 178
Cakile maritima, 176
Carmichaelia species, 273–275
Chamelaucium uncinatum, 12
Chenopodium fremontii, 81–84
Chionochloa species, 23
Chrysosplenium glechomaefolium, 113–116
Chrysosplenium valdivicum, 289
Cinchona species, 70–71
Cistus albidus, 41–43
Cnidoscolus spinosus, 117
Collinsia parviflora, 30
Cotoneaster wilsonii, 275–276
Cunila galioides, 70
Cymbopogon distans, 63
Cyperus rotundus, 253
Cypripedium calceolus, 179–180

Dactylis glomerata, 280–282
Dapsilanthus species, 185
Datisca species, 173–174
Datura metel, 61–62
Degeneria species, 262–263
Delphinium variegatum, 101–102
Dendroseris species, 264–265
Diplacus aurantiacus, 120
Dubautia species, 256
Duboisia myoporoides, 14
Dudleya species, 100

Echinacea angustifolia, 121
Empetrum nigrum, 44
Empetrum species, 276–277
Encelia canescens, 290
Encelia farinosa, 111
Ephedra species, 180–181
Erigeron species, 266–268
Eschscholzia californica, 291

Flyriella parryi, 94
Foeniculum vulgare, 52

Gaillardia pulchella, 99–100
Genista species, 53–54
Gunnera species, 268–271

Helianthus maximiliani, 94
Hesperomannia arborescens, 254
Hieracium, 265
Hippophae rhamnoides, 67–68
Hordeum species, 219
Hymenaea species, 76
Hymenaea verrucosa, 182
Hypochaeris, 265
Hypolaena fastigiata, 183

Isomeris arborea, 102
Isotachis lyallii, 19

Kramera species, 291–292
Kunzea ericoides, 22–23

Larrea species, 292–295
Larrea tridentata, 113
Lasthenia species, 88–92
Laurus azorica, 282
Lavandula stoechas, 58
Leptocarpus species, 185
Leptospermum scoparium, 20
Lippia dulcis, 122
Liriodendron tulipifera, 138–139
Lotus corniculatus, 28–30
Luetkea pectinata, 126
Lupinus sericeus, 103–105
Lycopersicon cheesmanii, 248–249
Lycopersicon pennellii, 72–74
Lycopersicon pimpinellifolium, 249
Lyonothamnus floribundus, 156

Marjorana syriaca, 57
Marshallia graminiifolia, 161
Melaleuca, 17, 19
Mentha suaveolens, 58–59
Menziesia, 134–136
Metrosideros polymorpha, 255
Mimulus aurantiacus, 120
Minthostachys andina, 78
Minthostachys verticillata, 77–78

Ocimum gratissimum, 12
Octomeles, 173
Origanum microphyllum, 43

Papaver aculeatum, 11
Pappobolus, 249
Passiflora colinvauxii, 248
Passiflora foetida, 247
Pearsonia species, 7
Peperomia species, 271–272
Pericome, 296
Perityle emoryi, 295–296
Phlomis lychnitys, 61
Phlox carolina, 86–88
Phytolacca dodecandra, 7
Piper methysticum, 259
Piper wichmannii, 260
Pistacia lentiscus, 56
Prunus lyonii, 156
Pulicaria dysenterica, 38

Quercus rubra, 116
Quercus tomentella, 156

Ramalina siliquosa, 24–25
Ranunculus species, 277–278
Rhizophora, 11
Rhodiola crenulata, 67
Robinsonia species, 265–266
Rubus chamaemorus, 44

Satureja douglasii, 105–108
Scalesia species, 249
Sedum lanceolatum, 101
Senecio angulatus varieties, 6
Senecio kleinia, 6
Senecio radicans complex, 5
Silene species, 31–36
Sophora species, 272–273
Stenotalis species, 185
Swertia franchetiana, 62
Symphonia globulifera, 4

Tagetes minuta, 76
Tetrameles, 173
Teucrium polium, 59
Thalictrum minus, 38–39
Thymus caespitosus, 47
Thymus praecox, 48–49
Thymus species, 56–57
Todaroa aurea, 282–284
Tolmiea menziesii, 126–128
Trachylobium species, 76
Tradescantia species, 296–297
Trifolium, 25, 28
Turnera, 248

Vaccinium calycinum, 257
Vaccinium reticulatum, 257
Valeriana officinalis, 40
Vanillosmopsis erythropappa, 68

Wilkesia gymnoxiphium, 255, 256
Withania somnifera, 51

Zanthoxylum species, 257
Zieria, 16

Bryophytes (liverworts)

Adelanthus decipiens, 228

Conocephalum conicum, 225–228

Isotachis lyallii, 19

Makinoa crispata, 233
Marchantia species, 231–232
Mastigophora diclados, 232–233

Plagiochila species, 229–231

Bryophytes (mosses)

Bazzania trilobata, 223–225

Ferns

Cheilanthes, 78

Gymnogramma triangularis, 108

Notholaena, 78–80

Pellaea, 78
Pentagramma triangularis, 108
Pityrogramma triangularis, 108
Pteridium aquilinum sens. lat., 219–221

Gymnosperms

Abies alba, 166–167
Abies species, 141–144
Araucaria araucana, 171
Austrocedrus chilensis, 170

Chamaecyparis, 64
Cupressus bakeri, 144–146
Cupressus, 64, 118

Fitzroya cupressoides, 167–170

Juniperus phoenicea, 60
Juniperus silicicola, 161
Juniperus virginiana, 161
Picea species, 146–147
Pinus albicaulis, 149–151
Pinus aristata, 152
Pinus balfouriana, 152
Pinus brutia, 164
Pinus cembroides, 157t
Pinus contorta, 147–148
Pinus flexilis, 157t
Pinus halepensis, 162–163
Pinus longaeva, 152
Pinus monophylla, 157t
Pinus monticola, 151
Pinus nigra, 157t
Pinus ponderosa, 150–151
Pinus quadrifolia, 157t
Pinus radiata, 157t
Pinus remota, 157t
Pinus sylvestris, 46, 164–165
Pinus torreyana, 155
Pinus uncinata, 49
Pinus, 46, 49
Pseudotsuga macrocarpa, 157
Pseudotsuga menziesii, 157
Pseudotsuga, 157–159

Sequoia sempervirens, 159–161

Thuja plicata, 137–138
Thujopsis dolabrata, 68

Lichens

Anomalographis madeirensis, 280

Cladonia cervicornis, 120

Jungermannia, vulcanicola, 65

Lobaria pulmonaria, 223

Pseudovernia furfuracea, 221

Ramalina siliquosa, 24–25
Rhizoplaca melanophthalma, 221–222

Thamnolia vermicularis, 223

Other organisms

Arion species (slugs), 28
Arvicola terrestris (vole), 27–28

Chelonia midas (turtle), 237

Eulachnus agilis (aphid), 46

Lygus rugulipennis (bug), 46

Schizolachnus pineti (aphid), 46

Subject Index

A
Acetophenones, 228–229
Acetylenes, 76, 122, 192, 287
Acetylenic amides, 122
Alerce, 167
Algae, brown, 233–234
  red, 235
Alkaloids, 5, 7, 38, 54, 70–71, 189, 203–204, 273
Alkaloid toxicity indices, 5
Alkanes, 8, 58, 100–101, 110, 171
Allelochemistry, 4–5
Aloins, 9
Angola, 10
Anthocyanidins, 215
Anthocyanins, 101, 112
  presumed, 85, 101, 112
Anthrone derivatives, 9
Antiherbivore activity, 46, 237
Arabian Peninsula, 8
Archipiélago de Colón, 245
Argentina, 75, 76
Aromatic esters, 19
Australia, 12, 16
Austria, 33
Azores, 47, 230, 279

B
Baja California, 113
Baker cypress, 144
Bark beetles, 158
Barley, 219
Basil, tree, 12
Bay, true, 282
Bees, odor sensing, 57–58
Beggar's ticks in Hawaiian Islands, 251–252
Benin, 10
Benzofurans, 111
Benzopyrans, 111
Betacyanins, 181
Bibenzyl derivatives, 231
Biflavonoids, 65
Big cone Douglas fir, 157
Blueberries in Hawaiian Islands, 257
Bracken carcinogens, 220
Brazil, coastal, 68–69
Bristlecone pines, 152
British Isles, 48
Brominated indoles, 243
Brown algae, 233–234
Buckthorn, 67

C
Caffeic acid derivatives, 121
California poppy, 291
Cameroon, 10
Canary Islands, 6, 8, 279
Cape Verde Islands, 11, 279
Carcinogens in bracken, 220
Caribbean Islands, 122
Central African Republic, 10
Chad, 10
Chile, 238
Chiloé Island, Chile, 169
China, southwestern, 67
Chromones, 290
Cloudberry, 44
Clover, 25
Cock's foot, 281
Conifer hybrids, 158
Corsica, 54
Coumarins, 191–192
Creosote bush, 113
Crete, 43
Crowberry, 44
Cyanogenesis, 25–28

Cyanogenesis, genetics, 26
Cyanogenesis and islands, 246, 248
Czechoslovakia, 33

D
Defense, immobile, 103
Defense, mobile, 103
Depsides, 120
Dispersal
  by ballast, 45, 178
  island rafting, 156
  land bridges, 156, 249
  long distance, 156
Diterpenes, 205, 233
Douala-Edea Forest, Cameroon, 3
Douglas fir, 157

E
East Africa, 6
Ecuador, 231
Edaphic factors, 3, 39, 41, 79, 90–91, 103, 131
Egypt, 56, 59
Elba, 54
Elevational variation, 26–27, 116
England, 48
Ephedrine isomers, 181
Estonia, 44
Ethiopia, 7

F
Falkland Islands, 276
Fatty acids, 144–145, 170, 244, 292
Feeding deterrent, 46, 237
Fennel, 52
Fiji Archipelago, 262
Finland, 33, 44–47
Fir, Douglas, 157
Firs (true, *Abies*), 141
Flavonoid
  genetics, 49
  glycosylation genetics, 32–33
Floral aroma, 179
France, southern, 41

G
Gabon, 11
Galapagos Islands, 245–249
Genetics
  flavonoid concentration, 49
  flavonoid glycosylation, 32–33
Geraldton waxflower, 12
Germany, 33
Ghana, 8, 10
Glucosinolates, 102, 176
Gough Island, 277
Grasses, C_3 and C_4, 74–75
Great Barrier Reef, 236
Greece, 59, 162
Greenland, 49, 223
Guyana, 11

H
Halogenated terpenes, 236–237, 240
Hawaiian Islands, 249
Herbivore pressure, 70
Hierba dulce, 122
Hops, 198
Human migration
  Europe, 36
  Pacific, 261
Hungary, 33, 36
Hydrocarbons, 58, 100–101, 110, 171

I
Iceland, 49
Inaccessible Island, 271
India, 62, 188
Ireland, 48
Islas Malvinas, 276
Israel, 51
Italy, northern, 40

J
Japan, 65
Juan Fernandez Islands, 263–272

K
Kanuka, 23
Kava, 259
Kavalactones, 259
Kenya, 6, 8
Kerguelen Islands, 277–279
Kibale Forest, Uganda, 3
Korea, 275

L
Latitudinal variation, 26, 71, 73, 105, 106, 111, 114, 122, 166

Lavender, 58
Leaf oils, 41, 43, 60
Leaf-pocket resins, 182
Leatherwood, 200
Lichen acids and esters, 25, 120, 222, 223, 224
Linamarase variation, 28
Lithuania, 38
Lodgepole pine, 147
Lord Howe Island, 273–275

M
Macaronesia, 279–284
Madagascar, 8
Madeira, 279
Malaysia, 232
Manuka, 20
Marquesas, 255
MBSD (Minimum biosynthetic distance), 87–88
Mediterranean coast, 234
Mediterranean region, 162
Megagametophyte fatty acids, 170
Melanesia, 188
Mexico, 70, 117
Micronesia, 189
Minimum biosynthetic distance (MBSD), 87–88
Mint, 58–59
Molluscicides, 7
Morocco, 59, 159
Mozambique, 10
Mustard oil glucosides, 102

N
Naphthalene derivatives, 9, 228–229
Negev Desert, 53
New Caledonia, 16
New Guinea, 14
New South Wales, 18
New Zealand, 19, 23, 223
Nigeria, 8
Norway, 49
Novaya Zemlya, 223

O
Oaks, 116, 156
Odor sensing by bees, 57
'Ohi'a lehua, 255
Oriental fruit fly, 257
Origanum, 43
Otters, defense against, 233–234
Oviposition effects, 46
Oxidation indices, 218

P
Pacific Islands, 253
Paperbark, 17
Papua New Guinea, 19
Peperina, 77
Peru, coastal, 72–73
Phaeophyta, 233
Phenolic ethers, 16, 17
Phenylpropanoids, 213, 258, 283
Phlorotannin, 233–234
Phlox, 86
Phytol derivatives, 8
Phytosterols, 20
Pines, European black, 157t
  bristlecone, 152
  limber, 157t
  lodgepole, 147
  Mexican piñon, 157t
  Monterey, 157t
  Parry piñon, 157t
  ponderosa, 150
  Scotch, 164
  single-leaf piñon, 157t
  Torrey, 155
  western white, 151
  whitebark, 149
Pistachio, 56
Ploidy levels, 88, 96, 126, 133, 203, 281, 292
Poland, 33
Polyacetylenes, 76, 122, 192, 287
Ponderosa pine, 150
Poppy, California, 291
Ptaquiloside, 220
Purple nut sedge, 253

Q
Qatar, 59
Queensland, 14, 18

R
Rain forests, 3
Red algae, 235
Redwood, coast, 159
Refuge, Chiloé Island, 168
Refugia, 36, 136, 139, 168
Relic species, 156
Rhodophyta, 235
Rift Valley, 9
Rock rose, 41

Romania, 36
Rooibos, 10
Russia, 33

S
Sallowthorn, 67
Sardinia, 54
Saudi Arabia, 59
Scotch pine, 164
Scotland, 29, 48
Sea level change, 141
Serbia, 38
Sesquiterpenes, 19, 65, 69, 76, 94, 95–99, 140, 167, 182, 227, 253
Silver fir, 166
Silversword alliance, 255–256
Snow tussock, 23
South Africa, 10
South West Africa, 6
Spain, 162
Spruces (*Picea*), 146
Stepping-stone dispersal, 263
Stewart Island, 202
Sudan, 10
Sugar esters, 72–73
Svalbard, 223

T
Tannins, condensed, 257
Tanzania, 10
Tasmania, 16, 202
Terpenes
  halogenated, 236–237
  simple, 1, 158
Tethys Sea, 179
Thiophenes, 76
Thyme, 47, 48–49, 56–57
Tibet, 67
Togo, 10
Torrey pine, 155
Tree basil, 12
Triacylglycerols, 44
Tristan da Cunha Islands, 271
Triterpenes, 11, 23, 53, 100
Tropolones, 118
Tulip tree, 138
Tunisia, 162
Turkey, 37

U
Uganda, 10
Ullung Island, Korea, 275

V
Valerian, 40
Vancouver Island, 30
Variation, altitudinal, 59, 62, 75

W
Wales, 228
Waxes, 11, 24, 171
Western Australia, 12
Western red cedar, 137
Western white pine, 151
Whitebark pine, 149
Withanolides, 53

X
Xanthones, 62, 218

Y
Yugoslavia, 36, 40

Z
Zambia, 8
Zimbabwe, 8

Printed in the United States
134747LV00002B/39/P

9 781402 090516